# Essentials of Oil and Gas Utilities

## Process Design, Equipment, and Operations

# Essentials of Oil and Gas Utilities

## Process Design, Equipment, and Operations

**Alireza Bahadori, PhD**

School of Environment, Science & Engineering
Southern Cross University, Lismore, NSW, Australia

ELSEVIER

AMSTERDAM • BOSTON • HEIDELBERG • LONDON
NEW YORK • OXFORD • PARIS • SAN DIEGO
SAN FRANCISCO • SINGAPORE • SYDNEY • TOKYO

Gulf Professional Publishing is an imprint of Elsevier

Gulf Professional Publishing is an imprint of Elsevier
50 Hampshire Street, 5th Floor, Cambridge, MA 02139, USA
The Boulevard, Langford Lane, Kidlington, Oxford, OX5 1GB, UK

**British Library Cataloguing-in-Publication Data**
A catalogue record for this book is available from the British Library

**Library of Congress Cataloging-in-Publication Data**
A catalog record for this book is available from the Library of Congress

ISBN: 978-0-12-803088-2

For information on all Gulf Professional publications
visit our website at http://store.elsevier.com/

*Dedicated to the loving memory of my parents, grandparents, and to all who contributed so much to my work over the years*

# Contents

# Biography

**Alireza Bahadori, PhD** is a research staff member in the School of Environment, Science & Engineering at Southern Cross University, Lismore, NSW, Australia. He received his PhD from Curtin University, Perth, Western Australia.

During the past 20 years, Dr. Bahadori has held various process and petroleum engineering positions and been involved in many large-scale projects at NIOC, Petroleum Development Oman (PDO), and Clough AMEC PTY LTD.

He is the author of around 250 articles and 12 books. His books have been published by multiple major publishers, including Elsevier.

Dr. Bahadori is the recipient of the highly competitive and prestigious Australian Government's Endeavour International Postgraduate Research Scholarship (IPRS) as part of his research in the oil and gas area. He also received the Top-Up Award from the State Government of Western Australia through the Western Australia Energy Research Alliance (WA:ERA) in 2009. Dr. Bahadori serves as a reviewer and member of editorial boards for a large number of journals.

# Preface

A typical oil and gas refinery or a petrochemical plant requires enough utilities to support a successful operation. The supply of water, steam, fuel, and air is the assignment of the utilities department. All refineries produce steam for use in process units. This requires water-treatment systems, boilers, and extensive piping networks. In addition, clean, dry air must be provided for many process units, and large quantities of cooling water are required for condensation of hydrocarbon vapors.

A gas processing, refinery, or a petrochemical plant utilizes a large number of utilities to support the manufacturing at production units. A comprehensive utilities complex exists to meet the requirements of utilities, steam, condensate, cooling water, instrument/plant air, and fuels. This consists of water treatment plants, condensate recovery plants, high-pressure steam boilers, induced draft cooling towers, a number of instrumentation/plant air compressors, and some units for refinery fuel gas and fuel oil system. This all operates round the clock for smooth operation of the entire refinery or a petrochemical plant.

This book specifies the essential requirements for the process design of many utilities facilities including fuel systems used in the oil and gas industries. There are major design parameters and guidelines for process design of fuel systems. This book covers gaseous and liquid fuel systems.

Fuel is utilized to provide heat for power generation, steam production, and process requirements. The fuel system includes facilities for collection, preparation, and distribution of fuel to users. Alternative fuels (as required) should be made available at all consuming points. The commonly used ones are liquid fuel and gas fuel. The selection of fuels used in the system shall be on the basis of the cost, availability, and dependability of supply; convenience of use and storage; and environmental regulations.

This book also covers:

- cooling water circuits for internal combustion engines
- cooling water circuits for reciprocating compressors
- cooling water circuits for inter cooling and after cooling facilities.

Thermal liquids are used for process heating and cooling in the form of liquid, vapor, or a combination of both. In addition to steam, thermal liquid include, hot and tempered water, mercury, molten salt mixtures, hot oils, and many others, each of which can be used for a specified field of application and can operate in different temperature ranges.

This book covers the minimum requirements and recommendations deem necessary to be considered in process design of hot oil and tempered water systems. The scope is covered in two parts as:

- process design of hot oil system
- process design of tempered water system.

Also, this book covers the requirements for the protection of process and utilities and all associated equipment and flow lines and instruments against the temperature which would cause congealing or freezing of contents, interfere with operation, or cause damage to equipment or pipelines and for heat conservation requirements as would be determined by process conditions.

This book also covers a preparation and storage of strong and dilute caustic solution, including storage facilities for the regenerated caustic from LPG Caustic Treating Section.

Transferring the proper caustic solution strength to the plants or process units where required with available pumps.

Supplying strong caustic solution for the unit by means of solid caustic or available strong caustic solution from existing petrochemical plants.

This book is intended to cover requirements for process design of field erected water tube boilers. Boiler shall be designed for guaranteed net steam capacity, for each boiler at superheated outlet excluding all blow down, heat losses and all steam used for boiler auxiliaries such as fans, soot blowers, burners, etc.

This book specifies the essential requirements for the process design and selection of various water supply systems, used in oil and gas Industries, and consists of the following systems:

- water treatment system
- raw water and plant water system
- fire water distribution and storage facilities

The object of installing a compressed air system is to provide air supply to the various points of application in sufficient quantity and quality and with adequate pressure for efficient operation of air tools or other pneumatic devices.

This book covers the process requirement and design guide for plant and instrument air systems as applicable to the oil and gas industries. This book also covers solid handling systems and dryers in oil, gas, and petrochemical processing industries and many other issues related to utilities in the oil and gas industries are discussed in this book.

**Dr. Alireza Bahadori**

*School of Environment, Science & Engineering,*
*Southern Cross University, Lismore, NSW, Australia*

# Acknowledgments

I would like to thank the Elsevier editorial and production teams, Ms. Katie Hammon and Ms. Kattie Washington of Gulf Professional Publishing for their editorial assistance.

# FUEL SYSTEMS

The increasing demand for energy and the continuing increases in prices for standard fuels demand greater flexibility in the use of fuels in gas turbines.

Besides the standard fuels natural gas (typical heating values between 39 and 46 MJ/kg) and Diesel No. 2 fuel oil (42 MJ/kg), there is an increasing interest in low-BTU gases, synthetic gases (syngas here), and even liquid fuels (eg, heavy fuel oil, Naphtha, and condensates).

Low-BTU gases refer to fuels with heating values between 10 and 35 MJ/kg. Syngas denotes synthetically produced gases that generally have even lower heating values (LHVs), between 4 and 12 MJ/kg. This chapter specifies the essential requirements for process design of fuel systems used in oil- and gas-processing industries and outlines the major design parameters and guidelines for process design of fuel systems. It covers both gaseous and liquid fuel systems.

## 1.1 FUEL-SUPPLYING SYSTEMS

As the need for higher efficiency power plants increases, a growing number of combined-cycle power plants are incorporating performance gas fuel heating as a means of improving overall plant efficiency. This heating, typically increasing fuel temperatures in the range of 365°F/185°C, improves gas turbine efficiency by reducing the amount of fuel needed to achieve desired firing temperatures. For fuel heating to be a viable method of performance enhancement, feedwater has to be extracted from the heat recovery steam generator at an optimum location. Boiler feedwater leaving the intermediate pressure economizer is commonly used. Using gas-fired, oil-fired, or electric heaters for performance gas fuel heating will not result in a power plant thermal efficiency improvement.

Proper design and operation of the gas fuel heating system is critical to insure reliable operation of the gas turbine. Improper selection of components, controls configuration, and/or overall system layout could result in hardware damage, impact plant availability, and create hazardous conditions for plant personnel.

This chapter addresses the critical design criteria that should be considered during the design and construction of these systems. Fuel should be used to provide heat for power generation, steam production and process requirements.

Fuel systems are necessary to provide the proper fuel for a variety of users in the plant.

A fuel system should include facilities for collection, preparation, and distribution of fuel to users.

Alternative fuels (as required) should be made available at all consuming points. The commonly used ones are liquid fuel and gas fuel. One liquid fuel supplies at least one pump and its standby should be steam driven or available via other reliable power sources. Standby pumping units should be arranged for instantaneous start-up on failure of the operating unit.

Low-pressure users are items such as boilers, fired heaters, and reciprocating engines. High-pressure users are gas turbines. All fuel systems need to be kept free of solid contaminants that can plug instrumentation and fuel nozzles. In addition, fuel streams need to be maintained above the hydrocarbon dew point to prevent any liquid slugs in the fuel users.

The following section identifies general system requirements that apply to all gas fuel heating systems. These requirements, in addition to those described in the combustion specific system requirements section shall be followed during the design and development of the system. Gas fuel supplied to the gas turbine shall meet the particulate requirements. If the components in the gas fuel heating system are constructed of materials susceptible to corrosion, a method of final filtration upstream of the gas turbine interface is required. Particulate carryover greater than that identified in the standards can plug fuel nozzle passages, erode combustion hardware and gas valve internals, and cause damage to first stage turbine nozzles. The new gas piping system must be properly cleaned prior to initial gas turbine operation. Additional design considerations are those related to gas fuel cleanliness. The fuel delivered to the gas turbine must be liquid free and contain a specified level of superheat above the higher of the hydrocarbon or moisture dew points.

Saturated fuels, or fuels containing superheat levels lower than specified, can result in the formation of liquids as the gas expands and cools across the gas turbine control valves. The amount of superheat provides a margin to compensate for temperature decrease due to pressure reduction, and is directly related to incoming gas supply pressure. (Note: within this document, gas fuel heating strictly for dew point considerations is still considered to be in a "cold" state. Heating for performance purposes is considered "heated" fuel.) The design of the gas fuel heating system shall prevent carryover of moisture or water to the gas turbine in the event of a heat exchanger tube failure. Water entrained in the gas can combine with hydrocarbons causing the formation of solid hydrocarbons or hydrates. These hydrates, when injected into the combustion system, can lead to operability problems, including increased exhaust emissions and mechanical hardware damage. Proper means of turbine protection, including heat exchanger leak detection, shall be provided.

Gas being supplied to the gas turbine interface point shall meet the minimum gas fuel supply pressure requirements as defined in the proposal documentation. These minimum pressure requirements are established to insure proper gas fuel flow controllability and to maintain required pressure ratios across the combustion fuel nozzles. The gas fuel heating system shall be designed to insure that these requirements are met during all modes of operation over the entire ambient temperature range. The design of the gas fuel heating system shall insure that the design pressure of the gas turbine gas fuel system is not exceeded.

Overpressure protection, as required by applicable codes and standards, shall be furnished. In addition to minimum and maximum pressures, the gas turbine is also sensitive to gas fuel pressure variations. Sudden drops in supply pressure may destabilize gas pressure and flow control. Sudden increases in supply pressure may potentially trip the turbine due to a high temperature condition. Limitations on pressure fluctuations are defined in the gas turbine proposal documentation.

The gas fuel heating system shall be designed to produce the desired gas fuel temperature at the interface with the gas turbine equipment. Guaranteed performance is based on the design fuel temperature at the inlet to the gas turbine gas fuel module. The gas fuel heating and supply systems shall compensate for heat losses through the system. Compensation shall include but not be limited to elevated heater outlet temperatures, use of piping and equipment insulation, and minimization of piping length from heater outlet to turbine inlet. The gas fuel heating system shall be designed to support specified gas fuel temperature set points required by the gas turbine. These set points include high and low temperature alarms, gas turbine controls permissives, and gas turbine controls functions. These set points are derived by GE Gas Turbine Engineering and are based on operability requirements and/or design limitations of components within the gas turbine gas fuel system. During specified cold and hot gas fuel turbine operating modes, the gas fuel heating system shall attain and maintain the fuel at a temperature that corresponds to a Modified Wobbe Index (MWI) within ±5% of the target value.

The MWI is a calculated measurement of volumetric energy content of fuel and is directly related to the fuel temperature and LHV. The MWI is derived as follows:

$$MWI = \frac{LHV}{T_g \times SG}$$

Where

MWI, Modified Wobbe Index (temperature corrected);
LHV, lower heating value of fuel (BTU/SCF);
$T_g$, absolute temperature (°R);
SG, specific gravity of fuel relative to air at ISO conditions.

The ±5% MWI range insures that the fuel nozzle pressure ratios are maintained within their required limits. If gas fuel constituents and heating value are consistent, the 5% tolerance can be based strictly on temperature variation. If the heating value of the fuel varies, as is the case when multiple gas suppliers are used, heating value and specific gravity must be considered when evaluating the allowable temperature variation to support the 5% MWI limit.

For the use of gas fuels having a significant variation in composition or heating value, a permanent gas chromatograph shall be furnished in the plant's main gas supply line. LHV and specific gravity readings from the gas chromatograph are used to regulate the amount of fuel heating so that the ±5% MWI requirement is satisfied. This control function shall be performed automatically by the plant control system.

### 1.1.1 FUEL SELECTION

Materials produced in the plant which cannot be sold for the least monetary value should possibly be used as fuel. Diverted to plant fuel oil system include visbreaker tar, lube extracts, waxes, and atmospheric residue. The selection of fuels used in the system should be based on the cost, availability and dependability of supply, convenience of use and storage, and environmental regulations.

All the previously mentioned materials should be used as liquid fuel, in a manner so as to maintain the threshold limits of the "Air Pollution Control" Standard. $H_2S$ content of the fuel gas main header should comply with local regulations. Provision(s) should be made for using the liquefied petroleum gas (LPG) and/or natural gas to supplement the gaseous fuel.

Gaseous materials diverted to refinery fuel are those which cannot be processed to saleable products economically, and frequently include, $H_2$, $CH_4$, $H_2S$, and $C_2H_6$ and should consist essentially of $CH_4$ and/or $C_2H_6$.

### 1.1.2 GASEOUS FUEL

The main source of fuel gas should be the gas produced by process units and treated by the treating unit (if necessary). All fuel gas streams should be routed to a mixing drum where entrained liquid is

**Table 1.1 Example Turbine Fuel Gas Specifications**

| Specifications | Maximum | Minimum |
|---|---|---|
| Lower heating value, MJ/m³ ᵃ | 44.7 | 33.5 |
| Fuel pressure. kPa (abs)ᵇ | 2580 | 1900 |
| Supply temperatureᶜ | 93°C | 5°C or 2°C above hydrocarbon dew point whichever is greater |
| Liquid hydrocarbons/dust, mg/kg | 20 | |
| Liquid droplet size, μm | 5 | |
| Dust particle size, μm | 10 | |
| Water | No free water | |
| Sodium content, mg/kg | 1 | |

ᵃThe turbines can be designed to handle higher or lower values of LHV, but are designed for a particular range by the vendor.
ᵇFeed gas pressure is entirely dependent on the type of turbine used. Most smaller turbines and industrial turbines require 2100–2800 kPa (abs). Large aeroderivative turbines can require fuel in the 4100 kPa (abs) range. Fuel systems are usually designed to provide fuel at 700 kPa greater than the compressor exit pressure of the turbine.
ᶜVirtually all turbine manufacturers require 0–10°C of superheat in the fuel stream.

separated from the gas and where good mixing is ensured before distribution. In order to enable the balancing of gas production and gas consumption, necessary provisions for installation of LPG vaporizer and natural gas supplying systems to the fuel gas-mixing drum should be considered.

The liquid from the knockout drum and mixing drum should be drained to a closed recovery system or flare header. The fuel gas supply system should be equipped with enough controls and alarms, such as a ow system pressure and high knockout drum liquid level alarms, to assure a safe fuel gas supply.

The fuel gas supply system should be designed to provide the consumers with liquid-free gas at constant pressure [about 350 kPa (g) or 3.5 bar (g)] and reasonably constant heating value.

The system should include, collecting piping, mixing drum controls, and distribution piping.

Location of fuel gas mixing drum should minimize collection and distribution piping. In the event of high pressure in the mixing drum (abnormal condition), the excess gas should be released to the flare on pressure control.

Alarms should be fitted to the pilot gas system to warn of low pressure/low flow. Gas turbines require more rigorous fuel specifications than do lower pressure systems. Table 1.1 shows an example of a fuel gas specification for a commercial gas turbine. Specifications for a particular turbine can be obtained from the manufacturer.

The pressure controlling system should be provided to the fuel gas-mixing drum, which actuates from the fuel gas main header and responses to the steam control valve of the LPG vaporizer and/or natural gas control valve to supply the required pressure.

If main fuel gas header pressure drops to its preset value, LPG and/or natural gas should be used to supplement the make-up gas. The mains and all fuel gas lines should be steam traced and insulated to hold a temperature of at least 49°C to prevent condensation and hydrate formation.

A liquid knockout drum near the gas-consuming furnace (or group of furnaces) should be provided to prevent liquid from entering the burners.

To counteract the tendency of butane to recondense in the mixing drum, a steam coil in its base should be provided. Provision for installation of a relief valve to the flare header should also be considered.

The superheating of the gas can be accomplished in several ways. One way is simply to heat the feed stream directly. If the fuel is compressed before entering the fuel header, the temperature of the gas from the discharge cooler can be controlled to ensure proper superheat.

If the fuel is supplied from a higher-pressure system by pressure let down, the stream often needs some processing to prevent condensation. This can be provided by chilling the stream, scrubbing, and then reheating. Fuel gas conditioning systems based on Low Temperature Separation schemes are commercially available to accomplish this fuel conditioning.

### 1.1.3 LPG VAPORIZER

Liquefied petroleum gases (LPG) should be used as a fuel in gaseous form. A vaporizer system should be provided for this purpose.

The system should consist of the following:

1. one LPG surge drum
2. two LPG fuel pumps, one in operation (motor driven) and one stand-by (turbine driven)
3. one LPG vaporizer
4. all necessary controllers.

Various streams of LPG and butane should be received in the LPG surge drum and will be pumped into the fuel gas vaporizer.

Pressure in the LPG surge drum should be uncontrolled and will fluctuate with composition and temperature.

Levels in LPG surge drum should be controlled and recorded and provisions for high and low liquid level alarms should be installed.

Provision should be made to cut incoming LPG streams to the surge drum and pump LPG.

### 1.1.4 LIQUID FUEL

The ultimate aim in liquid fuel supply system design should be to ensure that the supply of suitable fuel to each fired heater/furnace will not fluctuate with load changes. All liquid fuels lighter than fuel oil should be filtered through mesh of about 0.3 mm aperture.

### 1.1.5 FUEL OIL SYSTEM

A typical system includes tankage from which the circulation pumps take suction, pumping the fuel oil through the heaters and strainers to the main circulation system. This serves all units that are potential users of fuel oil and returns to the tank, through a back-pressure controller.

The system should be designed to supply fuel oil to the furnaces at constant pressure and at the required viscosity. The required pressure depends on the type of burners used in the furnaces. The viscosity requirement should be met by means of temperature control.

The system should be designed so that from the fuel oil tanks, one supply and return header serves the processing units while a separate supply and return header serves the boiler plant.

In the system design, particular attention should be paid to the following:

1. *Piping System.* In the case of heavy fuel oil, measures should be taken to prevent plugging of lines. These may include heat tracing, insulation, and a separate flushing oil system (low pour point fuel). The flushing oil system will facilitate furnace starting up and shutting down operations and flushing out of lines, filters, and fuel oil heaters.

2. *Circulation Pumps.* To provide a reliable supply of fuel oil at least three pumps should be used. Typically, at least one pump should be turbine driven (upon availability of steam) and the others motor driven.

   Automatic cut-in of the stand-by pump should be provided when pressure in the fuel system becomes low. Loss of one pump may nevertheless result in a considerable pressure transient in the fuel oil supply system, which may cause furnaces to trip. By having three pumps each of about 70% capacity this effect is reduced considerably.

3. *Instrumentation.* The system should be equipped with a low-pressure alarm for supply header, located in each control house. Heaters on each fuel oil tank should be able to keep the content at about 65°C. This temperature should be limited to a maximum of 115°C to minimize the possibility of boil-over due to vaporization of water in the tanks. The fuel oil supply header temperature should be maintained at a temperature consistent with burner supply viscosity requirements.

   To obtain the required fuel oil supply temperature adequate heat exchangers (fuel oil heaters) heated by 2000 kPa (g) [20 bar (g)] medium pressure steam should be provided. These heaters should be installed in parallel arrangement, and all will be required to be in service when maximum fuel oil consumption is experienced.

   The fuel oil supply temperature should be regulated by controlling steam flow to the heaters.

   By using fuel oil at each unit, provision should be made for a fuel oil return line with block valve.

   A fuel oil return meter should be provided on each unit that consumes fuel oil. The recirculating fuel oil should be returned at a substantially temperature difference with respect to the exchanger effluent, and it may be directed back to the tank through the small vapor disengaging drum. Smoother operation will result, if it is always directed into the pump suction while the tanks are only heated to about 65°C. The fuel oil lines should be steam traced.

   The fuel oil system should be designed such that at least two parts are supplied to the heater, one part burned, and one part returned. Unless otherwise specified the size of the return header should be the same as the size of the supply header.

   Separate nozzles should be provided on storage tanks for the make up of fuel oil, recirculation, and the withdrawal of oil. The arrangement of nozzles should minimize any short-circuiting of oil that has recirculated.

   The fuel oil supply header should be controlled at a minimum pressure of 1000 kPa (g) [10 bar(g)], unless otherwise specified for process requirements.

   Relief valves should be located on the discharge of the pumps and on the fuel oil heaters. Relief valve discharges should be piped back to the fuel oil storage tank.

4. *Strainers.* To prevent plugging of the burners, parallel strainers should be installed in the discharge and suction of fuel oil distribution pumps, with the mesh sizes of 0.75 and 1.5 mm respectively (unless otherwise specified by the pump manufacturer).

### 1.1.6 **REFINERY GASOLINE FUEL**

Refinery gasoline fuel (visbreaker gasoline) may be considered as an alternative liquid fuel in steam boilers. A gasoline fuel system should have its own facilities for storage, pumping, and filters.

To accommodate variations in gasoline fuel demand, pressure control spillbacks should be considered to allow excess fuel to be returned to the storage tanks as required.

The following instruments should be provided in a boiler house control room:

1. Visbreaker gasoline storage tank low-level alarm.
2. Visbreaker gasoline supply header, pressure indication, and low-pressure alarm.

   Surge drum. Provision should be made to pump LPG directly into the flare header (if necessary).*

   LPG surge drum. Provision should be made to cut incoming LPG streams to the surge drum and pump LPG directly into the flare header (if necessary).

By-pass line for LPG fuel pumps should be provided to transfer LPG from the LPG surge drum to the fuel gas vaporizer, in the event of high pressure in the LPG surge drum.

LPG fuel pumps should have a minimum flow by-pass line to protect them at times of low consumption of LPG.

Instrumentation should be provided to start the spare LPG fuel pump automatically in case of failure of the main pump.

Size of the vaporizer, that is, whether a heat exchanger is required depends upon the following factors:

1. maximum gas demand
2. size and location of LPG surge drum
3. minimum amount of gas carried in LPG surge drum
4. climatic conditions
5. gas pressure to be supplied by plant.

   Location of the safety valve on the fuel gas vaporizer should be in the vapor portion of the vaporizer to avoid the problem of having LPG going into the flare header.

   A vaporizer should be equipped with an automatic means of preventing liquid passing from vaporizer to gas discharge piping. Normally this should be done by a liquid level controller and positive shut-off liquid inlet line or by a temperature control unit for shutting off the liquid line at low temperature conditions within the vaporizer.

## 1.2 **FIRED HEATERS FUEL SYSTEM**

The fuel system should be in accordance with the following requirements.

The pilot gas, where practicable, should be taken from a sweet gas supply, independent of the main burner gas, or from a separate off-take on the fuel gas main, with its own block valve and spade-off

---

*This situation may occur temporarily due to low gas consumption.

position. Unless otherwise approved by the Company, the pilot gas pressure should be controlled at 35 kPa (0.35 bar) and the pressure-regulating valve should be the self-operating type.

Fuel manifolds around heaters should be sized such that the maximum pressure difference between individual burner off-takes should not exceed 2% of the manifold pressure at any time. In addition, account should be taken of the effect pipework sizes and arrangements of individual burners have on the distribution of fuel flow to each burner.

Individual burner isolation valves for the main fuels and steam should be located under the heater. The burner isolation valves, excluding pilots, should be located within an arm's length of the peep holes giving a view of the flames from those burners. Where possible, a standard disposition of valves for each burner should be used; namely: from left to right, gas, oil, and steam.

All burners and pilot isolation valves should be of the ball valve type meeting BS 5351 or equivalent, subject to the operating temperature and pressure, including any purge steam, being within the rating of the valve seat. All burner isolation valves should have some readily recognizable indication of the valve position.

Each burner isolation valve for pilot gas should be positioned safely away from the burner and so that an electrical portable ignitor, when inserted in the lighting port, can be remotely operated from the burner valve position. In the case of floor-fired heaters, the pilot burner valves should not be located under the heater and should be operable from grade.

## 1.3 MINIMUM DATA REQUIRED FOR BASIC DESIGN

The following data should be provided as a minimum requirement for basic design calculation of liquid fuel to be used for normal operation or alternative operations, including startup.

- Net heating value, in (kJ/kg)
- Gross heating value, in (kJ/kg)
- Sulfur, in mass, in (mg/kg)
- Vanadium, in mass, in (mg/kg)
- Sodium, in mass, in (mg/kg)
- Nickel, in mass, in (mg/kg)
- Iron, in mass, in (mg/kg)
- Conradson Carbon, in (mass %)Ash, in (mass %)
- Other impurities in (mass %) or mass, (mg/kg)
- API
- Viscosity: dynamic in (Pa.s) at 100°C or at specified temperature (°C)
- Vapor pressure, in (Pa) at specified temperature (°C)
- Flash Point, in (°C)
- Pour Point, in (°C)
- Supply header operating pressure, in [kPa (g)] or [bar (g)] (max., normal, min.)
- Return header operating pressure, in [kPa (g)] or [bar (g)] (max., normal, min.)
- Supply header operating temperature, in (°C) (max., normal, min.)
- System mechanical design pressure and temperature, in [kPa (bar)] and (°C).

The following data should be provided as a minimum requirement for basic design calculation of fuel gas to be used for normal operation and for alternate operations, including startup, if pilot gas is not supplied from the fuel gas header, its properties should be provided.

- Relative density (specific gravity) at 15°C
- Net heating value, in $(MJ/Nm^3)$ or (kJ/kg)
- Gross heating value, in $(MJ/Nm^3)$ or (kJ/kg)
- Flowing temperature, in (°C) (max., normal, min.)
- Header operating pressure, in [kPa (g)] or [bar (g)] (max., normal, min.)
- System mechanical design pressure and temperature, in [kPa (g)] or [bar (g)] and (°C)
- Total sulfur, in mass, (mg/kg)
- Chloride, in mass, (mg/kg)
- Other impurities, in (volume %) or mass, (mg/kg)
- Flow rate available, in (Nm3/h).

The valves for controlling the flow of foul or waste gases to the individual nozzles should not be located underneath floor-fired heaters but should be positioned near the pilot gas valves.

A flame trap of an approved type should be fitted in the main foul or waste gas lines leading to a furnace, with a high temperature alarm actuator installed immediately downstream of the trap. Cleaning of the traps should be provided.

Irrespective of any purging arrangements within the burners, steam purging of the oil lines between the burner valves and the burners should be fitted.

The gas lines between the burner isolation valves and the burners should be fitted with a purge connection.

The steam and purge valves should be located adjacent to the burner isolation valves.

Each fuel supply header to a heater and individual pilot gas supplies to each burner, excluding waste or foul gases, should be fitted with two filters in parallel or with dual filters. Where the latter incorporate two filter elements in one housing, individual elements should be removable whilst in service without interruption of fuel flow. There should be no leakage from the operating compartment to the open compartment when one element is removed.

The filter mesh sizes should be as specified by the burner supplier and approved by the Company. The mesh material on main gas and pilot gas should be Monel. For the pilot gas filter the mesh size should be approximately 0.5 mm. In the case of the pilot gas supply, the pipework between the filters and the pilots should be in 18/8 stainless steel.

Piping should be in accordance with relevant standards, except that where fuel atomizers or gas nozzles require positional adjustment within the burner for optimum combustion, flexible piping for all fuels and steam connections to individual burners should be provided. This flexible piping should be of the fireproof continuously formed stainless steel bellows type, protected by metal braiding and approved by the Company.

The fuel oil, atomizing steam, and gas piping to the burners should be arranged so that the oil, main gas, or pilot nozzles can be removed without isolating the other fuel supply to that burner.

Individual gas and oil burner off-takes should be from the top of the headers. The ends of oil and fuel gas headers should be flanged to allow access for cleaning.

Pilot gas pressure reducing valves should be of the self-operated type and in accordance with API PR 550. They should be provided with isolation and hand-operated bypass valves.

All fuel control valves and meters should be conveniently located at grade and a safe distance from the furnace.

## 1.4 ATOMIZING STEAM AND TRACING

The atomizing steam supply should be run from the main and separately from the steam tracing supply and should not be used as steam tracing. Additionally, where light distillate fuel firing is specified, the atomizing steam lines should be lagged separately from the fuel lines to prevent vapor locking.

Atomizing steam off-takes to the burners should be from the top of the header and adequate trapping arrangements should be provided to prevent the admission of condensate to the burners, including steam traps at the end of manifolds.

Unless otherwise specified by the burner Vendor, the atomizing steam pressure should be controlled by a steam/oil differential pressure controller capable of operating over the specified firing range, or by a steam pressure controller.

Tracing of the fuel lines should be separated from other tracing systems. The heavy fuel oil system, including instrument legs, is to be traced right through to the burner, but that section of the fuel line common to both low flash and heavy fuel oils should be traced separately from the rest of the heavy fuel oil system. Tracing may be by steam or electricity.

Arrangements should be made to ensure that traced lines and associated instrumentation are not over pressured due to overheating if the fuel oil becomes stationary in the lines for extended periods.

Unless otherwise approved, fuel gas lines upstream of the burner isolating valves should be traced.

## 1.5 SHUT-OFF SYSTEMS

To ensure the effective isolation of furnaces from remote control positions, a solenoid initiated shut-off valve should be installed in each main furnace fuel line in addition to the control valve, and in each waste or foul gas line. They should be installed next to the control valves. These valves will normally be shut by remote manual or automatic initiation, for example, by "Heat-Off Switch," but opened only by local manual operation. Additionally, these valves should be shut automatically when the main fuel pressure upstream of the control valves falls below the stable burning limit of the main burners, or the atomizing steam falls below a predetermined pressure.

Consideration should also be given to having the shut-off valve close automatically in the case of high liquid level in the fuel gas knockout drum. The shut-off valve should be operable from the control room. In addition a solenoid initiated shut-off valve should be installed in the pilot gas line to be operated only by the "Emergency Shutdown Switch."

All systems should have a fail safe, that is, in normal operating conditions sensor contacts should be closed, relays and solenoid valves should be energized, and in the trip conditions, air-operated valves should vent.

## 1.6 **GAS TURBINE FUEL ALTERNATIVES**

For a gas turbines fuel system, reference should be made to API standard 616.

### 1.6.1 **GASEOUS FUELS**

For LPG a liquid phase formation in the combustor should be avoided.

For natural gas/LNG boil off, the inlet gas temperature should be above the dew point of liquid hydrocarbons.

For sour gas the following consideration should be applied:

1. corrosion-resistant gas supply hardware
2. any heat recovery equipment should have cold end protection.

Process gas, due to the wide variation in composition should be considered on a case-by-case basis.

Practically, all types of gaseous fuels should be burned in heavy-duty gas turbines, but do not necessarily have to be interchangeable in the same machine.

The standard gas turbine should be designed for a natural gas specification. A fuel falling outside these requirements should be accommodated by suitable modifications to the turbine control system, gas-fuel components, rating, and fuel handling equipment. Fig. 1.1 represents the fuel system of standard turbine.

The liquid hydrocarbon content of natural gas should be reduced to a maximum of 12 L/Nm$^3$ "dry gas" before using it in a gas turbine.

Natural gas may have appreciable levels of hydrogen sulfide as a significant contaminant, which is known as sour gas. This hydrogen sulfide should be removed by fuel treatment. In some cases, it may

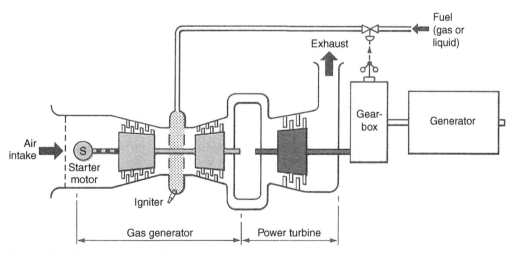

**FIGURE 1.1 Fuel System in Gas Turbine**

be burned directly in the gas turbine if the proper selection is made of materials and components in the gas turbine end fuel system.

## 1.6.2 LIQUID FUELS

Gas turbine liquid fuels have a wide range of properties, but for gas turbine application they should be divided into two broad classes:

1. True distillate fuels, which can normally be used without any change, but just as they are.
2. Ash-forming fuels, which generally require heating, fuel treating, and periodic cleaning.

Ash-forming fuels should require on-site fuel treatment to modify or remove harmful constituents. In addition, there should be provisions for periodically cleaning ash deposits from the turbine.

Liquid fuels, ranging from naphtha to residual fuels, should be successfully used in heavy-duty gas turbines.

True distillate fuels do not usually require heating for proper atomization, except for the heavy distillates and some light distillate used in cold regions. Heavy fuels should always require heating for proper fuel atomization; the temperature required being related to the type of fuel atomization.

For heavy residual fuels it may be necessary to heat the fuel to reduce the viscosity to the operating range of the fuel transfer and filter system. It may also be necessary to heat some crude and heavy distillates to keep wax dissolved.

A secondary and start-up/shut-down fuel should be considered for naphtha for safety reasons. A secondary fuel may need to be ready for heavy fuels, both for fuel system flushing and to provide fuel lightoff.

Explosion proofing of the gas turbine system may be required when used with low flash point fuels such as naphthas and some crude oils.

Gas turbines for heavy-fuel application may require a combustion liner designed for a more radiant flame.

## 1.6.3 CLASSIFICATION OF PETROLEUM FUELS

- *Gaseous fuels.* Gaseous fuels of petroleum origin that consist essentially of methane and/or ethane.
- *Liquefied gaseous fuels.* Gaseous fuels of petroleum origin which consist predominantly of propane–propane and/or butanes–butenes.
- *Distillate fuels.* Fuels of petroleum origin, but excluding LPG. These include gasoline, kerosenes, gas-oils, and diesel fuels. Heavy distillates may contain small quantities of residues. The products belonging to distillate fuels can be obtained not only by distillation, but also for example, by cracking, alkylation, etc.
- *Residue fuels.* Petroleum fuels containing residues of distillation processes.
- *Petroleum cokes.* Solid fuels of petroleum origin consisting essentially of carbon, mostly obtained by cracking processes.

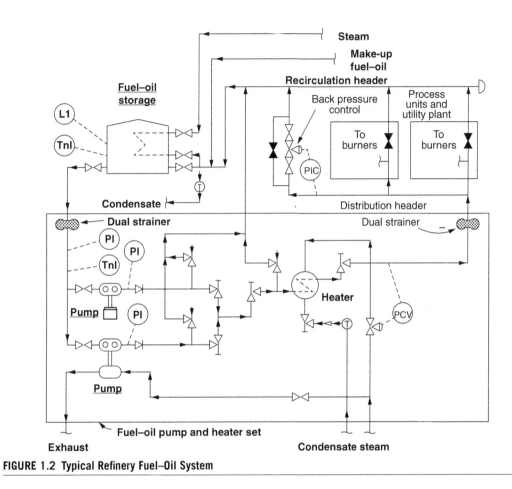

**FIGURE 1.2 Typical Refinery Fuel–Oil System**

Figs. 1.2 and 1.3 show typical refinery fuel–oil and fuel–gas systems.

## 1.7 **OPERATION OF HEAT-OFF AND EMERGENCY SHUTDOWN SWITCHES**
### 1.7.1 **HEAT-OFF SWITCH**

Heat-off switch operation should include one or more of the following:

1. shut-off all fuel supplies, with the exception of pilot gas supplies, to all fired process heaters
2. shut-off heat to reboilers and feed pre-heater
3. in certain cases stop the unit charge pumps. In such cases these should be agreed with the Company.

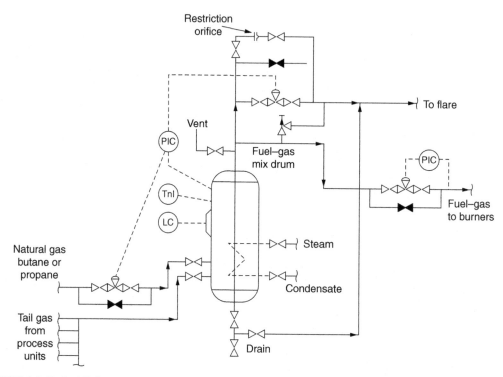

**FIGURE 1.3 Typical Refinery Fuel–Gas System**

## 1.7.2 EMERGENCY SHUTDOWN SWITCH

Emergency shutdown switch operation should include one or more of the following:

1.  perform all the operations listed in heat-off (D.1)
2.  initiate appropriate automatic devices
3.  shut-off all nominated feeds to the Unit
4.  fail safe critical control valves, a list of which is to be submitted to the Company
5.  shut-off pilots for gas supplied to heaters.

# COMPRESSED AIR SYSTEMS

## 2.1 COMPRESSED AIR SYSTEMS

In the oil and gas industrial sites air compressors with auxiliary equipment like dryers, filters, etc., are usually located in a central place, and this area is called the compressed air central plant. The overall energy efficiency of the compressed air system is greatly affected by the full load and part load compressor efficiencies, the control system that operates and sequences the compressors, and various other equipments in the central plant.

The plant air requirements should not be designed for coincident operation. Optimum schedule for overhaul, regeneration, decoking, etc. should be considered when arriving at the total capacity of the complex.

Determining the true demand in a compressed air system can be difficult, but it is a critical first step. Air demand often fluctuates significantly; however, if actual flow demand at any given time is known, storage and distribution systems can be designed to meet demand without installing additional compressors.

The most accurate method of establishing demand over time in an existing system is to monitor the air flow using a flow meter. Flow meters can be installed at various points in the system, but are typically installed in the main headers. Recorded data, such as airflow and pressure, can be evaluated to determine the flow pattern. It is especially important to note the peaks and valleys in demand and their duration.

Electronic data loggers offer an effective means to track compressor activity over time. While they don't directly measure as many system parameters as flow meters, they provide substantial information to make an accurate assessment of system dynamics.

For smaller, less complex systems, the ratio between loaded and unloaded compressor running time (on-line/off-line or start/stop times) can indicate average demand over a long period of time.

When establishing compressed air demand for a new system, we must consider operating pressure requirements and the duty cycle of individual equipment.

Equipment using compressed air is rated by the manufacturer for optimum performance at a certain pressure and airflow. To design a compressed air system that delivers uniform pressure, it is necessary to ensure that all tools and equipment work efficiently within a narrow pressure range. If this cannot be done, you have the option of operating the entire system at the higher pressure and regulating pressure down as required, adding a booster compressor to increase pressure for a particular application, or installing two independent.

A rough estimate of probable air requirements can be made based on Table 2.1 which represents the typical air requirement of one refinery.

| Table 2.1 Typical Refinery Air Requirement Data (Complexity Factor = 7) | | |
|---|---|---|
| @ 1325 Sm³/h (200,000 bbl/sd) crude charge | | |
| **A. Normal process consumption** | | |
| **Item** | **S m³/h** | |
| Instruments, all plants | 2875 | |
| Asphalt plant | * | |
| Plant air, all plants | 2550 | |
| | 5425 | |
| **B. Regeneration and decoking air requirements** | | |
| **Unit design capacity (Sm³/h)** | | |
| Crude heater | 663 | 480 |
| Vacuum heater | 324 | 186 |
| Platformer catalyst | 111 | 120 |
| Unifiner catalysts | 111 | 420 |
| Visbreaker heater | 126 | 247 |
| Isomax catalyst | 119 | 6800 |
| Nitrogen plant | 1700 (N₂ product) | 850 |
| Lube unit heater, i e, | | 231 |
| Propane deasphalting | 25 | |
| Furfural | 48 | |

*Installed three compressors each of 6796 Sm³/h.*
*\*Air requirement for asphalt plant oxidizer air would be produced by separate system, (about 51–85 m³ air per ton asphalt feed at 207 kPa (ga) like air blower, compressor, etc.*

Two items are involved in the load factor, the first is the time factor, which is the percentage of the total work time during which a device is actually in use. The second is the work factor, which is the percentage of air for maximum possible output of work per hour that is required for the work actually being performed by the device. The load factor is the product of time factor and work factor.

Most air compressors are controlled by line pressure. Typically, a drop in pressure signals an increase in demand, which is corrected by increased compressor output. Rising pressure, indicating a drop in demand, causes a reduction in compressor output.

The object of installing a compressed air system should be to provide air to the various points of application in sufficient quantity and quality and with adequate pressure for efficient operation of air tools or other pneumatic devices.

Compressors use various capacity control systems to monitor these changes in pressure and adapt the air supply to the changing air demand. One of the more efficient is a load/no load control which runs the compressor at full load or idle, accommodating the demand variations.

Total plant supply can be provided by either a single compressor or a multiple compressor installation which can be centralized or decentralized. Single compressor installations are best suited to smaller systems or systems which operate almost exclusively at full output.

Multiple compressor installations offer numerous advantages including the following:

- Application flexibility (the ability to efficiently adjust to shift demand variations)
- Maintenance flexibility

- The option of centralized or decentralized operation
- Floor space flexibility
- Backup capability

It is important that this study should include, in addition to immediate applications, anticipated future uses as well, since the availability of compressed air always leads to new applications.

The total air requirement is the sum of the plant air and instrument air inclusive of a contingency for each category.

The ratio of actual air consumption to the maximum continuous full-load air consumption, each measured in cubic meter of air per hour, is known as the load factor.

Pneumatic devices are generally operated only intermittently and are often operated at less than full load capacity. It is essential that the best possible determination or estimate of load factor be used in arriving at the plant capacity needed.

A compressed air system may consist essentially of one or more compressors with a power source, control system, intake air filter, after cooler and separator, air receiver, air dryer, and inter connecting piping, plus a distribution system to carry the air to points of application.

Before attempting to determine the amount of compressed air required, an investigation should be made of all likely as well as known air applications.

Sufficient air storage to meet short-term high demands should also be available. Determination of the average air consumption is facilitated by the use of the concept of the load factor.

Air for steam-air decoking requirement should be considered based on furnace tube size and as an estimation it can be 10% of steam flow or 2.93 kg/s.m2 of furnace tube.

The total air requirement for pneumatic devices should not be the total of the individual maximum requirement but the sum of the average air consumption of each.

A brief compilation of maximum air requirements of various tools is shown as a guide in Tables 2.2 and 2.3. It should be used for preliminary estimates, but tool and compressor manufacturers should be consulted in the final design.

A compressed air storage system consists of all the compressed-air containing vessels in your compressed air system. Sufficient storage is critical and represents available energy that can be released or replenished at any time as required.

The air receiver tank typically makes up the bulk of total storage capacity. Because some compressor controls (start/stop and on-line/off-line) depend on storage to limit maximum cycling frequency at demands less than 100% of supply, a properly sized receiver tank prevents excessive cycling.

If air requirements of a manufacturing process are evaluated on a basis of unit production in cubic meters of free air (standard air) per piece produced, they should then be combined on the basis of total production to arrive at the average rate of air required.

Flow controllers are also extremely important. Installation of a flow controller after the receiver tank is essential for providing additional compressed air when needed without downstream pressure fluctuations. The flow controller basically works like a precision regulator, increasing or reducing flow to maintain constant line pressure. It also provides the necessary pressure differential between the receiver tank and the system to create storage without changing system pressure downstream.

For determination of $m^3$/h required for various tools, the procedure should be as follows:

- Load factor (product of time factor and work factor) (B),
- Number of tools (A),
- $m^3$/h per tool when operating (C).

**Table 2.2  Typical Air Requirements of Various Tools**

| Tool | Free Air (m³/h) at 620 kPa (ga.) 100% Load Factor |
|---|---|
| Grinder 152.4 mm and 203.2 mm wheels | 85 |
| Grinder 50.8 mm and 63.5 mm wheels | 24–34 |
| File and burr machines | 31 |
| Rotary sanders, 228.6 mm pads | 90 |
| Rotary, sander, 177.8 mm pads | 51 |
| Sand rammers and tampers | |
| 25.4 mm × 101.6 mm cylinder | 42 |
| 31.75 mm × 127 mm cylinder | 48 |
| 38.1 mm × 152.4 mm cylinder | 66 |
| Chipping hammers, weighing 4.53–5.89 kg | 48–51 |
| Heavy | 66 |
| Weighing 0.9–1.81 kg | 20 |
| Nut setters to 7.93 mm weighing 3.62 kg | 34 |
| Nut setters 12.7–19.05 mm weighing 8.16 kg | 51 |
| Sump pumps 32.93 m³/h (a 15.24 m head) | 119 |
| Paint spray, average | 12 |
| Varies from | 3–34 |
| Bushing tools (monument) | 25–42 |
| Carving tools (monument) | 17–25 |
| Plug drills | 68–85 |
| Riveters 2.38 mm–25.4 mm rivets | 20 |
| Larger weighing 8.16–9.97 kg | 59 |
| Rivet busters 59–66 | |
| Wood borers TO 25.4 mm diameter weighing 1.81 kg | 68 |
| 50.8 mm Diameter weighing 11.79 kg | 136 |
| Steel drills, rotary motors | |
| Capacity up to 6.35 mm weighing 0.56–1.81 kg | 31–34 |
| Capacity 6.35–9.52 mm weighing 2.72–3.62 kg | 34–68 |
| Capacity 12.7–19.05 mm weighing 4.08–6.35 kg | 119 |
| Capacity 22.22–25.4 mm weighing 11.33 kg | 136 |
| Capacity 31.75 mm weighing 13.6 kg | 161 |
| Steel drills, piston type | |

**Table 2.2  Typical Air Requirements of Various Tools (*cont.*)**

| Tool | Free Air (m³/h) at 620 kPa (ga.) 100% Load Factor |
|---|---|
| Capacity 12.7–19.05 mm weighing 5.89–6.80 kg | 76 |
| Capacity 22.22–31.75 mm weighing 11.33–13.60 kg | 127–136 |
| capacity 31.75–50.8 mm weighing 18.14–22.67 kg | 136–153 |
| Capacity 50.8–76.2 mm weighing 24.94–84.80 kg | 170–187 |

**Table 2.3  Cubic Meters of Air per Hour Required by Sandblast**

| Nozzle Diameter (mm) | Compressed Air Gage Pressure (kPa) | | | |
|---|---|---|---|---|
|  | 414 | 483 | 552 | 689 |
| 1.58 | 6.8 | 8.5 | 9.3 | 11.0 |
| 2.38 | 15.3 | 18.7 | 20.4 | 25.5 |
| 3.17 | 28.9 | 32.3 | 35.7 | 44.2 |
| 4.76 | 64.6 | 73.0 | 79.8 | 98.5 |
| 6.35 | 113.8 | 129.1 | 144.4 | 175.0 |
| 7.93 | 178.4 | 202.2 | 226.0 | 273.5 |
| 9.52 | 256.5 | 290.5 | 324.5 | 394.1 |
| 12.7 | 455.3 | 516.5 | 577.7 | 700.0 |

Because not all compressors in a multiple compressor system remain on-line at all times, the actual air supply at any given time can be less than the total system capacity. During the time required to bring additional compressor capacity on-line, stored compressed air can be used to prevent any pressure drop in the system.

The amount of storage capacity needed depends on the amount of excess demand in cubic feet, available pressure differential between the compressor station and point of use, compressor start-up time, as well as the time available to replenish stored compressed air.

The capacity of a dry air system should be established according to requirement of the instrument system. The instrument air requirement customarily should be considered about 1.6 Sm³/h of dry air for each instrument pilot in process units.

A properly sized receiver tank also provides sufficient storage capacity for any peaks in demands. During peak demand periods, a poorly designed system will experience a drop in pressure as air in excess of the system's capacity is taken from the system.

The total requirement of instrument air would vary with types of process facilities and instruments installed.

The requirement of a typical refinery process for units of air is shown in Table 2.4.

Instrument air systems are critical to the proper operation of oil and gas processing facilities since all the instruments and controls depend on dry, pressurized instrument air for operation. A typical instrument air system is shown in Fig. 2.1.

Most systems are designed with 100% (or more) backup in the air compressor systems. Systems are generally designed to provide 700 kPa (ga) air pressure for users. While most instrumentation components do not require this high pressure, often, large valves with rapid closure requirements need this pressure to operate properly.

Drying of the instrument air is critical to prevent fouling and possible freezing in the air system and instruments. A −40°C dew point specification is common but specifications vary according to the climatic conditions of the site. Drying is usually accomplished with activated alumina or molecular sieve fixed bed driers similar to natural gas dehydrators. The second type of regeneration is the "heatless" regeneration. In this system the beds are regenerated by depressuring the bed and using some of the dry

**Table 2.4 Typical Air Requirement of Each Refinery Process Units 994 Sm³/h (150.000 bb1/sd) Crude Charge**

| Item | Air (Sm³/h) | |
| --- | --- | --- |
| | Instrument 6.9 (bar ga) | Plant 6.9 (bar ga) |
| Crude and vacuum | 422 | (1142)* |
| Heavy naphtha hydrotreater and CCR | | |
| Platformer | 836 | (1165) |
| Visbreaker | 133 | (441) |
| LPG recovery | 46 | (91) |
| Hydrocracker | 244 | (2854) |
| Hydrogen | 211 | — |
| Amine treating and sour water stripper | 85 | (91) |
| Sulfur recovery and solidification | 106 | (6737) |
| Asphalt blowing | 116 | (5300) |
| Nitrogen | (54) | (91) |
| Off–site | 73 | 11 |
| Utility | 528 | 666 |
| Flare | 14 | (91) |
| Waste water treatment | 21 | 43 (254) |
| Lpg tankage and loading | 4 | 91 |
| Total | 2893 | (6737) |
| *Parentheses indicates intermittent services.* | | |

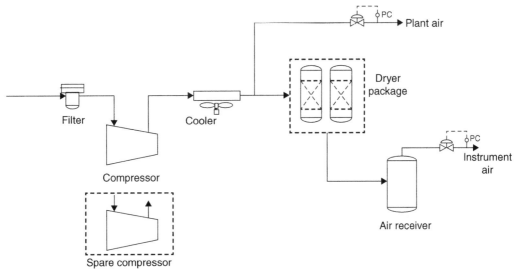

**FIGURE 2.1  A Typical Instrument Air System**

air to drive off the water. The "wet" air is vented. Instrument air is also sometimes dried by using refrigeration to chill the air and condense out the water. The air is then reheated to ambient user conditions. A number of different style compressors are used for instrument air systems. Since the air is needed at 700 kPa (ga), the compression ratio is over 7.5. If reciprocating compressors are used, multiple stages are necessary. Screw compressors (both oil flooded and oil free) are the most widely used style of compressors in this service.

## 2.2  SELECTING THE SIZE AND NUMBER OF COMPRESSORS

Air compressors of the rotary design usually require less downtime and should be considered when making a selection.

Air leakage of the compressed air system, customarily should be considered to be about 10% to the estimated amount of consumption.

In sizing and number of compressors, the following features should be considered:

- Total actual air requirement
- Estimated leakage of the air system
- Spare and operational philosophy
- Criticality of air supply
- Energy management
- Optimum size based on market availability

## 2.3 LOCATING THE COMPRESSORS

The centralized system has several inherent advantages. A good location may be selected where clean, cool air can be taken in. A separate compressor room should be justified in which the equipment can be protected from dirt and where control of maintenance should be relatively simple.

Water and electricity are needed only at the central compressor room rather than multiples throughout the plant.

The efficiency of larger compressors is generally higher than that of smaller machines. However, smaller air-cooled machines may be more economical and especially advantageous for variable demand situations.

In a centralized system noise may be better isolated or reduced. In a centralized system auxiliary equipment and controls may be provided, which could not be justified economically in smaller, multiple unit installations.

The cost of standby equipment should be weighed against the value of production losses, where compressed air is used for the regular operation of the plant and any shutdown may be very costly.

## 2.4 AUTOMATIC WARNING AND SHUTDOWN SYSTEMS

Protective control systems employing shutdown without indication, which merely drops out the holding coil of the motor starter, should usually be warranted only on very small compressors.

Some conditions, such as low oil pressure or excessive vibration, will demand immediate shutdown.

Compressor shutdown should be scheduled to occur at a preset point after a warning condition is reached. An automatic warning and shutdown system should be considered for installation with most air compressors for a completely reliable air system.

## 2.5 NONLUBRICATED AIR COMPRESSORS

Generally, nonlubricated compressors have a higher initial cost due to the special design and materials. Nonlubricated reciprocating air compressors also tend to have higher operating costs due to the higher maintenance cost associated with shorter valve and ring life.

Reciprocating, nonlubricated air compressors substitute low friction or self-lubricating materials such as carbon or teflon for piston and packing rings, where the piston is usually supported on wear rings of similar materials.

Oil free rotary screw and lobe-type compressors are available, having a design that does not require lubrication in the compression chamber for sealing and lubrication. Centrifugal air compressors are inherently nonlubricated due to their configuration.

## 2.6 COMPRESSED AIR DISTRIBUTION SYSTEM

For systems using only an oil-free compressor, it is strongly recommended that corrosion-resistant pipe be used. Unlike a system using lubricated compressors in which an oil film will form to protect the pipe from the corrosive effect of the moisture in the warm air, a nonlubricated system will experience corrosion. This corrosion can lead to contamination of products and control systems.

Each header or main should be provided with outlets as close as possible to the point of application. This permits the use of the shorter hose lengths and avoids large pressure drops through the hose. Outlets should always be taken from the top of the pipe line to prevent carry-over of condensed moisture to tools.

All piping should be sloped so that it drains toward a drop leg or moisture trap in order that condensation may be removed to prevent its reaching air-operated devices in which it would be harmful. The slope of lines should always be away from the compressor to prevent flow back into the compressor cylinder. A slope of about (2.0 mm/m) may be used, with drains provided at all low points. These may consist of a short pipe with a trap or drain at the bottom. All branches taken from the compressor discharge line should be from the top of the header.

Any drop in pressure between the compressor and the point of use is an unrecoverable loss. The distribution system is therefore one of the most important elements of the compressed air plant. In planning it, the following general rules should be observed.

Long distribution lines, including those in a loop system, should have receivers of liberal size located near the far ends or at points of occasional heavy use. Many peak demands for air are of short duration, and storage capacity near such points avoids an excessive pressure drop and may permit the use of a smaller compressor. Certain applications such as starting diesel engines or gas turbines are examples of this type of demand where the required rate may exceed the total compressor capacity.

Where possible, a loop system around the plant within each shop and building is recommended. This gives a two-way distribution to the point where air demand is greatest. The loop pipe should be made large enough that the pressure drop will not be excessive at any outlet regardless of the direction of flow around the loop.

Pipe sizes should be large enough that the pressure drop between the plot limits of air compression and consumer units does not exceed 10%.

## 2.7 AIR STORAGE (AIR RECEIVER)

A liberal "ASME" Air Receiver should be provided for the system.

The horizontal receiver should be sloped so that the collected water in the receiver moves toward the pot under the receiver for better drainage. The slope should be 2 mm/m.

The receiver serves several important functions. It damps pulsations from the discharge line, resulting in essentially steady pressure in the system. It serves as a reservoir to take care of sudden or unusually heavy demands in excess of compressor capacity. It prevents too frequent loading and unloading of the compressor. In addition, it serves to precipitate some of the moisture that may be present in the air, as it comes from the compressor or that may be carried over from the aftercooler.

## 2.8 INLET AND DISCHARGE PIPING

### 2.8.1 AIR INLET

The air inlet should always be located far enough from steam, gas, or oil engine exhaust pipes to ensure that the air will be free from dust, dirt, moisture, and contamination by exhaust gases.

A clean, cool, dry air should be supplied for a plant and instrument air compressor.

The compressor inlet should be taken from external air.

The filter should take air from at least 2 m or more from the ground or roof and should be located a few meters away from any wall to minimize the pulsating effects on the structure.

A silencer should be provided according to noise limitation.

Heating facilities should be provided where icing is expected.

## 2.8.2 INLET PIPING

It is frequently necessary to reduce the inlet-pipe diameter to match a centrifugal compressor inlet flange of a lesser diameter. Where such conversion is necessary, the transition should be gradual.

The inlet piping should be the full diameter of the intake opening of the compressor. If the inlet pipe is extremely long a larger size should be used.

The inlet piping should be as short and direct as possible, with long-radius elbows where bends are necessary. For centrifugal compressors, the air piping should be arranged for best performance to achieve uniform air velocity over the entire area of the compressor inlet. To attain this condition, there should be a run of straight pipe prior to the compressor inlet, with a length equivalent to about four diameters.

For large reciprocating compressors, the air inlet can be located on the bottom of the cylinder with the inlet piping located below floor level.

It is essential that underground piping be watertight, as the lower pressure within the pipe tends to draw leakage into the system.

## 2.8.3 AIR FILTERS

Air compressors handle large volumes of air over a period of time. Air-borne contamination that would otherwise seem insignificant can accumulate to a significant amount, the proper location of the inlet air source or inlet pipe can, to a great extent, minimize the amount of debris in the inlet air. However, it should be the job of the air filter to keep the quantity of abrasive materials that the air compressor would normally take in within acceptable limits.

For the best air filter selection a compromise among a number of variables such as filter design, compressor requirements, and atmospheric conditions should be made. Dry type filters should be selected.

Filters should provide a high degree of efficiency, to remove approximately 99% of particles larger than 10 $\mu$m (micron) and 98% or better of particles larger than 3 $\mu$m (micron).

Dry type filters should have some means of monitoring the air pressure drop through the element as an indicator of element contamination.

Depending on housing configuration, size, and element material, a clean element pressure drop of 75–200 mm of water (735–1960 Pa) is typical.

## 2.8.4 DISCHARGE PIPING

Unnecessary pockets should be avoided. The discharge pipe should be the full size of the compressor outlet or larger, and it should run directly to the aftercooler if one is used.

The discharge piping is considered to be the piping between the compressor and the aftercooler, the aftercooler separator, and the air receiver.

The discharge pipe should be as short and direct as possible, with long-radius elbows where bends are necessary, and should have as few fittings as possible. If the design cannot avoid pockets between the compressor and the aftercooler or receiver, it should be provided with a drain valve or automatic trap to avoid accumulation of oil and water mixture in the pipe itself.

For a centrifugal compressor, if it is necessary to increase the pipe diameter just beyond the compressor discharge flange, this transition should be gradual.

The installation of a safety valve between the aftercooler and the compressor discharge piping should be considered.

If an aftercooler is not used, the discharge pipe should run directly to the receiver, the latter should be set outdoors, and as close to the compressor as is practical.

Piping after the air receiver will have accessories dictated by the application (dryers for oil-free air) preseparators, afterseparators for the dryers, and so on.

If the discharge line is more than 30m long, pipe of the next size up diameter than that calculated should be used throughout. If the designer wants to install a shut off valve between the separator and the receiver, installation of a suitable safety valve between the compressor and the valve shut off point is mandatory.

A method of bleeding the air pressure from the system between the shutoff valves and the compressor discharge is required. This may consist of a simple plug valve located in the piping between these two points.

To detect possible clogging of aftercooler tubes, a means of monitoring the discharge pressure between the aftercooler and compressor discharge should be provided.

The main header size for plant air should not be less than DN 80 (3 in.).

## 2.9 **REFINERY AIR SYSTEM**

Discharge check valves on the air compressor should be installed as near to the compressor as possible. The moisture content of air should be considered in plant and instrument air design.

In the design of a refinery air system the required quantities and pressures for plant, instrument, and other air systems must be determined, and standby and crossover provision should be provided.

Air outlets should be provided in all areas requiring air such as for air-driven tools, exchangers, both ends of heaters, the compressor area, the top platform of reactors and on columns, so that each man way can be serviced with a 15 meter hose.

Specifications and capacities should be established for compressors, inter- and aftercoolers, receiver vessels, and air dryers when required. An air-distribution system between compressor locations and offsite users must be provided. These air facilities are frequently included within refinery utility plant limits to simplify air distribution systems.

### 2.9.1 **INSTRUMENT AIR**

Instrumentation should be provided so that the supply of instrument air has always the first priority and the spare unit(s) will start automatically in the event of failure of running unit(s) and/or excessive demand.

Instrument air pressure typically should be kept at 700 kPa (ga).

Instrumentation should be provided so that the spare unit will start automatically in the event of the failure of the running unit, and/or excessive demand.

### 2.9.2 PLANT AIR

Units using air in the process should have their own supply of air. Tying in to the refinery air header may be also considered for small users.

Air compressor(s) should be provided in addition to the available main air compressor(s) as an emergency spare, to supply separately air for the user in the process. Additionally, a separate compressor should be supplied for regeneration of hydrocracker and similar units. The compressor will also be used as a spare for the main air compressors.

## 2.10 AIR DRYERS

This equipment will dry compressed air having 100% humidity relative to the stipulated dew point, and should be durable for continuous operation under the specified pressure, humidity, and flow rate. The design dew point for plant air should be the lowest ambient recorded outside temperature under the line pressure.

The design dew point for instrument air should be more than 10°C below the lowest recorded outside temperature under the line pressure.

### 2.10.1 DESICCANT TYPE DRYER

The dryer has two desiccant drums, regeneration equipment, instrumentation, piping, and other accessories. While one desiccant drum is being operated, the other drum will be automatically regenerated.

The continuous operating time of one desiccant drum should not be less than 8 h under the specified conditions. However, based on ultimate life and other parameters, the operating time should be optimized.

When using an electric heater for regeneration integrated into the desiccant drum, this heater should be capable of being maintained without affecting the dryer operation.

### 2.10.2 INSTRUMENTATION

Operation control of the dryer should be fully automatic. Switching valves should not block the air flow if they fail unless an automatic back-up bypass valve across the dryer is installed.

Control panels should be provided as follows:

1. A local control panel for the air dryer should be employed.
2. The control panel should be equipped with a necessary instrument for monitoring operation of the air dryer.

### 2.10.3 AIR FILTER

An air after-filter with a drain valve should be installed at the exit of the dryer system. For the desiccant type dryer, the filter mesh should be less than 3 micrometers (microns).

An air pre-filter with a drain trap should be installed at the entrance to the desiccant drum.

Measures should be provided so that the flow of air will not stop even in the case of replacement or cleaning of the filter elements.

### 2.10.4 PERFORMANCE CHARACTERISTICS

Whenever use of packaged units are to be made, API Standards 672 and 680, should be considered for centrifugal and reciprocating compressors.

The following characteristics should be guaranteed:

- Inlet/outlet air-flow rate
- Dew point of outlet air
- Drying and regeneration cycle time
- Pressure drop through the dryer including air filters.

## 2.11 MOISTURE CONTENT OF THE AIR

Condensed water vapor can have corrosive effects on metals and wash out protective lubricants from tools and other pneumatic devices. To protect against such undesirable effects in a compressed air system, the use of various types of air drying systems has become increasingly popular.

All air contains moisture, the amount being influenced by pressure, temperature, and proximity to oceans, lakes, and rivers.

Relative humidity and dew point are two methods of indicating the amount of moisture in the air.

### 2.11.1 EFFECT OF PRESSURE AND TEMPERATURE ON RELATIVE HUMIDITY

Pressure has a major effect on the vapor content in air. The capacity of air at a given temperature to hold moisture in vapor form decreases as the pressure increases. Table 2.5 lists the water content of saturated air (relative humidity 100%) at given temperatures and pressures.

Table 2.6 lists the water content of air in milliliters per one cubic meter at various temperatures and relative humidities.

Fig. 2.2 illustrates the ability of air to hold water vapor at various temperatures while maintaining a constant pressure. It will be noted that as the temperature decreases the quantity of water vapor that can be held decreases, but the relative humidity remains constant. However, when the volume is reheated, its ability to hold moisture is increased; but since the excess moisture has been drained off, no additional moisture is available. The relative humidity therefore, decreases.

An illustration of pressure effect on vapor content and relative humidity is shown in Fig. 2.3.

Temperature itself has a significant effect on the ability of air at a given pressure to hold moisture. The higher the air temperature is, the greater its capacity to hold water vapor. Conversely, as the air temperature is lowered, its capacity to hold water vapor is decreased, measured in terms of relative humidity.

**Table 2.5  Water Content of Saturated Air in Milliliters per One Standard Cubic Meter**

| kPa (ga) | Temperature (°C) | | | | | | | | | |
|---|---|---|---|---|---|---|---|---|---|---|
|  | 2 | 5 | 10 | 15 | 20 | 25 | 30 | 35 | 40 | 45 |
| 0 | 5.37 | 6.71 | 9.38 | 13.20 | 18.24 | 25.03 | 33.93 | 45.67 | 61.04 | 80.94 |
| 68.9 | 3.18 | 3.97 | 5.56 | 7.79 | 10.74 | 14.69 | 19.83 | 26.52 | 35.16 | 46.15 |
| 137.9 | 2.25 | 2.81 | 3.94 | 5.53 | 7.61 | 10.40 | 14.00 | 18.68 | 24.69 | 32.27 |
| 206.8 | 1.75 | 2.18 | 3.06 | 4.29 | 5.90 | 8.05 | 10.83 | 14.43 | 19.02 | 24.81 |
| 275.8 | 1.43 | 1.79 | 2.50 | 3.49 | 4.82 | 6.56 | 8.82 | 11.75 | 15.47 | 20.15 |
| 344.7 | 1.22 | 1.51 | 2.11 | 2.95 | 4.06 | 5.53 | 7.44 | 9.90 | 13.04 | 16.97 |
| 413.7 | 1.05 | 1.30 | 1.83 | 2.56 | 3.52 | 4.79 | 6.43 | 8.56 | 11.26 | 14.65 |
| 482.6 | 0.93 | 1.14 | 1.61 | 2.25 | 3.09 | 4.22 | 5.67 | 7.54 | 9.92 | 12.90 |
| 551.6 | 0.82 | 1.04 | 1.44 | 2.01 | 2.77 | 3.77 | 5.06 | 6.73 | 8.86 | 11.51 |
| 620.5 | 0.75 | 0.93 | 1.31 | 1.82 | 2.50 | 3.42 | 4.58 | 6.09 | 9.00 | 10.39 |
| 689.5 | 0.68 | 0.85 | 1.19 | 1.66 | 2.29 | 3.11 | 4.18 | 5.55 | 7.29 | 9.47 |
| 758.4 | 0.62 | 0.78 | 1.09 | 1.52 | 2.10 | 2.87 | 3.85 | 5.10 | 6.70 | 8.71 |
| 827.4 | 0.58 | 0.72 | 1.02 | 1.41 | 1.95 | 2.64 | 3.55 | 4.72 | 6.21 | 8.05 |
| 896.3 | 0.54 | 0.67 | 0.95 | 1.32 | 1.81 | 2.46 | 4.22 | 4.39 | 5.77 | 7.48 |
| 965.3 | 0.50 | 0.63 | 0.88 | 1.23 | 1.69 | 2.30 | 3.09 | 4.11 | 5.40 | 7.00 |
| 1034.2 | 0.48 | 0.59 | 0.83 | 1.16 | 1.58 | 2.16 | 2.90 | 3.86 | 5.07 | 6.57 |
| 1103.2 | 0.45 | 0.55 | 0.77 | 1.09 | 1.50 | 2.04 | 2.74 | 3.63 | 4.77 | 6.19 |
| 1172.1 | 0.42 | 0.53 | 0.73 | 1.04 | 1.42 | 1.93 | 2.59 | 3.43 | 4.51 | 5.85 |
| 1241. | 0.40 | 0.50 | 0.69 | 0.97 | 1.34 | 1.83 | 2.45 | 3.26 | 4.28 | 5.55 |
| 1310 | 0.38 | 0.47 | 0.67 | 0.93 | 1.28 | 1.73 | 2.33 | 3.10 | 4.07 | 5.28 |
| 1379 | 0.37 | 0.45 | 0.64 | 0.87 | 1.21 | 1.66 | 2.22 | 2.95 | 3.87 | 5.03 |

## 2.12  EFFECT OF TEMPERATURE AND PRESSURE ON DEW POINT

It should be noted that as air leaves a compressor it is affected by both an elevated pressure and elevated temperature. A delicate balance exists under this condition since air under pressure has less capacity for water vapor, whereas air at elevated temperatures has a greater capacity for water vapor.

A more useful term than relative humidity for indicating the condition of water vapor in a compressed air system is dew point.

The dew point is the temperature at which condensate will begin to form if the air is cooled at constant pressure. At this point, the relative humidity is 100%.

The air leaving the compressor is generally saturated and any reduction in air temperature will cause water to begin to condense inside the downstream piping.

**Table 2.6  Water Content of Air in Milliliters per One Cubic Meter at Atmospheric Pressure**

| RH (%) | Temperature (°C) | | | | | | | | | |
|---|---|---|---|---|---|---|---|---|---|---|
| | **2** | **5** | **10** | **15** | **20** | **25** | **30** | **35** | **40** | **45** |
| 5 | 0.26 | 0.33 | 0.47 | 0.65 | 0.89 | 1.21 | 1.62 | 2.16 | 2.84 | 3.67 |
| 10 | 0.53 | 0.66 | 0.92 | 1.30 | 1.79 | 2.43 | 3.25 | 4.32 | 5.68 | 7.37 |
| 15 | 0.79 | 0.99 | 1.39 | 1.95 | 2.68 | 3.65 | 4.90 | 6.52 | 8.56 | 11.11 |
| 20 | 1.07 | 1.33 | 1.86 | 2.60 | 3.57 | 4.87 | 6.55 | 8.71 | 11.46 | 14.90 |
| 25 | 1.34 | 1.67 | 2.33 | 3.26 | 4.48 | 6.10 | 8.21 | 10.92 | 14.38 | 18.71 |
| 30 | 1.60 | 2.00 | 2.80 | 3.91 | 5.38 | 7.34 | 9.86 | 13.14 | 17.32 | 22.57 |
| 35 | 1.87 | 2.33 | 3.26 | 4.56 | 6.28 | 8.57 | 11.54 | 15.37 | 20.29 | 26.47 |
| 40 | 2.11 | 2.67 | 3.73 | 5.22 | 7.19 | 9.81 | 13.21 | 17.62 | 23.27 | 30.41 |
| 45 | 2.41 | 3.00 | 4.20 | 5.89 | 8.11 | 11.06 | 14.90 | 19.88 | 26.29 | 34.39 |
| 50 | 2.67 | 3.34 | 4.67 | 6.54 | 9.01 | 12.31 | 16.59 | 22.16 | 29.32 | 38.40 |
| 55 | 2.94 | 3.67 | 5.13 | 7.20 | 9.93 | 13.56 | 18.30 | 24.44 | 32.38 | 42.46 |
| 60 | 3.21 | 4.01 | 5.60 | 7.86 | 10.84 | 14.82 | 20.00 | 26.75 | 35.47 | 46.55 |
| 65 | 3.48 | 4.35 | 6.07 | 8.52 | 11.75 | 16.08 | 21.71 | 29.06 | 38.58 | 50.69 |
| 70 | 3.75 | 4.69 | 6.55 | 9.19 | 12.67 | 17.35 | 23.43 | 31.39 | 41.71 | 54.88 |
| 75 | 4.02 | 5.02 | 7.02 | 9.85 | 13.60 | 18.62 | 25.16 | 33.74 | 44.87 | 59.11 |
| 80 | 4.28 | 5.36 | 7.48 | 10.52 | 14.52 | 19.89 | 26.90 | 36.10 | 48.05 | 63.38 |
| 85 | 4.56 | 5.70 | 7.97 | 11.19 | 15.45 | 21.17 | 28.64 | 38.46 | 51.26 | 67.71 |
| 90 | 4.83 | 6.04 | 8.47 | 11.86 | 16.38 | 22.46 | 30.41 | 40.86 | 54.50 | 72.07 |
| 95 | 4.93 | 6.37 | 8.90 | 12.52 | 17.31 | 23.74 | 32.16 | 43.25 | 58.03 | 76.49 |
| 100 | 5.37 | 6.71 | 9.38 | 13.20 | 18.24 | 25.03 | 33.93 | 45.67 | 61.04 | 80.94 |

**EXAMPLE 2.1**

The amount of water vapor in 0.1 m$^3$ of air at 80%relative humidity and 25°C was determined from Table 2.6:

Now, if air is compressed from 0.1 to 0.01 m$^3$ (Fig. 2.6b), the pressure is increased to 912 kPa(ga) (from relation $P_1.V_1 = P_2.V_2$). That $P_1.V_1$ equals original pressure and volume, and $P_1,P_2$ must be stated as absolute pressures for calculation purposes. At 912 kPa(ga) and 25°C, the 0.01m$^3$ volume of air can hold only 0.0242 mL (refer to Table 2.5).

Since there were 1.989 mL in the air but it can now hold only 0.0242 mL, the excess moisture will condense. If the condensed water is not removed (see Fig. 2.3) and pressure is reduced to atmospheric, the excess water will gradually evaporate back into the air until an equilibrium is established. This will happen because the air under this condition can again hold 1.989 mL of water vapor. If the condensed water is removed as shown in Fig. 2.3d and the pressure is again reduced as shown in Fig. 2.3e, the excess water will not be available for evaporation back into the air. The water vapor content of 0.1 m³ of air will then be 0.0242 mL, which was the maximum vapor content that 0.01 m³ of air can hold at 25°C and 912 kPa (ga).

The 0.0242 mL per 0.1 m³ thus determined is 0.242 mL per 1 m³. Referring to Table 2.6, it will be seen that the 0.242 mL per 1 m³ is less than any quantity listed for 25°C. Therefore the relative humidity is less than 5%.

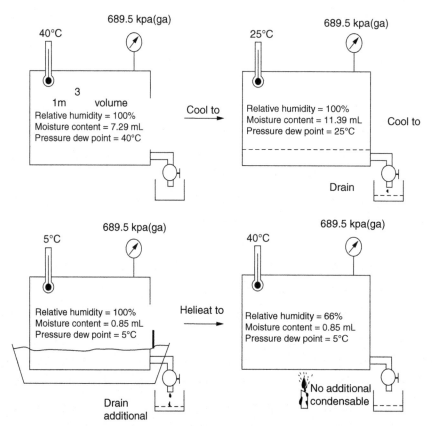

**FIGURE 2.2 How Temperature Influences the Capacity of Air to Hold Water Vapor, the Pressure Remaining Constant**

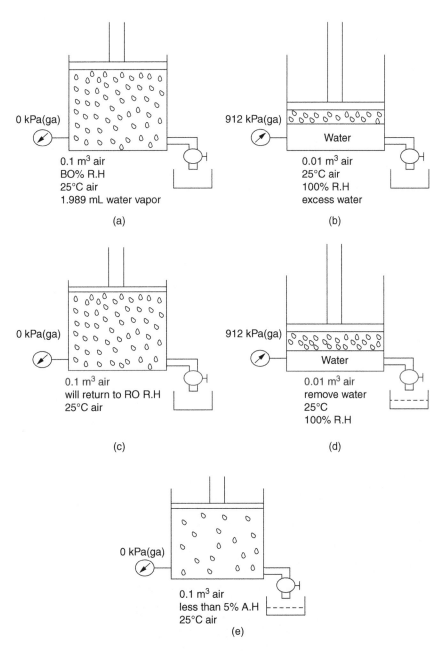

**FIGURE 2.3  How Change in Pressure Affects Moisture Content of Air at Constant Temperature**

**EXAMPLE 2.2**

As the air temperature is increased from 5 to 40°C, it again has the ability to hold 7.29 mL. However, since the excess moisture has been drained away, only 0.85 mL of moisture remain. The relative humidity is therefore reduced to (0.85/7.29) × 100 or 11.66%.

The 7.29 ml moisture content for saturated air (see Fig. 2.2) can be determined from Table 2.5 at the intersection of the 40°C column and 689.5 kPa(ga) row. The moisture content for 25 and 5°C air at 689.5 kPa(ga) can also be determined in the same manner.

## 2.13 **PRESSURE DEW POINT**

A pressure dew point temperature is more meaningful, since it indicates the temperature at which water vapor will begin to condense inside a pipeline at a given pressure.

For reference purposes, pressure dew points can be converted to atmospheric dew points by use of the graph shown in Fig. 2.4.

Refrigerant dryers are rated at pressure dew points, whereas desiccant dryers have been traditionally rated at atmospheric dew points. However, there is a trend toward rating desiccant dryers at the pressure dew point as well.

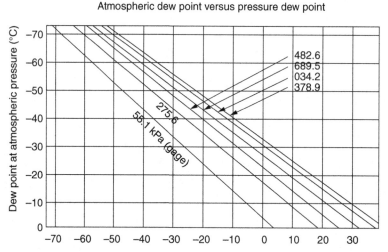

**FIGURE 2.4  Dew Point at Elevated Pressure in Degree Celsius**

## 2.14 **EFFECT OF MOISTURE ON AIR COMPRESSOR INTAKE CAPACITY**

It is not always necessary or desirable to assume that the air is saturated, however, this is the maximum condition with respect to water content.

When establishing intake capacity for air compressors, the moisture content of the air should be taken into account.

Intake air temperature should be selected with some recognition of maximum-minimum-normal for summer conditions. Fig. 2.5 is convenient for reading the air condition. Example 2.3 represents the sample calculation of using Fig. 2.5.

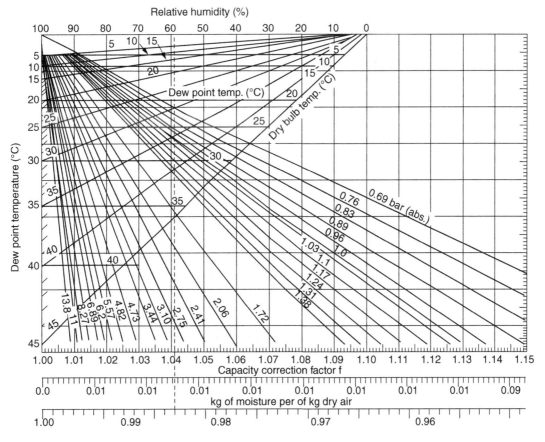

**FIGURE 2.5 Air Properties Compression Chart**

## EXAMPLE 2.3

A plant air system requires 17000 $Sm^3/h$ of dry air. The air is required at 6.89 bar abs. Intake conditions are: Atmospheric pressure at location = 0.89 bar (abs.) (89 kPa abs)

Temp. = 27°C
Relative humidity = 92%
Aftercooler: Water used can cool air to 23.9°C
Pressure at this point = 6.89 bar abs. (689 kPa abs).

Solution
Referring to Fig. 2.5 follow the dashed line starting at the left hand scale which shows the dry bulb temperature 27°C. Follow the line up and to the right to the intersection with vertical relative humidity of 92%, follow across to the intersection with an inlet pressure of 0.89 bar (abs) and read vertically down:

| | |
|---|---|
| Moisture capacity correction (F) | = 1.0412 |
| Water vapor (kg)/dry air (kg) | = 0.026 |

Relative density (specific gravity) of air–water vapor mixture = 0.985. In aftercooler conditions, the dry bulb equals the wet bulb temperature (air is saturated):

| | |
|---|---|
| Wet bulb = dry bulb temperature | = 23.9°C |
| Pressure = 6.89 bar (abs), (690 kPa abs) water (kg)/dry air (kg) | = 0.003 |
| Dry air capacity at inlet | $= 17000 \dfrac{(1.013)}{0.89} \times \dfrac{(300)}{288.5} = 20120\ ^3/h$ |
| Atmospheric air required | $= (1.0412)(20120) = 20949\ m^3/h$ |

Where,
 $R = 8.314$ kJ/kmol.K
 $P = $ kPa (abs)
 $T = $ K
 $MW = 28.69$ kg/kmol

Or,

$Vm = 0.286\ (T_1)/(Sp.\ Gr)(P_1)$
  $= 0.286(300)/(0.985)(89)$
  $= 0.978\ m^3$ moist air/kg

Mass of moist air = 20949/0.978 = 21420.2 kg/h

$$\text{Mass of dry air} = \frac{21420.2}{(1.0412)(0.985)} = 20885.9\ \text{kg/h}$$

Mass of water vapor entering compressor = 21420.2 − 20885.9 = 534.3 kg/h
Leaving:
Mass of water leaving in air from aftercooler = 20885.9 (0.003) = 62.7 kg/h
Mass of water condensed = 534.3 − 62 .7 = 471.6 kg/h

## 2.15  PIPE COMPONENTS DEFINITION OF NOMINAL SIZE

Nominal size (DN): A numerical designation of size which is common to all components in a piping system other than components designed by outside diameters or by thread size. It is a convenient round number for reference purposes and is only loosely related to manufacturing dimensions.

*Notes*:

1. It is designated by DN followed by a number.
2. It should be noted that not all piping components are designated by nominal size, for example steel tubes are designated and ordered by outside diameter and thickness.
3. The nominal size DN cannot be subject to measurement and should not be used for the purposes of calculation.

The purpose of this Table 2.7 is to present an equivalent identity for the piping components nominal size in the SI System and the Imperial Unit System.

## 2.16  INSTRUMENT AIR QUALITY AND QUANTITY

The instrument air should be dust free, oil free and dry.

Under normal operation the instrument air should have a pressure of at least 8.0 barg. in the buffer vessel, and a pressure of 7.0 barg. in the supply piping.

**Table 2.7  Pipe Component – Nominal Size**

| Nominal Size | | Nominal Size | | Nominal Size | | Nominal Size | |
| --- | --- | --- | --- | --- | --- | --- | --- |
| DN (1) | NPS (2) | DN | NPS | DN | NPS | DN | NPS |
| 6 | ¼ | 100 | 4 | 600 | 24 | 1100 | 44 |
| 15 | ½ | 125 | 5 | 650 | 26 | 1150 | 46 |
| 20 | ¾ | 150 | 6 | 700 | 28 | 1200 | 48 |
| 25 | 1 | 200 | 8 | 750 | 30 | 1300 | 52 |
| 32 | 1¼ | 250 | 10 | 800 | 32 | 1400 | 56 |
| 40 | 1½ | 300 | 12 | 850 | 34 | 1500 | 60 |
| 50 | 2 | 350 | 14 | 900 | 36 | 1800 | 72 |
| 65 | 2½ | 400 | 16 | 950 | 38 | | |
| 80 | 3 | 450 | 18 | 1000 | 40 | | |
| 90 | 3½ | 500 | 20 | 1050 | 42 | | |

*(1), Diameter nominal (mm); (2), nominal pipe size (inch).*

To prevent condensation in the supply piping or in the instruments, the dew point of the air at operating pressure after having been dried should always be at least 10°C lower than the lowest ambient temperature ever recorded in the area.

For the minimum allowable pressure during compressor failure, the required quantity of instrument air should be estimated as accurately as possible, taking into account the requirements for the following:

1. Pneumatically operated instrumentation, based on the data stated by the manufacturers or suppliers of such equipment.
2. Pressurizing the enclosures of electrical instruments located in hazardous areas.
3. Continuous dilution for enclosures of process stream analyzers, etc.
4. Regeneration of air drier, especially for heatless type as the required quantity is about 15–20% of drier outlet.

The consumption thus obtained should be multiplied by 1.3 to account for uncertainties in the data used for the estimate and for the installation of additional instruments during the first years of plant operation. Fig. 2.6 is a dew point conversion chart.

## 2.17 SEGREGATION

Where required for reasons of plant operation, the air supply system should include provisions for segregating certain plant sections or certain groups of users from others. This segregation should primarily be based on the importance of continued operation of a particular plant section of selected instruments in the case of partial or complete failure of the air supply plant.

In this context utility supply plants, such as electric power plants, boilers (with related de-aerators and boiler feed water pumps), fuel systems and cooling water pumps, are usually considered to be more essential than processing units. The latter may be segregated depending on the probability of calamities or the financial consequences of a sudden shutdown and/or the relative ease with which such a unit can be started up again after resumption of the instrument air supply.

The aforementioned segregation is achieved by installing one or more priority control valve(s) in the supply piping. Each priority control valve should consist of a pneumatically operated control valve with a local pneumatic pressure-indicating controller operating as a backpressure controller, that is, throttling the valve in the case of low air pressure in the upstream piping. Each priority valve should have block valves and a by-pass valve.

Where an emergency air compressor is installed, the distribution piping should be so arranged and where necessary provided with non-return devices that in the case of complete failure of the main compressors, only selected sections in the plant and the control center remain connected to the emergency compressor, with no possibility of back-flow to other sections.

Where no emergency compressors are installed, the consumers which must stay in operation after an air supply failure (such as depressurizing valves and pneumatically operated instrumentation in safeguarding systems) should be supplied from a buffer vessel which is connected to the distribution piping via a non-return device.

*Note*: Consideration should also be given to the installation of manually operated isolating valves in the distribution piping for segregating certain sections, for example, to allow the commissioning of these sections when plant construction is not yet complete.

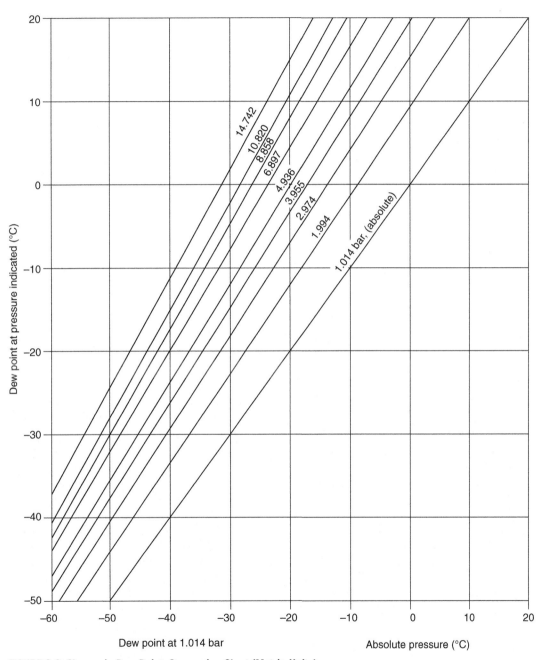

**FIGURE 2.6  Change in Dew Point–Conversion Chart (Metric Units)**

## 2.18 **TOOL AIR SUPPLY**

Where tool air is permanently required, the tool air supply system should be completely separate from the instrument air supply system, but consideration may be given to taking tool air from the instrument air compressors.

This is, however, only allowed if all the following requirements are satisfied:

*   The tool air is used for driving pneumatic tools only, and not for process applications such as blowing of asphalt or in-tank product blending, or for blowing-out of (plugged) process lines.
*   The compressors are adequately sized to provide the required quantity of tool air without detrimental effects on the instrument air supply.
*   The branch-off connection for tool air is upstream of the instrument air drier, and is provided with a non-return device and a safeguarding device ensuring priority for the instrument air.
*   At no other place are connections made between the tool air system and the instrument air system.

All cases where instrument air compressors are used for purposes other than supplying instrument air require the written agreement of the user.

## 2.19 **AIR SUPPLY PLANT**

The main air supply line should be provided with a flow-measuring element, and on the main panel a pressure recorder and a low-pressure alarm.

The air supply plant should be located in a non-hazardous area.

All piping interconnecting the compressors, buffer vessel, and air drier should be so arranged that each major piece of equipment can be taken out of operation without interrupting the air supply.

The piping between the compressor discharge, buffer vessel, and drier inlet should have automatic condensate draining facilities at all low points. In cold climates this piping, as well as the bottom part of the buffer vessel, should be (steam) traced and insulated.

A by-pass line with an automatic pressure control valve should be installed between the inlet and the outlet of the air drier. The valve should open at a low downstream pressure, eg, 5 barg, and should have a valve position switch which should initiate an alarm on the main panel when the valve starts to open.

The humidity should be measured with a water-content analyzer of the lithium chloride type (or equivalent), with a local indication and with a high-humidity alarm on the main panel.

Safety/relief valves should be provided when required by statutory regulations and/or by the relation between maximum compressor discharge pressure and the maximum allowable working pressure of vessels and piping.

An instrument air supply plant should be provided, comprising:

*   compressors.
*   buffer vessel.
*   air drier.

In addition, independent facilities may be required to ensure the continuation of instrument air supply to the utility supply plants and/or essential process instrumentation in emergency cases.

These facilities should then comprise an automatically starting emergency compressor with associated buffer vessel and air drier.

*Notes*:

1. This emergency compressor should be driven by a diesel engine, a petrol engine, or an electric motor; the latter only if an independent emergency electric generator or independent supply system is available. If an electric motor is used, it should be in accordance with Induction Motors.
2. Consideration may also be given to obtaining air for essential consumers from an outside plot instrument air supply system (if present).

## 2.20 **AIR COMPRESSORS**

To ensure maximum reliability of the instrument air supply, at least two compressors should be installed. These should be driven by two different and independent utilities, eg, steam and electricity. Each compressor should be arranged for normal operation and for stand-by, and should be capable of supplying the designed quantity of instrument air, plus the required quantity of tool air (4.4), and, if applicable, the required quantity of regeneration air.

Where it is essential to have stand-by also if one of the two compressors is not operational, eg, because of repairs or maintenance, the installation of a third compressor should be considered.

The installation of more than two compressors may also be considered for other reasons, eg, where the fluctuations in air consumption are greater than the range of one compressor, or where purchasing and maintaining a number of compressors each with a relatively low capacity is more attractive than a (small) number of compressors each with relatively large capacity.

In any case, the total capacity of the compressors driven by the most reliable utility should be sufficient to supply the design quantity of instrument air.

*Note*: In addition to the aforementioned compressors for normal plant operation, an independent emergency air compressor may be required.

### 2.20.1 **COMPRESSOR SPECIFICATION**

The compressor should be of the dry type cylinder and should supply oil-free air, and be complete with non-return valves, intercoolers, aftercoolers, condensate draining facilities, etc.

The compressors and their drives should satisfy the requirements for running equipment as specified by the user.

### 2.20.2 **COMPRESSOR PIPING**

The inlet of the compressors should be so located that the instrument air is free from toxic, obnoxious, or flammable gases, and is free from dust.

The inlet opening should be fitted with a wire mesh cage. The cage should be of adequate size to prevent flying papers, etc. from completely blocking the compressor inlet; the wire mesh should be adequate to prevent flying objects from entering the compressor and to prevent plugging by frost or hoar-frost.

Where the compressor inlet cannot be located in a completely dust-free area, consideration may be given to dust filters in the inlet piping.

To reduce the load on the air drier, the air from the compressors should be cooled to a temperature of 5–10°C above the cooling medium inlet temperature. Where the aftercoolers supplied as an integral part of the compressors are suspected of having only marginal capacity, the installation of additional aftercoolers should be considered.

### 2.20.3 COMPRESSOR CONTROLS

Each compressor should have the facilities for manual and automatic starting in the case of failure of the other compressor(s).

The automatic starting system should be so arranged that stopping of a compressor is only possible by manual control.

Automatic starting of the stand-by compressor(s) should be as quick as possible; in addition to an initiator on the piping downstream of the air drier. Initiators should be provided at each compressor discharge upstream of the non-return valve and/or on the compressor oil system. Starting of each stand-by compressor should be indicated by an alarm on the main panel.

The electric motor(s) should have local start/stop controls and be protected against repetitive starting. Electric controls supplied as integral parts of the compressor, as for oil filter, oil pump, oil heater, should be interlocked with the start/stop controls and should be located in a weatherproof housing on, or close to, the compressor.

## 2.21 AIR DRIER

The air drier should reduce the dew point of the air under operating pressure to at least 10°C below the lowest ambient temperature.

The air drier should normally be of the twin-vessel adsorption type, with regeneration. Switching of the vessels should be either manual or automatic.

The air drier should have a sight glass for indication of outlet air humidity.

The selected methods of drying, regeneration and switching are usually specified by the user; where this has not been done the contractor should submit a proposal for approval by the user.

### 2.21.1 THE DESICCANT

The desiccant should normally be activated alumina or silica gel or a molecular sieve in beaded form. When silica gel is used, a bottom layer (approx. 10%) of activated alumina should be provided to achieve a better resistance to entrained water. The quantity of desiccant should be such that adequate drying capacity is still available after the desiccant activity has deteriorated.

### 2.21.2 REGENERATION

The regeneration for a heater type drier should be at elevated temperature either at atmospheric pressure or at operating pressure.

Regeneration for a heatless air drier should be at ambient temperature and atmospheric pressure. The heat required for regeneration should be supplied by electric heaters or steam heaters. *Notes*:

1. When selecting electric heaters, it should be realized that these are large consumers of electric energy (approximately 30 kW for a drier of 0.5 m$^3$/s capacity at 15°C and 1.013 bar abs.), and this power must be available from emergency generators during prolonged power failures, for example, for more than 1 h.
2. Heatless regeneration is preferable.

Steam heaters should be of good mechanical construction to avoid leakage into the desiccant vessel.

For regeneration of a heater type drier at atmospheric pressure the water vapor is removed by means of air which can either be taken from the outlet of the drier (2–3 wt%) or be provided by a separate blower.

Where a separate blower is used, the heater should be external to the vessel, otherwise each vessel may be internally heated.

Where internal electric heaters are applied, these should preferably be removable during operation of the drier.

Where regeneration at atmospheric pressure is used, the vessel should be depressurized slowly to prevent blowing out and/or fragmentation of the desiccant and to reduce exhaust noise. After the desiccant has been regenerated, the vessel should be pressurized slowly before switch over.

For regeneration of a heater type drier at operating pressure, the regeneration air is taken upstream of a restriction in the drier inlet piping, heated by an electric or steam heater, passed through the desiccant to be regenerated, cooled, and (after separation of condensed water) returned to the drier inlet piping downstream of the restriction.

The cooler and water separator should be adequately sized. Quantity control for the regeneration air should be by means of a local flow-indicating controller with low-flow interlock on the heater and a pneumatically operated control valve with mechanical limit stop.

After the desiccant has been regenerated, it should be cooled by a flow of cold air (for heater type only).

For regeneration of a heatless type drier, the water vapor is removed by means of air which should be taken from the outlet of the drier (15–20 wt% as purge air to desorb the desiccant and carry the moisture to atmosphere).

## 2.21.3 **SWITCHING**

For switching the desiccant vessels from the drying stage to the regeneration stage and vice versa, the drier should be provided with a number of valves in vessel inlets and outlets. These valves may be four-way plug valves or four-way ball valves with mechanical interlocks, or individual valves with pneumatic operators which are interlocked via an automatic control system. Especially for a 4-in. and larger pipe , four-way valves may cause mechanical problems and individual valves are then preferred. *Notes*:

1. Valve bodies should preferably be made of cast steel, but cast iron may be used if agreed by the user.
2. Four-way valves on driers with regeneration at operating pressure should be of the opening-before-closing type, but carbon steel valves of this type are sometimes difficult to obtain.

**3.** Closing-before-opening may then be acceptable if there is an automatically controlled by-pass around the drier.

**4.** Where four-way plug valves or ball valves are used, consideration should be given to PTFE linings in order to reduce maintenance (greasing), air leakages and the force required for turning.

Switching should be initiated either manually or automatically.

*Note*: Manual switching is not applicable for heatless regeneration, as the time of regeneration is normally 10–20 m.

For a heater type regeneration switching should be on a once-per-shift (8 h) basis. Each drying vessel should then have a drying capacity equal to the design quantity of instrument air, during 10 h (minimum); the regeneration (including cooling) should not last more than 6 h.

Automatic switching should be integrated with the automatic controls for the regeneration cycle and should be either on a fixed time schedule or be initiated by a humidity instrument.

Pneumatic actuators for automatic switching should be suitable for an air pressure of 7.0 barg., but should still operate satisfactorily at 2.5 barg.

### 2.21.4 FILTERS

Prefilters may be necessary to prevent rust particles from settling on the desiccant.

Afterfilters (3 micron) should always be provided to prevent desiccant particles from entering the air supply piping and the instruments (see: ANSI/ISA-S 7.3).

All filters should be in duplicate and have isolating valves.

### 2.21.5 DRIER SPECIFICATION

The specification of the drier should contain all data necessary to ensure the supply of a suitable unit. Wherever possible a construction in accordance with the manufacturer's standard should be accepted.

For the design and construction of the vessels, BS 1515-Part I or the ASME Boiler and Pressure Vessel Code, Section VIII, Div. 1 or any other approved standard of equivalent authority is usually acceptable.

For the design, fabrication, erection, and testing of piping ANSI B16.5 and ANSI B31.3 are usually acceptable.

Where local regulations are more stringent than or conflicting with the requirements of the codes aforementioned, the former should prevail.

### 2.21.6 AFTERCOOLER

Because of the adsorption heat generated during the drying cycle, the outlet temperature of the drier may rise to 60°C. If the air cannot cool down to approximately 40°C before reaching the consumers, an aftercooler should be installed.

## 2.22 BUFFER VESSEL (AIR RECEIVER)

The buffer vessel should be of adequate size to serve:

- as condensate separator and draining vessel;
- as buffer volume during compressor failure;
- as fluctuation damper if compressors are on-load/off-load control.

The buffer vessel should be sized to maintain the air supply between the moment of compressor failure caused by mechanical failure of one compressor or failure of one utility supply for the compressor(s) normally in operation, and the moment that the stand-by compressor(s) is or are operating.

The period between these moments should be taken as the time required for starting the stand-by compressor(s) manually if automatic starting is unsuccessful, and should be determined by plant operations in connection with mechanical engineering and utility engineering, but should be at least 15 min.

During this period, the instrument air pressure should not drop below the minimum value required for proper operation of the instruments (especially control valves) and other services depending on instrument air. This minimum pressure can usually be taken as (3.0 barg), but may be higher for some special cases.

*Note*: If special equipment requiring an air pressure higher than 3.0 barg is used (eg, cylinder actuators for damper drives, or pressure repeaters), special devices such as volume chambers connected to the supply system via non-return devices may be considered to ensure that these individual consumers do not suffer from an unacceptable pressure drop.

Alternatively, consideration may be given to bottled high-pressure air or nitrogen as an emergency supply for such equipment.

The sizing of the buffer vessel should be based on the design quantity of instrument air plus the tool air consumption until the safeguarding device closes.

The buffer vessel should have automatic draining facilities. The wall thickness should have a 3 mm corrosion allowance, and the lower part of the vessel should be provided internally with a protective coating.

The vessel should be installed between the compressors and the drier. Where limited space makes it impossible to install the buffer vessel in this place, part of the required buffer volume may be located downstream of the drier, provided the buffer vessel between compressor and drier remains of sufficient size for condensate separation. For the design and construction of vessels, BS 1515 part 1 or the ASME Boiler and Pressure Vessel Code, Section VIII, Div. 1 or any other approved standard of equivalent authority is usually acceptable.

## 2.23 **AIR SUPPLY PIPING**

The piping system for instrument air supply should be designed in close cooperation between instrument engineering, utility engineering and mechanical engineering, taking into account, the:

- segregation
- plant lay-out
- pipe sizes

The complete lay out of the piping system, including the take-off points, pipe sizes, etc. should be shown on a drawing or on a set of drawings. These drawings should also clearly indicate the demarcation points between mechanical engineering and instrument engineering.

In general, all piping in pipe tracks and pipe bridges, and all piping in sizes 2 in. and larger in plant sections (including branch-off points and valves) and the piping to the air filter/reducer station in the control centre form part of mechanical engineering. All piping smaller than 2 in. in the plant and the air filter/reducer station(s) with downstream piping in the control centre forms part of the nstrument engineering.

### 2.23.1 LAY OUT

The lay out of the supply piping depends on the following:

- The lay out of the plant and plant sections
- The location of the air supply plant
- The location of pipe bridges, cable trunking, etc.
- The location of the instruments

Piping for instrument air supply should be completely separated from that for tool air supply.

The lay out drawings should include all instrument air supply piping in pipe tracks, pipe bridges, and plant sections up to and including the branch-off points for individual instruments or groups of instruments; piping for the latter need not, however, be shown in detail on these drawings.

Piping in the plant sections should run close to the trunking for instrument signal cables to facilities supporting pneumatic signal lines.

### 2.23.2 PIPE SIZES

The pipe sizes should be determined in accordance with the following Table 2.8, and thereafter in proportion to the aforementioned table, for example, 400 users for nominal 4-in. pipe.

*Note*: A user is considered to be a typical instrument using approximately 0.015 m³ (0.5 SCF) of air per minute.

Piping in pipe tracks and pipe bridges should have a minimum size of 1½ in.

For exceptionally long piping, a calculation should be made to ensure that the decrease in pressure between the outlet of the air drier and the most remote consumer does not exceed 1 bar.

### 2.23.3 PIPING DETAILS

The air supply piping to the control center may run underground with protection against corrosion. The location should be such that when required in the future the piping can be excavated for repairs, etc.

As an alternative the air pipe to the control center may run through prefabricated concrete trenches and in the open atmosphere.

Piping forming part of mechanical engineering should be in accordance with Piping Standards.

**Table 2.8 Determination of Pipe Size**

| Pipe Headers | No. of Pilots | Nominal Pipe Size | |
|---|---|---|---|
| | | in. | mm |
| | 1–5 | ½ | 15 |
| Branch | 6–20 | 1 | 25 |
| | 21–50 | 1½ | 40 |
| | 51–100 | 2 | 50 |
| Main | 101–300 | 3 | 80 |

Piping forming part of instrument engineering should normally be made of galvanized materials, however, for those plants where the use of galvanized materials is not allowed, the piping forming part of instrument engineering should be made of carbon steel.

All main piping should be provided with drain valves at low points and at dead ends.

Branch-off points for future extensions, etc. should be provided with an isolating valve and blind flange. Branch-off points from piping in pipe tracks and pipe bridges should be 1 in. minimum, be located on the top of the (horizontal) piping and be provided with an isolating valve.

Branch-off points from piping in process sections should be ½ in. minimum and be provided with a steel globe valve.

Groups of up to five instruments located close together may be supplied by a common ½ in. take-off; a ¼ in. brass or bronze globe valve (bronze is preferred) should then be provided close to each individual consumer. Such a ¼ in. valve should also be provided in individual supply piping if the isolating valve at the take-off point is not easily accessible.

At least 15% spare ½ in. valved connections should be evenly distributed through the plant.

## 2.24 AIR SUPPLY FOR PLANT-MOUNTED INSTRUMENTS

All plant-mounted instruments including final control elements requiring air should be provided with an individual air supply set, consisting of a filter, a pressure reducer with drain valve and a pressure gage. If the instrument has an integral supply pressure gage, the pressure gage on the reducer may be omitted. The variety in type of air supply sets should be kept as small as possible.

Instruments in local panels should have individual air supply sets, unless they are for larger panels where a common filter/reducer station is found. For more information see Figs. 2.7–2.16 and Table 2.9.

## 2.25 AIR SUPPLY IN THE CONTROL CENTER

In the bottom corner of the control panel one or more filter/reducer station(s) should be installed for reducing the incoming air to the required pressure, which is normally 1.5 barg.

Panel instruments for integrated processing units, ie, which cannot be operated separately, may be supplied from a common central filter/reducer station.

Separate stations and air headers should be provided for panel sections serving essential units, eg, boilers, which are expected to stay in operation during maintenance shut-downs of the processing units or during failure of the normal air supply plant.

Separate stations may also be required for consumers requiring higher air pressures; eg, for direct operation of depressurizing valves.

Each filter/reducer station should consist of at least two filters in parallel followed by two high-quality pressure reducers in parallel, one acting as stand-by for the other.

Each pressure reducer should be fitted with a gage indicating its downstream pressure.

The capacity of each pressure reducer should be such that with an upstream pressure of 2.0 barg. and only one reducer in operation, a downstream pressure of 1.5 barg. will be maintained.

*Note*: Excessive oversizing of reducers should be avoided as their operation may become unstable when in the nearly closed position.

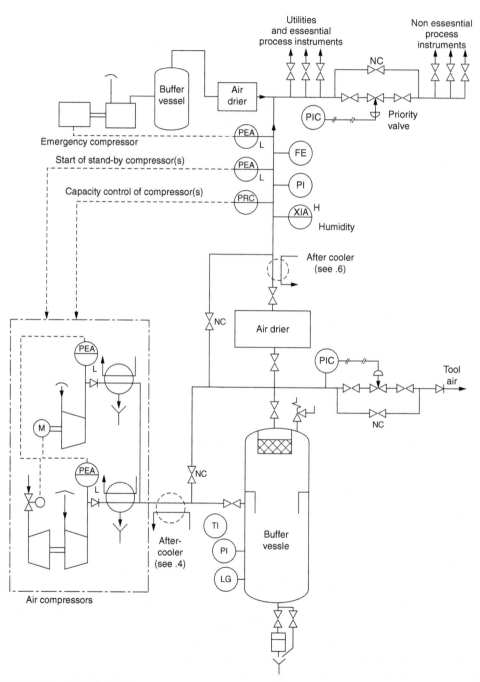

**FIGURE 2.7 Typical Air Supply Plant**

NC, normally closed.

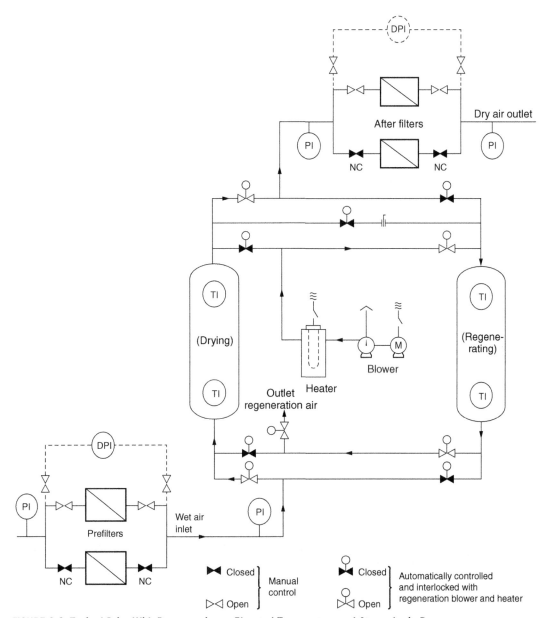

**FIGURE 2.8  Typical Drier With Regeneration at Elevated Temperature and Atmospheric Pressure**

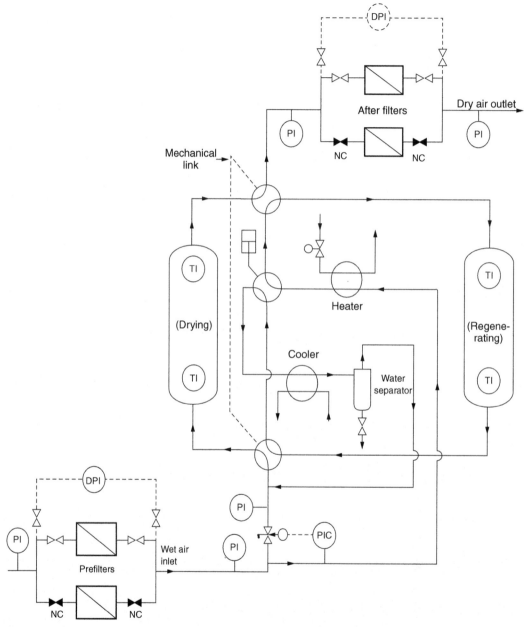

**FIGURE 2.9 Typical Drier With Regeneration at Elevated Temperature and Operating Pressure**

NC, normally closed.

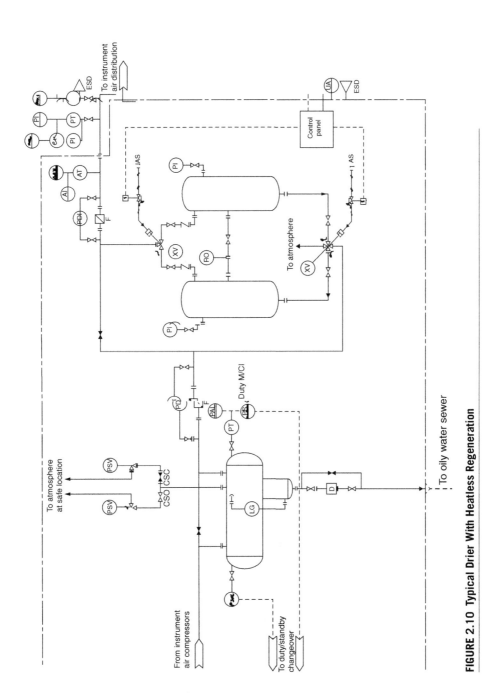

**FIGURE 2.10 Typical Drier With Heatless Regeneration**

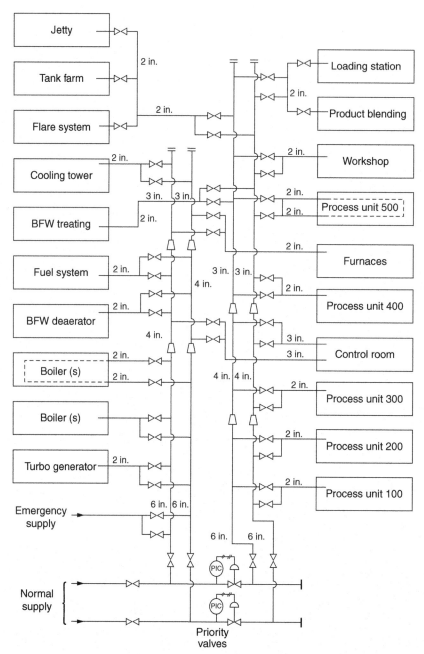

**FIGURE 2.11 Typical Arrangement of Air Supply Piping**

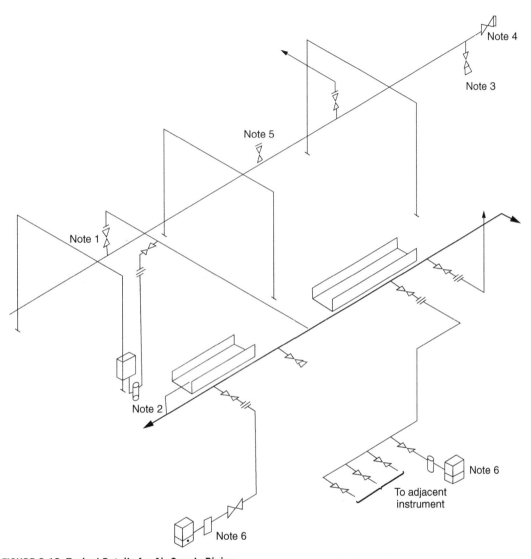

**FIGURE 2.12  Typical Details for Air Supply Piping**

(1) Branch-off points form horizontal piping in pipe bridges located on the top of the piping. (2) Supply piping close to trunking for instrument cables. (3) Drain valves at low points and dead end of piping. (4) Valve at the end of main piping for future extension. (5) Spare connection.

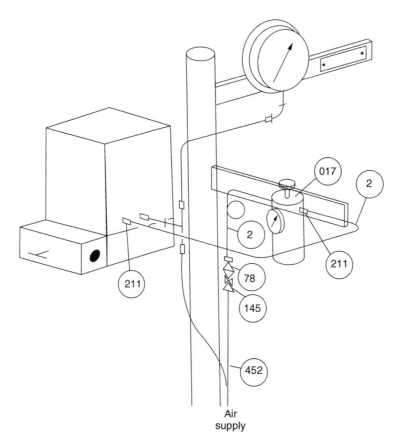

Air
supply

| TEM | SIZE | DESCRIPTION | MATERIAL |
|---|---|---|---|
| 017 | (¼ inch) | FILTER/REDUCER | |
| 452 | (½ inch) | LINE PIPE | CARBON STEEL, GAL V. |
| 2 | (¼ inch) | TUBING | COPPER |
| 145 | (½ inch) | GLOBE VALVE | BRASS |
| 78 | (½) (¼ inch) | REDUCER | Mal. IRON, GALV. |
| 211 | (¼ inch × ¼ inch) | STUD COUPLING (CONNECTOR, MALE SCRD. API COMPR. TYPE) | BRASS |

**FIGURE 2.13 Typical Air Supply for Plant-Mounted Instrument Typical Arrangement in Basement**

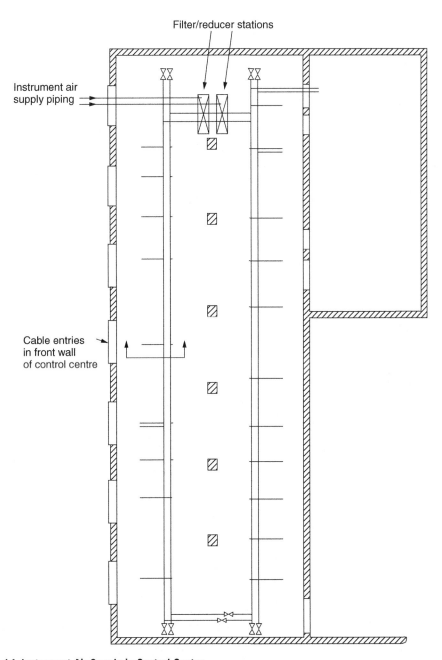

**FIGURE 2.14 Instrument Air Supply in Control Center**

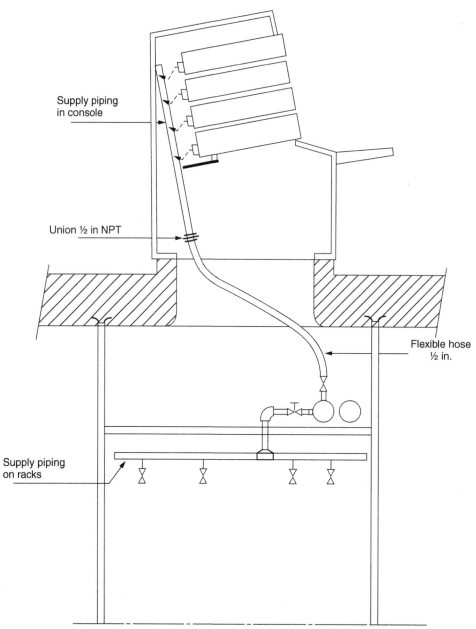

Supply piping
in console

Union ½ in NPT

Flexible hose
½ in.

Supply piping
on racks

**FIGURE 2.15 Instrument Air Supply in Control Center Typical Arrangement in Console Where Basement Used**

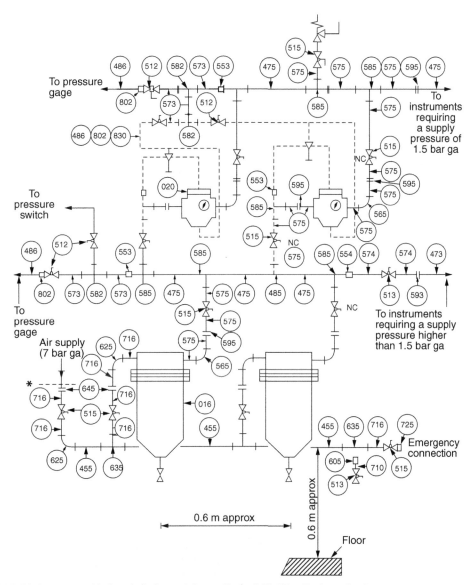

**FIGURE 2.16 Instrument Air Supply in Control Center Typical Air Filter/Reducer Station**

A third air filter is required if one filter in operation would cause too great a drop in pressure. NC, normally closed; * demarcation point between mechanical engineering and instrument engineering.

**Table 2.9 Instrument Air Supply in Control Center Typical Air Filter/Reducer Station**

| Item | Size (inch) | Description | Material |
|------|-------------|-------------|----------|
| 016 | 2 | Air filter | Doulton/Aerox |
| | | | Norgern |
| 020 | 2 | Reducer (with | Norgern |
| | | regular) | Galv. steel |
| 18 | 2 | Line pipe | |
| 473 | 1 | Line pipe | Al. brass |
| 475 | 2 | Line pipe | Al. brass |
| 1 | ¼ | Tubing | Copper |
| 512 | ½ | Globe valve | Brass |
| 513 | 1 | Globe valve | Brass |
| 515 | 2 | Globe valve | Brass |
| 553 | 2 × ½ | Bushing | Bronze |
| 554 | (2 × 1) | Bushing | Bronze |
| 565 | 2 | Elbow | Bronze |
| 242 | ½ × 2 | Nipple | Al. brass |
| 243 | 1 × 2 | Nipple | Al. brass |
| 575 | 2 × 3 | Nipple | Al. brass |
| 582 | ½ | Tee | Bronze |
| 585 | 2 | Tee | Bronze |
| 593 | 1 | Union | Bronze |
| 595 | 2 | Union | Bronze |
| 605 | (2 × 1) | Bushing | Mal. iron |
| 625 | 2 | Elbow | Mal. iron |
| 635 | 2 | Tee | Mal. iron |
| 645 | 2 | Union | Mal. iron |
| 710 | 1 × 2 | Nipple | Galv. steel |
| 716 | 2 × 3 | Nipple | Galv. steel |
| 725 | 2 | Plug | Galv. steel |
| 802 | ¼ OD × ½ NPT | Stud coupling | Brass |
| 235 | ¼ OD | Tee, compression | Brass (fem) |

A panel-mounted pressure gage and a low-pressure alarm switch should be connected to the air supply piping between the filters and the reducers. The alarm switch should be set at 0.5 bar below the normal operating pressure.

The outlet of each filter/reducer station should be provided with a safety relief valve of adequate capacity to prevent the downstream pressure from rising above 2.0 barg in the case of a complete failure of one reducer.

For a typical arrangement, see Figs. 2.7–2.16 and Table 2.9.

All piping downstream of the filter/reducer stations should be of ample size, and made of brass, bronze, or stainless steel.

Isolating valves should be installed in each branch, and be of the diaphragm or ball type, and should have a brass or bronze body.

The drawings for the air supply system in the control center should show details of the arrangement such as the following:

- The main filter/reducer station
- The supply piping for 1.5 barg
- Segregation between normal and essential duties
- The take-off points to console sections and auxiliary instruments

# FRESH AND SPENT CAUSTIC UNITS AND CHEMICAL INJECTION SYSTEMS

# 3

A significant investment in a chemical feed system can often be justified when compared with the high cost of these control problems. When a chemical feed system is not properly engineered, chemical levels are often above or below program specifications.

The chemical injection system usually controls the chemistry of some process fluids in order to reduce or to prevent corrosion or deposition. Oxygen scavenger is continuously injected into the feedwater stream in order to reduce the amount of $O_2$ remaining downstream, the deaeration treatment of feedwater and into the condensate stream to reduce the corrosion problems.

This chapter sets forth the content and the extent of the essential process and control system requirements of fresh and spent caustic units and "chemical injection systems."

A well-engineered feed system is an integral part of an effective water treatment program. If a feed system is not designed properly, chemical control will not meet specifications, program results may be inadequate, and operating costs will probably be excessive. Some of the costly problems associated with poor chemical control include the following:

- High chemical costs due to overfeed problems.
- Inconsistent product quality, reduced throughput, and higher steam and electrical costs due to waterside fouling.
- High corrosion rates and resultant equipment maintenance and replacement (ie, plugging or replacing corroded heat exchanger tubes or bundles).
- High labor costs due to an excessive requirement for operator attention.
- Risk of severe and widespread damage to process equipment due to poor control or spillage of acid into cooling towers.

The use of a proper feed system can prevent difficult operational situations. Chemical feed systems can be classified according to the components used, the type of material to be fed (powder or liquid), the control scheme employed, and the application. Chemical storage treatment chemicals are usually delivered and stored in one of three ways: bulk, semi bulk, and drums. The choice among these three depends on a number of factors, including usage rate, safety requirements, shipping regulations, available space, and inventory needs.

The scavenger solution is intermittently injected into the closed cooling water system by a common scavenger injection hardware. The dosing pumps are automatically controlled by a flow signal from the feedwater system and by residual $O_2$ concentration signal. Ammonia is continuously injected into the condensate extraction pumps discharge.

The dosing pumps are automatically controlled by a flow signal from the condensate system and by pH signal. Sodium phosphate is injected to transform the silica content of the boiler water into sodium silicates. In the case of raw water inlet into the condenser, calcium and magnesium salts are transformed into nonadherent precipitates.

Installed in a petrochemical complex the coagulant dosing package is supplied to prepare and feed solutions of coagulant (aluminum sulfate) and water to assist the coagulation of solids in the raw water prior to filtration in the raw water filters.

Three packages are usually supplied for a chemical injection system for a power station and desalination plant. One is for the washwater recovery system injecting polyelectolyte to aid settlement of washwater.

The second is for the seawater filtration system injecting liquid polyelectrolyte as a filter aid and the third is a caustic dosing skid to inject variable quantities of caustic solution to provide pH correction of the carrier water. Features include automatic proportional control of metering pumps by speed and stroke length through a control panel. A polyelectrolyte powder-wetting system and solution eductor system are used.

The boiler treatment injection package forms part of the steam generation system in an aromatics complex. Polymer or chelant, diluted with steam condensate is injected into the boiler steam drums, oxygen scavenger and neutralizing amine, diluted with steam condensate is injected into the deaerator storage drum.

*Bulkstorage.* Large users often find it advantageous to handle their liquid chemical delivery and storage in bulk. Liquid treatments are delivered by vendor tank truck or common carrier. A large tank, often supplied by the water treatment company for storing the liquid treatment, is placed on the property of the user near the point of feed. Service representatives often handle all inventory management functions.

Treatment can be drawn from these storage vessels and injected directly into the water system or added to a smaller, secondary feed tank, which serves as a day tank.

Day tanks are used as a safeguard to prevent all of the material in the main storage tank from accidentally being emptied into the system. They also provide a convenient way to measure daily product usage rates. This is known as semi bulk storage. Where chemical feed rates are not large enough to justify bulk delivery and storage, chemicals can be supplied in reusable shuttle tanks.

Usually, these tanks are designed in such a way that they can be stacked or placed on top of a permanent base tank for easy gravity filling of the base tank. This is known as drum storage. Although 40 and 55-gallon drums were widely used for chemical delivery only a few years ago, increasing environmental concerns have sharply reduced drum usage.

The restrictions on drum disposal and drum reclamation have reduced the popularity of this delivery and storage method in favor of reusable or returnable containers.

For design of any equipment inside the subject units/systems, reference should be made to the relevant standards.

## 3.1 PROCESS REQUIREMENTS OF FRESH AND SPENT CAUSTIC UNITS
### 3.1.1 FRESH CAUSTIC UNIT

Fresh caustic unit should involve but not be limited to the following:

- Transferring the proper caustic solution strength to the plants or process units where required with available pumps.

- Supplying strong caustic solution (typically 50% by mass) for the unit by means of solid caustic or available strong caustic solution from existing petrochemical plants.
- Preparation and storage of strong and dilute caustic solution, including storage facilities for the regenerated caustic from a LPG caustic treating section.

### 3.1.2 DESIGN REQUIREMENTS

Regenerative caustic solution should be used in nonregenerative batch caustic wash treating for $H_2S$ removal throughout the refinery/plant.

The overflows and drains of all caustic tanks in the fresh caustic unit area and drains of nonregenerative caustic wash treaters should be drained in a closed loop drainage system which is fed into caustic sump pit in the spent caustic treating unit.

Connections of cold condensate or demineralized water addition should be considered for diluting the contents of the caustic dissolving tank and each of the dilute caustic tanks, to prepare the desired concentration.

The capacity of dilute caustic tanks should be enough to provide continuous make-up caustic, during catalyst regeneration of the catalytic unit. Provision of the tank heaters (typically by use of LP steam) should be considered for caustic tanks to maintain the solutions above their freezing points.

Caustic tanks should be provided with temperature and level indicators as well as air spargers for homogenizing the solutions and/or to maintain constant bulk temperature.

The regenerated caustic that is purged continuously from the regenerative caustic treating section in the LPG unit should be stored in a regenerative caustic tank in the fresh caustic unit area.

A pump(s) should be provided to pump a constant supply of fresh caustic from dilute caustic tanks, to the regenerative caustic treating section to maintain the content of $Na_2S$ and other impurities with a value less than the specified value (typically 2% by mass).

The temperature of the solution to the suction of caustic transferring pump during transfer from the caustic dissolving tank should not exceed 98°C.

Dilute caustic pumps should be provided to pump the regenerative caustic where required. In addition these pumps should be used to transfer fresh dilute caustic from appropriate tanks under the following circumstances:

1. To supply fresh caustic to the catalytic unit.
2. To supply fresh caustic in the absence of regenerated caustic.

Pump(s) should be provided to pump constantly fresh caustic from dilute caustic tanks into the line carrying sour water produced in the regenerative caustic treating section, to the sour water plant.

## 3.2 DESIGN REQUIREMENTS

The utility air injection should be flow controlled to the bottom of each reactor, through a convenient distributor.

Dual filters and cooler should be provided to filter and cool (typically up to 35°C) the treated liquid effluent from the effluent separator.

A mixer should be provided for mixing the hydrochloric acid solution (typically 30% by mass) with effluent liquid from cooler.

A storage tank for hydrochloric acid with appropriate transferring pumps should be provided for the neutralization system mentioned above. A degassing vessel should be provided to enter the neutralized solution.

A level indicator controller should be provided to discharge the neutralized solution (treated spent caustic) to the water sewer.

In the case of existence of oil in the spent caustic drains sump, an oil separator with internal baffle should be installed to remove the oil from the spent caustic before transfer to spent caustic surge tank.

An automatic level control system should be provided to transfer the entrained oil to the API separator.

Provision of cold condensate addition should be made to dilute the spent caustic feed line to the desired concentration of contaminants, before entering the spent caustic surge tank.

Provision of tank heater (typically by use of LP steam) should be considered for the spent caustic surge tank to maintain the contents of the tank above its freezing point.

A series of stirred reactors should be provided to convert the sulfide contained to thiosulfate/sulfate with atmospheric oxygen at the correct temperature and pressure.

The size and numbers of series reactors should be chosen to reduce the sulfide content of liquid effluent from the last reactor to an acceptable standard figure.

The temperature of the reactors could be logged and controlled (TIC) by the injection of live steam to the reactor through the air distributor line to supplement the heat of reaction.

Vent gases and effluents from the reactors should be discharged to the effluent separator, from where the gas (essentially air) should be sent to the atmosphere at a safe location.

The pressure in the reactors and effluent separator should be monitored and maintained by a pressure indicator controller (PIC) in order to control the gas discharge to the atmosphere.

## 3.3 SPENT CAUSTIC TREATING UNIT

The purpose of the spent caustic treating unit should be to improve the quality of spent caustic up to the point at which it is not harmful to the environment, before sending it to the water sewer.

The spent caustic unit as a typical unit may involve but not be limited to the following equipment:

- Spent caustic drains sump
- Spent caustic drains pump
- Spent caustic surge tank
- Spent caustic feed pump(s)
- Spent caustic oxidizer(s)
- Hydrochloric acid tank
- Caustic oxidizer effluent separator
- Spent caustic filter(s)
- Spent caustic cooler
- Static HCl acid mixer
- Spent caustic degassing vessel.

## 3.4 **CHEMICAL INJECTION SYSTEMS**

Chemical feed systems should be designed to ensure high reliability and have flexibility enough to cover contingencies that might arise. The required volume of chemical as well as its physical characteristics should also be considered in the feed system design.

### 3.4.1 **FEED CONCEPTS**

The method by which a chemical is added should be suited to both its intended use and the system into which the product is being added. Feeding mechanisms should be categorized to: continuous feed, intermittent feed, slug feed, and shock feed:

1. Continuous feed. Continuous feed is the method most commonly encountered. It may be manual, providing a constant rate of chemical addition, or it might be automatic, the feed rate being automatically adjusted in response to some measured variable such a pH or flow rate. Feeders which cycle on and off over short time spans should also be considered continuous.
2. Intermittent feed. Intermittent feed is on/off feed, over an extended time span, with chemicals added at fixed intervals to a threshold level of treatment.
3. Slug feed. Slug feed involves the addition of chemical in excess of the amount required to produce a desired concentration after a specific time interval. As make-up is added to compensate for system losses over a period of time, the residual is gradually lowered to an unacceptable level, therefore requiring another slug.
4. Shock or shot feed. Shock feed is a specialized form of slug feed as applied to the introduction of micro-biocides to the recirculating cooling system. Shock feed is utilized to provide maximum benefit from the "kill" effect on microbiological growths afforded by a high level of treatment.

### 3.4.2 **CHEMICAL FEED EQUIPMENT USED IN WATER TREATMENT**

Most cooling water and boiler water treatment products are liquids or solutions prepared from powders. The pump is therefore the most frequently encountered feed device. Different feed equipment is briefly mentioned in the following:

- Shot feeders. The shot feeder typically shown in Fig. 3.1 consists of a pressure-rated chemical tank installed across a pressure differential such as the feedwater or circulation pump.
- Miscellaneous feeders
  - Bypass feeders
    Treatment chemicals which dissolve slowly, or are made in special briquette form can be added, using the bypass feeder, typically shown in Fig. 3.2.
  - Acid/Alkali feeding
  - Acids and alkalis require suitable feed mechanisms and control. Acid feed is recommended to regulate automatically by a pH controller. The following safeguards should commonly be considered:
    - Day tanks: feeding from a day tank in lieu of a bulk tank limits the total volume which might enter the system in the event of an accident.
    - Dilution: acid is frequently diluted to improve mixing and pH control. Alarms indicating loss of dilution flow should be considered to stop further acid flow, if necessary.

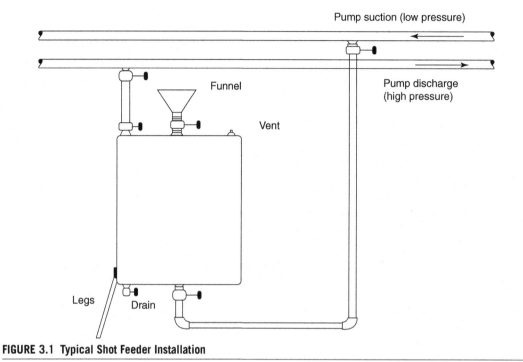

**FIGURE 3.1 Typical Shot Feeder Installation**

- Safeguard: the pH-sensing element should be provided with a no-flow switch in the sample line to indicate if sample flow is lost. Duration timers should also be used to provide the alarm if acid is fed for an unusually long period of time.
- Chlorine feeders (chlorinator)
  The water jet eductor uses the kinetic energy of water under pressure to entrain chlorine gas, mix the two and discharge the mixture to water in a flow line or in a treating basin. Chlorine should not flow if a vacuum is not produced. If hypochlorite solutions are to be used, pumps and shot feeders can be used.

## 3.4.3 ACCESSORIES

- Injection nozzles. Specialized nozzles are often needed when injecting chemicals into the pipeline. Fig. 3.3 typically shows a nozzle for adding liquid chemicals into a steam line or other gaseous stream. This nozzle mechanically atomizes the liquid by force of the gas.
- Mixers
  - An agitator or mixer should be used whenever a powdered chemical is dissolved or if a heavy or viscous liquid product is to be diluted.
  - The mixer should not be run continuously except for slurries.
  - Dissolving polymers and other viscous materials requires a larger propeller and slower speed.
  - Direct injection of air or steam is satisfactory for dissolving chemicals.

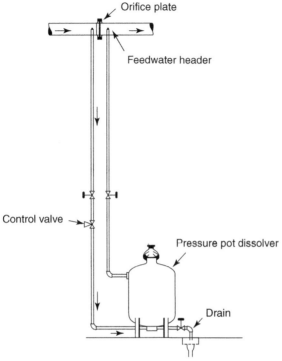

**FIGURE 3.2  Typical Bypass Feeder**

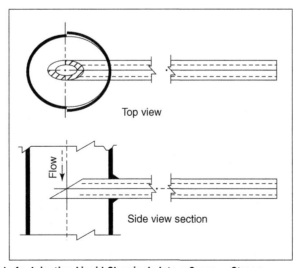

**FIGURE 3.3  Typical Nozzle for Injecting Liquid Chemicals Into a Gaseous Stream**

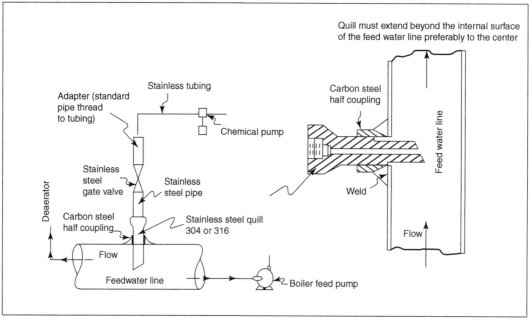

**FIGURE 3.4 Typical Boiler Water Treatment Feeding Systems**

- Level alarm
  - A level alarm should be installed on a feed tank when the chemical feed should not be interrupted or when the pump will be damaged if it runs dry.
  - Chemical feed tanks usually use an electrode-type level control operating on the conductivity of the chemical solution.
  - The level control package will automatically perform a variety of functions, such as shutdown of a chemical feed pump and alarm at a low level, energizing a pump or opening a valve at a low level to fill the tank and an alarm at a high level.

### 3.4.4 CHEMICAL ADDITIVE FEEDING SYSTEMS

Figs. 3.4 and 3.5 are presented to show a typical feeding system in boiler water treatment and corrosion inhibitor, respectively.

Important factors for slipstream inhibitor feeding:

1. The slipstream line must be large enough for structural strength, yet not so large as to keep fluid from filling it and entering the overhead vapor line under pressure. Recommended size is DN 20 to DN 25.
2. A rotameter is important to ensure that the inhibitor is flowing through the line and is thoroughly mixed.

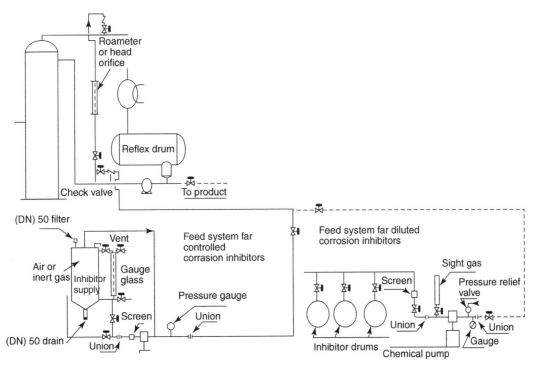

**FIGURE 3.5  Typical Feed System for Concentrated and/or Diluted Corrosion Inhibitors**

### 3.4.5 SYSTEM DESIGN AND OPERATION OF CHEMICAL FEEDERS FOR COOLING TOWERS

Fig. 3.6 shows the equipment, piping, and auxiliaries to be included in the vendor's scope of supply.

Wet chemical feed systems should be designed to hold a minimum of 72 h supply of chemicals based on design flow and design raw water analysis, or on peak treatment dosages. For metaphosphate inhibitors use 24 h hold up.

Solution strengths for phosphates should not exceed 3% (by mass).

All equipment should be suitable for unsheltered outdoor installation for the climatic zone specified.

Units should be designed for continuous service and an uninterrupted operation for a period of 2 years.

Controlled volume pumps should have a capacity of two times the specified design feed rate to allow flexibility (increase or decrease) in dosing. A minimum of two pumps (one spare) should be included.

## 3.5 EQUIPMENT DESIGN AND SELECTION

Mechanical agitators should be designed to operate continuously.

Each positive displacement pump should be furnished with a calibration pot. A gage glass should be furnished with a gage board calibrated in cm.

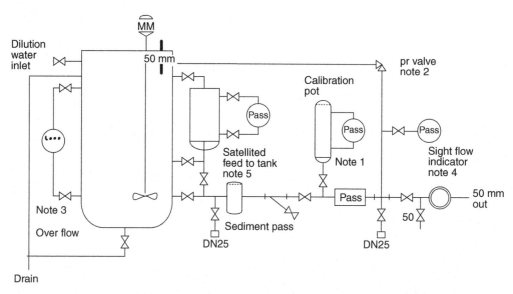

**FIGURE 3.6 Controlled Volume Pump Feeder (Diagrammatic Layout of Equipment and Piping)**

Notes: 1, Gage glass with calibrated gage board; 2, PR valve to be furnished even when built-in relief valves are provided with pumps. Set pressure to be provided by the pump manufacturer; 3, Gage glass not required if tank is constructed of transparent plastic; 4, Sight flow indicator with flapper; 5, Satellite feed tank required when chemical mixing and storage are done in the same tank. The satellite tank is used when the main tank is being used for preparing new batch.

Tanks and small vessels used for storage and handling of water and solutions not subject to pressure or vacuum should be designed and fabricated to the requirements of API 650.

Dispersion equipment for dispersing concentrated chemicals should be furnished to assure complete solution of soluble chemicals or a complete dispersion of suspended solids.

Enclosures for electrical equipment should be appropriate for the specified area classification and environmental exposure.

## 3.6 GENERAL REQUIREMENTS FOR DESIGN AND CONSTRUCTION OF CHEMICAL DOSING UNITS IN A WATER TREATING SYSTEM

1. The chemical dosing unit should be capable of withstanding continuous operation.
2. The accuracy of the chemical feeds should be ±5% through the range of 20–100% of the design maximum feed rate of the unit.
3. Adjustment for changing the feed rate from 0 to 100% of design should be possible.

### 3.6.1 CONSTRUCTION

1. The chemical dosing unit should consist of a chemical tank (and measuring tank, if necessary), feed pump (or eductor), and mixer.

2. The equipment containing chemicals should be fabricated of suitable materials for each respective chemical service.
3. All equipment should be installed on the common base plate at the shop so as to constitute a packaged unit.
4. All equipment should be suitable for unsheltered outdoor installation.

### 3.6.2  MATERIALS

1. Each item of equipment of the chemical dosing unit should be of materials, which are sufficiently resistant to corrosion and erosion by the chemicals.
2. Table 3.1 "Material Selection Guide for Dosing Unit" is applicable to typical chemicals used in the water treatment unit.

### 3.6.3  INSTRUMENTATION AND CONTROL

1. Control system
   a. The control system should be integrated into the control and instrumentation system of the water treating system, unless the equipment is isolated.
   b. Flow control should be automatic except in the case of systems in which fluctuations in the chemical dosing rates can be disregarded.
2. Instruments
   The instrumentation should include, but not be limited to the following:
   a. Local level indicators on all tanks.
   b. Low-level sensor that will initiate an alarm on the control panel.
   c. Measurement of charge (as required).

### 3.6.4  MECHANICAL
#### 3.6.4.1  Chemical Tanks
Chemical tanks should have sufficient capacity on the basis of the maximum operating rate.

Where the chemical solution is to be prepared by dissolving powder chemicals or by diluting concentrate chemicals, the chemical tanks should be provided with motor-driven mixers which should be capable of performing continuous operation. The mixers should be of the magnetic type where the content of the tanks are toxic or hazardous materials.

Where the chemicals handled are physically nonhazardous and nontoxic, and for tanks smaller than 200 L in capacity, hand-operated mixers may be used.

Chemical tanks should be provided with instruments capable of measuring the quantities in the tanks.

The powder chemicals are to be dissolved. For this purpose, dissolving baskets should be provided in the tank.

#### 3.6.4.2  Pumps
Safety valve discharge lines should be connected to the chemical tanks.

Proportioning pump stroke should be capable of being changed manually even during operation.

**Table 3.1 Material Selection Guide for Dosing Unit**

| Chemical | Solution Concentration (Mass %) Max. Temp. | Material |
|---|---|---|
| Aluminum Sulfate $Al_2(SO_4)_3 \cdot 14H_2O$ | 20% Ambient | Plastic or natural rubber lined carbon steel or Alloy 20 |
| Ammonia $NH_4OH$ | 40% Ambient | Carbon steel (copper alloys not to be used) |
| Calcium Hypochlorite $Ca(OC_{12})_2$ | 10% Ambient | Plastic or natural rubber lined carbon steel or Alloy 20 |
| Copper Sulfate $CuSO_4 \cdot 5H_2O$ | 10% Ambient | Natural rubber lined carbon steel or type 316 stainless steel or Alloy 20 |
| Cyclohexy-Amine $C_6H_{11}NH_2$ | 40% Ambient | Carbon steel |
| Disodium Phosphate $Na_2HPO_4 \cdot 12H_2O$ | 3% Ambient | Carbon steel |
| Hydrazine $N_2H_4 \cdot H_2O$ | 50% Ambient | Type 304 or type 316 stainless steel |
| Hydrochloric Acid HCl | 40% Ambient | Natural rubber lined carbon steel or FRP or Hastelloy B |
| Monosodium Phosphate $NaH_2PO_4 \cdot H_2O$ | 3% Ambient | Type 304 stainless steel or rubber lined carbon steel |
| Potassium Permanganate $KmnO_4$ | 20% Ambient | Carbon steel |
| Sodium Chloride NaCl | 25% Ambient | Fiber-glass reinforced plastic (FRP) or FRP lined carbon steel or natural rubber lined carbon steel or bronze |
| Sodium Hydroxide NaOH | 50% Ambient | Carbon steel |
| Sodium Hypochlorite NaOCl | 10% Ambient | Natural rubber lined carbon steel or plastic |
| Sodium Metaphosphate $(NaPO_3)\times$ | 3% Ambient | Type 304 stainless steel or natural rubber lined carbon steel |
| Sodium Sulfite $Na_2SO_3$ | 30% Ambient | Type 304 stainless steel |
| Sulfuric Acid $H_2SO_4$ | 80–100% Ambient | Type 316 stainless steel or Alloy 20 for pump and mixer, carbon steel for tank and piping (Velocity less than 1.0 m/s) |
| Trisodium Phosphate $Na_3PO_4 \cdot 12H_2O$ | 3% Ambient | Carbon steel |

Pumps other than proportioning pumps should be provided with flow indicators and control valves.

Chemical pumps into which slurry may pass should be of an open impeller centrifugal type and should be provided with flow indicators and control valves.

Provisions should be made to isolate the pump during maintenance period.

All types of pumps should be furnished with spare parts.

### 3.6.4.3 Other Items

For dosing of acid and caustic soda for regeneration in the ion exchange system, eductors may be used instead of pumps. In this case, flow indicators and control valves should be provided.

Ladders and platforms required for operation, inspection, and maintenance should be provided.

Enclosures for electrical equipment should be appropriate for the specified area classification and environmental exposure.

### 3.6.4.4 Inspection and Testing
#### 3.6.4.4.1 Shop Inspection and Testing
The following inspections and tests should be performed on each respective part or equipment:

1. Visual and dimensional check.
2. Material check against the mill test certificate.
3. Mechanical running test on each respective equipment in the unit.

Field inspection and testing

1. A running and performance test should be performed.

## 3.7 GENERAL DESIGN REQUIREMENTS OF CHEMICAL FEED EQUIPMENT
### 3.7.1 COOLING TOWER WARM LIME SOFTENER
The equipment should be designed to feed the chemicals in direct proportion to the amount of water entering the softener.

The equipment should accurately proportion the chemicals at all rates and ranges of flow within the specified capacity of the softener.

The equipment should also include means of adjusting the dosage of chemicals without increasing or decreasing the strength of solution.

The equipment should include wet or dry chemical feeders with level indicators, tank agitators, proportion devices, and chemical feed pumps.

Two chemical pumps (one as spare) should be furnished for each chemical feed system.

The vendor should guarantee that the chemical proportioning and feeding equipment should automatically deliver the respective chemicals in proportion to the rate of water flow.

An acid feed system should also be provided in the inlet line to the pressure filters for pH control to prevent clogging of filters. The feed system should include a solution tank with support stand, two acid feed pumps and all necessary accessories.

### 3.7.2 CLARIFIER
The equipment should be designed to feed the chemical in direct proportion to the amount of raw water entering the reactivator.

The equipment should accurately proportion the chemical at all rates and ranges of flow within the specified capacity of the clarifier.

The equipment should also include a means of adjusting the dosage of chemicals without increasing or decreasing the strength of solution.

The equipment should include dry chemical handling facilities for storage and conveyance of chemicals to the central chemical dissolving tank.

One central chemical dissolving tank with air agitation, dry chemical handling facilities and its own chemical transfer pumps to pump the chemical solution to the clarifiers' chemical feed tanks, level indicator, and all interconnecting piping should be provided.

## 3.8 GENERAL DESIGN REQUIREMENTS OF METERING PUMP IN CHEMICAL INJECTION SYSTEMS

The following requirements should be considered in the design of metering pumps that are used in chemical injection systems.

### 3.8.1 INSTALLATION

All factors and considerations of sound hydraulic practice, including freedom from air and foreign matter, accurate and reliable seating of valves, proper size and length of piping, liquid vapor pressure, viscosity, and temperature should be considered for successful metering pump installation.

The application of basic hydraulic principles during planning, installation (as shown in Fig. 3.7), and operation is essential.

The installation should be made with careful attention to all instructions regarding handling of corrosive, toxic, or hazardous chemicals to assure personnel safety.

### 3.8.2 LOCATION

The preferred location of metering pump is indoors, although pumps can be installed outdoors. Manufacturer's recommendations for ambient operating temperatures should be followed.

All pumps used outdoors where temperature can fall below 0°C should be provided with a means of heating the pump, as well as being sheltered for protection from precipitation, blowing sand, dust, or other possible contamination.

All pump installations should allow sufficient room all around for operator access for adjustments or servicing.

- The pump should not be installed under tanks or other equipment where possible overflow would damage the pump.
- The pump foundation should have sufficient height to accommodate system piping.

### 3.8.3 INSTALLATION TIPS

1. A strainer should be employed to prevent foreign matter or undissolved lumps of chemicals from entering the pump that may interfere with check valve operation.
2. Strategically located shutoff and check valves should be incorporated to permit servicing the pump without draining the entire system.
3. Drain valves should be installed at the lowest point in the discharge line.
4. If the pump is not provided with check valves that are removable without disconnecting the piping, unions should be installed near the pump suction and discharge valves to facilitate removal of the pump head.

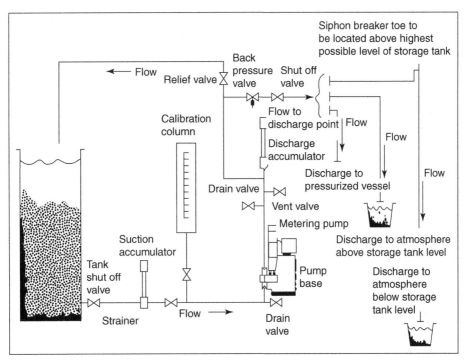

**FIGURE 3.7 Typical Pump, Tank, and Piping Arrangement**

5. Suction and discharge piping runs should be as straight and short as possible.
6. Piping should be sloped, if necessary, to eliminate vapor pockets.
7. A manual vent on the pump discharge line is desirable to facilitate removal of entrapped air particularly during pump start-up.
8. The shut-off valve should not be placed between the pump discharge and the system relief valve; to do so would make the relief valve ineffective.
9. A nipple or pipe should not be welded to valve bodies without first being removed from the pump head. If the pipe is welded, be sure to use flanges near valve bodies for easy disassembly.

### 3.8.4 SUCTION LINE

The suction line is a critical part of the system. To assure proper operation of the metering pump, the following recommendations should be followed:

Keep suction lines short; locate the pump as close to the chemical supply tank as possible.

Locate both pump and tank as close to the application point as possible, long lines may require large-diameter pipe.

If possible, locate the pump and chemical supply tank so that high-positive suction does not coexist with low discharge pressure, as liquid will siphon through the pump.

Long lines may result in poor performance, notably under feeding, nonlinearity, noisy operation, and vibration of the piping.

When a number of individual pumps are connected to a common supply, the suction line and/or header should be sized to accommodate the total flow required by all pumps running simultaneously.

A pump calibration column should be installed in the suction line, suitably valved to shut off flow from the supply tank. The calibration column should provide sufficient volume for at least a 30-s test run.

To minimize problems inherent in long suction lines, a day tank or an accumulator may be located close to the pump. Installation of an accumulator at the pump suction can act as a day tank. Essentially, the flow will be continuous in the long line between the supply tank and the accumulator and discontinuous between the accumulator and the pump.

In installations where a suction line of over 10 m in length is needed, and use of a day tank is impractical, a suction accumulator should be installed. The accumulator should be installed close to the suction connection, within 0.3 m if possible.

### 3.8.4.1 Suction pressure
The pump selected should be capable of operating against a specified pressure, and it should supply the pump with liquid at a certain minimum suction pressure (NPSH).

## 3.8.5 RELIEF VALVE
A process line relief valve is required for chemical injection system protection and should always be installed in the discharge line close to the pump. This valve will protect the line from damage due to plugging or accidental valve closure.

It is recommended relief valve discharge is piped back to the supply tank above the fill level (refer to Fig. 3.8).

If the distance and/or cost of the relief valve return line preclude its being piped to the tank, this line may be piped into the pump suction.

The relief valve, whether in the hydraulic fluid line or process line, should be set for a pressure 10–20% higher than the operating pressure of the system.

## 3.8.6 BACK PRESSURE
All reciprocating metering pumps require some amount of positive system pressure or back pressure to assure accurate metering. This required pressure prevents overfeed from the internal force of the suction line liquid due to the hydraulic characteristic of the pump design.

Adjustable in-line or nonadjustable internal pump-mounted back-pressure valves should be provided for this supplementary pressure.

A back-pressure valve should not be used to prevent a positive liquid level from draining or siphoning through the pump to an atmospheric discharge.

A back-pressure valve should not be used to prevent the siphoning of fluid into a below-grade normally pressurized main that has been depressurized. The sole purpose of the back-pressure valve should be to assure accuracy of pump delivery.

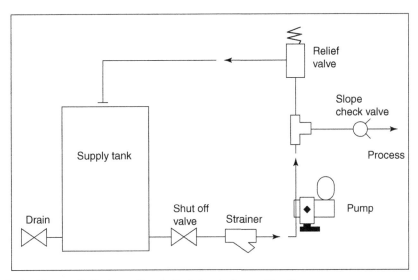

**FIGURE 3.8  Typical System Relieving to Supply Tank Due to Excess Line Pressure**

Most back-pressure valves cannot be used in a line handling slurry. The pump manufacturer should be consulted for appropriate back-pressure valves.

Pressure loss created by long discharge lines should not be considered as required back pressure. Similarly, a throttling valve or other fixed orifice will not be usable.

In order to overcome inertial pressure in the lines to and from the pump, a minimum of 207 kPa (ga), and/or [2.07 bar (ga)] back pressure should be required.

### 3.8.7 SIPHONING

Static siphoning may occur in situations where suction pressure is high relative to discharge pressure (typically when pumping into open tanks).

An antisiphon device is required in any chemical injection system that has a positive suction head in excess of the pressure at the discharge of the system. Without such a device, flow will pass from the tank through the pump to the end of the pipe.

Antisiphon valves are usually spring-loaded valves whose long-term reliability depends upon the frequency of operation. If the antisiphon valve fails, the system may be subject to overfeed or uncontrolled flow.

The pump back-pressure valve should not be used as an antisiphon valve. A separate antisiphon device should be installed and should provide greater capability than the system differential between supply tank and discharge point.

Drainage should be prevented by locating the antisiphon device at the end of the discharge line.

Antisiphon set pressure value should be equal to the static pressure in the discharge system as seen by the pump.

It is strongly recommended that in place of a mechanical valve, a vented riser be used in the manner shown in Fig. 3.9.

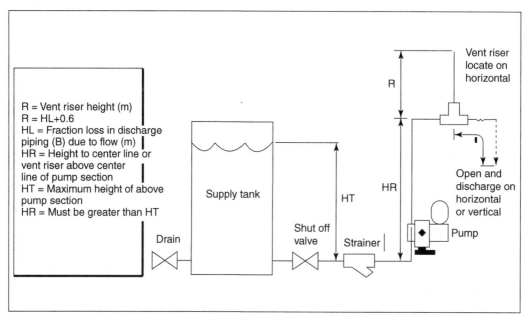

R = Vent riser height (m)
R = HL+0.6
HL = Fraction loss in discharge piping (B) due to flow (m)
HR = Height to center line or vent riser above center line of pump section
HT = Maximum height of above pump section
HR = Must be greater than HT

Vent riser locate on horizontal

R

Open and discharge on horizontal or vertical

HR

HT

Supply tank

Drain

Shut off valve

Strainer

Pump

**FIGURE 3.9 Guidelines for Proper Venting**

## 3.8.8 SLURRIES

- Pump selection
  - A high-speed pump should be selected for slurries. The higher liquid velocities aid in maintaining a slurry suspension.
  - For fast settling materials, such as slaked lime, speeds less than 2.4 strokes per second should be avoided. For slurries such as hydrated lime, speeds down to 1.6 strokes per second may be used.
- Piping system layout
  - Consideration should be given to the piping layout of slurry materials in a chemical injection system, to avoid slurry settling out.
  - Vertical runs should be minimized in slurry systems.
  - Any 90-degree direction changes should be accomplished by using plugged tees or crosses. These fittings permit prodding out deposits and also provide temporary flushing connections.
  - If deposits are likely (that is, calcium carbonate scaling from lime slurries), flexible plastic tubing or rubber hosing should be used rather than rigid pipe. Normal flexing from pump pulsations will dislodge scale. A flexible discharge line also allows long radius bends and direction changes with a few fittings.
  - If the water used for slaking the lime is softened, flanged steel pipe may be used.
- Valves
  - A relief valve should be used to protect the pump against dead ending or from severe plugs in long, vertical runs.

- To prevent siphoning, the slurry should be pumped to an elevated atmospheric break, from which it flows to the application point by gravity.
- Flushing
  - As settling during shutdowns is unavoidable, a flushing connection should be provided between the chemical feed tank and the suction check valve.
  - The flushing systems can be manual or automatic. If automatic a timed sequence flushing cycle will be established and consequent plugging will be materially reduced.

## 3.9 GENERAL DESIGN REQUIREMENTS OF PACKAGE TYPE CHEMICAL INJECTION SYSTEMS

### 3.9.1 SCOPE OF SUPPLY

A packaged type chemical injection system should typically consist of, but not be limited to the following composite items:

Each system should be shop assembled as much as possible and skid-mounted complete with chemical solution tank, tank level gage, mixer, diaphragm and plunger chemical feed pump, piping dosing device, complete necessary accessories and instruments for proper operation.

Each system should be painted, lined, skid-mounted and pretested at the vendor's shop so that it should be shipped to site ready to operate

### 3.9.2 DESIGN REQUIREMENTS

- Mixing tanks
  - Mixing tanks should be cone bottomed and mounted on legs of sufficient length to insure satisfactory pump operation, and allow clearance for drain connection.
  - Tanks should be complete with gage glass, hinged cover, connections for drain, pump suction, and discharge.
  - All tanks should have mixer supports:
    - Polymer phosphate mixing tanks should have a dissolving basket.
- Mixers
  - A portable mixer should be provided for each tank.
  - Mixers should be stable while agitating contents of tank from 1/3 to full.
- Pumps
  - Chemical feed pumps should be piston type (diaphragm and plunger, as required by the service), and should have facilities to permit adjustment of capacity from 0 to 100% of maximum specified.
  - Accessories should include coupling guard, floor stand, back-pressure valve, relief valve, and strainer.
- Relief valves
  A relief valve should be provided for each pump for the purpose of protecting the pump and piping from excessive pressure. The relief valves should have an internal construction adequate for the pressure, temperature, and material being pumped and should be sized to pass the maximum output of which the pump is capable of the relieving pressure.

- Strainers

  Each pump should be provided with a suction strainer adequate to protect the pump against damage from insoluble materials, which may enter the suction line. The construction materials should be suitable for the material to be pumped.

## 3.10 SCALE CONTROL

Scale formation involves the deposition of insoluble salts on heat transfer surfaces. The most common method of controlling scale is to precipitate potential scale-forming ions as nonadhering solids, or sludge, or as loosely adhering scales. Calcium ions are preferably precipitated as calcium hydroxyapatite [$3Ca_3(PO_4)_2 \cdot Ca(OH)_2$]; magnesium is preferably precipitated as serpentine [$2MgSiO_3 \cdot Mg(OH)_2 \cdot H_2O$]. These sludges are more flocculent or fluid when precipitated at a pH above about 9.5. Caustic soda, soda ash, or a blend of phosphates can be fed to provide this alkalinity if there is inadequate natural alkalinity in the feedwater. Proper control of phosphate and silicate residuals avoids the formation of magnesium phosphate (a sticky precipitate) and calcium silicate (normally a dense, hard scale).

In higher pressure boilers and some lower pressure boilers with high purity feedwater, coordinated phosphate-pH control is practiced. This control method provides both the phosphate residual and the pH desired in the boiler by feeding a combination of disodium and trisodium phosphates. Its purpose is to avoid the presence of free hydroxide, thus eliminating the potential for caustic attack of boiler surfaces.

Chelating agents provide an alternative approach to scale control that may be attractive for some low-pressure boiler systems. These chemicals form soluble complexes with ions such as calcium and magnesium. Some chelating agents will also solubilize iron and copper ions. Chelating agents should be supplemented with an antifoam agent and an oxygen scavenger.

The boiler feedwater must be low in hardness (1–2 mg/kg or less) for chelating agents to demonstrate an economic advantage over the precipitation scale control methods. Chelating agents have not been successfully utilized in high-pressure boilers.

## 3.11 SLUDGE CONDITIONING

Various organic materials are often used to condition the boiler precipitates or sludges to make them fluid or free flowing for easier removal by blowdown. These are usually derivatives of tannin or lignin, synthetic materials, or, in some cases, derivatives of seaweed. Starch is sometimes used in high silica waters. Sludge conditioners are frequently combined with phosphates and chelating agents. Antifoam materials, for smoother boiler operation, are sometimes also incorporated in these formulations.

### 3.11.1 CAUSTIC EMBRITTLEMENT

Caustic embrittlement is intercrystalline cracking of boiler steel, which may occur in the presence of all of the following factors:

- The metal must be subjected to a high level of stress.
- There must be some mechanism (a crevice, seam, leak, etc.) permitting concentration of the boiler water on the stressed metal.

| Table 3.2 Recommended NaNO3/NaOH Ratio for Boilers | |
|---|---|
| **Boiler Operating Pressure (KPa)** | **NaNO₃/NaOH Ratio** |
| Up to 1725 | 0.20 |
| 1725–2750 | 0.25 |
| 2750–4825 | 0.40 |

- The concentrated boiler water must possess embrittling characteristics and chemically attack the boiler metal.

Of these three factors, the embrittling characteristics of the boiler water generally can best be shown to be present or absent in a boiler. An Embrittlement Detector developed by the U.S. Bureau of Mines can be used to determine this water characteristic. As an alternative, since there are no simple chemical tests to measure embrittlement and there is always the possibility of embrittlement occurring, a chemical embrittlement inhibitor, generally sodium nitrate, is often added to the boiler. A definite ratio of sodium nitrate to caustic alkalinity in the boiler water is required for inhibition according to the formula:

$$\frac{NaNO_3}{NaOH} \text{Ratio} = \frac{(\text{Nitrate as NO}_3, \text{mg}/\text{kg})(2.14)}{(\text{M alkalinity as CaCO}_3, \text{mg}/\text{kg}) - (\text{Phosphate as PO}_4, \text{mg}/\text{kg})}$$

This ratio depends upon the operating pressure of the boiler as recommended by the U.S. Bureau of Mines in Table 3.2.

# STEAM BOILERS

Steam generators, or boilers, use heat to convert water into steam for a variety of applications. Primary among these are electric power generation and industrial process heating.

The boiler design should be proven in practice to be rugged and reliable, and the tenderer should provide a list of similar installations which have already been built and which are in operation. The boilers and auxiliary equipment should be designed and erected in accordance with the latest edition of section I of the American Society of Mechanical Engineers (ASME) boiler and pressure vessel code, including all published addenda and interpretations thereto. The boiler and all auxiliary equipment listed as being supplied by the vendor should be suitable for outdoor installation.

Fossil fuels are used in factories to provide heat and some electricity. The fuels vary in the amount of acid gas they produce. Natural gas contains only a very small amount of sulphur, whereas fuel oils can contain up to 3% sulphur. Some fuels such as coal vary in their composition from one region to another.

Ease of operation, safety, inspection, maintenance, repair, and cleaning should be of major concern when considering the design and arrangement of boilers.

A boiler is a closed vessel in which water or another fluid is heated. The fluid does not necessarily boil. (In North America the term "furnace" is normally used if the purpose is not actually to boil the fluid.) The heated or vaporized fluid exits the boiler for use in various processes or heating applications, including water heating, central heating, boiler-based power generation, cooking, and sanitation.

Boiler and ancillaries should be capable of continuous 24 h a day operation between a 36 month statutory shut-down period.

The source of heat for a boiler is combustion of any of several fuels, such as wood, coal, oil, or natural gas. Electric steam boilers use resistance- or immersion-type heating elements. Nuclear fission is also used as a heat source for generating steam, either directly or, in most cases, in specialized heat exchangers called "steam generators." Heat recovery steam generators (HRSGs) use the heat rejected from other processes such as a gas turbine.

Most boilers produce steam to be used at saturation temperature; that is, saturated steam. Super-heated steam boilers vaporize the water and then further heat the steam in a superheater. This provides steam at a much higher temperature, but can decrease the overall thermal efficiency of the steam generating plant because the higher steam temperature requires a higher flue gas exhaust temperature. There are several ways to circumvent this problem, typically by providing an economizer that heats the feedwater, a combustion air heater in the hot flue gas exhaust path, or both. There are advantages to superheated steam that may, and often will, increase overall efficiency of both steam generation and its utilization: gains in input temperature to a turbine should outweigh any cost in additional boiler complication and expense. There may also be practical limitations to using wet steam, as entrained condensation droplets will damage turbine blades.

Superheated steam presents unique safety concerns because, if any system component fails and allows steam to escape, the high pressure and temperature can cause serious, instantaneous harm to anyone in its path. Since the escaping steam will initially be completely superheated vapor, detection can be difficult, although the intense heat and sound from such a leak clearly indicates its presence.

Superheater operation is similar to that of the coils on an air conditioning unit, although for a different purpose. The steam piping is directed through the flue gas path in the boiler furnace. The temperature in this area is typically between 1300 and 1600°C (2372 and 2912°F). Some superheaters are the radiant type; that is, they absorb heat by radiation. Others are the convection type, absorbing heat from a fluid. Some are a combination of the two types. Through either method, the extreme heat in the flue gas path will also heat the superheater steam piping and the steam within. While the temperature of the steam in the superheater rises, the pressure of the steam does not and the pressure remains the same as that of the boiler. Almost all steam superheater system designs remove droplets entrained in the steam to prevent damage to the turbine blading and associated piping.

Boilers should be designed to operate under the following operating conditions:

- Guaranteed net steam capacity, for each boiler at superheater outlet excluding all blow down, heat losses, and all steam used for boiler auxiliaries such as fans, soot blowers, burners, etc., as specified on the data sheet.
- Over design capacity for 4 h continuous operation with an interval of not less than 20 h between periods, the vendor to specify (minimum acceptable 15%). The vendor should also specify the time required for increasing the load of the boiler from ¼ of maximum continuous rating (MCR) to a full load of MCR.
- Guaranteed turn down ratio.
- Minimum thermal efficiency, as specified in the project specification, on fuel gas firing based on LHV.

The previously mentioned boiler efficiency should be guaranteed to operate under the following conditions:

1. final superheater/desuperheater pressure
2. final superheater/desuperheater temperature
3. feedwater temperature at the economizer inlet
4. ambient air temperature at Fine Dust Filter (FDF) inlet
5. ambient air relative humidity
6. barometric pressure
7. fuel gas temperature
8. based on ASME PTC 4.1 heat loss abbreviated method
9. continuous blow down rate.

- The superheater (in case a desuperheater is not considered) outlet temperature should be uncontrolled and the vendor should guarantee the maximum variation in superheater steam outlet temperature throughout the operating range from 40% to full load.
- Guaranteed operation under the superheater/desuperheater outlet pressure specified on the data sheet.
- Feedwater inlet temperature specified on the data sheet.
- Purity of produced steam as specified in the project specification in ppm mass (mg/kg) with consideration of ppm mass (mg/kg) of total solids in the boiler water. The boiler manufacturer

should state the maximum total dissolved solids (TDS) in the boiler water at which the required steam purity can be obtained.

- The steam as measured at drum outlet should have an impurity not greater than 0.02 mg/kg (ppm mass) silica. The wetness of the steam leaving the drum should not exceed 0.02%.
- Heat release (maximum and average) per cubic meters of furnace volume (volume enclosed by the effective heating surface) when firing specified fuels, the vendor to specify subject to the company's approval.
- Maximum intensity of heat flow rate in kW/m$^2$ released in the furnace based on the effective heating surface (as defined later), the vendor to specify subject to the company's approval.
- Effective heating surface is defined as the flat projected area of tubes and extended surface integral.
- With tubes exposed to direct radiation (only the first row of tubes in flue gas passes exposed to direct radiation should be counted for calculation of flat projected surface and the refractory covered surface should not be counted).
- Superheater, drums, and boiler tube system maximum design pressure, the vendor to specify, but a minimum of 300 kPa (3 bar) above operating pressure is required.
- Steam drum and boiler tube system maximum design temperature specified on data sheet, superheater design temperature per code.
- Total continuous and intermittent blowdowns should not be more than the allowable figure specified in the boiler data sheet of steam generated, the vendor to specify capacity at 100% rating.
- Steam for fan drivers specified on the data sheet.
- Exhaust steam from fan drivers specified on the data sheet.
- Stack temperature, the vendor to specify.
- Boiler feedwater and chemical injection specified on the data sheet.

Supercritical steam generators are frequently used for the production of electric power. They operate at supercritical pressure. In contrast to a "subcritical boiler," a supercritical steam generator operates at such a high pressure (over 3200 psi or 22 MPa) that the physical turbulence that characterizes boiling ceases to occur; the fluid is neither liquid nor gas but a super-critical fluid. There is no generation of steam bubbles within the water because the pressure is above the critical pressure point at which steam bubbles can form. As the fluid expands through the turbine stages, its thermodynamic state drops below the critical point as it does work turning the turbine, which turns the electrical generator from which power is ultimately extracted. The fluid at that point may be a mix of steam and liquid droplets as it passes into the condenser. This results in slightly less fuel use and therefore less greenhouse gas production. The term "boiler" should not be used for a supercritical pressure steam generator, as no "boiling" actually occurs in this device.

## 4.1 SPECIAL DESIGN FEATURES

Boilers can be classified into the following configurations:

- "Pot boiler" or "Haycock boiler". a primitive "kettle" where a fire heats a partially filled water container from later. Eighteenth century Haycock boilers generally produced and stored large volumes of very low-pressure steam, often hardly above that of the atmosphere. These could burn wood or most often, coal. Efficiency was very low.

- Fire-tube boiler. Here, water partially fills a boiler barrel with a small volume left above to accommodate the steam (*steam space*). This is the type of boiler used in nearly all steam locomotives. The heat source is inside a furnace or *firebox* that has to be kept permanently surrounded by the water in order to maintain the temperature of the *heating surface* below the boiling point. The furnace can be situated at one end of a fire-tube which lengthens the path of the hot gases, thus augmenting the heating surface which can be further increased by making the gases reverse direction through a second parallel tube or a bundle of multiple tubes (two-pass or return flue boiler); alternatively the gases may be taken along the sides and then beneath the boiler through flues (three-pass boiler). In the case of a locomotive-type boiler, a boiler barrel extends from the firebox and the hot gases pass through a bundle of fire tubes inside the barrel which greatly increases the heating surface compared to a single tube and further improves heat transfer. Fire-tube boilers usually have a comparatively low rate of steam production, but high steam storage capacity. Fire-tube boilers mostly burn solid fuels, but are readily adaptable to those of the liquid or gas variety.
- Water-tube boiler. In this type, tubes filled with water are arranged inside a furnace in a number of possible configurations, often the water tubes connect large drums, the lower ones containing water and the upper ones, steam and water; in other cases, such as a mono-tube boiler, water is circulated by a pump through a succession of coils. This type generally gives high steam production rates, but less storage capacity than the earlier. Water tube boilers can be designed to exploit any heat source and are generally preferred in high-pressure applications since the high-pressure water/steam is contained within small diameter pipes, which can withstand the pressure with a thinner wall.
- Fire-tube boiler with water-tube firebox. Sometimes the two earlier types have been combined in the following manner: the firebox contains an assembly of water tubes, called thermic siphons. The gases then pass through a conventional fire-tube boiler. Water-tube fireboxes were installed in many Hungarian locomotives, but have met with little success in other countries.
- Flash boiler. A flash boiler is a specialized type of water-tube boiler in which tubes are too close together and water is pumped through them. A flash boiler differs from the type of mono-tube steam generator in which the tube is permanently filled with water. In a flash boiler, the tube is kept so hot that the water feed is quickly flashed into steam and superheated. Flash boilers had some use in cars in the 19th century and this use continued into the early 20th century.
- Sectional boiler. In a cast iron sectional boiler, sometimes called a "pork chop boiler" the water is contained inside cast iron sections. These sections are assembled on site to create the finished boiler.

## 4.1.1 DRUMS AND STEAM GENERATORS

The boiler designer should state the minimum and maximum water levels between which the boiler should be allowed to continue operating.

Furnace wall, floor, and roof tubes should not incorporate bends or sets sufficiently small in radius to interfere significantly with water circulation. In particular, roof tubes exposed to radiant heat should be free from bends and sets as far as possible, so as not to upset the division of flow between tubes or bring to a doubtful value the head available to promote circulation in any part of the circuits.

The boiler designer should state the holding time provided by the reserve of water in the steam drum, between "low liquid level" and "low-low liquid level", and the company will approve this time against that required to introduce effectively the standby boiler feed pump. The size of the steam drum may have to be increased to provide a longer period in which to recover the water level without incurring the automatic shut down of the boiler.

Downcomers supplying the furnace wall, etc. with feedwater should preferably be outside the flue gas path. If the downcomers are in contact with the flue gases, the heat transfer should not significantly affect the circulation head.

Tube banks should be arranged, as far as practicable, to permit access for tube renewal with minimum cutting out of serviceable tubes.

Provision for acid cleaning of boilers and nitrogen blanketing of (idle) boilers should be provided.

Adequate provision should be considered for inspection and cleaning of waterwall headers (minimum of two, one at each end of the header).

Steam drums should be equipped internally with steam separators and scrubbers to ensure that the carryover of total solids from the boiler water should not exceed the following:

- ppm mass (mg/kg) up to 65 bar (ga)
- ppm mass (mg/kg) from 65 bar (ga) to 135 bar (ga).

In order that drum stability may be evaluated, the boiler designer should indicate the steam drum water content (effective) at normal, low-level, and MCR loading.

Necessary drum connections should be provided for chemical cleaning and nitrogen sealing of boilers in addition to connections as required, that is, steam outlet, safety valves, continuous blowdown, chemical feed, water column level controls, level alarm, feedwater, vent and bottom blowdown, etc. In case of necessity for winterization of the boiler, the vendor should provide steam coils in each lower drum.

Suitable internals should be provided for the distribution of the incoming feedwater to ensure a proper distribution of the incoming water along the length of the drum, suitably placed to feed the downcomer tubes but not to interfere with the correct function of the water level gages, and also for the chemicals and for the collection of the continuous blowdown.

## 4.1.2 DOORS AND OPENINGS

Observation ports should be furnished to permit visibility of furnace and flame conditions, the furnace floor and the superheater space during operation of the boiler. Observation ports of pressurized boilers should be furnished with an aspirating type air interlock to prevent opening if the seal air is not turned on. Seal air should be sufficient to prevent pressurized furnace gases from blowing out through the observation port.

## 4.1.3 DUCTS AND STACKS

The stack should be designed as an individual self-supporting steel stack with a minimum height specified for each boiler, but in any case not less than 76 meters. Stack lining should be the vendor's standard design, subject to the company's approval.

- Each boiler should have a separate stack, unless otherwise specified.
- Stacks should be equipped with aircraft warning lights per relevant job specification.

- Ducting for air and flue gases should be air tight and sufficiently stiffened.
- Steel stacks should have a minimum of 3 mm allowance for corrosion.

For air and flue gas ducting the air and gas velocities in ducting should not exceed 13.7 m/s and 15.2 m/s respectively, taking all internal bracing and stiffeners into account.

The gas outlet damper for control of furnace pressure should be supplied by the vendor and fitted with an extended shaft.

All ductwork should be designed for fabrication in flanged sections. Particular attention should be paid to the design of the ducting from the boiler fans to the stack to ensure proper performance of one or both fans at all loads.

### 4.1.4 BURNERS

Two main flame detectors should be fitted to each burner, with any one detector signal arranged to give an alarm and the two signals together to cause lockout of the fuels to the burner.

The system should be complete, without any areas of split responsibility, especially regarding furnace purging and boiler safety.

Separate buttons should be provided to initiate purge and individual burner start up, and also for individual burner and boiler shut down.

Local and control room panels should provide all the information necessary to enable the operators to ascertain the condition of each burner and all the associated functions of fans, purging, register positions, fuel valve positions, and safety interlocks.

The boiler designer should justify any atomizing steam consumption greater than 0.5% of boiler MCR.

Burner minimum turndown ratio should be 33% for liquid fuels and 10% for fuel gas, with the boiler supplier's guaranteed low $O_2$ in the flue gas maintained over the ranges mentioned in the standard.

Boilers having four or more burner assemblies for use with fuel gas or commercial grade liquid fuel should operate satisfactorily with combustion conditions as near stoichiometric as practicable. The excess air should not exceed 3% for liquid fuels and 5% for gaseous fuels. Over the full operating range of the boiler the following $O_2$ vol. percentage in flue gases should be achieved with liquid fuels:

- 0.5% $O_2$ between 70 and –100% MCR
- 1.0% $O_2$ between 25 and 70% MCR
- 5.0% $O_2$ between 0 and 25% MCR

Carbon monoxide in the flue gas should not be greater than 0.01% by volume at the specified $O_2$ content in flue gases.

Duplex type filters, or two filters in parallel, of 125 μm (0.005 in.) mesh in monel, should be provided in the gas supply for each convenient group of pilot burners. The pipework from the strainers to the pilot burners should be in stainless steel.

Horizontal distance between the main burners and the vertical distance between rows of burners should be such as to facilitate discrimination between individual flames by the proposed flame detectors.

Burner viewing ports should be fitted to each burner assembly front plate in such a position as to afford an adequate visual examination of the pilot burner and the root of the flame.

Provision should be made for the steam purging of burner guns to remove all liquid fuels. It should not be possible to withdraw a gun from the burner assembly unless the fuel is shut off, the purging carried out, and steam shut off. It should also not be possible to turn on fuels or steam with the gun withdrawn. This mechanism must only be capable of being overridden by a locked "defeat" switch with a removable key.

Before the first burner on a boiler can be ignited, an adequate purge of the furnace and gas passes should be automatically carried out. Indication of the unpurged condition should be visible from the firing floor and boiler control panel. The purging sequence should be initiated by a local push-button control by the operator.

The air-flow rate and duration of this purge procedure should be specified and this should be based upon the shape of the furnace and complexity of the flue gas passes. However, the airflow should not be less than 25% MCR airflow for a period of at least 5 min with all air registers open, or for such a length of time as to give at least five volume changes of the plant combustion chamber and gas passages up to the exit of the flue, whichever is greater.

The purge procedure should be an inescapable action on every start up and one which the operator cannot override, reduce in flow rate, or shorten in duration.

The start-up and shut-down sequence should be automatic with push buttons to start and stop the sequence for each burner. Coloured lamps on the panels should indicate the status of the burners.

It should not be possible for the fixed periods of fuel admission to be extended or overridden by the operator before the flame is established.

Interlocks should be provided to prevent burner start up if the furnace conditions are not satisfactory. These should initiate the master fuel trip system to shut-off the main fuel trip valves to the boiler at any time during operation, if they are not continuously satisfied. Conditions producing this trip should at least include the following:

1. Low-low water level in the steam drum
2. Low-low pilot fuel gas supply pressure (shut off pilot gas at start up only)
3. Low-low supply pressure for the relevant fuel
4. Loss of forced draft
5. Loss of main burner flames (individual burner fuel cut off)
6. Loss of atomizing steam pressure (on liquid fuel firing)
7. Low-low pressure of instrument air (start up conditions only, "fail-locked" would operate when on load).

*Note*: All level, temperature, and pressure transmitters/switches which are supposed to shut down the boiler, should be triple and function as a voting system (two of three). Actuation of one transmitter/ switch should give an alarm and the action of two transmitters/switches should shut down the boiler.

Following a main fuel valve trip, the FDF and tripping equipment should be so arranged that the furnace should not be unacceptably pressurized.

While burners may be arranged for control from a remote control room, the start up of a boiler, and every additional burner thereafter, should be initiated and observed by an operator at the boiler-firing floor. The control and indicating equipment should, therefore, be arranged accordingly. On large boilers having two or more burner platform levels, the local control panel should be divided into sections positioned appropriately at each platform level (eg, upper burner start-up panel, lower burner start-up panel).

To ensure the effective isolation of all fuels to a furnace, solenoid-operated valves should be inserted in the air lines to pneumatically operated ball valves placed immediately upstream of the control valves, these isolating valves should be arranged for remote manual activation in an emergency and to work, automatically, in conjunction with the safety interlocks, when unacceptable conditions arise.

The general physical arrangement of pipes, valves, and control equipment, etc. at each burner and in the firing floor area as a whole, should be given specific attention so as to provide a neat, uncluttered, and logical layout, capable of being readily identified by the operator and facilitating easy access for operation and maintenance.

Gas off-takes for individual burners should be from the top of the header.

Platforms at each burner level should be provided, together with stairways and necessary escape ladders. The platforms should be wide enough to enable burner guns to be withdrawn without difficulty and to be safely handled by the operator.

Fuel pipework should have blanked-off connections to which temporary steam lines may be attached for purging before maintenance. They should be located close to, and downstream of, the shut-off valves.

Atomizing steam lines should be lagged separately from fuel lines.

The atomizing steam pressure should be controlled to give a constant value, or a constant differential pressure from that of the relevant fuel, as the particular type of burners may require.

Each burner should be sized for 110% of its design load or such that the boiler MCR can be maintained with one burner out of use, whichever is greater. Burner flames should be horizontal and not parallel with the steam drum. The flame should not impinge on the wall or any metal parts.

The boiler supplier should state the heat input of the proposed pilot burners.

Combination-forced draft burners for firing all specified gas, all gasoline, all gas oil, all oil, or any combination thereof should be provided. Atomizing steam facilities for fuel and gas oil (if required) burners should be provided. The fuel on which boiler performance should be guaranteed should be specified. Provision should be made for changing fuel or any one burner without affecting boiler operation in any way. One liquid fuel burner gun and one gas fuel burner should be provided per each burner, separate tips for fuel oil and gasoline fuel should be provided. The vendor should specify the possibility of using a single tip for both liquid fuels. Provision should be made for preventing flue gas leakage when the oil burner gun is removed for tip change or cleaning by use of air purge, etc.

A fixed gas-fired pilot burner, removable for maintenance while the boiler is in operation should be provided at each burner assembly. It must be suitable to ensure safe and efficient ignition of all fuels specified. Each pilot burner should be permanently lit when its main burner is in use. The pilot flame should be visible through the burner peephole, at least prior to the ignition of the main flame. The pilot burner should be proven capable of igniting the main fuels efficiently and of remaining lit under all windbox and furnace conditions likely to be experienced.

Each furnace should be supplied with burners with insulatable forced draft registers with steam atomizing oil units, center fired-type gas units, flexible metallic oil and steam hoses, flexible stainless steel gas hose, oil and steam shut-off valves and oil burner fittings. The oil units should be equipped with swing check valves and air seals for use in a pressurized type steam-generating unit. The registers should be equipped with all necessary seals. Where gasoline fuel is specified, each burner should be equipped with an interlocking device on the gasoline and atomizing steam supply complete with valves, interlock discs, piping, and flexible joints on gasoline supply lines.

Each burner should be provided with an electric gas igniter with complete flame protection system with flame scanners, flame protection relays, interlocks, purge cycle timer, operating switch indicating lights, transformer, safety shut-off valves, stainless steel gas hose, air hose, stainless steel gas strainer, and air seal.

One burner holder and wrench assembly should be provided for each boiler. Steam atomizing oil units for gasoline and one spare gun complete with tip end assembly should be provided for each boiler. Fuel oil return lines should be equipped with nonreturn valves. The liquid fuel oil lines should be large enough with a low friction factor.

The boiler manufacturer should furnish a burner windbox cut and drilled to accommodate the burners described herein. This windbox should be complete with necessary supports, division plates, and access door, if required.

Specification of fuels to be burnt will be indicated for each installation.

Number of burners and arrangement of burners should be submitted to the company for approval.

A complete flame monitoring and safety control system to perform the functions should be furnished. The system should be clearly described in the proposal and should be guaranteed for safe and efficient operation of the boiler. Any logic circuity being proposed should be included in the vendor's proposal. Burner management systems should be installed locally to the burners, the system should jointly monitor the burner and boiler to ensure safe start up and shut down of burners and boiler. The system should, on the pressing of push buttons, arrange for the whole sequence of burner light up or shut down to be automatically carried out with a high degree of safety and reliability. It should also automatically shut down the burners on identification of a fault condition serious enough to warrant such action, or raise alarms to indicate faults of a less serious nature. Reset facilities should be provided for both boiler and individual burner trips.

The system should ensure that the agreed logic on which the sequence of operating functions is based should not be capable of being interchanged or abridged. It should be of proven reliability, operating on the stop-check principle where the system can only proceed if the preceding sequence has been completed. A failure to complete a sequence should operate an alarm and a fault location system will identify the area of malfunction, and where practicable, the fault itself. Separate circuits should be used so that only those required for actual operation are retained in service, the others maintaining a passive, but energized role so that should component failure occur, it can be identified and repaired with the burner in service.

## 4.2 INSULATION AND CASINGS

Insulation material should be applied in sufficient thickness to prevent casing distortion, to reduce radiation losses to an economic minimum and to ensure personnel protection. The percentage radiation loss for the whole boiler should be stated in the proposal.

Casings should be designed to prevent escape of the flue gases or the circulation of gases into cool sections of the casings or structural steelwork, thereby creating conditions for internal or external corrosion. The temperature of the casing plus attachment should be maintained above the dew point of gases by the installation of adequate external insulation.

## 4.3 AIR HEATER

Air heaters, utilizing the flue gas sensible heat to raise the temperature of the combustion air, will be accepted when there is a need to obtain higher thermal efficiencies than can be attained by an economizer alone. When an air heater is proposed the boiler designer should satisfy the company concerning the advantages of higher efficiency, considering the increased capital cost, increased maintenance costs, effectiveness of soot blowing, expected operating time, efficiency, and likely problems.

Air and gas bypasses should be provided with proven soot-blowing equipment.

The flue gas exit temperature should not be less than that recommended by the air heater manufacturer who must consider the air/flue gas temperature differential in relation to the possibility of corrosion.

Air heaters using surplus low-pressure steam should be considered. The condensate discharge should be returned to the deaerator. On-line cleaning facilities for the finned tubes should be provided.

## 4.4 FURNACE

The maximum temperature on the outside of the boiler casing should be low enough so as not to constitute a hazard to personnel.

Where the boiler design incorporates a refractory front wall around the burner area, the boiler manufacturer should obtain the company's approval on the suitability of the method of attachment and anticipated life of this refractory.

Drains should be furnished at the low point of the boiler furnace and bank areas to permit removal of the flue gas deposits by water washing. Drain openings should be effectively sealed against flue gas bypassing and casing overheating.

Cavities above the furnace roof tubes should be designed to prevent the accumulation of gases that might form explosive mixtures. This may be achieved by purging with air, reliable sealing, or some other proven method. Access to such areas for the examination of tubes, penetration seals, and hangers, etc., should be possible during boiler surveys.

The furnace width should enable sufficient spacing of burners to ensure burner flame discrimination by individual viewing heads and also to make certain that there is no flame impingement on the sidewalls. The furnace depth should be sufficient to ensure that burner flames do not impinge on the rear wall or penetrate the screen tube arrangement.

Floors should not be utilized for heat transfer and should be effectively shielded from furnace radiant heat by refractory tiles and insulation as necessary.

The angle to the horizontal of floor tubes should not be less than 15 degree and the angle of roof tubes should not be less than 5 degree. Where the roof tubes have to be offset for any reason, continuity of drainage should be ensured, but such offset should be avoided where possible.

## 4.5 SUPERHEATER/ATTEMPERATOR

On boilers that are required to produce a specified degree of superheat to the steam over a wide range of boiler operation, or on units where superheat temperatures are expected to be near maximum design temperature of the tube metal concerned, attemperators should be provided. In the former case attemperation may be performed at the final outlet before the stop valve, but in the latter case the attemperator should be positioned between the primary and secondary stages of the superheater.

Spray-water type attemperators should normally be used, provided that there is a supply of demineralized water of suitable quality for this duty. Spray-type attemperators must not be capable of blocking the steam flow through the superheater in the event of mechanical failure.

Where interstage desuperheating is used, the downstream construction material should be capable of temporarily withstanding the resulting higher temperature, should the spray water supply fail.

Metal temperatures should not be high enough to allow corrosion to take place in the presence of vanadium and sulfur compounds, or other corrosive constituents resulting from the use of the specified fuels.

To monitor metal temperatures during boiler start up, skin thermocouples should be secured to the tubes at appropriate points. These thermocouples and their connecting leads should be so positioned and protected that they will not suffer rapid deterioration by exposure to the flue gases or radiant heat of the furnace.

Air vents should be provided where necessary.

The flow of steam through the superheater should create such a pressure drop over the entire operating range as will ensure an adequate distribution of steam through all tubes and thereby prevent overheating of any element. The boiler designer should state the pressure drop across the superheater at 40, 70, 100, and 110% of MCR.

## 4.6 VALVES AND ACCESSORIES

The vendor should furnish the following valves and accessories, all as per ASME codes, details of which will be specified on the data sheet:

1. Safety valve (on boiler)
2. Blow-off valve (tandem)
3. Steam drum vent valve
4. Chemical feed stop valve
5. Chemical feed check valve
6. Continuous blowdown micrometer valve
7. Continuous blowdown line stop valve
8. Steam gage
9. Water gage stop valve
10. Water gage drain valve
11. Steam gage line stop valve
12. Water gage
13. Water wall drain valves (as required)
14. Superheater safety valves (according to ASME Section I PG-71)
15. Superheater inlet header drain valve
16. Superheater outlet header drain valve
17. Steam sampling connection from superheater inlet header
18. Superheater vent valve
19. Feedwater stop and check valves

**20.** Nonreturn and stop valves on superheater outlet (nonreturn valve means combined stop/check valve)

**21.** Two DN25 (1 in.) ID (RF flanged) openings for the installation of the thermowells at the following locations:
   **a.** boiler air inlet
   **b.** boiler outlet (full gas)

**22.** Double valves in series should be used for blowdown facilities, steam drum venting, sample connection (boiler water, saturated, and superheated steam) and drain connections (lower drum, water walls, superheater). Isolating facilities required for maintenance of the boiler or equipment without shutting down the plant should also be equipped with double isolation valves and a vent valve.

## 4.7 INTEGRAL PIPEWORK

The feedwater system arrangement should be such that the feed regulating valve, isolating valves and bypass valves can be manually operated from the floor level in the event of an emergency.

Drains from the boiler, superheater, economizer, and soot blowers, etc. should be operable from the floor level where practicable. Pipework should not be positioned where it may possibly obstruct or trip operators.

## 4.8 ECONOMIZER

Economizers should generally be an integral part of the boiler but may be supplied as separate units when the boiler, as a standard model design, cannot readily incorporate an economizer, or the physical limitations of the proposed site make it necessary or desirable to position the economizer away from the boiler.

Feedwater operating pressure and temperature in the economizer should at no time permit the possibility of steam being generated.

When economizers can be isolated on the waterside, a safety relief valve should be fitted.

## 4.9 INSTRUMENTATION

The vendor should submit the proposed instrument and control schematic drawings including combustion and feedwater controls, adequate to fulfil the requirements of his process and mechanical guarantees for Company's approval.

Retractable thermocouples for superheater flue gas inlet temperature and superheater skin temperature thermocouples should be included for use during start-up periods.

There should be local indication of instruments for each boiler and all controls and indicators should be brought to the central control room.

An external float type water column with switches for high and low level alarms and three external float type water column with switches for low-low alarm with low water cut off should be separately mounted on the steam drum. Low water cut off (low-low water level) should be separate from low level alarm and should cause the fuels to the boiler to be cut off and an emergency alarm to be raised, both visual and audible in the boiler control room. The point at which this switch operates should be at a water level high enough to protect all pressure parts from overheating and to be still visible in the gage glasses. It must not be so near to the normal-low water alarm level that there would be insufficient time for an operator to make adjustments for the first condition before the second arises, causing the shut down of the boiler. High and low level alarms should have different sounds and be sufficiently loud to be heard in the boiler area.

Identical direct reading water level gage glasses should be provided on each side of each boiler. Each gage glass should be fed by independent feed lines. Gage glass and water column drain valves should be furnished. Water columns should be fitted with flanged shut-off valves installed between the column and the drum. Each gage should be capable of being blown-down or isolated for removal and repair without taking the boiler off-load. Lighting should be incorporated in the design of the gage glasses and the gage glasses should be constructed by suitable materials such that the detection of the level in the gage glass to be easily traced and additionally the material of the gage glasses should be resistant against thermal shocks and their color should not be changed upon any changing in temperature which causes any turbidity on the gages.

Where the drum elevation above operating floor level prevents the operator from viewing the direct water level gages, a remote direct-reading gage of a proven type should be provided in addition to the two gages local to the drum. It should be located at operating floor level and positioned so as to be easily seen by the operator standing at the feedwater regulating and bypass valves. This remote direct-reading gage should be of sturdy construction to overcome vibration.

Design of low level steam drum switches should be such as to avoid tripping due to vibration.

A pressure tap connection and thermowell connection at superheater outlet header should be provided.

Connection of gas sampling should be provided at the boiler outlet, burner air inlet, furnace, and should be DN25 (1 in. NPS) size. Their location should be such that representative samples are obtained.

## 4.9.1 FANS AND DRIVERS

Fans, drivers, and gearboxes should be at an easily accessible location (for maintenance and motor removal) at the rear of the boiler and mounted on foundation blocks at grade level. Unless otherwise specified, each boiler should have two 75% capacity fans. The forced draft fans on all boilers should be identical. Fans should be of the centrifugal type, providing stable operation under all conditions of boiler load. Fans efficiency should not be less than 80%.

The fan performance guarantee should be in accordance with the standard test code of "National Association of Fan Manufacturers." Fan performance should also be guaranteed to meet all operating conditions specified on the data sheets. A characteristic fan performance curve should be submitted in the vendor's proposal for the company's approval.

Fans and associated drivers should be designed, sized, (and capable of handling all air requirements for all fuels) as follows:

1. Fan rating should be established on the maximum ambient temperature for the location. Unless otherwise specified, each fan should be sized to provide 75% capacity at the rated load.
2. Fans should be turbine driven with a gear reducer, one fan for one boiler (at least one, total number of dual drive fans to be specified by the company), and should have dual drive for

stand-by purpose. One end of this fan should be coupled to a steam turbine and the other end to an electric motor. Each driver should be connected through a one-way clutch for automatic disconnection when either driver is disengaged.

3. Fans should be supplied with an inlet screen, cleanout door, and a silencer. The air intake main connection for the silencer should be flanged and a minimum of 3.2 mm corrosion allowance should be considered. Pressure drop across each silencer should not exceed 12.7 mm $H_2O$ (1.25 kPa).

4. The vendor should furnish for each draft fan inlet a suitable air intake device to reduce sand and dust intake.

5. The control of combustion air should be accomplished by a controlling fan inlet damper. The damper operator should be furnished by the vendor and should be equipped with a pneumatic positioner. The damper operator should be provided with a continuously connected hand wheel.

6. The boiler Vendor should provide, in the air duct between the fan and the burners, a primary air measuring element for the purpose of measuring total airflow rate.

7. Turbines' steam inlet and exhaust conditions (maximum/normal/design pressure and temperature) should be as per relevant Standard Specification.

8. The elevation of FDF air intake point should be at least 5 meters above the high point of paving.

## 4.9.2 BOILER CONTROLS

The vendor should guarantee that the proposed control system will permit sound and reliable operation of boilers over the whole range of operation.

Boiler controls should be designed to fire all types of fuels or combination of them as indicated on the boiler data sheet. The turn down ratio under all firing conditions should be as defined on the boiler data sheet. However, this should be qualified to the extent that: any minimum firing range for a single fuel or a combination of fuels is determined from actual field tests when placing the equipment in operation.

### 4.9.2.1 Steam Header Pressure Control (General Plant Control)

The plant master should control the main header pressure to a set point determined by the operator. The plant master should control this pressure by changing the firing rate of the boilers. The plant master controls a boiler when its combustion controls are in full automatic.

If the load demand increases, the main steam header pressure will decrease and the plant master should sense this and increase its output until the main header pressure is back to the set point. If the load demand decreases the opposite is true.

### 4.9.2.2 Combustion Controls

The plant master output should change to meet load demand. The change in plant master output should act as an air-rich control as follows:

1. Boiler load up
   a. Boiler master signal increase
   b. Airflow demand signal increase
   c. Fuel flow demand signal increase after actual airflow increase

**2.** Boiler load down
   **a.** Boiler master signal decrease
   **b.** Fuel flow demand signal decrease
   **c.** Airflow demand signal decrease after actual fuel flow decrease.

### 4.9.2.3 Feedwater Control System

The feedwater flow should be controlled by a three-element feedwater control system.

The drum water level should be measured by a differential pressure transmitter. The level signal from this transmitter should be compared with the fixed set point in the master drum level controller. The output signal of the master level controller, summed with the steam flow signal, is fed to the (slave) boiler feedwater flow controller as the set value of the feedwater flow. The output signal of this controller (ie, feedwater flow controller) should change the boiler feedwater flow control valve opening. For the start-up period and low load operation a single element control should be provided, the transfer between single element and three element controls should be carried out according to boiler load automatically, and also manually through distributed control system selector switches.

### 4.9.2.4 Boiler Master Control System

Boiler master control should provide two kinds of control mode, one is the common master mode and the other one is the individual boiler master mode.

**1.** Common master control: When the boiler is operated in parallel with the other boiler units, this control mode must be selected. In this mode the boiler is controlled so as to maintain the common high-pressure steam header pressure.
**2.** Individual boiler master mode: In this mode, the boiler should be controlled so as to maintain the individual superheater outlet steam pressure. This control mode is used for boiler start up and shut down and/or special isolated operation. After the steam pressure rises up to the same value as the high-pressure steam header pressure, a controlled signal must be transferred from the superheater outlet line pressure to the common high-pressure steam header pressure.

### 4.9.2.5 Oxygen Compensation Control

Excess air ratio control should be achieved by measuring the flue gas oxygen content and trimming the airflow. The $O_2$ content in the flue gas is measured by an oxygen analyzer. The output signal of the analyzer is to be fed to the flue gas oxygen controller. The desired value of oxygen content is to be set according to the steam flow signal. Also, it should be possible to change the set value of the oxygen controller by using the bias function from a hand controller.

Comparing the signal from the oxygen analyzer and the desired value, the controller should alter the output signal until both signals are the same. The output signal of this controller should be fed to the airflow control loop to trim the airflow signal.

The following equipment will be mounted on a lighted, freestanding, rear-closed, weatherproof local instrument panel, boiler control panel, or locally as appropriate:

**1.** Draft gages [Pressures at FD fan outlet, air heater outlet (air side), burner windbox, furnace at burner level/after superheater/boiler outlet, economizer outlet, air heater outlet (flue gas side)]
**2.** Flame safety equipment
**3.** Soot blower control equipment

**4.** The following indicators should be easily and significantly visible from the local panel or located on the local panel:
  **a.** Atomizing steam pressure
  **b.** Combustion airflow
  **c.** Pressure of fuels and pilot gas downstream of control valves (also local instruments)
  **d.** Fuel supply pressure (also local instruments)
  **e.** Boiler water conductivity
  **f.** Steam purity analyzer
  **g.** Air and flue gas temperature at:
    – Air heater inlet (flue gas)
    – Air heater outlets (air and flue gas)
    – Flue gas after secondary stage of superheater
    – Flue gas, primary stage of superheater
    – Flue gas at boiler outlet
  **h.** Flue gas $O_2$ analyzer/controller/recorder (also local indication)
  **i.** Feed flow recorder
  **j.** Drum level recorder
  **k.** Fan speed indicator
  **l.** Steam pressure recorder
  **m.** Drum and superheater outlet pressure (also local gage)
  **n.** Steam flow indicator/recorder
  **o.** Instrument air pressure gage
  **p.** Fuel flow indicator/recorder with integrator for each fuel
  **q.** Fuel temperatures at burners (also local instruments)
  **r.** Burner and pilot ON/OFF indication
  **s.** Final steam temperature
  **t.** Desuperheater/attemperator sprays water supply pressure
  **u.** Conductivity indication for saturated steam leaving the steam drum, including sample cooler, with facilities for conductivity recording if specified
  **v.** Conductivity indication for superheated steam leaving the superheater, including sample cooler, with facilities for conductivity recording if specified
  **w.** All low/high and shut/down alarms.

Boiler lock-up system: each boiler should lock up on loss of instrument air to the boiler control panels. A master relay should trip all lock-up valves on control drives and valves. This should allow the boiler to operate at a fixed rate. Trimming of bypass valves may be necessary to maintain drum level and steam pressure. An alarm window should indicate that a boiler is locked up.

All instrumentation should be suitable for continuous working in the conditions of their location. Provision should be made for local tripping of critical equipment.

The boiler supplier should be responsible for the satisfactory design and operating capability of the instruments, controls, and safety equipment associated with the boiler, and he should submit details to the company for approval.

### 4.9.2.6 *Control Valves*

a. Control valves should be specifically selected for the full dynamic turndown of the system, that is, for start up and over the full firing range

b. The type of valve should be selected according to the service. Special valves should be used where cavitation, noise, flashing, or erosion may occur.

The boiler designer should be responsible for the auxiliary equipment necessary to raise and control the temperature and/or pressure of the liquid fuels to the boiler if the conditions at which fuels are to be supplied are not satisfactory for the burners he intends to use.

Provision should be made to prevent fuel supply pressure from falling when additional burners are lit.

Under normal operating conditions, including the load fluctuation specified by the company, the water level in the steam drum should not rise or fall to the point of operating the level alarms, which should be normally set to operate at not more than 100 mm (4 in.) higher/lower than the design level.

The connections to the drum for mounting the water level transmitters should be separate from those for the direct reading level gages.

For boiler drum level control applications, a water column should be used, designed to reduce errors due to temperature effects to a minimum.

Feedwater to the boiler should be controlled by a regulating valve in the feed line to the economizer, or to the steam drum direct, if no economizer is supplied.

The regulating valve should be supplied by the boiler supplier and installed in the integral pipework associated with the unit.

The operation of the regulating valve may also have to take into account a preset pressure differential across the valve in order to avoid excessive wear of the valve seating.

The following instruments should be provided and mounted on the boiler control panel:

1. Feedwater supply pressure, indication, and recording
2. Feedwater flow, indication, and recording
3. Feedwater temperature at inlet to the economizer
4. $O_2$ in boiler feedwater, indication, and recording
5. pH of feedwater.

Key-operated override switches should be provided for all shut-down functions. These switches should also override those start "permissives" which are also shut-down functions. The override switches should normally be located on the front of the main control panel. If located on the rear of the panel, then indication of override condition should be given on the panel face.

All shut-down systems should be capable of full function testing from primary sensor up to final actuation device while the plant is on-line. Test key-operated override switches should be provided for this function. These should override the minimum number of function components. Alarms should be provided to show automatically when the trip circuit is being overridden for test. Final element trip testing on a single fuel basis should be provided where more than one fuel is used.

All override test facilities should be mechanically protected and should be accessible only to the personnel authorized to carry out testing.

### 4.9.3 **BOILER FEED AND BOILER WATER QUALITY AND CHEMICAL CONDITIONING**

Company will specify the quality of the boiler feedwater available, including condensate if intended to be used.

The boiler designer should notify the company of any objection or any difficulties he may foresee in using the specified water, and should recommend to the company any further treatment or conditioning of the feedwater he considers necessary or advisable.

The recommended maximum TDS in the boiler water should be stated by the boiler designer.

Any necessary chemical mixing and injection equipment should be included by the boiler supplier and the company will specify the required extent of duplication of equipment such as injection pumps and chemical mixing tanks.

Experts will specify the type of container to be used for delivery of chemicals. All equipment necessary for the safe handling and storage (in a closed system) of hazardous chemicals should be provided locally to the injection pumps.

Water sampling points for both boiler feedwater and boiler water complete with coolers, should be provided.

### 4.9.4 **SOOT BLOWERS**

An adequate number of soot blowers should be provided to keep each boiler clean in service (free of soot). Soot blowers may be of the retractable or rotary type per operational requirement and should come complete with electric motor drive. Retractable soot blowers should be operated automatically. A control for automatic sequential operation of soot blowers should be mounted on the local panel provided by the vendor.

Each soot blower should also be capable of independent manual operation. The system should be designed for automatic removal of condensate to avoid water shock to the tubes and keeping steam temperature above saturation upstream of the lance tubes.

All headers, branches, fittings, valves, drain valves, control valves, pipe hangers, and guides as required for soot blowing systems to be supplied.

The boiler supplier should support his proposal of soot blowers with details of steam flows, jet angle, extent of effective penetration, etc. Suitable stops should be fitted to the tracks inside the boiler, to prevent lances coming off the rails due to overtravel of the drive mechanism.

The supervisory controls of soot blowers should ensure that soot blowing does not commence until all the soot blower steam distribution system has reached its working temperature and all condensate has been removed.

On completion of the operation, complete shut-off of the steam supply should be assured and drains opened. The drains should not be connected to other systems from which a blow-back might occur.

The automatic sequence and system management control should monitor and indicate all stages of operation. Facilities to interrupt the sequence or obtain selective operation of soot blowers should be included.

It should not be possible to interrupt the supply of steam to a retractable soot blower until it is in the fully retraced position.

Means of manually retracting a soot blower should be provided, and it should be possible to remove all soot blowers completely from the boiler, for maintenance, while the boiler is on load.

Sealing of wall boxes, lances, and nozzles should be provided.

### 4.9.5 **TOOLS**

One complete set of pneumatic tube rollers with two complete sets of spare tools for each size of tube to be expanded should be furnished. Any other special tools required for maintenance of operation should be provided. These furnished tools are to be used only for Owner's maintenance.

Any other special tools required for carrying out normal maintenance should be specified and provided.

### 4.9.6 **CHEMICAL CLEANING**

The internal surfaces of the boiler and economizer should be mechanically cleaned as necessary and then chemically cleaned before being put into service.

All boilers should be given an alkaline boil-out, followed by acid cleaning if specified, depending upon the amount of mill scale and iron oxide to be removed and the pressure at which the boiler will operate.

The temporary recirculating system used during chemical cleaning should be hydraulically tested to 1.5 times the pressure at which the cleaning process is to be carried out. Circulation should be arranged in such a manner as to ensure that no part of the system will be short circuited.

Superheaters should not be acid cleaned but steam blown at velocities calculated to provide a forward momentum greater than the service momentum.

The entire cleaning procedure should be the responsibility of the boiler supplier and be agreed by the company.

## 4.10 **PERFORMANCE CONDITIONS**

Performance data for the equipment supplied will be submitted herewith and made part of this section in the form recommended by the American Boiler Manufacturers Association and affiliated industries as follows:

It is recognized that the performance of the equipment supplied cannot be exactly predicted for every possible operating condition. In consequence, any predicted performance data submitted are intended to show probable operating conditions which may be closely approximated to, but which cannot be guaranteed except as expressly stated in the guarantee clauses.

The general arrangement of the equipment furnished by the boiler manufacturer and the general design and arrangement of related equipment furnished by the purchaser should be as shown on the drawings submitted by the boiler manufacturer.

### 4.10.1 **PERFORMANCE TESTS**

Prior to acceptance, the boiler manufacturer should conduct such operating tests as are necessary or required by the purchaser such as:

1. The steam dryness fraction
2. That the purity of the steam meets the guarantee under the declared conditions of boiler water, TDS, boiler load, steam drum level, and normal water level fluctuations

3. The limits of adverse conditions such as high TDS, high water level, high silica content, and severe load swings, which might cause a deterioration in steam quality beyond that acceptable, to demonstrate satisfactory functional and operating efficiency. Boiler manufacturer should be responsible for furnishing all instruments and equipment, such as portable analyzers, which are required when making the specified performance and efficiency tests.

Performance tests and performance calculation should be made in accordance with the applicable short test form in the latest edition of the ASME Test Code for Stationary Steam Generating Units and the measure of performance should be the results of such tests.

In addition to the mechanical guarantee required, the vendor should guarantee in writing that each boiler will produce from ¼ of load to full load of steam rating as specified on the data sheet and "Design Data" section without detrimental carryover into the superheater tubes and without flame impingement upon any boiler tubing when burning any combination of the gas, gasoline, and fuel oil specified herein. The vendor should also guarantee that each boiler will be capable of producing the overload requirements of the 4 h overdesign capacity specified in "Design Data" section and the minimum load for a continuous period of 24 h.

This equipment should be guaranteed on the basis of steam output (MCR) specified in the data sheet with steam and feedwater conditions described when burning the specified fuels and taking an amount of the blowdown percentage specified by the company. The supplier guarantees the purchaser that the equipment furnished is free from fault in design, workmanship, and material and is of sufficient size and capacity and is of proper material to fulfill satisfactorily the operating conditions specified.

Should any defect in design, material, workmanship, or operating characteristics develop after 12 months of operation or after 36 months following shipment from the supplier's plant, whichever occurs later, the supplier agrees to make all necessary or desirable alterations, repairs, and replacements free of charge and, if the defect or failure to function cannot be corrected, the supplier agrees to replace promptly, free of charge, said equipment or to remove the equipment and refund the full purchase price.

The following aspects of performance should also be guaranteed by the boiler supplier:

1. Steam quality at MCR and specified part loads at the maximum allowed boiler water TDS.
2. Steam temperature at MCR and specified part loads.
3. Excess combustion air at MCR and specified part loads.
4. Steam consumption of steam-atomizing burners.
5. Contaminants, for example, $NO_x$, in flue gases released to the atmosphere when firing the specified fuel(s).
6. Overall thermal efficiency when burning the specified fuels at MCR.
7. Electrical power or steam consumption at MCR of ancillary equipment in the boiler supplier's scope.
8. Generated noise level in accordance with the relevant company's noise specification.
9. The specified time for increasing the load of boiler from 25% to MCR.

Predicted performance data should be filled in by the vendor as per the boiler data sheet. It is understood that this data is predicted only and should not be considered as being guaranteed except where the conditions given coincide with those stipulated elsewhere in the standards.

### 4.10.2 **AIR AND GAS**

The unit should have sufficient forced draft fan capacity available to provide the necessary air for combustion at a pressure in the burner windbox required to overcome all of the resistance through the unit including the ducting and stack. Means should be provided to control the furnace pressure and the supply of air throughout the operating range.

The $CO_2$ or excess air in gas leaving the furnace should be determined by sampling uniformly across the width of the furnace where the gases enter the convection-heating surface. There should be no delayed combustion at this point nor at any point beyond.

The fuel-burning equipment should be capable of operation without objectionable smoke.

### 4.10.3 **WATER**

The boiler water concentration in the steam drum should be specified in the job specification. Samples of water for testing should be taken from the continuous blowdown. Samples should be taken through a cooling coil to prevent flashing. Sampling and determination of boiler water conditions should be under the methods contained in ASTM Special Technical Publication No. 148.

Test procedure for solids in steam: samples of condensed steam for determination of solids should be obtained in accordance with the method specified in the latest edition of ASTM D-1066 entitled "Tentative Method for Sampling Steam." The electrical conductivity method should be used to determine the dissolved solids in the steam. The test should be made in accordance with ASTM D-1125-50T.

Sample collector for checking of boiler TDS should be provided with sample cooler and pressure reducing device.

### 4.10.4 **INFORMATION REQUIRED WITH QUOTATIONS**

- Data sheet should be completely filled out. Additional data normally supplied by the vendor should be given separately. The following data must also be included:
- Outside radiant heating surface in square meters.
- Outside convection heating surface in square meters.
- Outside superheater surface in square meters.
- Overcapacity rating and maximum time allowed to run at this overcapacity rating and the time for increasing the load of the boiler from 25% to MCR.
- Efficiency at full, 75, 50, and 25% load on fuel gas firing based on LHV. Excess air and ambient temperature used in calculating these efficiencies should be included.
- Heat release in megajoule per cubic meters of firebox volume.
- Individual specification data sheets for fans and other ancillary equipment, indicating turbine steam rate/power, fan test block conditions at this power (kW) should be indicated.
- Material, type, size, and wall thickness of water tubes and superheaters. Material, size, and wall thickness of headers. Material, size, and design pressure of drums.
- Steam and mud drum dimensions including wall thicknesses.
- Stack height, diameter, and wall thickness.
- Capacity capability using natural draft when draft fans fail.
- Proposed field-testing procedures including dry-out, start-up sequencing, time intervals, etc.

- Overall dimensions for layout purpose, ladders, stairways, and platforms supplied should be shown.
- Drawings showing the general layout of the boiler including firebox, tubes, burners, air ducts, soot blowers, inspection and access doors, peepholes, observation windows, piping and valve arrangement, stack ducts, and all connections.
- Vendor should specify maximum anticipated noise levels at full boiler capacity.
- The proposed design should be proven in practice, rugged and reliable. The tenderer should provide a list of similar installations already in satisfactory operation for a period of at least 2 years.
- The proposal should either state compliance with the specifications or list the exceptions taken, exceptions mentioned are subject to the company's approval.
- Superheater steam inlet and outlet temperatures and velocities on the basis of boiler load.
- Hold-up time and capacity of steam drum between normal level and low level shutdown at the rated capacity of boiler.
- Heat balance.
- Quantity of fuel at MCR, heat supplied by fuel, rate of flue gas produced, heat to steam, heat losses due to dry flue gases/moisture in air/atomizing steam/radiation/others, flue gas temperature and pressure leaving furnace/entering and leaving superheater/entering and leaving convection and evaporation sections, flue gas velocity at superheater and boiler passes.
- Specification of burner type, layout, model, size, and manufacturer.
- Soot blower type, location, manufacturer, and arrangement.
- Statement of maximum permitted boiler water TDS.
- Statement of furnace positive and negative pressures.
- Method of fuel consumption and thermal efficiency tests, with correction formulae, curves, etc. used in the calculations.
- Approximate dimensions, layout, and location of local and remote control panels.
- List of the major control loops, general statement of local and remote panel instrumentation.
- Details of steam drum internals.
- Design and construction of flue gas dampers.
- Burner fuel gas flow rate and automatic control.
- Breakdown of total feedwater pressure requirement.
- Specifications, individually completed data sheets (including process performance, constructional, and test data) and cross-sectional drawings for all machinery.
- Specifications, data sheets, sizing data, and proposed suppliers of control equipment.
- Schematic and hook-up drawings of emergency shutdown systems and automatic trip systems, burner management and controls, accompanied by a detailed description of operation.
- Design details and mode of operation for isolating plates in common ducting or stack.
- Burner flame detection equipment.
- Design of stack duct entries.
- Winterization proposals for plant protection.
- Flue gas ducting internal lining.
- Block logic diagrams of the burner management.
- Design details, specification, and data sheets of level gages, transmitters, and switches.

Approval of final drawings and detail design is required as listed here:

1. Flame failure system with a list of material
2. Sufficient copies, as required by the company, of operating manuals containing procedures for dryout, boil out, safety valves floating, commissioning, etc.

Spare parts should be considered in three categories as follows:

1. Precommissioning
2. Commissioning
3. Permanent (for 24 months' operation).

The supplier should specify in his proposal all the previously mentioned spares separately with the required lists.

The supplier should submit, prior to precommissioning work, a complete spares manual, to include all spares recommended as permanent stock.

In general all plant and equipment, including auxiliaries, should be protected against damage or inability to operate under the winter conditions specified/agreed for the location concerned. Detailed requirements for winterization should be as per the relevant Company's winterization specification.

## 4.11 **FIRE-TUBE PACKAGED BOILERS**

This section covers the minimum requirements for materials, design, fabrication, inspection, testing, preparation for shipment, and guarantees of fire-tube and packaged boilers.

In the case of packaged type, the unit should meet the following requirements. The unit supplied should be completely fabricated, assembled, tested, and dismantled only to the extent necessary for inspection and practical shipping. The unit should be supplied piped, wired, and ready for operation with a minimum of field tie-ins.

### 4.11.1 **MATERIALS**

The materials used in the manufacture of pressure parts and also the materials for plates, tubes, bars, and forgings, should comply with BS 2790: Section 2.

For installations operating in sour environment, materials of construction selected should be least affected when subjected to atmospheric corrosion attack.

Under no circumstances should valves, regulators, etc., contain any copper or brass components.

Superheater tubes should be carbon steel or alloy steel.

The quality of refractory material covering furnace floor tubes should be at least high-duty brick.

Flue gas and air ducts should be carbon steel.

All insulation should be covered with metal jackets, either zinc coated (galvanized) steel, aluminum-coated steel, or aluminum.

Burner piping should be carbon steel. Except for the example mentioned by experts.

## 4.11.2 DOCUMENTATION

The following documentation is usually necessary for approval:

### 4.11.2.1 With the Proposal

- General specification
- Statement of design code compliance
- Outline arrangement and cross-sectional drawings of boiler, ducting, and stack, showing burner and platform location. Preliminary P&ID diagram of boiler fuel and controls.
- Specification of burner type, layout, model, size, and manufacturer.
- Soot blower type, location, manufacturer, and arrangement.
- Guarantees as required by the Standard.
- Noise information for "Noise Control."
- Statement of maximum permissible boiler water TDS.
- Statement of furnace positive and negative pressures.
- Furnace maximum and average heat flux density.
- Heat flux to cause "Departure from Nucleate Boiling."
- Furnace flue gas exit temperature.
- Method of fuel consumption and thermal efficiency tests, with correction formula, curves, etc. used in the calculations.
- Statement of percentage radiation loss for the whole boiler.
- Over capacity rating, indicate maximum time allowed to run.
- Heat release $Mj/m^3$ firebox volume.
- Efficiency at full, 75, 50, and 25% load. Include excess air and ambient temperature used in calculating these efficiencies.
- Specification of accessories and optional equipments.

### 4.11.2.2 Prior to Order

- Individual specification data sheets for fans and other ancillary equipment.
- Recommended spares list for 2 years' operation.
- Specifications for instrument piping and cable.
- Design and construction of flue gas dampers.
- Burner fuel gas flow rate and automatic control.
- Breakdown of total feedwater pressure requirement.
- Capacity capability using natural draft when draft fans fail, if exist.

### 4.11.2.3 During Design and Prior to Manufacture

- Valve data sheets for HP steam, feedwater, safety relief, and pressure-reducing over 14 bar.
- Details of connections and conditions at supply limits.
- Specifications, individually completed data sheets (including process performance, constructional, and test data) and cross-sectional drawings, for all machinery.
- Local control panel layout and location.
- List of instrument makes and models.
- Schematic and P&ID diagrams of boiler fuel and controls and hook-up drawings of emergency shutdown systems, automatic trip systems, burner management and controls, accompanied by a detailed description of operation.

- Foundation general arrangement and details.
- Cable trench layout and details.
- Structure loading under various design conditions, together with design stresses.
- Piping specifications.
- Welding procedures with supporting PQRS.
- Design details and mode of operation for isolating plates in common ducting or stack.
- Burner flame detection equipment.
- Lighting levels at burner and access platforms and stairways.
- Design of stack duct entries.
- Winterization proposals for plant protection.
- Any special type screwed pipework fittings.
- Flue gas ducting internal lining.
- Stack height, diameter, and wall thickness.
- Overall dimensions for layout purpose, showing ladders, stairways, and platforms supplied.
- Spare parts interchangeability record list as completed by vendor.

### 4.11.3 PIPING, FITTINGS, VALVES, AND APPLIANCES WATER GLASSES

#### 4.11.3.1 Feed Piping

When a horizontal-return tubular boiler exceeds 1000 mm in diameter, the feedwater should discharge at about three-fifths the length from the end of the boiler, which is subjected to the hottest gases of the furnace (except a horizontal-return tubular boiler equipped with an auxiliary feedwater heating and circulating device), above the central rows of tubes. The feed pipe should be carried through the head or shell farthest from the point of discharge of the feedwater and be securely fastened inside the shell above the tubes. In accordance with ASME PFT-48.2.

#### 4.11.3.2 Auxiliary Piping Component-Connections

Minimum size of piping should be DN15 unless otherwise specified.

Minimum bore of piping should be 6.35 mm.

Draft gage connections should be provided and located: one in the burner windbox, one in the furnace zone, and one at the boiler outlet.

A flue gas sampling connection, DN25 should be provided at the boiler outlet. Its location should be such that a representative sample is obtained.

Boiler drain, level gage, and sample connections should be piped to grade.

Pressure Relief valves discharging materials such as hot water, steam, etc. to atmosphere, should be furnished with outlet piping to direct the flow away from areas where personnel may be present.

#### 4.11.3.3 Valves

For steam and feedwater shut-off duties, parallel slide valves should be used. All valves should be of steel construction suitable for the pressure and temperature concerned and cast iron should not be used for any valve or fitting. The main steam stop valve may be an angle screw-down type mounted directly on the outlet nozzle of the boiler.

Boiler isolation from the range should be to double-isolation standard. As a minimum, a block valve and screw down nonreturn valve and a drain should be provided.

All gate type and screw down valves should have rising spindles with handwheels rotating clockwise to close, and marked accordingly. Stainless steel or alloy nameplates should be fitted to each valve to indicate valve duties and item number.

Two blowdown valves, of a size depending upon the size of the boiler, should be fitted at the bottom of the boiler water space.

An air cock should be fitted to the top of the boiler shell.

An antisyphon valve should be provided to prevent the boiler filling with water as the internal pressure falls during shutdown operations.

Soft sealed type valves should not be used.

All valves should be suitable for the line service classification. Single or double valves at a classification change should be suitable for the more severe line classification on either side of the valve(s) location.

Materials for the first block and check valve and the connective piping through the first check valve should be suitable for the more corrosive condition of the process or utility service for maximum metal temperature of the connective piping.

Material used for valve packing and seals which are preferably asbestos free should be suitable for the maximum and minimum fluid design temperatures to which these components will be exposed.

### 4.11.4 DESIGN

Wetback units should normally be used. A dryback design may be considered when a superheater is required, provided that the refractory does not present a potentially severe maintenance problem.

The flow of steam through the superheater should create such a pressure drop over the entire operating range so as to ensure an adequate distribution of steam through all tubes, and thereby prevent overheating of any element. The boiler designer should state the pressure drop across the superheater at 40, 70, 100, and 110% of MCR.

Facilities should be provided to enable inspection of internal surfaces without recourse to cutting and rewelding.

The manufacturer should show by calculation that the attachment of tubes to tube sheets is satisfactory in relation to the pressure and heat transfer rates envisaged for the conditions stated.

Superheater tubes should be to BS 3059 or equivalent standard.

Internal pipework, which is the responsibility of the boiler supplier, should be based generally on ANSI B 31.3 for fuel piping, ANSI 31.1 for steam piping, or other equivalent standards. Exceptions may be made where necessary to meet special requirements and any such exceptions should be stated.

Terminal flange connecting to external pipework should be of raised face type complying with ANSI B 16.5 or BS 1560, adopting the pressure/temperature ratings of ANSI B 16.5 latest issue.

In the case of package type, the unit should meet the following requirements.

All necessary equipment such as ladders and platforms, guards for moving parts, etc., should be supplied as part of the package.

All boiler heaters should be installed inside the buildings.

Indoor equipment should be suitably protected against damage by infiltration of moisture and dust during plant operation, shutdown, washdown, and the use of fire protection equipment.

Outdoor equipment should be similarly protected, and in addition, it should be suitable for continuous operation when exposed to rain, snow or frost, high winds, humidity, dust, temperature extremes, and other severe weather conditions.

The unit should be laid out such as to make all equipment readily accessible for cleaning, removal of burner, replacement of filters, controls and other working parts, and for adjustment and lubrication of parts requiring such attention.

For similar reasons, the boiler front and rear doors should be hinged or davitted. The boiler should be installed with the following minimum clearances:

| | |
|---|---|
| Vertical | 1200 mm |
| Sides and rear | 1200 mm |
| Front | 1200 mm |

Clearances should be increased to take into account for the front and/or the rear enclosures to swing open and for the tube removal.

### 4.11.4.1 Design of Welded Joints
Longitudinal, circumferential, and other joints, uniting the material used for channels, shells, or other pressure parts should be butt-welded with full penetration.

### 4.11.4.2 Joints Between Materials of Unequal Thickness
A tapered transition section having a length not less than three times the offset between the adjoining surfaces, as shown in Fig. 4.1, should be provided at joints between materials.

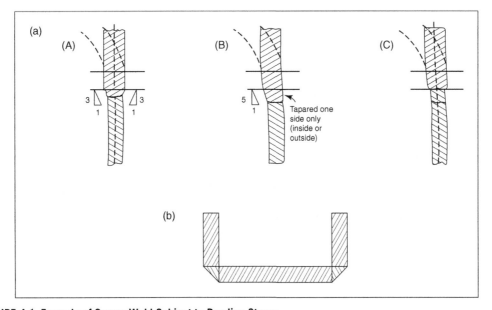

**FIGURE 4.1  Example of Corner Weld Subject to Bending Stress**

(a) Butt welding of plates of unequal thickness: (A) preferred method (center lines coincide); (B) permissible (circumferential joints only), and (C) not permissible; (b) Butt welding of plates of equal thickness.

The transition section may be formed by any process that will provide a uniform taper. The weld may be partly or entirely in the tapered section or adjacent to it as indicated in Fig. 4.1.

This paragraph is not intended to apply to joint design specifically provided for elsewhere in the Standard or to joints between tubes, between tubes and headers, and between tubes and tubesheets.

### 4.11.4.3 Welded Joints Subject to Bending Stress

The design of welded shells and heads should be such that bending stresses are not brought directly on the welded joint.

No single-welded butt joint or fillet weld should be used where a concentrated bending stress will occur at the root of the weld due to the bending of the parts joined, as in the corner weld shown in Fig. 4.1, unless the parts are properly supported independently of the welds.

### 4.11.4.4 Thickness Requirements

- Shell and dome. The minimum thickness of shell plates and domeplates, after forming, should be as follows in Table 4.1.
- Tubesheet. The minimum thickness of tubesheets of fire-tube boilers should be as follows in Table 4.2, but it should not be less than 0.75 times the thickness of the shell to which it is attached.

### 4.11.4.5 Requirements for Inspection Openings

All openings should meet the requirements of BS 2790: Section 3.6. Manholes may be substituted for handholes.

Where washout plugs are used the minimum size should be 38 mm.

**Table 4.1  Minimum Thickness of Shell Plates and Domeplates, After Forming**

| ID of Shell or Dome | Minimum Thickness, in. (mm) |
|---|---|
| 900 mm or under | 6 |
| Over 900–1370 mm | 8 |
| Over 1370–1800 mm | 10 |
| Over 1800 mm | 13 |

**Table 4.2  The Minimum Thickness of Tubesheets of Fire-Tube Boilers**

| Diameter of Tubesheet | Minimum Thickness, in. (mm) |
|---|---|
| 1060 mm or under | 10 |
| Over 1060–1370 mm | 11 |
| Over 1370–1800 mm | 13 |
| Over 1800 mm | 14 |

### *4.11.4.6 Opening Between Boiler and Safety Valve*

The opening or connection between the boiler and the safety valve should have at least the area of the valve inlet.

After the boiler manufacturer provides for the opening required by the Code, a bushing may be inserted in the opening in the shell to suit a safety valve that will have the capacity to relieve all the steam that can be generated in the boiler and which will meet the Code's requirements.

No valve of any description should be placed between the required safety valve or safety relief valve or valves and the boiler, or on the discharge pipe between the safety valve or safety relief valve and the atmosphere. When a discharge pipe is used, the cross-sectional area should be not less than the full area of the valve outlet or of the total of the areas of the valve outlets discharging and should be as short and straight as possible and so arranged as to avoid undue stresses on the valve or valves.

## 4.11.5 **BURNERS**

Fuels to be used and conditions of supply will be specified by the company. Heating for fuel oil should be provided as necessary so that the viscosity required at the burners can be achieved at all times.

Liquid fuels should be filtered through a mesh of nominal 0.25 mm aperture for heavy fuel oils and 0.18 mm for light fuel oils, or as specified by the burner manufacturer. Duplex type filters or two filters in parallel should be provided, allowing change-over to take place without interruption of flow. A differential pressure gage should be provided across the filters.

Steam-atomized, pressure-jet, or rotary-cup type burners may be used, provided that they are fully proven for use with the particular furnace and fuels to be burnt. Works tests should be carried out, using samples of the fuels concerned, on a boiler similar to that under consideration. If steam-atomized burners are used, the boiler designer should justify any steam consumption greater than 0.5% of the boiler MCR.

During the works tests the opportunity should be taken to measure the noise levels of the combustion equipment, etc.

Each main burner assembly should be equipped with a fixed gas-fired pilot burner suitable for a gas supply pressure of 0.2–0.35 bar, for which the gas supply should be from a source independent of the main fuel.

Each pilot burner should incorporate an electric igniter as part of its assembly and must be suitable to ensure safe and efficient ignition of all fuels specified. They should also be permanently lit when their respective main burners are in use, but should be removable for maintenance while the boiler is in operation.

Both main and pilot burner should have proven flame detection equipment responsible for controlling fuel admission and cut-off to their respective burners.

Fuel gas burners should be of the multispud or gun type.

When both liquid and gaseous fuels are specified, all burners should be capable of satisfactorily burning any of the fuels separately or simultaneously.

Where waste fuels are to be burned, they should be considered as intermittent supplies and the reliable operation of the boiler should not depend on their use.

Burner minimum turndown should be 3:1 for liquid fuels and 10:1 for fuel gas. The percentage of $O_2$ in the flue gas should be in the range of 1½–2% for liquid fuels and 3–3½% for gaseous fuels, over the turndown range of 50–100% MCR.

For boilers above 4.5 ton/h capacity, the burner turndown should not be less than 4:1 for liquid fuels provided that where steam superheaters are fitted, the steam flow should always be adequate to prevent overheating of any element.

Unburnt carbon in the flue gas should not be greater than 0.05% wt. of the fuel.

Carbon monoxide in the flue gas should not be greater than 0.01% by volume at specified $O_2$ content in flue gases.

The pilot flame should be visible through the burner peephole, at least prior to the ignition of the main flame, and should be monitored by a reliable flame detector, preferably of the ionization probe type, at all times. The pilot burner should be proven capable of igniting the main fuels efficiently and of remaining lit under all windbox and furnace conditions likely to be experienced.

Separate combustion air to pilots must be arranged if the main windbox supply, under all pressure changes normally experienced, cannot be relied upon to maintain the flame in a satisfactory condition.

Duplex type filters, or two filters in parallel, of 125 μm mesh in monel, should be provided in the gas supply for each convenient group of pilot burners. The pipework from the strainers to the pilot burners should be in stainless steel.

Burner viewing ports should be fitted to each burner assembly front plate in such a position as to afford an adequate visual examination of the burner stabilizer and the root of the flame.

Provision should be made for the automatic steam purging of burner guns to remove all liquid fuels. It should not be possible to withdraw a gun from the burner assembly unless the fuel is shut off, the purging carried out and steam shut off. It should also not be possible to turn on fuels or steam with the gun withdrawn. This mechanism must only be capable of being overridden by a locked "defeat" switch with a removable key. When a burner trips out on default of flame, or any other essential condition, the burners should not be automatically purged. Indication of the unpurged condition should be visible from the firing floor and boiler control panel. The purging sequence should be initiated by local push-button control by the operator when he is satisfied that it is safe to so purge the fuel from the guns into the furnace. Under these conditions, the pilots must be in operation.

Where automatic valves are proposed for the "on" and "off" control of the fuels and steam to individual burners, separate manually operated valves should also be provided at the boiler front. All these valves, both automatic and manual, should be specifically selected to give reliable operation, tight shout-off, and no external leakage over the full operation period between boiler overhauls which should be taken as 36 months. Valves should preferably be of the ball valve type, Volume 2, Par II or equivalent, subject to the operating temperature and pressure being within the rating of the valve seat, etc. Overtravel on automatic valves should be sufficient to operate limit switches satisfactorily.

After purging, the sequence of events of start up should embody the following principles:

1. Ensure all interlocks are in the correct condition to proceed
2. Prove fuel valves closed and correct the airflow rate
3. Prove the required number of air registers are open to ensure the minimum boiler airflow
4. Start pilot igniter
5. Open pilot gas valves and closed bleed valve.

The start-up and shut-down sequence should be automatic with push buttons to start and stop the sequence for each burner. As specified by the company, colored lamps on the panels should indicate the status of burners.

Interlocks should be provided to prevent burner start up if the furnace conditions are not satisfactory. These should initiate shut-off of the main fuel trip valve to the boiler at any time during operation, if they are not continuously satisfied. Conditions producing lock-out or trip should include the following:

1. Extra-low water level
2. Low pilot fuel gas supply pressure (shut off pilot gas at start up only)
3. Low supply pressure for the relevant fuel
4. Loss of forced drought
5. Loss of induced drought
6. Loss of main burner flames (individual burner fuel cut off)
7. Loss of atomizing steam pressure (on liquid fuel firing)
8. Low pressure of control air/instrument air (start up conditions only; "fail-locked" would operate when on load)
9. Loss of electric power supply (start-up conditions only; "fail-locked" would operate when on load).

Fuel pipework should have blanked-off connections to which temporary steam lines may be attached for purging before maintenance. They should be located close to, and downstream of, the shot-off valves.

Fuel oil and fuel gas pipework should have tracing, thermostatically controlled.

Atomizing steam lines should be lagged separately from fuel lines.

The atomizing steam pressure should be controlled to give a constant value, or constant differential pressure from that of the fuel, as the particular type of burners may require.

The manufacturer should specify a purging procedure in the boiler manual. This should include at least a five-volume air change of the furnace before the first burner is started up.

## 4.11.6 BOILER FEED PUMPS

Boiler feed water pumps should be protected from excessive temperature rise under minimum flow conditions, for example, by the provision of a minimum leak-off path.

For a boiler operating on modulated feed control, the feed pump should normally run continuously. For the ON/OFF system, the pumps may be arranged to start up as required, unless supplying several boilers.

Feedwater pumps on modulated feed control should have a capacity of 110% of MCR.

Feedwater pumps on an ON/OFF system of feed control should have a capacity in the range 125–175% of MCR.

## 4.11.7 FANS, DUCTING, AND STACKS

### 4.11.7.1 Forced Draft Fans

The fan should meet the following requirements at its rated point:

- Rated airflow should be 120% of the airflow required by the unit operating at MCR and firing design fuel at 20% excess air for oil fuel and 15% excess air for gas fuel.
- Rated static head should be 145% of the head required at MCR and firing design fuel at 20% excess air for oil fuel and 15% excess air for gas fuel.

- Fan should be rated at summer design air temperature plus 14°C, summer design relative humidity, and altitude at site of installation.

  Fan-blading design should be either of the airfoil or backward curved non overloading type.

### 4.11.7.2 Ducting

Ducting for flue gas and air should be of continuous seal welded construction to insure air tightness with flanged connections for field assembly. Ducts should be reinforced and stiffened for the operating air pressure and temperature under all conditions within the guaranteed operation.

Minimum thickness of flue gas ducts should be 6 mm.

Minimum thickness of air ducts should be 5 mm.

Expansion joints in ducting should be designed and furnished by the Manufacturer. Designs should be submitted to the company for approval.

The dampers should be designed for tight shutoff and should be braced sufficiently to withstand maximum forced draft fan discharge pressure.

### 4.11.7.3 Stacks

The stack should be designed as individual self-supporting steel stack with minimum height specified for the boiler. Stacks should be checked for dynamic and static wind loadings.

Aircraft warning lights may be required by local regulations.

Stacks may require facilities for flue gas sampling and smoke and temperature measurement if suitable locations in the flue ducts cannot be provided.

Steel work external surface should be given protective treatment on the ground, before erection.

## 4.11.8 INSTRUMENTS, CONTROLS, AND SAFETY EQUIPMENT

Where "fail-locked" control circuits are specified, digital signals should be used. Provision should also be made for local tripping of critical equipment.

Unless otherwise specified, automatic control of the following functions should be provided:

1. Fuel supply to burners
2. Combustion conditions
3. Boiler water level
4. Feedwater supply
5. Burner management
6. Furnace purge before first burner light off.

Controls may be pneumatic or electronic, or a combination of both, as the company may specify.

Solid-state burner management equipment should be used unless the company specifies or agrees otherwise.

The boiler manufacturer should be responsible for the satisfactory design and operating capability of the instruments, controls, and safety equipment associated with the boiler, and he should submit details to the company for approval before placing purchase orders.

The flow rate of fuels to the burners should be controlled by the pressure of the steam in the boiler or discharge header common to other boilers, unless specified otherwise.

For boilers of up to 4.5 ton/h capacity the fuel may be controlled at two firing rates, one high and one low, related to boiler steam pressures, the low firing rate also being used for the warming up of the boiler. The steam pressure band over which the burner is controlled should normally be no more than 5% of the operating pressure, unless otherwise agreed with the company.

Above 4.5 ton/h capacity, modulating control of the fuel flow rate should be employed, with the flow inversely proportional to steam pressure existing at the boiler or common steam header. An override should be incorporated in each boiler control system to shut down the burners of a boiler whose steam pressure is about to lift the safety relief valve. The setting of the override should be adjustable between normal operating pressure and the safety relief valve set pressure.

All automatic burners should be provided with management equipment to control the sequences of operation during start up and shut down and to monitor the conditions of the burner flame at all times, in accordance with the technical requirements of BS 799: Parts 3 and 4.

The steam pressure should actuate a control signal, which will position a modulating motor connected to the air damper and fuel control valve. The air-to-fuel ratio should be mechanically set by means of levers, links, and an adjustable cam, or similar positive device, over the modulating range to give satisfactory combustion conditions and thereafter require little attention.

For boilers of 4.5 ton/h capacity and above, feedwater to the boiler should be modulated according to actual boiler water level, using a level transmitter.

For boilers below 4.5 ton/h capacity, control of the feedwater to the boiler should either be by modulation as earlier, or by means of an intermittent ON/OFF switch at predetermined levels. The feed regulating valve should be provided with isolating and bypass valves for emergency manual operation.

Two direct-reading water level gages should be fitted on each end of the boiler shell and preferably diagonally opposed. Audible water level alarms should be incorporated, located at high, low, and extra low positions. At the low-level alarm position, the fuel supply should be cut off but may be automatically restored if the level recovers. At the extra-low level position, the firing should be shut down and a lock-out condition introduced which should require manual re-setting, including sequence purging for restart.

### 4.11.9 FABRICATION

The rules in the following paragraphs apply specifically to the fabrication of boilers and parts thereof that are fabricated by welding and should be used in conjunction with the general and specific requirements for fabrication in the applicable Sections of BS 2790 that pertain to the type of boiler under consideration.

### 4.11.10 WELDING PROCESSES

The welding processes that may be used under this section should meet all the test requirements of BS 2790 and are restricted to the following:

- In accordance with ASME PW-27.1
- In accordance with ASME PW-27.2.

The manufacturer should submit copies of welding procedures, before production commences for comments and approval by experts.

The proposed welding procedures should include the following information:

1. Welding process or combination of processes
2. Name and designation of welding consumables.
   *Note*: Where no related material specifications are available, all chemical and mechanical data should be supplied with the welding proposals.
3. Dimensioned sketch of joint design
4. Method of making weld preparations
5. Details of welding techniques ie, diameter of electrodes or filler wire and sketch showing sequence of welding
6. Welding position
7. Range of production thicknesses to which welding specification applies.
8. The welding proposal should include details for the removal of deleterious weld defects and the subsequent re-welding operation
9. Pre and postweld heat treatment
10. Shielding and purging composition and flow rates (for gas shielded arc welding)
11. Type of power sources, amperage, speed of travel
12. Base metal chemical composition.

All root pass on tubes or pipes including nozzles and branches must be made with the Tungsten inert gas shielded arc welding method with a suitable filler wire.

All filler runs on tubes and pipes should be made with the electric arc coated electrode on the butt and fillet welds.

The outside surface of the weld should be free from undercuts, abrupt edges or valleys. The weld reinforcement should not exceed 2.4 mm.

Every precaution should be taken to avoid excessive penetration of root runs, on the inside of the tubes.

All longitudinal and circumferential welds on the pressure vessel including tube membrane walls should be made by the submerged arc method.

Permanent back-up rings are only permitted on the closing circumferential seams of vessels without an external man way with the approval of experts. Back-up rings, when permitted, should match the analysis of the base material. If back-up rings are removed, the area should be dressed and examined for cracks by the magnetic particle method.

Welds which are deposited by procedures differing from those recommended by properly qualified to approve should be rejected and completely removed from the equipment.

### 4.11.10.1 Assembly

Parts that are being welded should be fitted, aligned, and retained in position during the welding operation within the tolerance specified in the Standards.

Tack welds used to secure alignment should either be removed completely when they have served their purpose, or their stopping and starting ends should be properly prepared by grinding or other suitable means so that they may be satisfactorily incorporated into the final weld. Tack welds, whether removed or left in place, should be made using a fillet weld or butt weld. Tack welds to be left in place should be made by qualified welders and should be examined visually for defects and, if found to be defective, should be removed.

### 4.11.10.2 *Tube to Tubesheet Joints*

To obtain sound tube/tubesheet welded joints absolute cleanliness should be maintained until all joints have been proved to be sound and all repair welds complete. The following stages of the manufacturing process should be observed:

1. Initial cleanliness of component parts and the removal of oil, moisture, shop dirt, etc.
2. Expansion of tubes before welding should not be permitted since air trapped between tube and tube sheet causes weld porosity.
3. To avoid contamination of welds by water, no hydraulic testing is to be carried out until joints have been proven sound and all weld repairs have been completed.

The basic requirements for cleanliness should be as follows:

1. Tube holes must be free from scale, shop dirt, oil, and grease before welding commences. These clean conditions should be maintained until all welding, including weld repair, is concluded.
2. Welding consumables should be stored prior to use in accordance with the manufacturer's recommendations and filler wire should be cleaned immediately before use with a suitable solvent.
3. The work pipes should be maintained above an ambient condensing temperature for the duration of the welding operation and for the periods when the work piece is standing without work being carried out on it.
4. For some welding processes it is essential to mechanically position tubes in the tubeholes before welding. When this is required; expansion of the tube with the tubehole should not be allowed. Instead a light expansion with a drift pin can be used to produce a line contact. The positioning contact line should be such that it forms part of the subsequent weld root run. No lubricants should be used in the positioning operation.
5. Weld should be examined by any nondestructive testing means, but after test all dye and oil, should be completely removed from the tube sheet before proceeding.
6. A pneumatic pressure test should be applied to the tube/tube sheet weld before any hydrostatic test is carried out. The medium used may be air or air freon mixtures, using soapy water or a halogen leak detector to determine faulty welds.
7. All welds should be pneumatically tested again after any weld repairs have been carried out.
8. Carry out the full hydraulic test on the tube side of the boiler.
9. When all welds are proven satisfactory, the tubes should be lightly expanded onto a tube sheet to minimize cyclic bending stresses on the weld joint.
10. Final hydrostatic proving tests on the shell side and tube side of the boiler should proceed after all the foregoing steps have been completed.

### 4.11.10.3 *Heat Treatment*

All postweld or postoperation heat treatment whether carried out in a furnace or by local induction heating, should have pyro-metric control, with automatic chart recorders. Thermocouples must be attached to the metal. It is not sufficient to measure only furnace atmosphere. No heat treatment is permitted without this instrumentation.

## 4.11.11 PREPARATION FOR SHIPMENT

The boiler manufacturer should properly prepare the boiler for shipment to the jobsite.

Outside exposed metal surfaces should be prepared and painted in accordance with painting project specification.

Machined surfaces and flange faces should be coated with heavy rust preventive grease.

All threads of bolts, including exposed parts, should be coated with a metallic base waterproof lubricant to prevent galling in use and corrosion during shipment and storage.

To prevent damage, all flange facings should be protected with gaskets and 6 mm thickness plates, and all couplings should be protected by steel pipe plugs.

Suitable bracing and supports should be provided to prevent damage during shipment.

## 4.11.12 INSPECTION AND TESTING

All materials and work including the work of subsuppliers should be subject to inspection as indicated in the conditions of the contract.

Certificates of shop inspection and all related test reports plus material mill test certificates should be furnished by the boiler manufacturer.

### 4.11.12.1 Hydrostatic Test

After a boiler has been completed, it should be subjected to pressure tests using water at not less than ambient temperature, but in no case less than 7°C.

All welded pressure parts should be subjected to a hydrostatic test pressure of not less than 1.5 times the design pressure. The hydrostatic test may be made either in the manufacturer's shop or in the field.

- In accordance with ASME PG-99.3.3
- In accordance with ASME PW-54.2
- In accordance with ASME PG-99.4.1.

The test pressure should be applied and maintained for a sufficient length of time to permit a thorough examination to be made of all seams and joints, but not less than 60 min, excluding inspection time.

After completion of the hydrostatic tests, the water should be immediately drained and the equipment tested, and dried by blowing with dry compressed air.

## 4.12 WATER-TUBE BOILERS

This section covers the minimum requirements for engineering, material, installation, fabrication, inspection, and preparation for shipment of water-tube boilers and accessories.

Boilers are intended to be suitable for heavy duty uses in, oil and gas refineries, petrochemical plants, and other oil industry applications where necessary.

These boilers should be designed for outdoor uses, which automatically can be satisfactorily used for indoor applications, when required.

This section specifies general requirements for design, fabrication, cleaning, testing, and painting of water-tube type steam-generating plants. This includes integral steel tube economizers and superheaters, and all parts connected to the pressure parts of the boiler without the interposition of a shut-off valve and other ancillary equipment. Natural, forced, assisted, controlled-circulation, and once-through boilers are included. These may be for use either in a central power station or in an installation for supplying steam directly to a process plant.

Boilers should be constructed to ASME Boiler & Pressure Vessel Code Section 1, BSEN 12952 and should be subject to prior approval by experts.

All conflicting requirements should be referred to the purchaser in writing, the purchaser will issue a conflicting document if needed for clarification.

- Steam boiler data sheets showing expected performance, fuel requirements and characteristics, materials of construction etc. (as specified in appendices) form a part of this specification.
- The boiler unit as a whole should be designed, fabricated, and mounted in a manner which will facilitate ease of maintenance.
- Outdoor equipment should be suitably protected against damage by infiltration of moisture and dust during plant operation, shutdown, washdown, and the use of fire protection equipment, and in addition, it should be suitable for continuous operation when exposed to rain, snow or ice, high winds, humidity, dust, temperature extremes, and other severe weather conditions.
- The vendor should supply all equipment completely fabricated, assembled, tested, and dismantled only to the extent necessary for inspection and practical shipping.
- The boiler unit should be selected to allow maximum interchangeability of spare parts consistent with proper performance characteristics.
- Spare parts must be readily available. If a stock of parts is not maintained by the vendor, strategic spare items should be furnished with the unit.
- The boiler unit and its auxiliary equipment should be thermally and mechanically designed by the Vendor. All materials, fabrications, testing and inspection should be included in the Vendor's scope of supply.

## 4.12.1 BASIC DESIGN

### 4.12.1.1 Steam Quality

Steam drums should be equipped internally with steam separators and scrubbers to ensure that the carry over of total solids from the boiler water should not exceed the following:

1 ppm up to 65 bar
5 ppm from 65 bar up to 135 bar

The boiler manufacturer should state the maximum TDS in the boiler water at which the required steam quality can be obtained.

The steam as measured at the drum outlets should have an impurity not greater than 0.02 ppm silica. The wetness of the steam leaving the drum should not exceed 0.02%.

### 4.12.1.2 Circulation

In order to evaluate the drum stability, the boiler designer should indicate the drum water content (effective) at normal low level and MCR loading.

The boiler designer should state the minimum and maximum water levels between which the boiler should be allowed to continue operating.

Furnace wall, floor, and roof tubes should not incorporate such bends or sets sufficiently small in radius to interfere significantly with water circulation. In particular, roof tubes exposed to radiant heat should be free from bends and sets as far as possible, so as not to upset the division of flow between tubes or bring to a doubtful value the head available to promote circulation in any part of the circuits.

Downcomers supplying the furnace wall etc., with feedwater should preferably be outside the flue gas path. If the downcomers are in contact with the flue gases, the heat transfer should not significantly affect the circulation head.

### 4.12.1.3 Pressure Parts

All pressure parts, wherever reasonably possible, should be constructed in carbon steel of a quality such that neither pre nor post weld heat treatment is required. The boiler supplier should inform company in his proposal where higher grade material is proposed. Drums of high-tensile, low-ductility steel should not be used where the thickness of the drum would otherwise be less than 75 mm.

Steam drums should not be less than 0.61 m internal diameter.

Strength calculation of boiler pressure parts should be made in accordance with relevant codes and standards and with due account of the following:

1. Combination of temperature and pressure at design and hydrostatic test conditions
2. Maximum static head of contained fluid under design and hydrostatic test conditions
3. Weight of equipment (if applicable)
4. Wind or earthquake loading whichever is greater (if applicable)
5. Special consideration may be required for critical items of equipment for the effect of the following:
   a. Local stresses due to supporting lugs, ring girders, saddles, internal structures, or connecting pipes
   b. Shock loads due to water hammer or surging of vessel content
   c. Temperature difference and differences in coefficients of thermal expansion
   d. Fluctuating pressure and temperature
   e. Fatigue loading.

### 4.12.1.4 Furnace

Furnaces should be water-cooled and of membrane-wall (also referred to as mono-wall or panel-wall), or skin-cased, construction.

Pressurized furnaces should be of membrane-wall construction.

Where the boiler design incorporates a refractory front wall around the burner area, the boiler manufacturer should obtain the company's approval on the suitability of the method of attachment and anticipated life of this refractory.

Special attention should be given to the design of burner throat refractory, whether brick, cast, or plastic material applied to studded or plain tubes. Evidence of satisfactory service experience should be submitted.

At least one furnace access door should be provided with at least 600 mm diameter and designed to give efficient sealing and ease of operation over the useful life of the boiler. The sizing of the doors should be adequate for the passage of necessary maintenance materials including internal access equipment.

There should be a clear space in front of the access door to facilitate handling and entry of all materials and equipment.

Observation ports should be furnished to permit visual checking of furnace and flame conditions, the furnace floor and the superheater space during operation of the boiler. Observation ports for pressurised boilers should be furnished with a sight glass type which is resistant against the heat and pressure.

Drains should be furnished at the low point of the boiler furnace and bank areas to permit removal of flue gas deposits by water washing. Drain openings should be effectively sealed against flue gas bypassing and casing overheating.

Cavities above furnace roof tubes should be so designed as to prevent the accumulation of gases that might form explosive mixtures. This may be achieved by purging with air, reliable sealing, or some other proven method. Access to such areas for the examination of tubes, penetration seals, and hangers, etc. should be possible during boiler inspections.

For site-erected boilers, the average heat release rate in the furnace, based on the net calorific value, should not exceed the following:

for fuel oil firing: 380 kW/m$^2$ (120,000 Btu/ft$^2$h)
for fuel gas firing: 470 kW/m$^2$ (150,000 Btu/ft$^2$h)

Where high availability is required, lower heat release rates may be specified by the company.

The maximum heat flux density at any localized "hot spots" should not exceed 300 kW/m$^2$ (95,000 Btu/ft$^2$h). The boiler designer should present with his proposal the predicted furnace heat flux distribution along the length and height of the furnace at MCR and state the maximum and average heat flux and the furnace flue-gas exit temperature.

*Note*: The heat flux density is the quantity of heat passing through unit surface area in unit time and is expressed as kW/m$^2$ (Btu/ft$^2$h). The surface area should be that of the flat projected area of exposed tubes and exposed extended surface, integral with the tube. Refractory covered surfaces should not be included.

The furnace width should enable sufficient spacing of burners to ensure burner flame discrimination by individual viewing heads and also to make certain that there is no flame impingement on the sidewalls. The furnace depth should be sufficient to ensure that burner flames do not impinge on the rear wall or penetrate the screen tube arrangement.

The depth of furnace should not be so disproportionate to the width that a significant volume of furnace is not occupied by the flame, thereby making the total projected surface area not truly reckonable for the radiant heat rate calculations.

Floors should not be utilized for heat transfer and should be effectively shielded from furnace radiant heat by refractory tiles and insulation as necessary.

The angle to the horizontal of floor tubes should not be less than 15degree, and the angle of roof tubes should not be less than 5degree. Where the roof tubes have to be offset for any reason, continuity of drainage should be ensured, but such offset should be avoided where possible.

### 4.12.1.5 Generation Tube Banks

Tubes should normally be plain, positioned in line. However, finned tubes may be used where fuel gas or distillate fuel is to be fired and if an additional surface is needed which cannot be provided by installing further plain tubes. This may be the case, particularly in supplementary, fired waste heat boilers associated with gas turbines.

If finned tubes are proposed they should not be used in zones where the flue gas temperature exceeds 670°C (1240°F). Fins should be in carbon steel, continuously welded to the tubes and should have a minimum spacing of 5 fins per 25 mm, a maximum height of 19 mm and a minimum thickness of 1.25 mm.

Finned tubes should not be placed at any angle that would facilitate the accumulation of deposits. Vertical tubes may be acceptable, but their use should be approved by experts.

Tube banks should be arranged, as far as practicable, to permit access for tube renewal with minimum cutting out of serviceable tubes.

If low-grade fuels are to be employed and regular soot blowing is envisaged, the company will specify whether provision is to be made for the off-load water washing of tube banks, with facilities for collection and disposal of effluent.

Flue gas baffles should preferably be water-tube walls or constructed of refractory tiles securely positioned, easily replaced, and with ready access.

Where refractory baffles or protective screens are supported or reinforced with metal parts, the boiler designer should demonstrate that the materials selected are suitable for the intended service life.

### 4.12.1.6 Superheater

1. Superheater tubes material should be to BS 3059 or ASME Section II or Company Approved Equivalent.
2. Self-draining type superheaters, integrated with the boiler, are preferred if these can be conveniently arranged, provided the boiler type is in all other respects suitable for the particular duty.
3. Where horizontal sections of a superheater are proposed, the span between supports should not be so great, or the angle of inclination so near to the horizontal, that sagging of tubes will occur with the consequent collection of a condensate in pockets.

On boilers that are required to produce a specified degree of superheat to the steam over a wide range of boiler operation, or on units where superheat temperatures are expected to be near maximum design temperature of the tube metal concerned, attemperators should be provided. In the former case attemperation may be performed at the final outlet before the stop valve, but in the latter case the attemperator should be positioned between the primary and secondary stages of the superheater.

Surface-type attemperators should not be used unless approved by expert.

Spray-water type attemperators should be used, provided that there is a supply of demineralized water of suitable quality for this duty. Spray-type attemperators must not be capable of blocking the steam flow through the superheater in the event of mechanical failure.

Where interstage desuperheating is used, the downstream construction material should be capable of temporarily withstanding the resulting higher temperature, should the spray water supply fail.

If the proposed tube metal temperature or fuel constituents require higher grades of steel to be used, a nominal 9% chromium, 1% molybdenum steel is preferred to the austenitic steels.

Metal temperatures should not be high enough to allow corrosion to take place in the presence of vanadium and sulphur compounds, or other corrosive constituents resulting from the use of the specified fuels.

To monitor metal temperatures during boiler start up, skin thermocouples should be secured to the tubes at appropriate points. These thermocouples and their connecting leads should be so positioned and protected that they will not suffer rapid deterioration by exposure to the flue gases or radiant heat of the furnace. Any requirement for tube skin thermocouple details will be specified by experts.

Junction boxes should be located against the insulation of the external plating of the boiler or some other convenient position where they will be protected against the heat of the furnace.

To prevent overheating of tubes during steam pressure raising, superheater drains should be arranged to facilitate removal of condensate from the appropriate points and the final superheater drain should be large enough, together with any additional blow-off vent, to ensure sufficient flow of steam through the superheater to prevent overheating of tubes. Such drains and vents should be equipped with silencers, unless routed to a blowdown drum.

Hangers, which should be cooled where possible, and spacers, should be of such material and design that overheating will not occur during start up or normal operation of the boiler, up to 110% MCR. Where tubes or supports pass through baffles on the furnace roof, efficient and durable seals should be provided. The materials should be selected to give the intended service life.

Air vents should be provided where necessary.

The flow of steam through the superheater should create such a pressure drop over the entire operating range as will ensure an adequate distribution of steam through all tubes, and thereby prevent overheating of any element. The boiler designer should state the pressure drop across the superheater at 40, 70, 100, and 110% MCR.

A steam superheater should be protected from furnace radiation by a boiler tube screen or by a boiler wall.

The steam superheater and access to it should be designed such that individual elements or tubes can be removed or repaired with minimum disturbances.

## 4.12.2 **REFRACTORY, INSULATION, AND CASINGS**

Refractory should only be used where it is absolutely necessary. Where used, it should be a proven design feature and of a quality of material and construction appropriate to the duty involved. It should not be of lower quality than 42% alumina.

The furnace wall and floor refractory should be suitable for use at the temperatures concerned, subject to a minimum of 1450°C (2640°F). The boiler designer should state the thermal conductivity of the materials he proposes for use.

Where brickwork is installed, the mortar joints should not be thicker than 2.5 mm.

Insulation should be applied in sufficient thickness to prevent casing distortion, to reduce radiation losses to an economic minimum and to ensure personnel protection. The percentage radiation loss for the whole boiler should be stated in the proposal.

Outer casings should be of metal, not less than 1.5 mm thickness, of adequate stiffness and be capable of being easily removed and replaced without damage or requiring extensive dismantling of equipment, pipework, etc. The method of fixing should, whenever economical, be suitable to withstand repeated removal and replacement without deterioration, cutting, and rewelding. Each sheet of cladding should be independently supported.

Outer casings should be corrosion-resistant and should not support combustion in the event of fire. Aluminum and its alloys should not be used.

Casings should be designed to prevent escape of the flue gases or the circulation of gases into cool sections of the casings or structural steelwork, thereby creating conditions for internal or external corrosion. The temperature of the casing plus attachment should be maintained above the dew point of gases by the installation of adequate external insulation. The earlier is particularly important on pressurized furnaces.

The design of casings for balanced draught furnaces should preclude the influx of air, so as to ensure maintenance of the required combustion excess air throughout the entire setting.

### 4.12.2.1 Integral Pipework

All integral pressure pipework should be designed and constructed in accordance with one of the following: BS 1113, ASME Boiler and Pressure Vessel Code Section 1, ISO 1129, or to the national design requirements.

The pipework should be welded, with flanges provided only at valves and fittings used for pressures up to 45 bar (ANSI Class 300). Above this pressure, flanges should be eliminated altogether unless the company specifies or agrees otherwise. Control valves may be an exception to this requirement, as specified or agreed by experts.

For pressures above 45 barg (ANSI Class 300), only integral or weldneck flanges should be used for those connections where flanges cannot be eliminated.

Flanges should comply with ASME/ANSI B 16.5, or equivalent national standards. All flanges should be raised face type and no screwed flanges should be used.

Flanges on pipework used for fuels should be a minimum of ANSI Class 150.

Screwed connections and unions should not be used unless the fittings are of a special type approved by experts.

Socket-welding fittings and screwed fittings, where permitted, should be in accordance with ASME/ANSI B 16.11, as applicable.

The feedwater system arrangement should be such that the feed regulating valve, isolating valves and bypass valves can be manually operated from the floor level in the event of an emergency.

Drains from the boiler, superheater, economizer, and soot blowers, etc., should be operable from floor level where practicable. Pipework should not be positioned where it may possibly obstruct or trip operators. All clean drain pipework should terminate at the clean drains tank. No drain or atmospheric trap should discharge into any gulley under, or leading to, the covered areas. Any potentially contaminated stream should be led to the blowdown system.

Pipework supports should be capable of accepting the additional weight of water during the hydrostatic test.

When differential settlement between items of equipment may be expected, sufficient flexibility should be provided in the connecting pipework.

Pipework to fans, pumps, and similar equipment, which may be required to be periodically removed for overhaul should be self-supporting when the item is removed. Such supports must, whenever possible, be provided by the overhead pipe track. Individual pipe supports from grade must be minimized.

Valves should be line size, except that pump discharge valves (check and block) should be the same size as the pump discharge or greater:

- Screwed and socket-welding valves should be of the bolted bonnet type unless otherwise specified.
- Bolting should be in accordance with ASTM A-193.
- Compressed asbestos fibre jointing may be used for classes 150 and 300 duties up to 400°C (752°F). Above this temperature, and in Classes 600 and 900, spiral wound gaskets should be used.

- Where pipework is laid on overhead structures, in tracks, or in permanent trenches, the arrangement and clearance should permit removal of any pipe fitting or valve without the necessity to disturb adjacent pipework.

Piping systems should be designed, and equipment laid out, to provide adequate access by mobile maintenance equipment. Access ways should be planned accordingly and minimum headroom clearances for this purpose should be:

- 4.5 m over railways and clearways for mobile equipment
- 5.5 m over access roads and where specified over clearways or heavy equipment
- 2 m over walkways and platforms.

The minimum clearance for walkways around pumps, fans, motors, compressors, etc. is 0.9 m. This clearance should be measured from the furthest projection on the equipment including associated pipework, filters, valves in their open position, drains, cabling, instruments, etc. between grade and 2 m above grade.

Pipe smaller than 15 DN (½ in.) should not be used without agreement, except for instruments. The use of steel pipe in 32 DN (1¼ in.), 65 DN (2½ in.), 90 DN (3½ in.), 125 DN (5 in.), 175 DN (7 in.), 225 DN (9 in.) sizes should be avoided, except for alloy steel pipe, which may include 65 DN (2½ in.) size.

Valves 200 DN (8 in.) and above should be fitted with double valved integral type bypasses, unless otherwise specified.

### 4.12.2.2 Steam Trapping

Steam traps should be selected and agreed between the boiler supplier and trap supplier, according to their required duty and should be of cast or forged steel. The trap supplier should provide data sheets for all duties.

A trap size should be based on the maximum quantity to be discharged at the minimum pressure difference between inlet and outlet.

No trap should be connected to more than one steam line or to more than one section of the same steam line.

Where condensate may collect downstream of closed valves, drain valves should be provided. Traps should not be fitted.

Normally, traps operating on different steam pressures should discharge to a separate condensate collecting header but, providing the condensate line is sized to accommodate the flash steam, the discharges may be taken into a common line. Where there is condensate back pressure, a nonreturn valve should be installed in the individual condensate return lines.

Traps should be supported whenever possible by the lines to which they are attached.

In locations specified as subject to exceptionally low temperatures, steam traps should, where possible, be grouped together and installed on racks in steel cupboard-type enclosures so that satisfactory operation of each trap can readily be checked and frost damage to any inoperative traps prevented.

### 4.12.2.3 Economizer

Economizers should generally be an integral part of the boiler but may be supplied as separate units when a boiler, as a standard model design, cannot readily incorporate an economizer, or the physical limitations of the proposed site make it necessary or desirable to position the economizer away from the boiler.

Construction should be steel tubes, either plain or finned. Fins may be of steel or cast iron according to the existing flue gas conditions.

With fuels containing no sulphur, either steel or cast iron fins may be used. With fuels containing sulphur, the metal operating temperature in contact with the flue gases should preferably not fall below 150°C (302°F), and fins may be of steel or cast iron. The company may specify or agree flue gas temperatures down to 105°C (220°F), when cast iron fins should be used. Designs involving any metal temperatures below 105°C (220°F) should not be used.

Tube bends should be all-welded except where feedwater conditions make it necessary for frequent cleaning or inspection, in which case, for boiler pressures up to 38 barg (ANSI Class 300), steel bends should be flanged.

In-line tube arrangement is preferred. Water velocity should be of the order of 1 m/s for water upflow circuits and 2 m/s for downflow. Headers should preferably be external to the casing and tubes efficiently sealed where they pass through to the bends and headers.

Feedwater operating pressure and temperature in the economizers should at no time permit the possibility of steam being generated. Recirculating type economizers should not be included unless agreed by experts.

When economizers can be isolated on the waterside, a safety relief valve should be fitted.

Flue gas bypass and/or recirculating facilities should only be provided with the agreement of the company.

Means for off-load water washing of economizers should be provided on boilers firing residual fuel oils or unsweetened gas.

### 4.12.2.4 Air Heater

Air heaters, utilizing the flue gas sensible heat to raise the temperature of the combustion air, will be accepted when there is a need to obtain higher thermal efficiencies than can be attained by an economizer alone. When an air heater is proposed the boiler designer should satisfy the company concerning the advantages of higher efficiency, considering the increased capital cost, increased maintenance costs, effectiveness of soot blowing, expected operating time, efficiency, and likely problems.

Air and gas bypasses should be provided, together with proven soot-blowing equipment.

The flue gas exit temperature from the air heater should not be less than 120°C (250°F), but may need to be considerably higher to meet the environmental requirements. In any case the gas exit temperature should not be less than that recommended by the air heater manufacturer who must consider the air/flue-gas temperature differential in relation to the possibility of corrosion.

Air heaters using surplus low-pressure steam should be considered. The condensate discharge should be returned to the deaerator. On-line cleaning facilities for the finned tubes should be provided.

### 4.12.2.5 Feed Heaters

Experts will specify, or agree, the temperature at which the feedwater will be available to the boiler from the deaerator, and the cost and condition of the steam which may be used for further heating of the feedwater.

The boiler supplier should provide a feed heater within the feed circuit of the boiler integral pipework, if it is necessary or desirable to raise the water temperature above that specified or agreed.

The heater should be a shell and U-tube type, unless otherwise agreed.

Normally, the feed heater should be of the high-pressure type receiving water direct from the boiler feed pump. The water should be on the tube side and the steam on the shell side.

A burst tube should be considered as a design case, and safety relief provided if necessary. A nonreturn valve should be fitted on the steam side at the steam inlet to prevent water entering the steam main.

The heater should be provided with a bypass and isolating valves. The assembly should be arranged for the easy removal of the shell or tube bundle for inspection, as necessary.

### 4.12.2.6 Draught Equipment

Draught fans should be of the centrifugal type, and preferably of aerofoil construction, designed for a maximum operating speed not exceeding 1500 rpm for forced-draught (FD) fans, and not exceeding 1050 rpm for induced-draught (ID) fans, having characteristics particularly suited to the duties concerned. Fans should have design margins of +15% on capacity and +32% on pressure for an MCR basis, or +10% on capacity, and +21% on pressure for a peak-load basis.

Fans should provide stable operation under all conditions of boiler load, and should have nonoverloading power characteristics. Flue gas fans should have, where necessary, self-cleaning impellers, and be resistant to abrasion. For ID and FD fans, overall efficiencies should not be less than 80%.

Specific attention should be given to the fan head/volume characteristics, especially to the maximum negative head that can be experienced in the furnace and/or flue, imposed by the ID fan under emergency operating conditions. A main fuel trip, or a restriction occurring in the flue gas paths, must be considered in the design to avoid unacceptable negative pressures.

If the possible draught excursion under any of the conditions thought likely to arise with the selected centrifugal fan cannot be accommodated in the design, then an axial flow fan with variable pitch-angle blades should be used. The stall line for this fan should be so paced as to ensure that no unacceptable negative pressures could be applied to the furnace and/or duct. The selection of an axial flow fan will require special attention to achieve an acceptably low noise level.

For boilers supplying steam to process units and production plant, fan drivers should normally be steam turbines, with the turbines exhausting to an LP steam system serving deaerators. If electric motors are used to drive the fans of such boilers, continuity of power supply should be arranged to cover the situations where the normal power supply might fail. This may take the form of an alternative source of power supply, or a dual-drive facility at the fan shafts such as clutch-connected steam turbines or diesel engines.

In circumstances where boiler start up from cold, or a hot re-start, may involve temporary loss of auxiliary steam pressure, the dual-drive arrangement of motor/steam turbine or steam turbine/diesel engine should be considered for at least one or two of the boilers in the installation.

Draught regulation should preferably be effected by the use of adjustable inlet guide vanes on each fan. Regulation may alternatively be by speed control if the response rate is demonstrated to be satisfactory and should be agreed by experts.

Dampers in the ducts may be regarded as satisfactory for combustion air control to burner registers. However, single or multiblade dampers located in the flue gas ducts should only be used for isolating duties, or where chimney natural draught has to be modulated at low operating loads. Dampers positioned in the flue gas ducting should be arranged to prevent complete closure during boiler operation.

On boilers of 200 ton/h capacity and greater, or where a high dependence is to be placed on boiler reliability, two separate fans for both FD and ID functions should be provided unless otherwise specified. The part-duty rating, between 50% and 75%, will be considered by experts in relation to the steam load reduction, which can be readily made in the event of the failure of one fan. On-load isolating facilities should be provided for each draught unit.

Rotor shafts of fans should be stiff, and bearings sleeve type with forced or oil ring lubrication. Bearings should have spherical seatings in their housings and should be mounted on individual pedestals independent of the fan casing.

Where speed-regulated fans are specified, a positive means of setting the minimum speed should be provided.

The lubrication of the driver, gearbox, and fan bearings should be adequate at speeds of 90% of the minimum speed.

Fans should have integral thrust bearings and transmission of axial thrust to the prime mover bearings should not be permitted.

ID fan bearings should be water-cooled. Internal flexible water connections should not be used.

Fan casings should be sufficiently heavy gage material, stiffened as necessary, to prevent "drumming" over the entire range of operation.

Shaft couplings and seals should be capable of continuous and efficient service without maintenance, for periods of not less than 26 months.

Fans and their drivers should be freely accessible for maintenance and rotor removal, and should be mounted on foundation blocks at grade level.

Geared drivers should preferably be avoided but, where gears are used, they should be totally enclosed. Parallel shaft gearing should meet the requirements of:

AGMA 420.03. "AGMA Standard Practice for Helical and Herringbone Gear Speed Reducers and Increasers"
AGMA 421.05. "AGMA Standard Practice for High Speed Helical and Herringbone Gear Units"

whichever is applicable.

For high-speed transmissions an epicyclic type is preferred.

Couplings should be of the metastream spacer type, or equal, capable of operating for not less than 26 months without attention.

Couplings should be capable of accommodating misalignments, both angular and lateral, and should allow both the driving and driven shafts to take up their individual positions against their thrust bearings without the transmission of unacceptable forces due to end-float of the shafts concerned.

Safety guards must be fitted to cover all exposed moving parts. They should be easily removable for machinery maintenance.

### 4.12.2.7 Air and Flue-Gas Ducting

The air and gas velocities in ducting should not exceed 13.7 m/s (45 ft./s) and 15.2 m/s (50 ft./s) respectively, taking all internal bracing and stiffeners into account.

Guide vanes should be fitted internally to ducts, for example, at elbows and severe bends, to ensure a satisfactory air/gas flow characteristic.

Ducting should be of welded construction from carbon steel, not less than 5 mm in thickness for air duties and not less than 6 mm for flue gases.

Heavier plate should be used, with stays and angle or tee stiffeners where duct surface areas might be conducive to drumming, or if strengthening is necessary against ID fan maximum negative head. Stays and stiffeners should only be used internally when there is agreed economic advantage and the effect on air/gas flow is insignificant. Expansion joints should be provided to prevent unacceptable forces and moments being transmitted to the boiler windbox, fan casing or stack, etc. and should be capable of operating for not less than 26 months without attention. Packed gland type joints should not be used.

When there is the possibility of flue gases reaching dew point temperature of any corrosive constituent, ducting should be in metal having high resistance to corrosion or be metal coated internally, with zinc or other corrosion resistant metal sprayed onto the duct metal. The whole method and procedure is to be to the approval of the company.

Duct anchor points should be incorporated in the support structure to ensure that the expansion joints accommodate the duct movements.

Sliding supports and guides should be designed to provide free movement without attention over a period of not less than 26 months between boiler surveys.

Insulation should be applied externally to duct walls, anchor supports and any other attachments, to maintain air and flue gas temperatures and for the protection of operators.

Access doors giving a minimum access of 600 mm × 600 mm should be provided, where duct size permits, to facilitate access to all parts of the ducting. For smaller ducting, inspection doors should be provided, as agreed by experts. Side entry from grade level is preferred, but the erection of temporary scaffolding to gain entry is acceptable unless otherwise specified by experts.

Painting of steelwork external surfaces below a metal temperature of 425°C should be in accordance with standards. Any supplementary requirements above this temperature will be specified or agreed by experts.

Dampers should preferably be of the multiblade type, set in a channel frame and provided with indicators and means of positive operation from grade level. Where ducting is elevated to heights at which it is impracticable to operate the dampers from grade level by direct mechanical/hydraulic means, a caged ladder and platform should give access to the damper operating mechanism, including the whole operating drive unit.

The design, material, and construction of dampers should be subject to approval by experts.

Materials for damper blades, shafts, and all damper components exposed to the flue gas should be limited to a maximum service temperature as follows:

| | |
|---|---|
| Carbon steel | 343°C |
| 1.25 Cr–0.5 Mo | 454°C |
| Type 321 stainless steel | 760°C |
| Type 310 stainless steel | 927°C |

Special attention should be given to ensure adequate blade clearances in the frame and adjacent ducting, under the operating and upset conditions. A suitable allowance should be made for material expansion, differential expansion, and fouling likely to occur in service.

Damper shafts should be supported in self-aligning dry sleeve bearings, mounted externally to the ducting. Adequate allowance must be made for relative expansion of the components.

When motorized operation of dampers is to be used, control push buttons should be located both at grade and damper levels. A friction clutch should be fitted to the motor shaft, and reliable limit switches incorporated. Position of the damper opening should be clearly indicated at points of operation.

For boilers connected to a common duct or stack shared by other boiler or process heaters, isolating plates should be provided in each flue connection.

Isolating plates should be arranged so that they operate vertically downwards to shut. Facilities to permit locking in the open position should be provided. Where it is considered that alternative arrangements to the use of isolating plates would permit significant economies to be made, these alternatives should be approved by experts. In any case, full design details and proposed mode of operation should be stated.

### 4.12.2.8 Boiler Support Structure

The boiler support structure should be minimized by placing equipment such as fans at ground level wherever practicable and locating stacks separately.

The design of steel-framed structures should comply with the requirements of BS 449, ASME/AWS.

In meeting the requirements of the standard, allowance should be made for all applicable dead and live loadings including:

1. Weight of boiler including test fluids
2. Lifting equipment and maintenance requirements
3. Dynamic loads resulting from vibrating equipment
4. Loads resulting from wind-excited oscillations
5. Wind loading
6. Earthquake loading.

Earthquake loads should not be considered as acting simultaneously with wind loads.

The boiler main load supporting steelwork should be fireproofed to give 2-h rating in accordance with standards. Unless company specify otherwise, it should be fireproofed up to a level of 7.5 m (25 ft.) above the highest fuel line.

Painting and the preparation for painting should be in accordance with standards.

### 4.12.2.9 Boiler Mountings, Valves, Gages, and Safety Fittings

Mountings, valves, gages, and safety fittings should be in accordance with ASME Boiler and Pressure Vessel Code Section 1.

For steam and feedwater shut-off duties, parallel slide valves should be used. All valves should be of steel construction. Cast iron should not be used for any valve or fitting.

Boiler isolation from the range should be to double-isolation standard. As a minimum, a block valve and nonreturn valve and a drain should be provided.

All gate type valves should have rising spindles with handwheels rotating clockwise to close, and marked accordingly. Stainless steel or alloy nameplates should be fitted to each valve to indicate valve duties and item number.

Double valving should be fitted in all drain and blowdown lines connected to boiler pressure parts. Isolating facilities required for maintenance of boiler or equipment without shutting down the plant should also be equipped with double isolation valves and a vent valve.

Such internal fittings to boiler drums as are necessary to ensure control of circulation and the production of steam quality as specified, such as nozzles, baffles, separators, and scrubbers, etc. should be provided, all of which should be of robust design, suitable material and, although securely fixed, must be easily removable for inspection purposes.

At least two direct-reading level gages. One installed at each end of the drum, should be provided, and in addition one level gage easily visible from the operating platform.

Safety relief valve vent piping should be complete with drains, expansion chambers, and exhaust mufflers, the latter meeting any noise level requirements of the local authorities. Safety relief valve drains should be open-ended. They should not be connected into a collecting system or present a hazard to personnel.

The safety valves should be of the direct spring-loaded type with the springs exposed to the open air, that is, with open bonnets. They should be provided with lifting gear.

All safety valves should have flanged connections. They should be adequate to meet the requirements of the service, but should have inlet and outlet flange ratings of at least ANSI Class 300 RF and ANSI Class 150 RF respectively. Welded connections are not allowed.

Rating and adjustment of the safety valves should be in accordance with the ASME Boiler and Pressure Vessel Code, Section 1. Blowdown pressure should be not more than 4% of the set pressure.

The set pressure of any boiler drum safety valve should be at least 5% in excess of the maximum operating pressure in the drum or 2.5 bar in excess of the maximum operating pressure in the drum, whichever is the higher.

Vertical outlets, at least 2000 mm high, should be provided for the safety valves. They should blow off to a safe location.

All valve outlets should be adequately supported to take care of the reaction forces. They should have safe drainage facilities that should prevent accumulation of water in the outlets.

### 4.12.2.10 Platforms, Stairways, and Ladders
Platforms, stairways and ladders should comply with the requirements of the standard.

### 4.12.2.11 Burners
The boiler designer should state the viscosity required at the burners so that any additional heating can be installed.

Liquid fuels should be filtered through a mesh of nominal 0.25 mm aperture for heavy fuel oils and 0.18 mm for light fuel oils, or as specified by the burner manufacturer. Duplex type filters or two filters in parallel should be provided, allowing changeover to take place without interruption of flow. A pressure gage should be provided and should be connected to the inlet and outlet of the filters.

Liquid fuel burners should normally be of the steam atomizing type. The boiler designer should justify any atomizing steam consumption greater than 0.5% of boiler MCR. The company will state whether there will be a reliable supply of steam for use with steam atomizing burners under cold-start conditions, or if the boiler supplier is to provide a special facility to meet this situation.

Fuel gas burners should be of the multispud or gun type.

When both liquid and gaseous fuels are specified, all burners should be capable of burning satisfactorily any of the fuels separately or simultaneously.

Where waste fuels are burned, they should be considered as intermittent supplies and the reliable operation of the boiler should not depend on their use.

Multifuel firing arrangements should provide an even distribution of the waste fuel burners in the burner-firing pattern. The boiler designer should recommend the arrangement for agreement by experts.

Burners for augmenting the exhaust heat should be one of the following types, listed in order of preference:

1. Conventional register burners
2. Sidewall burners with flame protection channels

3. Intertube burners of fabricated construction with integral gas supply chambers
4. Grid burners of cast construction in high-grade alloy steel.

Liquid fuels or contaminated gas fuels should only be burned in these burners when burner tip cleaning can readily be effected with the other burners remaining in operation.

Burner minimum turndown should be 3:1 for liquid fuels and 10:1 for fuel gas, with the boiler supplier's guaranteed low $O_2$ in the flue gas maintained over the ranges mentioned in the standard.

Burner air registers may be of parallel or venturi shape. Boilers with only a single burner should have burners equipped with two guns, one concentric and the other in an angular position, to enable either gun to be withdrawn without reducing the boiler-firing rate.

Boilers having four or more burner assemblies for use with fuel gas or commercial grade liquid fuel should operate satisfactorily with combustion conditions as near stoichiometric as practicable. The excess air should not exceed 3% for liquid fuels and 5% for gaseous fuels. Over the full operating range of the boiler the following $O_2$ vol. percentage in flue gases should be achieved with liquid fuels:

0.5% $O_2$ between 70 and 100% MCR
1.0% $O_2$ between 25 and 75% MCR
5.0% $O_2$ between 0 and 25% MCR

Excess $O_2$ for boilers with less than four burners should not be greater than 1% over the load range 70–100% MCR, with up to 2% at loads 40–70% MCR.

Carbon monoxide in the flue gas should not be greater than 0.01% by volume at specified $O_2$ content in flue gases.

Unburnt carbon in the flue gas should not be greater than 0.05% wt. of the fuel, or such limits.

Each burner should be sized either for 110% of its design load, or such that the boiler MCR can be maintained with one burner out of use, whichever is greater.

A fixed gas fired pilot burner, removable for maintenance while the boiler is in operation, should be provided at each burner assembly. It must be suitable to ensure safe and efficient ignition of all fuels specified. Each pilot burner should be permanently lit when its main burner is in use.

When a permanent clean gas supply is not available, or in other exceptional circumstances, the use of discontinuous pilots may receive the approval of the company.

Each pilot burner should be fitted with an electrically operated igniter as an integral part of its assembly. The pilots should operate from a sweet gas supply independent of the main gas supply to the boilers, and be suitable for a gas supply pressure of 0.2–0.35 bar (3–5 psig).

The pilot flame should be visible through the burner peephole, at least prior to the ignition of the main flame, and should be monitored by a reliable flame detector, preferably of ionization probe type, at all times. The pilot burner should be proven capable of igniting the main fuels efficiently and of remaining lit under all wind box and furnace conditions likely to be experienced.

The boiler supplier should state the heat input of the proposed pilot burner.

The combustion air supply to pilots must be arranged separately if the main wind box supply, under all pressure changes normally experienced, can not be relied upon to maintain the flame in a satisfactory condition.

Duplex type filters, or two filters in parallel, of 125 microns ($5 \times 10^{-3}$ in.) mesh in monel, should be provided in the gas supply for each convenient group of pilot burners. The pipework from the strainers to the pilot burners should be in stainless steel.

Horizontal distance between main burners, and the vertical distance between rows of burners should be such as to facilitate discrimination between individual flames by the proposed flame detector. Each main burner flame should be monitored by flame detection equipment.

Burner viewing ports should be fitted to each burner assembly front plate in such a position as to afford an adequate visual examination of the burner stabilizer and the root of the flame.

Provision should be made for the automatic steam purging of burner guns to remove all liquid fuels. It should not be possible to withdraw a gun from the burner assembly unless the fuel is shut off, the purging carried out and steam shut off. It should also not be possible to turn on fuels or steam with the gun withdrawn.

This mechanism must only be capable of being overridden by a locked "defeat" switch with a removable key. When a burner trips out on default of flame, or any other essential condition, the burners should not be automatically purged. Indication of the unpurged condition should be visible from the firing floor and boiler control panel. The purging sequence should be initiated by local push-button control by the operator when he is satisfied that it is safe to so purge the fuel from the guns into the furnace. Under these conditions, the pilots must be in operation.

Where automatic valves are proposed for the "on" and "off" control of the fuels and steam to individual burners, separate manually operated valves should also be provided at the boiler front. All these valves, both automatic and manual, should be specifically selected to give reliable operation, tight shut-off and no external leakage over the full operation period between boiler overhauls, which should be not less than 26 months. Valves should preferably be of the ball valve type subject to the operating temperature and pressure being within the rating of the valve seat, etc. Overtravel on automatic valves should be sufficient to operate limit switches satisfactorily.

Flame traps should be provided on all gas vent pipework.

Nonretracting type guns should be used, provided that the gun nozzles and stabilizers are adequately cooled when not in use. This should not significantly increase the excess air in the furnace.

Where retracting guns are proposed, connection between fuel pipe, etc. and burner inlets should be by flexible hose of three-ply construction; inner hose of close pitch corrugated monel tubing; a middle layer of type 321 stainless steel braid, and an outer layer of interlocked galvanized armor. The hose should be proven for the duties. Different hose connectors should be provided for each hose duty so that they cannot be wrongly connected. The fuel hose connectors should be self-sealing on disconnection from the burner.

The company may specify that, where no more than four burners are proposed for a boiler, the air duct may be divided into individual branches in which modulating dampers are fitted, enabling a strict fuel/airflow ratio to be provided to each burner.

Where gas only will be fired, the system should be so arranged as to reduce automatically the fuel gas flow rate to each burner, or column of in-duct burners associated with gas turbine waste-heat recovery, during the ignition and proving stages.

The boiler designer should inform the company of the flow rate proposed and how he intends to provide the automatic control of the fuel flow without affecting those burners already in operation.

Before the first burner on a boiler can be ignited, an adequate purge of the furnace and gas passes should be automatically carried out. The airflow rate and duration of this purge procedure should be agreed between the boiler designer and company, and this will depend upon the shape of the furnace and complexity of the flue gas passes. However, the airflow should not be less than 25% of MCR airflow for a period of at least 5 min with all air registers open, or for such a length of time as to give at least five volume changes of the plant combustion chamber and gas passages up to the exit of the flue, whichever is greater.

The purge procedure should be an inescapable action on every start up, and one which the operator cannot override, reduce in flow rate or shorten in duration.

The start-up and shut-down sequence should be automatic with push buttons to start and stop the sequence for each burner. As specified by experts, colored lamps on the panels should indicate the status of burners.

It should not be possible for the fixed periods of fuel admission to be extended or overridden by the operator before the flame is established.

If fully reliable flame detection equipment is not installed, individual flame detection bypass switches should be provided on boilers having more than two burners, in order to make it possible to maintain a boiler on full load. Such bypasses should be operable only by a special key, which should be removable. The bypasses should be arranged to be ineffective for a minimum of the first two burners lit and the last two burners in service.

Interlocks should be provided to prevent burner start up if the furnace conditions are not satisfactory. These should initiate shut-off of the main fuel trip valve to the boiler at any time during operation, if they are not continuously satisfied. Conditions producing lockout or trip should include the following:

1. Extra-low water level in steam drum
2. Low pilot fuel gas supply pressure (shut-off pilot gas at start up only)
3. Low supply pressure for the relevant fuel
4. Loss of FD
5. Loss of ID
6. Loss of main burner flames (individual burner fuel cut off)
7. Loss of atomizing steam pressure (on liquid fuel firing)
8. Low pressure of control air/instrument air (start up conditions only; "fail-locked" would operate when on load)
9. Loss of electric power supply (start-up conditions only; "fail-locked" would operate when on load).

Following a main fuel valve trip, the ID and FD equipment and tripping equipment should be so arranged that the furnace should not be unacceptably pressurized.

While burners may be arranged for control from a remote control room, the start up of a boiler, and every additional burner thereafter, should be initiated and observed by an operator at the boiler-firing floor. The control and indicating equipment should, therefore, be arranged accordingly. On large boilers having two or more burner platform levels, the local control panel should be divided into sections positioned appropriately at each platform level.

To ensure the effective isolation of all fuels to a furnace, solenoid-operated valves should be inserted in the air lines to pneumatically operated ball valves placed immediately upstream of the control valves. These isolating valves should be arranged for remote manual activation in emergency and to work, automatically, in conjunction with the safety interlocks, when unacceptable conditions arise.

The general physical arrangement of pipes, valves, and control equipment, etc. at each burner and in the firing floor area as a whole, should be given specific attention so as to provide a neat, uncluttered, and logical layout, capable of being readily identified by the operator and facilitating easy access for operation and maintenance.

Gas off-takes for individual burners should be from the top of the header. Each header supplying a horizontal row of burners should connect to a main vertical header that should be connected at its base to the outlet of a knock-out pot. The gas main to the boiler should be connected to the knock-out pot, with provision for flexibility of pipework to accommodate all boiler relative movements due to thermal expansion and vibration.

Platforms at each burner level should be provided, together with stairways and escape ladders as necessary. The platforms should be wide enough to enable burner guns to be withdrawn without difficulty and to be safely handled by the operator.

Drip trays should be fitted where necessary, and racks for spare guns, with facilities for gun maintenance, should be provided in agreed positions.

Fuel pipework should have blanked-off connections to which temporary steam lines may be attached for purging before maintenance. They should be located close to, and downstream of, the shut-off valves.

Fuel oil and fuel gas pipework should have electrical tracing, thermostatically controlled.

Atomizing steam lines should be lagged separately from fuel lines.

The atomizing steam pressure should be controlled to give a constant value, or constant differential pressure from that of the fuel, as the particular type of burners may require.

Expansion bellows should be avoided and should not be used in fuel gas lines of 50 mm (2 in.) diameter and above without the approval of the company.

## 4.12.3 **SOOT BLOWERS**

Boilers that will be fired on residual fuel oils or sour gas should be equipped with soot blowers to enable the unit to be kept in operation continuously for not less than 26 months without loss of thermal efficiency.

Where the future possibility of a conversion to residual fuel oil firing is specified, boilers should have the necessary soot blower openings and wall boxes fitted and blanked off. Additionally, any brackets and supports, etc. which cannot readily be inserted after boiler erection, should be provided at the necessary spacing between tube banks.

Steam blowing soot blowers are preferred, using steam from the boiler, where the pressure is acceptable, that is, up to approximately 14 bar (200 psig). A blanked-off pipe branch should also be included for connecting to an alternative steam supply when specified.

Retractable lance or nozzle type soot blowers should be used in the high temperature zones, with calorized tube rotary blowers installed in the zones where flue gas temperature permits. Economizer banks may be provided with rake or lance type soot blowers. The boiler supplier should support his proposals with details of steam flows, jet angle, extent of effective penetration, etc. Suitable stops should be fitted to the tracks inside the boiler, to prevent lances coming off the rails due to overtravel of the drive mechanism.

On boilers requiring frequent soot blowing, the soot blowers should be arranged for auto-sequence operation. The supervisory controls should ensure that soot blowing does not commence until all the soot blower steam distribution system has reached its working temperature and all condensate has been removed.

On completion of the operation, complete shut-off of the steam supply should be assured and drains opened. The drains should not be connected to other systems from which a blow-back might occur.

The automatic sequence and system management control should monitor and indicate all stages of operation. Facilities to interrupt the sequence or obtain selective operation of soot blowers should be included.

It should not be possible to interrupt the supply of steam to a retractable soot blower until it is in the fully retracted position.

Means of manually retracting a soot blower should be provided, and it should be possible to remove all soot blowers completely from the boiler, for maintenance, while the boiler is on load.

Automatic retraction of the lances should be at twice the speed of entry.

Sealing of wall boxes, lances, and nozzles should be provided.

### 4.12.3.1 Noise Limitations

Noise limits will normally be specified in detail in the inquiry. However, in the absence of such requirements, noise levels should not exceed 87 dB(A), at a distance of 1 m from equipment surfaces.

The boiler supplier should provide details of the noise emission in octave bands from his equipment.

The supplier should also provide details of any narrow-band or impulsive noise emitted by his equipment, which is noticeable to the ear, and the octave band or bands in which it occurs.

When the boiler supplier cannot meet the noise limits without the addition of noise attenuation measures, the levels with and without these measures should be stated in the proposal. Any noise attenuation measures proposed by the supplier should not conflict with the other requirements of the standard.

### 4.12.3.2 Corrosion

Corrosion and other types of failures of boiler tubes, drums, parts, should be fully investigated. Full attention should be paid for preparing procedures that evaluate the formation of waterside deposits (to control under deposit corrosion), fire side ash (for oil-firing), weld detail failures, thermal fatigue (thermal sleeves should be provided if required), overheating, creep failure, embrittlement (hydrogen damage and/or graphitization), dew point corrosion in low temperature zones of flue-gas passages, erosion, etc. Water-side corrosion control should be affected by removal of dissolved solids, oxygen, and control of pH. For idle time periods components should be drained, and filled with dry inert gas. Reduction of fire-side corrosion should be accomplished by one or more of these methods; fuel analysis recommendation, combustion control boiler design and construction, periodic ash removal, and use of fuel additives.

### 4.12.3.3 Instruments, Controls, and Safety Equipment

Where boilers are required for critical duties or specified as such by experts, "fail-locked" shutdown systems should be provided on loss of electrical power or instrument air supply. In other applications fail-safe, shut-down systems should be employed. The shut-down system should be subject to approval by experts.

All instrumentation should be suitable for continuous working in the conditions of their location. The burner management systems should be reviewed in detail and approved by the boiler designer.

Where "fail-locked" control circuits are specified, digital signals should be used. Provision should also be made for local tripping of critical equipment.

Unless otherwise specified, automatic control of the following functions should be provided:

1. Fuel supply to burners
2. Combustion conditions

3. Steam superheat
4. Steam drum water level
5. Feedwater supply.

The boiler supplier should be responsible for the satisfactory design and operating capability of the instruments, controls, and safety equipment associated with the boiler, and he should submit details to the company for approval before placing orders.

Local indicators should be provided, except where local panels are used for boiler start up. In the latter case, the local indicators may be located on the start-up panel.

For extra-high-pressure steam lines requiring welded-in primary elements, flow nozzles should be specified by experts.

The instrumentation technology employed should be based on single loop digital controllers or distributed control system as indicated in the purchase order.

Feedwater control of the boiler system should be according to ISA S 77.42 "Fossil Fuel Power."

### 4.12.3.4 Control Valves

Control valves should be specifically selected for the full dynamic turndown of the system, that is, for start up and over the full firing range. For high turndown, split-range valves or special start-up valves should be used.

The type of valve should be selected according to the service. Specialist valves should be used where cavitation, noise, flashing, or erosion may occur. Cage-guided valves may be considered for these specialist services, except on dirty or erosive applications.

### 4.12.3.5 Combustion Controls

An increase in demand of fuel, caused by falling steam pressure, or increase in steam flow in some cases, should increase the supply of combustion air. Measured increase in airflow should then initiate an increase in fuel flow, maintaining the correct fuel/air ratio. Falls in demand of fuel should initiate the reduction in fuel flow, followed by a corresponding reduction in airflow.

A reliable $O_2$ analyzer should be supplied. The airflow should be constantly "trimmed" within adjustable limits dependent on load, to maintain the specified excess $O_2$ in the flue gases.

Supply of combustion air from the FD fans should be measured and regulated to meet the precise needs of the burners, according to the quantity of fuel that the steam pressure (or flow) controller demands. The control signal thus obtained should vary the position of the fan inlet guide vanes, or change the speed of the fan, depending upon the system adopted.

In boilers having balanced draught, the ID fan inlet guide vane setting or fan speed should be varied to maintain a constant negative pressure in the furnace of minus 6 mm to minus 12 mm water gage.

Provision should be made to prevent a change in windbox pressure, when additional burners are introduced or withdrawn, from significantly affecting the combustion condition of those burners already in service.

The following indications and recordings should be included, unless specified otherwise, mounted on the boiler control panel or locally as appropriate:

1. Pressures at:
   a. FD fan outlet
   b. Air heater outlet (air side)

    **c.** Burner windbox
    **d.** Furnace at burner level
    **e.** After superheater
    **f.** Boiler outlet
    **g.** Economizer outlet
    **h.** Air heater outlet (flue gas side)
    **i.** ID fan outlet

**2.** Air and flue gas temperature at:
    **a.** Air heater inlet (flue gas)
    **b.** Air heater outlets (air and flue gas)
    **c.** Flue gas after secondary stage of superheater
    **d.** Flue gas after primary stage of superheater
    **e.** Flue gas at boiler outlet

**3.** Flue gas $O_2$ analyzer/controller/recorder (also local indication)
**4.** Smoke density indicator with audible alarm
**5.** Feed flow recorder
**6.** Drum level recorder
**7.** Fan speed indicator
**8.** Airflow indicator recorder
**9.** Steam pressure recorder
**10.** Drum and superheater outlet pressure (also local gage)
**11.** Steam flow indicator recorder
**12.** Instrument air pressure gage
**13.** Battery charger failure alarm of instrumentation power supply (where applicable)
    **a.** Fuel flow indicator recorder with integrator for each fuel
    **b.** Fuel supply pressure (also local instruments)
    **c.** Fuel pressure after control valve (also local instruments)
    **d.** Fuel temperatures at burners (also local instruments)
    **e.** Atomizing steam pressure (also local instruments)
    **f.** Burner and pilot ON/OFF indication.

### *4.12.3.6 Fuel Supply to Burners*

The boiler designer should be responsible for the auxiliary equipment necessary to raise and control the temperature and/or pressure of the liquid fuels to the boiler if the conditions at which the fuels are to be supplied are not satisfactory for the burners he intends to use.

The flow rate of fuels to the burners should be controlled by the pressure of the steam in the boiler steam drum or the discharge header common to other units with flow-limiting override where necessary. Alternatively, the company may specify a preference for the fuel flow to be controlled by steam flow from the unit and modulated by pressure compensator.

Provision should be made to prevent the fuel supply pressure from falling when additional burners are lit.

Individual burner and main fuel trips should be arranged as described in the standard, dealing with burners.

### 4.12.3.7 Steam Superheat

Where attemperators are proposed to control the degree of steam superheat, the temperature of the steam at the boiler stop valve should be the control criterion. The steam temperature leaving the attemperator should also be used in the control loop to ensure maximum response rate and accuracy.

Control of surface-type attemperators should be by regulating the amount of steam passed to the cooling surfaces. The spray-water type should be controlled by regulating the amount of spray water injected into the steam.

The final temperature of the steam should not vary more than ±5°C from the specified figure at any load, for steady state operation of the boiler, that is, ±3% MCR.

For rapid load changes at a rate of 10% MCR per minute, limited to a total of 10% MCR, the variation in final steam temperature should not be greater than ±8°C, above 60% MCR.

Instruments should be provided for the following duties on the boiler control panel, or locally mounted, as appropriate:

1. Final steam temperature (also recorded)
2. Spray-water supply pressure (where applicable)
3. Conductivity indication for saturated steam leaving the steam drum, including sample cooler, with facilities for conductivity recording if specified
4. Conductivity indication for superheated steam leaving the superheater, including sample cooler, with facilities for conductivity recording if specified.

Skin thermocouple installations on boiler tubes should be designed on an individual basis, having regard to the accuracy of the measurement required and protection against burnout of the thermocouple for each installation.

### 4.12.3.8 Steam Drum Water Level

Under normal operating conditions, including the load fluctuation stated by experts, the water level in the steam drum should not rise or fall to the point of operating the level alarms, which should be normally set to operate at not more than ±100 mm (4 in.) from the design level.

The control elements for the operation of the valve regulating the supply of feedwater to the boiler should be the measurement of the actual level of the water in the drum, the measurement of the steam flow from the boiler and the measurement of the feedwater flow to the boiler. On boilers subject to significant load fluctuation all control elements should be used.

For boilers in the small output range, that is, up to approximately 60 ton/h and subject to only gradual load changes, the company may specify the use of only two control elements: the water level and the steam flow.

Two direct-reading water level gages, or other primary level devices as approved by the company, should be provided on each steam drum, preferably one at each end. Each gage should be capable of being blown down or isolated for removal and repair without taking the boiler off load.

The connections to the drum for mounting the water level transmitter should be separate from those for the direct-reading gages.

The transmitter used for water level control for boilers in the smaller output range of up to approximately 100 ton/h, may also be used for providing the signal to the remote level indicator on the control panel. For larger boilers there should be a separate transmitter.

Where the drum elevation above operating floor level prevents the operator from viewing the direct water level gages, a remote direct-reading gage of a proven type should be provided in addition to the two gages local to the drum.

It should be located at operating floor level and positioned so as to be easily seen by the operator standing at the feedwater regulating and bypass valves. Bicolor gages equipped with mirrors to reflect a view of the drum gages down to operating floor level should not be used, although the bicolor gage itself is acceptable to improve identification of the water column.

A displacement-type "high" and "low" water alarm should be separately mounted on the steam drum, and operated by the steam. The "extra-low" water level switch should be similarly mounted and should cause the fuels to the boiler to be cut off and an emergency alarm to be raised, both visual and audible, in the boiler control room. The point at which this switch operates should be at a water level high enough to protect all pressure parts from overheating and to be still visible in the gage glasses. It must not be so near to the "normal-low" water alarm level that there would be insufficient time for an operator to make adjustments for the first condition before the second arises, causing the shut down of the boiler.

The boiler designer should state the holding time provided by the reserve of water in the steam drum between "low" level and "extra-low" level, and the company will approve this time against that required to introduce effectively the standby boiler feed pump. The size of the steam drum may have to be increased to provide a longer period in which to recover water level without incurring the automatic shut down of the boiler.

For boiler drum level control applications, a water column should be used, designed to reduce errors due to temperature effects to a minimum.

Boiler drum water level gages should be selected to suit individual applications. Proprietary level gages of established design should be used.

### 4.12.3.9 Feedwater Supply

Feedwater to the boiler should be controlled by a regulating valve in the feed line to the economizer, or to the steam drum direct if no economizer is supplied.

The control signal for positioning the opening of the regulating valve should be derived from the three elements (two in some cases).

The regulating valve should be supplied by the boiler supplier and installed in the integral pipework associated with the unit.

The operation of the regulating valve may also have to take into account a preset pressure differential across the valve in order to avoid excessive wear of the valve seating.

The boiler supplier should inform the company of the pressure of the feedwater required at the inlet to that section of the feed pipework in his supply. He should provide the company with the breakdown of the total pressure requirement, indicating maximum operating pressure of the boiler and the various pressure losses in the feed system, including static head to be overcome.

Within the steam drum, a feedwater distribution pipe should be arranged to ensure a proper distribution of the incoming water along the length of the drum, suitably placed to feed the downcomer tubes but not to interfere with the correct function of the water level gages.

The following instruments should be provided and mounted on the boiler control panel:

1. Feedwater supply pressure, indication, and recording
2. Feedwater flow, indication, and recording
3. Feedwater temperature at inlet to economizer

**4.** $O_2$ in boiler feedwater, indication, and recording
**5.** pH of feedwater.

### 4.12.3.10 Burner Management

Burner management systems should be installed local to the burners.

The systems should use solid-state logic, unless specified otherwise by experts. Shutdown systems only may use relay logic if the required functions cannot be achieved with solid-state logic.

The systems should jointly monitor the burner and boiler to ensure safe start up and shut down of burners and boilers.

Unless otherwise required by experts, the system should, on the pressing of push buttons, arrange for the whole sequence of burner light up or shut down to be automatically carried out with a high degree of safety and reliability. It should also automatically shut-down burners on identification of a fault condition serious enough to warrant such action, or raise alarms to indicate faults of a less serious nature.

Separate buttons should be provided to initiate purge and individual burner start up, and also for individual burner and boiler shut down.

Reset facilities should be provided for both boiler and individual burner trips.

Unless otherwise approved by experts, flame monitors should be:

**1.** The Babcock dual-signal type, or similar, monitoring both the high frequency flicker signals generated at the root of the flame, and the brightness of the flame, or
**2.** More advanced equipment when available, but subject to agreement by experts.

Where a very high degree of operating availability is specified, two main flame detectors should be fitted to each burner, with any one detector signal arranged to give an alarm and the two signals together to cause lockout of the fuels to the burner.

Automatic self-checking flame detectors are preferred and are mandatory if ultra-violet sensing is used.

The system should be complete, without any areas of split responsibility, especially regarding furnace purging and boiler safety. The boiler designer should ensure that the actions of the interlock equipment, in the event of a plant failure, are compatible with the actions of the analogue control equipment.

Local and control-room panels should provide all the information necessary to enable the operators to ascertain the condition of each burner and all the associated functions of fans, purging, register positions, fuel valve positions and safety interlocks.

The systems should be "fail-locked" or "fail-safe," and should lock out the boiler or individual burners if faults occur during the start-up sequence. After successful starting, the system should lock in and the boiler should shut down only by means of the manual or automatic trip system.

Key-operated override switches should be provided for all shut-down functions. These switches should also override those start "permissives" which are also shut-down functions. The override switches should normally be located on the front of the main control panel. If located on the rear of the panel, then indication of override condition should be given on the panel face.

All shut-down systems should be capable of full function testing from primary sensor up to final actuation device while the plant is on line. Test key-operated override switches should be provided for this function. These should override the minimum number of function components. Alarms should be provided to show automatically when the trip circuit is being overriden for test. Final element trip testing on a "single fuel" basis should be provided where more than one fuel is used. Where the loss of an individual burner can be tolerated in the steam system, the test facility should include tripping individual burners. Company will specify those trip functions where it is necessary to provide duplication for trips or a two-from-three voting system to give increased reliability.

All override test facilities should be mechanically protected and accessible only to personnel authorized to carry out testing.

All systems should preferably be energized during normal operation, but if systems that are deenergized during normal operation are used, they should be provided with power supply and trip circuit monitoring.

## 4.12.4 ELECTRICAL EQUIPMENT

Unless otherwise specified, all electrical equipment attached to, or closely associated with the boiler should be of the normal industrial standard suitable for "nonhazardous" areas, except for items on the fuel systems that have only one seal between the fuel and the electrical components, for example, motorized valves and pressure switches on fuel lines. Such electrical items should be at least to Zone 2 standard.

Lighting should be provided at all platforms, ladders, and stairways and around the boiler and its auxiliaries. The lighting levels should be approved by experts.

A separate system of emergency lighting should also be installed, with lights positioned at critical points, including lighting to facilitate the reading of the water level and other important gages, the easy identification of emergency valves, etc. and to permit safety of movement for personnel.

Electrical cables should be routed to avoid areas where there is a potential high risk of damage from fire, high temperatures, or any other cause. Where this cannot be achieved, suitable fire resistant cables should be used or a fire protection system should be installed.

## 4.12.5 STACKS

The stack should be designed as an individual self-supporting steel stack with minimum height specified for each boiler, but in any case not less than 30 m. Stack linings should be the vendor's standard design. Material for stack should be ASTM A-36 carbon steel.

The anchor bolts for stacks should have a ¼ in. minimum corrosion allowance. Stacks should be checked for earthquake, dynamic, and static wind loadings.

Stacks should be of welded construction and have a minimum thickness of 6 mm. A minimum allowance for corrosion from the inside of the stack of 3 mm should be specified.

Each boiler should have a separate stack, unless otherwise specified.

An access opening for internal inspection and cleanout should be provided at the base of each stack. Stacks should be equipped with aircraft warning lights.

Maintenance access for boiler stacks should be provided. This should enable two men, with paint spray or gunite machine to work inside or outside each stack. Inside access will be via a 3-point stack tip mounted stainless trolley bars and stainless pilot cables, and outside access via a trolley rail around stack tip with pilot cable. In both cases loading should be designed for a trolley capacity of a minimum of one ton. Stack ladders are not required, unless specified.

Stacks should be provided with sufficient protection against corrosion, subject to approval by the purchaser.

## 4.12.6 BOILER FEED AND BOILER WATER QUALITY AND CHEMICAL CONDITIONING

Boiler feedwater treatment should be as specified separately by experts.

Company will specify the quality of the boiler feedwater available, including condensate if intended to be used.

The boiler designer should notify the company of any objection or any difficulties he may foresee in using the specified water, and should recommend to the company any further treatment or conditioning of the feedwater he considers necessary or advisable.

The recommended maximum TDS in the boiler water should be stated by the boiler designer.

All blowdowns should be led to a blowdown drum with atmospheric venting and spray cooling as necessary. Each boiler should be provided with its own blowdown drum. Continuous and intermittent blowdown should be provided and separately routed to the blowdown drum.

Any necessary chemical mixing and injection equipment should be included by the boiler supplier, and the company will specify the required extent of duplication of equipment such as injection pumps and chemical mixing tanks.

The company will specify the type of container to be used for delivery of chemicals. All equipment necessary for the safe handling and storage in a closed system of hazardous chemicals should be provided local to the injection pumps.

Water sampling points for both boiler feed and boiler water, complete with coolers, should be provided.

For boilers intended to operate at low or zero solids in the boiler water, the quality of the boiler water should be continuously monitored and recorded, including pH, conductivity before and after an ion exchange vessel, residual ammonia level, and any other condition which might indicate a change in water quality of any serious consequence. Appropriate alarms at control position should be provided.

## 4.12.7 PURCHASING REQUIREMENTS

Steam boilers should be supplied in accordance with the contents of this specification that should include all the applicable sections of the codes and regulations etc. listed herein.

### *4.12.7.1 Basic Design*
#### 4.12.7.1.1 Terminal Points
The terminal points for the work of standard specification provided by the vendor should include but not be limited to the following:

- Outlet of main steam valve
- Boiler feedwater inlet, upstream of the stop and check valves
- Outlets of all safety valves
- Steam drum, economizer, and super heater vents, downstream of the tandem valves
- Boiler drain, downstream of the tandem valves

- Boiler blowoff, downstream of the tandem valves
- Chemical injection, upstream of the stop and check valves
- Fuel inlets at the burners and igniters upstream of the flexible hoses
- Motor electrical terminals, boxes
- Forced draft fan air ducting inlet
- Boiler flue gas duct at stack
- Stacks should be provided with sufficient protection against corrosion, subject to approval by the purchaser
- General drains to be collected to a single point local to the boiler at grade level
- One single terminal point for instrument air (instrument air lines harnessing by boiler manufacturer)
- Field-mounted instruments electronic and pneumatic junction boxes.
- All instruments for boiler local and remote control (including control room and control panels).

### 4.12.7.2 Boiler Mountings
#### 4.12.7.2.1 Safety Valves
The safety valves should be of the direct spring-loaded type with the springs exposed to the open air, that is, with open bonnets. They should be provided with lifting gear. All safety valves should have flanged connections. They should be adequate to meet the requirements of the service but should have inlet and outlet flange ratings of at least ANSI Class 300 RF and ANSI Class 150 RF respectively. Welded connections are not allowed.

Rating and adjustment of the safety valves should be in accordance with the ASME boiler and Pressure Vessel Code, Section 1. Blowdown pressure should not be more than 4% of the set pressure.

The set pressure of any boiler drum safety valve should be at least 5% in excess of the maximum operating pressure in the drum or 2.5 bar in excess of the maximum operating pressure in the drum, whichever is the higher.

Vertical outlets, at least 2000 mm high, should be provided for the safety valves. They should blow off to a safe location.

All valve outlets should be adequately supported to take care of the reaction forces. They should have safe drainage facilities that should prevent accumulation of water in the outlets.

#### 4.12.7.2.2 Water Level Gages
At least two direct-reading level gages. One installed at each end of the drum, should be provided, and in addition one level gage easily visible from the operating platform.

### 4.12.7.3 Fabrication Requirements
#### 4.12.7.3.1 Welding
Joints should be made by welding, wherever possible, unless otherwise specified. All welded connections to the steam and water drums should be of the full-penetration type. The connections should be of the set-on or set-through type, so as to obtain the minimum weld volume. If set-on nozzles are applied and wall thickness is greater than 30 mm, the material should have reduction of area, through thickness properties, established by means of a tensile test of 25%.

For set-on nozzles that cannot be back welded, the bore of the nozzle should be machined to sound metal in the root of the weld.

4.12.7.3.2 Performance Requirements

### *4.12.7.4 Performance/Acceptance Tests*

Performance and acceptance testing should start only after the installation has been operating satisfactorily at maximum continuous rating for a consecutive period of 5 days. The company may, however, stipulate a lower load and/or period to suit conditions prevailing at the time. Performance guarantees made by the supplier should be met.

At least the following tests should be carried out using such fuel as specified by the company:

at 100% of maximum continuous rating—for 8 h
at minimum load on automatic control—for 8 h
any additional performance test, as specified on the data/requisition sheets
tests for automatic control and load response
a test at 100% maximum continuous rating with one burner out of operation, if specified on the data/requisition sheets
tests for operation of the safeguarding system.

Tests will be done by the purchaser in the presence of the supplier who should give assistance, if necessary. Unless otherwise specified, the plant instruments may be used for the performance tests after agreement has been reached between the purchaser and the supplier on the calibration of the plant instruments.

If it is specified that the plant instruments may not be used for the performance tests, or if the supplier does not agree to the use of plant instruments, the supplier should provide the test instruments and apparatus for the tests.

Methods for the determination of steam quality should be agreed between the company and supplier. Efficiency tests and calculations should be carried out according to the "Losses Method" described in the ASME Performance Test Code PTC 4.1.

### *4.12.7.5 Air and Gas*

The unit should have sufficient forced draft fan capacity available to provide the necessary air for combustion at a pressure at the burner windbox required to overcome all of the resistance through the unit including the ducting and stack. Means should be provided to control the furnace pressure and the supply of air throughout the operating range.

The $CO_2$ or excess air in gas leaving the furnace should be determined by sampling uniformly across the width of the furnace where the gases enter the convection heating surface. There should be no delayed combustion at this point or at any point beyond.

The fuel burning equipment should be capable of operation without objectionable smoke.

### *4.12.7.6 Water*

The boiler water concentration in the steam drum should be specified into job specification.

Samples of water for testing should be taken from the continuous blowdown. Samples should be taken through a cooling coil to prevent flashing. Sampling and determination of boiler water conditions should be under the methods contained in ASTM Special Technical Publication No. 148.

Test Procedure for Solids in Steam: Samples of condensed steam for determination of solids should be obtained in accordance with the method specified in the latest edition of ASTM D-1066 entitled "Tentative Method for Sampling Steam."

The Electrical Conductivity Method should be used to determine the dissolved solids in the steam. The test should be made in accordance with ASTM D-1125.

### 4.12.7.7 Spares

Detailed requirements of spares required for 2 years of operation should be as specified by the vendor.

Spares should be considered in three categories as follows:

1. Precommissioning
2. Commissioning
3. Permanent.

The supplier should specify in his proposal all precommissioning spares [Category (a) earlier].

The supplier should submit, for company's approval prior to any order, a list of commissioning spares [Category (b) earlier]. These will include spares for installation at the first overhaul.

The supplier should submit, prior to precommissioning work, a complete spares manual, to include all spares recommended as permanent stock. This will also be subject to approval by experts who will also specify the manual format to be used.

### 4.12.7.8 Special Tools

All special tools required for maintenance and operation, such as tube expanders, special wrenches, etc. which are not normally found in a workshop should form part of the installation.

#### 4.12.7.8.1 Preparation for Shipment

Boiler manufacturer should properly prepare the boiler parts for shipments to the jobsite.

Machine surfaces and flange faces should be coated with heavy rust preventive grease.

All threads of bolts, including exposed parts, should be coated with a metallic base waterproof lubricant to prevent galling in use and corrosion during shipment and storage.

To prevent damage, all flange facings should be protected with gaskets and ¼ in. plates, and all couplings should be protected by steel pipe plugs.

Suitable bracing and supports should be provided to prevent damage during shipment.

Equipment must be suitably crated, packaged, and weather protected to guard against damage while in transportation. All pieces of equipment and spare parts should be identified by item number and services, and should be suitably marked inside and outside of boxes.

### 4.12.7.9 Guarantee

In addition to the mechanical guarantee required by the conditions of contract, the vendor should guarantee in writing that each boiler will produce from Load to Full ¼ Load of Steam Rating as specified on the data sheet without detrimental carry-over into the superheater tubes and without flame impingement upon any boiler tubing when burning any combination of the gas, gasoline, and oil fuels specified herein.

The vendor should also guarantee that each boiler will be capable of producing the overload requirements of the 4 h overdesign capacity specified in Paragraph 9.5 and the minimum load for a continuous period of 24 h.

### 4.12.7.10 Information Required with Quotations

The English language should be used throughout unless otherwise specified. However, descriptions on drawings may be in other languages, provided English translations are given.

The supplier should provide all drawings, design details, operation, and maintenance manuals, and other information necessary for the design assessment, erection, operation, and maintenance of the installation.

All information, especially the manuals for operation and maintenance should be clear and not open to misinterpretation and should apply specifically to the installation supplied.

### 4.12.7.11 Schedule of Vendor's Documentations

Specification sheets should be completed and the following documentation and information should be given by the supplier.

1. Drawings of:
   a. dimensioned general arrangement, front and side elevations, of complete installation showing boiler, burners, galleries and ladders, ducting, fan, and stack.
   b. dimensioned front and side sectional elevation of boiler, showing drum, casing, furnace, burners, access and observation ports, soot blowers, and all tube banks. The furnace in particular should be fully dimensioned including burner center lines.
2. Description of:
   a. extent of shop fabrication
   b. general description of installation
   c. boiler, indicating site fabrication required
   d. casing
   e. refractory, insulation, stack lining
   f. burners
   g. desuperheater
   h. aspirating, sealing, and cooling air system
   i. fan and drive
   j. soot blowing system
   k. mountings, valves and fittings, including safety valves
   l. graph showing superheated steam temperature against load
   m. control schemes and description of all controls especially combustion control scheme.
3. The capital costs of:
   a. boiler including furnace, superheater, economizer, drum
   b. combustion air supply (fans, air ducting, controls)
   c. fuel burning equipment
   d. flue gas ducting, stack and flue duct per meter run
   e. pipework
   f. mountings, valves, including safety valves and fittings
   g. refractories
   h. instrumentation
   i. miscellaneous items
   j. erection/supervision of erection.

**4.** Period of delivery:
   **a.** time from award of contract to arrival f.o.b. at port
   **b.** time from arrival at site to acceptance
   **c.** estimated man-hours and minimum time needed for erection.
**5.** Lists of:
   **a.** reference boilers of the same type, including location, capacity, superheated steam pressure and temperature and fuels fired, and if possible, feedwater quality
   **b.** any deviations from the requirements of this specification
   **c.** subsuppliers
   **d.** all instruments needed and their location (local, local panel, control room)
   **e.** major shipping weights and dimensions
   **f.** provisions made for safety and emergencies
   **g.** fabrication procedures, tests and inspection certificates.

## 4.13 PROCESS DESIGN OF STEAM TRAPS

This section is intended to cover minimum requirements and guidelines for process engineers to specify proper type and prepare data sheet for steam traps. It contains basic reference information, data and criteria for steam trap selection as mentioned previously.

### 4.13.1 TYPES OF TRAPS

Most steam traps used in the chemical process industries fall into one of the three basic categories:

Mechanical traps, which use the density difference between steam and condensate to detect the presence of condensate. This category includes float-and-thermostatic traps and inverted bucket traps. Thermostatic traps, which operate on the principle that saturated process steam is hotter than either its condensate or steam mixed with condensible gas. When separated from steam, condensate cools to below the steam temperature. A thermostatic trap opens its valve to discharge condensate when it detects this lower temperature. This category of trap includes balanced pressure and bimetal traps as well as wax or liquid expansion thermostatic traps. Thermodynamic traps, which use velocity and pressure of flash steam to operate the condensate discharge valve.

### 4.13.2 OPERATING CHARACTERISTICS AND SUGGESTED APPLICATIONS

The key to trap selection is understanding the application requirements and the characteristics of the steam and knowing which traps meet those requirements while handling the steam condensate. Table 4.3 summarizes the operating characteristics and suggested applications for each type of trap.

### 4.13.3 DESIGN CRITERIA

Surveys have found that only 58% of all steam traps are functioning properly. Other studies have found that almost half of all failures were not due to normal wear, but were, in fact due to misapplication, undersizing, oversizing, or improper installation.

**Table 4.3  Comparison Table to be Used to Identify Which Steam Trap to Consider for a Particular Application**

| Type of Steam Trap | Key Advantages | Significant Disadvantages | Frequently Recommended Services |
|---|---|---|---|
| F&T | Continuous condensate discharge<br><br>Handles rapid pressure changes<br><br>High noncondensible capacity | Float can be damaged by water hammer<br><br>Level or condensate in chamber can freeze, damaging float and body<br><br>Some thermostatic air vent designs are susceptible to corrosion | Heat exchangers with high and variable heat-transfer rates<br><br>When a condensate pump is required<br><br>Batch processes that require frequent start up of an air -filled system |
| IB | Rugged<br><br>Tolerates water hammer without damage | Discharges noncondensibles slowly (additional air vent often required)<br><br>Level of condensate can freeze, damaging the trap body (some models can handle some freezing g)<br><br>Must have water seal to operate, subject to losing prime<br><br>Pressure fluctuations and superheated steam can cause loss of water seal (can be prevented with a check valve) | Continuous operation where noncondensible venting is not critical and-rugged construction is important |
| Wax or liquid expansion thermostatic (TS) | Utilizes sensible heat of condensate<br><br>Allows discharge of noncondensibles at start up to the set point temperature<br><br>Not affected by superheated steam, water hammer, or vibration Resists freezing | Element subject to corrosion damage<br><br>Condensate backs up into the drain line and/or process | Ideal for tracing used for freeze protection<br><br>Freeze-protection, water and condensate lines and traps<br><br>Noncritical temperature control of heated Tanks |
| BP thermostatic | Small and light mass<br><br>Maximum discharge of noncondensible start up<br><br>Unlikely to freeze | Some types damaged by water hammer, corrosion and super-heated steam<br><br>Condensate backs up into the drain line and/or process | Batch processes requiring rapid discharge of noncondensibles at start up (when used for air vent)<br><br>Drip-legs on steam mains and tracing<br><br>Installations subject to ambient conditions below freezing |

*(Continued)*

**Table 4.3 Comparison Table to be Used to Identify Which Steam Trap to Consider for a Particular Application** *(cont.)*

| Type of Steam Trap | Key Advantages | Significant Disadvantages | Frequently Recommended Services |
|---|---|---|---|
| Bimetal thermostatic (BM) | Small and light mass<br><br>Maximum discharge of noncondensibles at start up<br><br>Unlikely to freeze, unlikely to be damaged if it does freeze<br><br>Rugged, withstands corrosion, water hammer, high pressure and superheated steam | Responds slowly to load and pressure changes<br><br>More condensate back-up than BP trap<br><br>Back-pressure changes operating characteristics | Drip legs on constant—pressure steam mains<br><br>Installations subject to ambient conditions blew freezing. |
| Thermodynamic (TD) | Rugged, withstands corrosion, water hammer, high pressure, and superheated steam<br><br>Handles wide pressure range compact and simple<br><br>Audible operation warns when repair is needed | Poor operation with very low-pressure steam or high back-pressure<br><br>Requires slow pressure build-up to remove air at start up to prevent air binding<br><br>Noisy operation | Steam mains drips, tracers<br><br>Constant-pressure, constant-load applications<br><br>Installations subject to ambient conditions below freezing |

That is why it is essential to follow these three steps (in addition to proper steam trap installation, checking and trouble shooting and correct steam trap maintenance) for successful steam trapping:

1. Application definition
2. Steam trap selection
3. Steam trap sizing.

### 4.13.3.1 Application Definition

Steam trap application fall into two categories:

1. Drip and tracer
2. Process.

#### 4.13.3.1.1 Drip and Tracer Traps

Drip traps drain condensate caused by natural heat loss that is formed in steam mains and steam driven equipment. If this condensate remained in the piping, water hammer, corrosion and damage to the piping, valving and equipment would occur. Tracer traps drain condensate from steam tracers, which is tubing or pipe strapped to a process pipeline, water line, or instrument to keep it warm. Winterization tracing protects against freezing, while process tracing maintains the temperature of process liquids. Both drip and tracer traps are for system "protection." The failure of these traps can cause severe and costly consequential damages.

#### 4.13.3.1.2 Process Traps

Process application falls into four categories based on the type of equipment, with steam either heating a liquid indirectly, air or gas indirectly, a solid indirectly, or a solid directly. Table 4.4 provides examples of each type of process application.

| **Table 4.4 Categories of Process Steam Trap Applications** | |
|---|---|
| **Type of Heating Equipment** | **Typical Examples of Equipment Being Heated** |
| 1. Steam heats a liquid indirectly | Submerged surfaces (batch still, evaporator, fuel heater, shell and tube exchanger, tank coil, vat water heater) |
| | Jacketed vessel (pan, kettle, concentrator) lift or syphon drainage (tilting kettle, sulfur pit, submerged pipe or embossed coil, shipboard tank) |
| 2. Steam heats air indirectly | Natural circulation (dry air: convector, pipe coil, moist air: blanket dryer, dry kiln, drying room). |
| | Forced circulation (air blast heating coil dry kiln, air dryer, pipe coil, process air heater, unit heater) |
| 3. Steam heats a solid or slurry indirectly | Gravity drained (chest-type ironer, belt press, chamber dryer, hot plate, platen) |
| | Syphon drained (cylinder ironer, cylinder dryer, drum dryer, dry can, paper machine) |
| 4. Steam heats a solid directly | Gravity drained (autoclave, reaction chamber retort, sterilizer) |

## 4.13.4 STEAM TRAP SELECTION

After defining the application, the next step is to select the correct type of steam trap based on performance criteria such as design failure mode (open or closed), speed of response, air handling capability, ease of checking, environment high or low temperature), potential for water hammer in the system, range of pressure operation, and the presence of superheat. With rare exception, a steam trap should always be selected for fail open service. Other criteria, which are more feature-oriented, include ease of maintenance, ease of installation (including flexibility of horizontal or vertical piping), and integral strainer and blowdown valve. Table 4.5 provides some selection guidance. Selected trap types are subject to the company's approval.

When choosing a steam trap, the following steam trap codes and standards which define steam trap design, performance, and manufacturing are applicable:

**1.** ANSI/ASME PTC 39.1, "Performance Test Code for Condensate Removal Devices for Steam Systems"
**2.** ANSI/FCI 69-1, "Pressure Rating Standards for Steam Traps"
**3.** ANSI/FCI 85-1, "Standards for Production and Performance Tests for Steam Traps."

## 4.13.5 STEAM TRAP SIZING

Once the correct trap type has been selected it must be sized. A steam trap must be sized based on the condensate load, not pipe size. Sizing a trap based on pipe size typically results in an oversized trap, which will cycle more frequently or operate with the valve too close to the seat, causing wear and short service life.

The procedure for sizing a steam trap is to first calculate the condensate load based on equations (or other means) subject to employer approval. Once the condensate load has been calculated, the trap should be sized using a reasonable safety factor. For drip and tracer applications, which typically

**Table 4.5 Table to be Used for Narrowing the Choice of Steam Trap Type**

| Type of Steam Trap | Failure Mode | Operates Over Wide Press Range | Easy to Check | Air Handling Ability | Designed for Superheat | Designed for Fast Response | Resistant to Water Hammer |
|---|---|---|---|---|---|---|---|
| Mechanical F&T | Closed or open | No | No | Excellent | No | Yes | No |
| IB Thermo-static | Closed | No | Yes | Poor | No | Yes | Yes |
| Bellows | Closed or open | Yes | No | Excellent | No | Yes | No |
| Bimetal | Open | No | No | Fair | Yes | No | Yes |
| Diaphragm | Closed or open | Yes | No | Fair | No | No | No |
| Thermody-namic | | | | | | | |
| Disk | Open | Yes | Yes | Poor | Yes | Yes | Yes |
| Piston | Open | Yes | Yes | Good | Yes | Yes | Yes |
| Lever | Open | Yes | Yes | Good | Yes | Yes | Yes |

operate at constant steam pressure and have relatively light condensate loads, a safety load factor of 2 to 3 should be considered.

Size of trap's condensate lines/subheaders/headers should be enough to avoid excessive backpressure on trap to facilitate proper trap functioning.

Process applications, however, typically operate with a controlled steam supply with variable pressure and have condensate loads that are both variable and heavy. The appropriate safety load factor will depend on such factors as the heating category (as defined in Table 4.2), the type of equipment, and whether the steam pressure is constant or variable (the latter of which implies the use of control valves to regulate steam pressure based on a signal from the process controller).

Values range between 2 and 5. Table 4.6 should be used to select the safety load factor.

In addition to the considerations mentioned previously, the following items should be strictly followed:

1. Impulse type steam traps should be used for general service such as headers, branches, and tracing as detailed in the relevant piping specifications.
2. Inverted bucket traps should not be used without written permission from the company in cases where these types apply.
3. Vacuum or lifting traps should be used for draining condensate from low-pressure systems where the available pressure differential is too low for other types of traps.
4. Automatic drain valves, either float or diaphragm type for draining condensate or liquid from air or gas lines and receivers should be used.

**Table 4.6 Chart to be Used for Selection of Safety Factor for Traps in Process Application**

| Heating Category | | Steam Pressure | |
|---|---|---|---|
| | | Constant | Variable |
| 1 | Drainage to trap: | | |
| | Gravity | 2 | 3 |
| | Syphon/lift | 3 | 4 |
| 2 | Ambient air: | | |
| | 0°C and higher | 2 | 3 |
| | Below 0°C | 3 | 4 |
| 3 | Drainage: | | |
| | Gravity | 3 | |
| | Syphon | 5 | |
| 4 | Warm-up: | | |
| | Normal | 3 | |
| | Fast | 5 | |

5. Ball float traps (continuous drainers) should be used for modulating service such as draining condensate from temperature controlled reboilers, for trapping liquid in gas or air streams and for venting air or gas from liquid streams.
6. Strainers should be installed in the piping upstream of all continuous drainers. Metallic gaskets should be used for steam pressure above 2000 kPa (ga) and/or 20 bar (ga). Integral strainers are preferred.
7. The body material for ball float traps and automatic drain valve should be as follows:
   **a.** 1700 kPa (ga) and/or 17 bar (ga) and lower, cast steel;
   **b.** over 1700 kPa (ga) and/or 17 bar (ga) forged steel or stainless as applicable.
8. End connections should conform to piping specifications, except for steam tracing traps that should be screwed type.
9. Trim material for traps and strainers should be stainless steel.
10. The body material for steam tracing traps should be stainless steel.
11. Minimum body size should be DN15 (½ in. NPS) for traps in steam tracing or unit heater services. Minimum size should be DN20 (¾ in. NPS) for all other traps.

For traps in winterizing and heat conservation services the items listed below should be strictly followed:

1. Condensate collecting piping for grouped tracer traps should be such as to avoid excessive back pressure on traps and trap discharge lines and should be based on the lowest expected steam supply pressure. Minimum size of condensate collecting piping for grouped tracer traps should normally be as follows:
   **a.** 1–2 traps DN20 (¾ in. NPS)

    **b.** 3–5 traps DN25 (1 in. NPS)

    **c.** 6–15 traps DN40 (1½ in. NPS)

**2.** Each tracer should have its own steam supply valve and steam trap.

**3.** For heat conservation service, each trap should have a block valve upstream and downstream of trap. Traps will have an integral strainer and plugged drain. In winterization service no blocks will be required at steam traps. Drains will be valved.

**4.** Steam trap should be impulse tilting disc type with DN15 (½ in.) or DN20 (¾ in.) threaded ends with integral strainers and blow off valves with removable internals as shown in Figure 4.4. Body should be forged steel, seat and disc should be stainless steel or stellited. Traps should be preferably installed with the flow down. If the trap is in a horizontal run, it should be installed on its side to prevent freezing.

**5.** The condensate discharge from the tracers should be carried out through one steam trap for each individual tracer. The steam trap may service two tracers only if they are tracing the same pipe in parallel for the same length and follow the same route. The steam trap may collect the discharge of more tracers in the particular cases of pumps and instrument tracing provided the tracers are completely self-draining with no pockets.

**6.** Valves and piping at trap should be the same size as the trap size.

**7.** Piping from the trap discharge to the header should normally be DN15 (½ in.) minimum piping. Condensate recovery should be 100%, however in exceptional approved cases where it is not practicable to recover, discharge piping should be short, without elbows and discharged into sewage or into a properly designed soakaway sump.

**8.** Instrument steam tracers should be supplied only from independent main headers which will not supply steam to any other facility.

### 4.13.6 COMMON PROBLEMS OF STEAM TRAPS

#### 4.13.6.1 Freezing

If subjected to ambient conditions below 0°C, condensate in the trap will freeze unless it is continuously replenished with hot, newly formed condensate. Generally, freezing is a problem only when the steam system is shut down (or idled) and a heel of condensate remains in the trap.

    Some traps, such as the float-and-thermostatic (F&T) and balanced pressure (BP) traps using conventional bellows, are more easily damaged by freezing than other types of traps. The inverted bucket (IB) trap can also freeze, subjecting its body to damage.

#### 4.13.6.2 Air Binding

Air and other noncondensibles in the steam system reduce heat-transfer rates and can confound steam trap operation. When subjected to excessive air removal requirements, thermodynamic traps can stay closed longer than normal and IB traps will release the air only very slowly. When these types of traps are used for frequently shut-down batch operations, the steam system should be fitted with an auxiliary vent for noncondensibles. Such a vent valve is usually similar in design to a BP trap. It should be installed at a high point in the steam system or parallel with the trap.

#### 4.13.6.3 Noise

Noisy operation is generally not a problem with steam traps that discharge condensate to a closed pipe. With the exception of the thermodynamic trap, most traps tend to operate relatively silently.

In some circumstances, however, a trap may cause a slight, audible "woosh" sound as condensate flashes into steam downstream of the trap valve. Noise in steam systems is usually caused by lifting condensate up vertical return lines, water hammer, or a failed trap that leaks live steam into a condensate line.

### 4.13.6.4 Steam Leakage

Like any valve, the valve seat in a steam trap is subjected to erosion and/or corrosion. When the seat is damaged, the valve will not seal completely and the trap may leak live steam. Some trap designs allow seat replacement without removing the trap body from the line, others may require replacement of the entire trap.

Steam trap valve seats are especially susceptible to a type of erosion called "wire drawing." This phenomenon is caused by high-velocity droplets of condensate eroding the valve seat in a pattern that looks like a wire has been drawn across the surface. Oversized traps are more susceptible to wire drawing.

### 4.13.6.5 Insufficient Pressure Difference

Steam traps rely on a positive difference between process steam pressure and the pressure downstream of the trap to remove condensate. If such a difference is not present, condensate will not drain from the trap and a pump will be required.

There are two circumstances where insufficient pressure difference will occur:

1. When the pressure downstream of the trap is too high because of overloading of a condensate return line (ie, high back pressure).
2. When the process steam pressure is too low. This condition frequently occurs in the modulating service where the process temperature controller throttles steam pressure in the exchanger to a pressure below that of the pressure downstream of the trap. In some circumstances the pressure in the exchanger will actually fall below atmospheric pressure if the exchanger calls for a heat source below 100°C.

When either circumstance occurs, condensate will back up into the exchanger, no matter which type of trap is used. In modulating service, the temperature controller will eventually increase the steam pressure and force the condensate out of the trap. This filling and emptying of the exchanger causes temperature cycling that cannot be controlled by any instrumentation system. Also, when the high-pressure steam forces condensate out of the exchanger at a high velocity, the exchanger will be subject to physical damage from water hammer. The tube bundle may also be corroded at the condensate-steam interface inside the exchanger.

(This interface is a point at which the corrosive effects of oxygen and $CO_2$ in the steam can be concentrated.)

The only way to stabilize condensate removal in such circumstances is to install a condensate pump in conjunction with the trap.

### 4.13.6.6 Dirt

Steam condensate often contains particles of scale and corrosion products that can erode trap valves. If the particles are large enough, they can plug the trap discharge valve or jam it in the open position. Levers in the F&T and IB traps can also can be jammed by particulates and valve movement in BM traps can be restricted by solids jammed between the bimetal plates.

To extend trap-life, a strainer should be installed immediately upstream of each trap. Strainers should be cleaned frequently when the system is first started up and when any steam piping is replaced so that any mill scale present is removed upstream of the trap. Subsequently, strainers should be cleaned on schedule consistent with how quickly they load up with particulate.

### 4.13.6.7 Maintenance

Most traps can be repaired rather than replaced. The repairs can usually be done in-line without removing the trap body from the connecting piping. Such repairs usually require less labor than replacement because removing the trap cover is easier than removing the trap from the line. Repairing the also eliminates the possibility of having to replace pipe if the trap piping is damaged when the trap is removed. Of course, it also costs less to buy trap parts than to buy an entire trap.

## 4.13.7 PROPER STEAM TRAP INSTALLATION

Certainly, steam traps should be installed according to the manufacturer's guidelines. There are, however, some basic considerations worth noting.

First, condensate gets into steam traps by gravity. Thus, a steam trap should always be installed below the equipment that is being drained by gravity flow.

However, there are applications where a steam trap can not be installed lower, such as buried tanks, drop-in (submerged) heating coils, or rotating drum dryers.

In these instances, special consideration should be given to providing tank coils with a lift fitting, as shown in Fig. 4.2 and utilizing steam traps with control orifices to vent flashed condensate.

A lift fitting is a loop into which condensate drains, a syphon tube is then placed down into it to syphon out the condensate. Control orifices are orifices in steam traps to vent flash vapor by providing continuous drainage (ie, the trap never shuts tightly). Control orifices are common to thermodynamic piston and lever traps, other types of traps may have internal or external orifices, but only if specified as an option.

Beyond gravity drainage, proper trap station and support should be considered. Fig. 4.3 shows a typical steam trap station with recommended piping and specialties.

Some key points to remember are:

1. Adequate drain legs should be provided to ensure collection and storage of condensate prior to the trap to permit operation free of water hammer. Size of drain legs should be the same as the equipment outlet connection and generally 460–610 mm long. Their length is generally limited based on the equipment installation and clearances to grade.

   Process equipment engineers should consider these necessary clearances when designing equipment support structures.

2. A Y-type strainer (integral or separate) with a blowdown valve is essential. Dirt is a major cause of steam trap failures. The strainer catches impurities and can then be flushed to remove them.

   In addition to protection from dirt, a strainer is also a good diagnostic tool. A cold trap can be checked by simply blowing down the strainer. If pressure is present, the trap is either failed closed or plugged with dirt. It is also possible that a downstream condensate valve is closed or some other restriction exists. And most important, the strainer blowdown valve depressurizes the trap station for safe maintenance.

3. A test tee should be installed in systems where condensate is collected and either returned to the boiler or some other location. A test after the trap provides quick visual examination of trap discharge for ease of checking and troubleshooting.

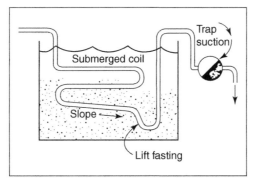

**FIGURE 4.2 A Lift Fitting Must be Installed to Drain Condensate From Equipment That is Lower Than the Steam Trap**

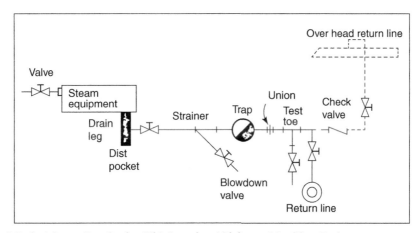

**FIGURE 4.3 A Typical Steam Trap Station With Associated Piping and Auxiliary Equipment**

**4.** Steam trap stations that include isolation block valves allow steam trap maintenance to be performed without having to turn off the steam supply at the root valve (ie, steam supply valve or the first valve in the system).

**5.** Flanges or pipe unions may be required for installations that use nonrepairable steam traps or repairable steam traps that require removal from the pipe for repair. In threaded pipe installations, only downstream unions or flanges are recommended, as upstream unions or flanges may leak and cause expensive high-pressure steam to be lost.

Flanges or unions will not be needed if in-line renewable steam traps are used for simplicity of installation and reduction of maintenance costs. (The term "renewable" is an alternative to "repairable"). Repairing implies changing a bad part.

In renewable steam traps, the maintenance results in a "new" trap, in that the valve, seat and (operating mechanisms are all replaced).

Bypasses around steam traps are to be installed when traps are needed to be removed or repaired, or when traps could not handle either the air or the heavy condensate load during start up.

In rare instances, where the process cannot even be shut down for quick in-line maintenance of the steam trap, installing a backup steam trap with the necessary valves, strainers and so on in parallel is the best alternative arrangement.

Simple instrumentation with pressure gages and thermometers on the upstream and downstream side of steam traps in critical process applications can provide valuable assistance in future trouble-shooting of system problems and trap performance. Such instrumentation is recommended for process heat exchangers where loss of temperature control may ruin a batch of material and cause significant monetary losses. Steam main drips and tracing do not require such monitoring.

Proper drip pot installation and general notes applicable to typical steam trap piping.

Insulation is an enemy to a good steam trap maintenance program and should not be used. To avoid problems, it is recommended that pipe insulation start approximately 300 mm upstream and downstream from the trap. Insulating steam traps makes them difficult to check and maintain because once insulated, a steam trap may never be accessed unless it is clearly affecting process operation. Additionally, the performance of a steam trap can be affected by insulation, thermostatic traps, for example, tend to be sluggish when insulated and bucket traps can lose their prime (ie, fail to open).

As for safety, the use of expanded metal screening wrapped around a trap, instead of insulation, can provide personnel protection where necessary.

### 4.13.7.1 Pipe Component-Nominal Size

The purpose of Table 4.8 is to present an equivalent identity for the piping components nominal size in SI System and Imperial Unit System, in accordance with ISO 6708-1980.

Figure 4.4 shows the details of thermodynamic steam tarp with removeable internals.

**Table 4.8 Pipe Component—Nominal Size**

| Nominal Size | | Nominal Size | | Nominal Size | | Nominal Size | |
|---|---|---|---|---|---|---|---|
| DN[a] | NPS[b] | DN | NPS | DN | NPS | DN | NPS |
| 6 | ¼ | 100 | 4 | 600 | 24 | 1100 | 44 |
| 15 | ½ | 125 | 5 | 650 | 26 | 1150 | 46 |
| 20 | ¾ | 150 | 6 | 700 | 28 | 1200 | 48 |
| 25 | 1 | 200 | 8 | 750 | 30 | 1300 | 52 |
| 32 | 1¼ | 250 | 10 | 800 | 32 | 1400 | 56 |
| 40 | 1½ | 300 | 12 | 850 | 34 | 1500 | 60 |
| 50 | 2 | 350 | 14 | 900 | 36 | 1800 | 72 |
| 65 | 2½ | 400 | 16 | 950 | 38 | | |
| 80 | 3 | 450 | 18 | 1000 | 40 | | |
| 90 | 3½ | 500 | 20 | 1050 | 42 | | |

[a]Diameter nominal, mm.
[b]Nominal pipe size, in.

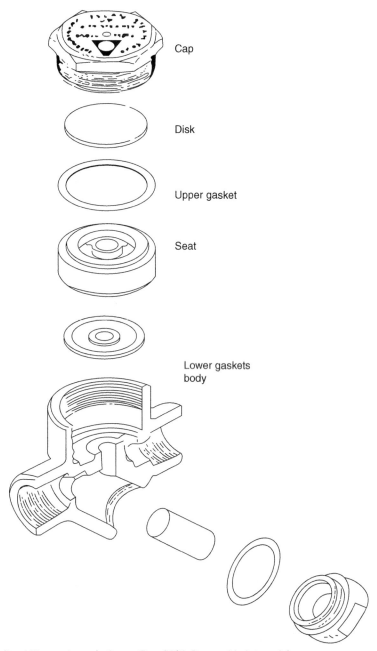

Cap

Disk

Upper gasket

Seat

Lower gaskets
body

**FIGURE 4.4  Details of Thermodynamic Steam Trap (With Removable Internals)**

# COOLING TOWER AND COOLING WATER CIRCUITS

## 5.1 INTRODUCTION

A cooling tower is a heat rejection device which extracts waste heat to the atmosphere through the cooling of a water stream to a lower temperature. Cooling towers may either use the evaporation of water to remove process heat and cool the working fluid to near the wet-bulb air temperature or, in the case of closed circuit dry cooling towers, rely solely on air to cool the working fluid to near the dry-bulb air temperature.

Common applications for cooling towers are providing cooled water for air-conditioning, manufacturing, and electric power generation. The smallest cooling towers are designed to handle water streams of only a few gallons of water per minute supplied in small pipes like those one might see in a residence, while the largest cool hundreds of thousands of gallons per minute supplied in pipes as much as 15 ft. (about 5 m) in diameter on a large power plant.

Cooling towers vary in size from small roof top units to very large hyperboloid structures (as in the adjacent image) that can be up to 200 m (660 ft.) tall and 100 m (330 ft.) in diameter, or rectangular structures that can be over 40 m (130 ft.) tall and 80 m (260 ft.) long. The hyperboloid cooling towers are often associated with nuclear power plants, although they are also used to some extent in some large chemical and other industrial plants. Although these large towers are very prominent, the vast majority of cooling towers are much smaller, including many units installed on or near buildings to discharge heat from air conditioning.

The generic term "cooling tower" is used to describe both direct (open circuit) and indirect (closed circuit) heat rejection equipment. While most think of a "cooling tower" as an open direct contact heat rejection device, the indirect cooling tower, sometimes referred to as a "closed circuit cooling tower" is nonetheless also a cooling tower.

The type of heat rejection in a cooling tower is termed "evaporative" in that it allows a small portion of the water being cooled to evaporate into a moving air stream to provide significant cooling to the rest of that water stream. The heat from the water stream transferred to the air stream raises the air's temperature and its relative humidity to 100%, and this air is discharged to the atmosphere. Evaporative heat rejection devices such as cooling towers are commonly used to provide significantly lower water temperatures than achievable with "air cooled" or "dry" heat rejection devices, like the radiator in a car, thereby achieving more cost-effective and energy efficient operation of systems in need of cooling. Think of the times you've seen something hot be rapidly cooled by putting water on it, which evaporates, cooling rapidly, such as an overheated car radiator. The cooling potential of a wet surface is much better than a dry one.

A direct, or open circuit-cooling tower is an enclosed structure with internal means to distribute the warm water fed to it over a labyrinth-like packing or "fill." The fill provides a vastly expanded air-water interface for heating of the air and evaporation to take place. The water is cooled as it descends through the fill by gravity while in direct contact with air that passes over it. The cooled water is then collected in a cold-water basin below the fill from which it is pumped back through the process to absorb more heat. The heated and moisture laden air leaving the fill is discharged to the atmosphere at a point remote enough from the air inlets to prevent its being drawn back into the cooling tower.

The fill may consist of multiple, mainly vertical, wetted surfaces upon which a thin film of water spreads (film fill), or several levels of horizontal splash elements which create a cascade of many small droplets that have a large combined surface area (splash fill).

An indirect, or closed circuit cooling tower involves no direct contact of the air and the fluid, usually water or a glycol mixture, being cooled. Unlike the open cooling tower, the indirect cooling tower has two separate fluid circuits. One is an external circuit in which water is recirculated on the outside of the second circuit, which is tube bundles (closed coils) which are connected to the process for the hot fluid being cooled and returned in a closed circuit. Air is drawn through the recirculating water cascading over the outside of the hot tubes, providing evaporative cooling similar to an open cooling tower. In operation the heat flows from the internal fluid circuit, through the tube walls of the coils, to the external circuit and then by heating of the air and evaporation of some of the water, to the atmosphere. Operation of the indirect cooling towers is therefore very similar to the open cooling tower with one exception. The process fluid being cooled is contained in a "closed" circuit and is not directly exposed to the atmosphere or the recirculated external water.

In a counter-flow cooling tower air travels upward through the fill or tube bundles, opposite to the downward motion of the water. In a cross-flow cooling tower air moves horizontally through the fill as the water moves downward.

Cooling towers are also characterized by the means by which air is moved. Mechanical-draft cooling towers rely on power-driven fans to draw or force the air through the tower. Natural-draft cooling towers use the buoyancy of the exhaust air rising in a tall chimney to provide the draft. A fan-assisted natural-draft cooling tower employs mechanical draft to augment the buoyancy effect. Many early cooling towers relied only on the prevailing wind to generate the draft of air.

If cooled water is returned from the cooling tower to be reused, some water must be added to replace, or make up, the portion of the flow that evaporates. Because evaporation consists of pure water, the concentration of dissolved minerals and other solids in circulating water will tend to increase unless some means of dissolved-solids control, such as blow-down, is provided. Some water is also lost by droplets being carried out with the exhaust air (drift), but this is typically reduced to a very small amount by installing baffle-like devices, called drift eliminators, to collect the droplets. The make-up amount must equal the total of the evaporation, blow-down, drift, and other water losses such as wind blowout and leakage, to maintain a steady water level.

*Some useful terms, commonly used in the cooling tower industry:*

Drift. Water droplets that are carried out of the cooling tower with the exhaust air. Drift droplets have the same concentration of impurities as the water entering the tower. The drift rate is typically reduced by employing baffle-like devices, called drift eliminators, through which the air must travel after leaving the fill and spray zones of the tower.

Blow-out. Water droplets blown out of the cooling tower by wind, generally at the air inlet openings. Water may also be lost, in the absence of wind, through splashing or misting. Devices such as wind screens, louvers, splash deflectors, and water diverters are used to limit these losses.

Plume. The stream of saturated exhaust air leaving the cooling tower. The plume is visible when water vapor it contains condenses in contact with cooler ambient air, like the saturated air in one's breath fogs on a cold day. Under certain conditions, a cooling tower plume may present fogging or icing hazards to its surroundings. Note that the water evaporated in the cooling process is "pure" water, in contrast to the very small percentage of drift droplets or water blown out of the air inlets.

Blow-down. The portion of the circulating water flow that is removed in order to maintain the amount of dissolved solids and other impurities at an acceptable level.

Leaching. The loss of wood preservative chemicals by the washing action of the water flowing through a wooden structure cooling tower.

Noise. Sound energy emitted by a cooling tower and heard (recorded) at a given distance and direction. The sound is generated by the impact of falling water, by the movement of air by fans, the fan blades moving in the structure, and the motors, gearboxes, or drive belts.

This chapter covers the minimum process design requirements, field of application, selection of types, design consideration, and thermal process design for cooling towers.

A water-cooling tower is a heat exchanger in which warm water falls gravitationally through a cooler current of air.

Heat is transferred from the water to the air in two ways:

1. by evaporation as latent heat of water vapor;
2. by sensible heat in warming the air current in its passage through the tower.

As a general measure, about 80% of the cooling occurs by evaporation and about 20% by sensible heat transfer. The transfer of heat is effected from the water through the boundary film of saturated air in contact with the water surface. This air is saturated at the water temperature. From this saturated air film, heat transfer occurs to the general mass of air flowing through the tower.

In the interests of efficiency, it is essential that both the area of water surface in contact with the air and the time of contact be as great as possible. This may be achieved either by forming a large number of water droplets as repetitive splash effects in one basic kind of tower packing, or by leading the water in a thin film over lengthy surfaces.

Airflow is achieved either by reliance on wind effects, by thermal draught or by mechanical means. The direction of air travel may be opposed to the direction of water flow giving counterflow conditions, or may be at right angles to the flow of water giving crossflow conditions. Although the methods of analysis may be different for counterflow and crossflow conditions, the fundamental heat transfer process is the same in both cases. In some designs mixed flow conditions exist.

The cooling range of the tower corresponds to the difference in temperature of the air-water film between entry to and exit from the tower. Air enters the tower having wet- and dry-bulb characteristics dependent on the ambient conditions. It is generally in an unsaturated state and achieves near-saturation in passing through the tower. It may be considered saturated at exit in all but very dry climates.

The performance characteristics of various types of towers will vary with height, fill configuration and flow arrangement crossflow or counterflow. When accurate characteristics of a specific tower are required the cooling tower manufacturer should be consulted.

Performance tests on a cooling tower should be done in accordance with the Cooling Tower Institute Acceptance Test Code and the American Society of Mechanical Engineers (ASME) test code.

## 5.2 TYPES OF COOLING TOWERS

There are many types of tower used in evaporating cooling; generally they tend to be divided into two groups depending upon the method used for moving air through the tower:

1. natural draught;
2. mechanical draught.

### 5.2.1 NATURAL DRAUGHT TOWERS

#### 5.2.1.1 Hyperboloidal tower

Airflow is affected by the reduction in density of the column of warm saturated air within the tower shell. Secondary effects of wind velocity may influence airflow but are not normally taken into consideration in tower design. Notes: See BS 4485: Part 4 for shell geometry (Fig. 5.1a).

Hyperboloidal tower is commonly known as hyperbolic tower.

The choice of counterblow, mixed flow or cross flow arrangements is dictated primarily by site and economic considerations.

The advantages are as follows:

1. It is suited to large water flow rates.
2. High-level emission of plume virtually eliminates fogging at ground level and recirculation.
3. It occupies less ground space than multiple mechanical draught towers for large thermal duty.
4. It is independent of wind speed and direction when compared with atmospheric towers.
5. There is no fan noise.
6. There is no mechanical or electrical maintenance.

The disadvantages are as follows:

1. The chimney effect of the shell diminishes as the humidity decreases and this may be a disadvantage in hot dry climates.
2. Close approach is not economical.
3. The considerable height of shell frequently arranged in multiple installations presents an amenity disadvantage.

#### 5.2.1.2 Atmospheric tower

Air movement through the tower is almost entirely dependent upon natural wind forces. Water falls in a vertical path through a packing while the air moves in a horizontal path, resulting in a crossflow arrangement to achieve a cooling effect. Wind speed is a critical factor in the thermal design and should always be specified. This type of tower is infrequently used in practice.

The advantage is that there is no mechanical or electrical maintenance.

The disadvantages are as follows:

1. Narrow construction results in considerable length of tower.
2. There is high capital cost due to low thermal capacity.
3. Unobstructed location broadside on to prevailing wind is required.
4. The recooled water temperature varies widely with changes in the wind speed and direction.
5. The drift loss may be substantial under high wind conditions.

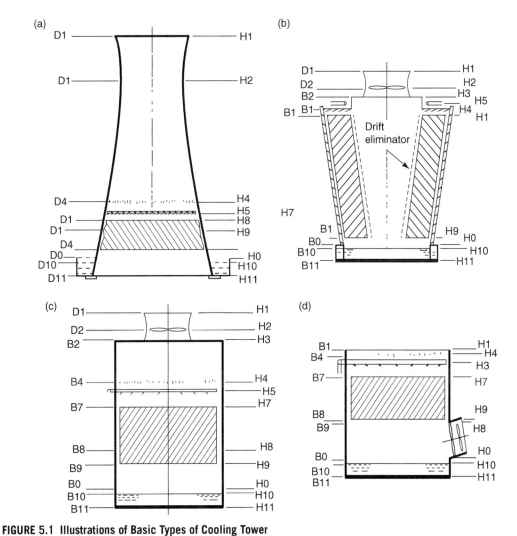

**FIGURE 5.1 Illustrations of Basic Types of Cooling Tower**

(a) Natural draught, mixed or counterflow cooling tower, (b) induced draught crossflow, (c) induced draught or counterflow cooling tower, (d) forced draught mixed or counterflow cooling tower.

## 5.2.2 MECHANICAL DRAUGHT TOWERS

Fans are used to produce air movement through the tower. This enables the airflow to be determined independently of other process conditions. Correct quantities and velocities of air may be selected to satisfy various design demands (Fig. 5.1b–d).

Several alternative ways of locating the fans in relation to tower structure are used to obtain specific advantages; also there are two basic flow arrangements for air-water flow, the counterflow and the crossflow.

Note: A standard reference sheet for physical dimensions is given in Table 5.1.

**Table 5.1 Standard Reference Sheet for Physical Dimensions (Fig. 5.1)**

| Item | Natural Draught (Fig. 5.1a) | | Crossflow (Fig. 5.1b) | | Induced Draught (Fig. 5.1c) | | Forced Draught (Fig. 1d) | |
|---|---|---|---|---|---|---|---|---|
| Top of air outlet | D1 | H1 | D1 | H1 | D1 | H1 | B1 | H1 |
| Throat | D2 | H2 | D2 | H2 | D2 | H2 | — | — |
| Fan deck | — | — | B3 | H3 | B3 | H3 | — | — |
| Eliminator screen | D4 | H4 | — | — | B4 | H4 | B4 | H4 |
| Distribution pipes | — | H5 | — | H5 | — | H5 | — | H5 |
| Distribution basins | — | — | B6 | H6 | — | — | — | — |
| Top of packing | D7 | H7 | B7 | H7 | B7 | H7 | B7 | H7 |
| Top of air inlet | D8 | H8 | B8 | H8 | B8 | H8 | B8 | H8 |
| Bottom of packing | D9 | H9 | B9 | H9 | B9 | H9 | B9 | H9 |
| Cold water basin | D0 | H0 | B0 | H0 | B0 | H0 | B0 | H0 |
| Water level | D10 | H10 | B10 | H10 | B10 | H10 | B10 | H10 |
| Bottom of cold water basin | D11 | H11 | B11 | H11 | B11 | H11 | B11 | H11 |

The advantages are as follows:

1. There is positive control of the air supply.
2. Minimum capital costs make it appropriate for low load factor applications.
3. High water loadings can be maintained regardless of the size of tower.
4. Difficult duties (long range combined with close approach) are more easily attainable than in natural draught.
5. It has a low height structure.

The disadvantages are as follows:

1. Power is required to operate the fans.
2. It requires mechanical and electrical maintenance.
3. Warm, moist discharge air may recirculate into the air intakes.
4. For large multitower installations, the total ground area required is greater than for natural draught hyperboloidal towers for equivalent duty. This is due to the spacing of towers to minimize recirculation.
5. Fogging and drift may create problems at low levels.
6. Fan noise may be a nuisance.

### 5.2.2.1 Forced draught tower

A forced draught tower is a mechanical draught tower having one or more fans located in the air intake, normally limited to capacities of up to 9.3 m³/s (Fig. 5.1d).

The advantages are as follows:

1. There is low vibration due to rotating components being located near the base of the tower.

2. Fan units are placed in a comparatively dry air stream; this reduces the problem of moisture condensing in the motor or gearbox.
3. Fan units located at the base of the tower facilitate inspection and maintenance.
4. Fans moving ambient air will absorb less power than in induced draught towers.

The disadvantages are as follows:

1. It may be more subject to recirculation than induced draught towers for equivalent duties.
2. Ice may form on fan inlets during operation in winter. This can be minimized by arranging the fan ducts at a slight angle for draining any water back into the storage basin.

### 5.2.2.2 Induced draught tower

An induced draught tower is a mechanical draught tower having one or more fans at the air discharge (Fig. 5.1b and c).
The advantages are as follows:

1. It has the ability to handle large water flow rates.
2. It is suitable for larger cell sizes and fan sizes as compared with forced draught. Larger fan sizes may result in greater efficiency and consequently lower power and sound levels.
3. It uses a more compact ground area than a forced draught tower of equivalent capacity due to the absence of fans on one side.
4. Fan equipment in warm exhaust air is less liable to icing up in winter operation.

The disadvantages are as follows:

1. Protection is required for mechanical equipment against corrosion and internal condensation.
2. Inspection and maintenance of mechanical equipment is relatively difficult due to fans being located 5 m to 20 m above the base.

### 5.2.2.3 Counterflow tower

A counterflow tower is a mechanical draught tower in which air and water flow in opposite, mainly vertical directions.
The advantages are as follows:

1. Normally it is an economical choice for difficult duties (long range combined with close approach).
2. It is less prone to icing than crossflow towers.

The disadvantages are as follows:

1. With induced draught arrangement the water distribution system (generally piping or troughs with spray nozzles) cannot be easily inspected and cleaned unless the tower is shut down.

### 5.2.2.4 Crossflow tower

A crossflow tower is a mechanical draught tower in which airflow is normally horizontal, in contact with falling water drops. It is normally associated with an induced draught arrangement (Fig. 5.1b).
The advantages are as follows:

1. It may be an economical choice for large water flows.
2. The plan area at basin level and the total power for fans and pumps can be less than for other mechanical draught towers.

**3.** The water distribution system of the open pan type is easy to clean without shut-down.
**4.** It may be designed to suit low-silhouette applications for small duties.

The disadvantages are as follows:

**1.** Prevention of icing during extreme weather conditions generally demands more care from the operator.
**2.** The exposure of water distribution basins to sunlight promotes growth of algae.

Cooling towers have two types of airflow: crossflow and counterflow. In crossflow towers, the air moves horizontally across the downward flow of water. In counterflow towers, the air moves vertically upward against the downward fall of the water. There are many types and sizes of cooling towers:

### 5.2.3 MECHANICAL DRAFT TOWERS

Fans are used to move the air through the mechanical draft tower. The performance of the tower has a greater stability because it is affected by fewer psychometric variables. The fans provide a means of regulating the airflow. Mechanical draft towers are characterized as either forced draft or induced draft.

Forced draft towers (Fig. 5.2). The fan is located on the air stream entering the tower. This tower is characterized by high air entrance velocities and low exit velocities, therefore, the towers are susceptible to recirculation thus having a lower performance stability. The fans can also be subject to icing under conditions of low ambient temperature and high humidity.

Induced draft towers (Fig. 5.3a and b). The fan is located on the air stream leaving the tower. This causes air exit velocities which are three to four times higher than their air entrance velocities. This improves the heat dispersion and reduces the potential for recirculation. Induced draft towers require about 1 kW of input for every 18 000 m$^3$/h of air.

Coil shed towers (Fig. 5.4). This application exists in many older cooling towers. The atmospheric coils or sections are located in the basin of the cooling tower. The sections are cooled by flooding the

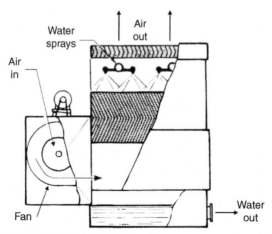

**FIGURE 5.2 Mechanical Forced Draft Counterflow Tower**

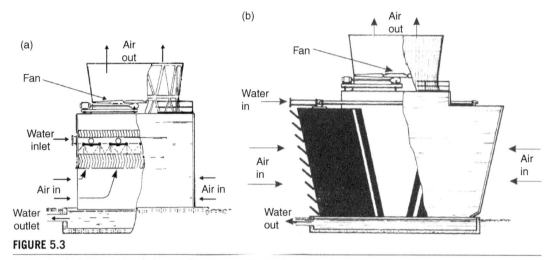

**FIGURE 5.3**

(a) Mechanical Induced Draft Counterflow Tower, (b) Mechanical Induced Draft Crossflow Tower.

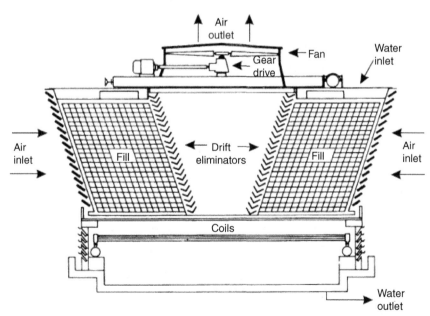

**FIGURE 5.4  Mechanical Draft Coil Shed Tower**

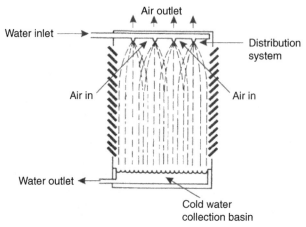

**FIGURE 5.5  Atmospheric Spray Tower**

surface of the coils with cold water. Reasons for discontinued use were scaling problems, poor temperature control, and construction costs. This type tower can exist both as mechanical or natural draft.

## 5.2.4 **NATURAL DRAFT TOWERS**

Atmospheric spray towers (Fig. 5.5). Cooling towers of this type are dependent upon atmospheric conditions. No mechanical devices are used to move the air. They are used when small sizes are required and when low performance can be tolerated.

Hyperbolic natural draft towers (Fig. 5.6). These towers are extremely dependable and predictable in their thermal performance. A chimney or stack is used to induce air movement through the tower.

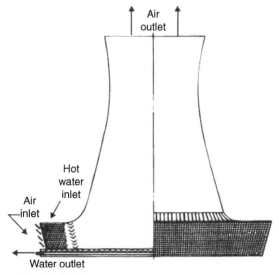

**FIGURE 5.6  Hyperbolic Natural Draft Tower**

## 5.3 DESIGN CONSIDERATIONS

### 5.3.1 DESIGN PARAMETERS

The parameters involved in the design of a cooling tower are:

1. ambient wet-bulb temperature;
2. approach;
3. cooling range;
4. circulating water flow;
5. altitude (considered if more than 300 m above sea level).

An additional parameter in the case of natural draught towers is ambient dry-bulb temperature or, alternatively, ambient relative humidity.

### 5.3.2 AMBIENT AIR TEMPERATURES

It is important that the correct design ambient conditions are chosen with care. Generally the hottest period of the year is selected as the critical area to be studied. For climatic conditions the atmospheric information covering the average 5 hot months period inclusive, i.e., last 2 months of spring and summer months (May to Sep. inclusive) are analysed and presented in the form of wet and dry bulb temperature isotherm maps for the different localities.

In general the tower should be designed for a wet-bulb temperature that will not be exceeded more than 2.5% of the time in five hot spring and summer months.

### 5.3.3 PACKINGS

The function of packing in a cooling tower is:

1. to increase the duration of contact between the air and the water;
2. to cause fresh surfaces of water to be formed, thus increasing the rate of heat transfer per unit volume.

#### 5.3.3.1 Types and selection

Packing may be of the two types namely splash packings and film packings. The intended situation of a tower should be considered in deciding on a particular type of packing. In general, the film packings will be more susceptible to fouling by suspended solids, fats and oils, biological growth, or other process contamination. Where fouling may become a problem, the spacing and configuration of the packing elements should be considered regarding the potential for cleaning.

#### 5.3.3.2 Height of packing

The height of cooling tower packing will vary considerably even within the various types of packing according to the design economics relating to any specified requirements. In general, it can be stated that for equivalent duties and fan power requirements the film or extended surface packings will be of lower height than the splash bar type of packing.

### 5.3.4 COOLING RANGE AND WATER QUANTITY

Cooling range and water quantity variations are usually considered in relation to a fixed heat load and are selected in conjunction with other plant conditions.

### 5.3.5 RECIRCULATION

The percentage of air recirculating on the leeward side of the cooling tower can vary between 3% and 20%. However, the higher figure is normally associated with installations of one or more large multicell mechanical draught cooling towers.

In general, therefore, recirculation of the warmed air discharged from the cooling towers is relatively insignificant in mechanical draught cooling towers under 0.5 m$^3$/s capacity. For other cases allowance should be made for the maximum anticipated recirculation.

Hydrocarbon detection facilities to be located in cooling tower basins should be provided to account for probable hydrocarbon leakage.

### 5.3.6 APPROACH

Approach is a very sensitive design parameter. Closer approaches are limited by practical difficulties such as minimum water loading on the packing.

The cooling tower supplier should be consulted before consideration is given to approaches closer than 3°C for mechanical draught towers or 7°C for natural towers.

At these levels an increase of 1°C in approach may result in a reduction of 20% in tower size and is therefore of considerable economic significance.

A 5.5°C approach between cold-water temperature and wet-bulb temperature should be used unless otherwise specified.

### 5.3.7 WATER LOADINGS

The maximum water loading on a packing is determined largely by the increase in resistance to airflow and by the risk of excessive drift.

Somewhat higher water loadings can in general be used in a cross flow-cooling tower irrespective of the type of packing.

Cooling tower water loadings do not approach the level at which flooding takes place. The only problem with high water loadings is in obtaining adequate airflow and cross flow towers will often therefore be found advantageous.

Water loading should not exceed 407 L/m$^2$ per min (10 gpm/ft.$^2$) of tower cross section area in the horizontal plane.

### 5.3.8 WINDAGE LOSSES

Typical windage losses, expressed as percentages of the total system water circulation rate, for different evaporative equipment are as follows:

| | |
|---|---|
| Spray ponds | 1.0–5.0% |
| Atmospheric draft towers | 0.3–1.0% |
| Mechanical-draft towers | 0.1–0.3% |

## 5.3.9 DRIFT LOSSES

Drift losses should not exceed 0.01% of design flow rate. (For further information reference is made to BS 4485:Part 3: 1988)

## 5.3.10 EFFECT OF ALTITUDE

Cooling tower calculations involve the use of published tables of psychometric data that are generally based on a barometric pressure of 1000 mbar (1 mbar = 100 N/m$^2$ = 100 Pa). Barometric pressure falls at a rate of approximately 1 mbar for each 10 m increase in altitude and, although this may be ignored for locations up to 300 m above sea level, appropriate corrections should be applied when designing for sites at higher altitudes.

## 5.4 WATER QUALITY

For the proper use of water resources, the consumption of high quality fresh water in an industrial cooling context should be discouraged, particularly if the quantities concerned are large as is usually the case in OGP industries. Any installation should use the lowest quality water suitable for the process concerned.

### 5.4.1 MAKE-UP WATER

A full mineral and biological examination of the make-up supply is essential. A typical analysis data sheets provided by the Company is shown in Table 5.2.

| Table 5.2 Typical Make-Up Water Characteristics |
| --- |
| **Sources** |
| Availability over use (dm$^3$/s)<br>Value (cent/1,000 dm$^3$)<br>pH<br>Total hardness as CaCO$_3$ (mg/kg)<br>Calcium as CaCO$_3$ (mg/kg)<br>Magnesium as CaCO$_3$ (mg/kg)<br>Total alkalinity as CaCO$_3$ (mg/kg)<br>Sodium as CaCO$_3$ (mg/kg)<br>Potassium as CaCO$_3$ (mg/kg)<br>Sulfate as CaCO$_3$ (mg/kg)<br>Chloride as CaCO$_3$ (mg/kg)<br>Nitrate as CaCO$_3$ (mg/kg)<br>Silica as SiO$_2$ (mg/kg)<br>Total iron (mg/kg)<br>Suspended solids (mg/kg)<br>Dissolved solids (mg/kg)<br>COD (as manganeses) (mg/kg)<br>Others |

Pretreatment for larger systems is usually limited to suspended solids removal by sedimentation and possibly flocculation. In the case of smaller systems, however, instances arise where base exchange softening of the make-up is economically favorable by virtue of enabling operation at a higher concentration factor with consequent savings in water charges and chemical treatment costs.

### 5.4.2 CIRCULATING WATER

An economical cycle of concentration (with the consent of the Company) should be selected and quality of circulating water should be maintained by making provisions for control of pH, hardness, biological growth, corrosion, etc.

## 5.5 COLD CLIMATE DESIGN CONSIDERATIONS

The necessity for antiicing devices to be incorporated in the tower design should be considered at the pretender stage, and the means to be employed preferably discussed between the Company and the prospective suppliers.

The methods available for providing antiicing facilities, and the need for such facilities, depend largely on the following factors:

**1.** The severity of weather conditions at the cooling tower site.
**2.** The thermal design conditions. For example, a close approach tower will mean a relatively higher airflow and a greater tendency to icing conditions in the air inlet.
**3.** The type of cooling tower used. A counterflow tower will normally have the packing protected by the tower shell, with a lesser tendency to icing on the packing than a mixed flow tower when part of the packing may be exposed.

When the tower is designed to be drained down in winter the system and tower should be designed so that no water pockets remain. In these circumstances antiicing devices are unnecessary.

### 5.5.1 AVOIDANCE OF ICING BY CONTROL OF WATER LOAD

The most commonly used device is a bypass of the tower producing a warm water curtain outside the exposed packing face. For efficient operation it is important that such a curtain should consist of jets of water rather than small particles, which might themselves become frozen.

Generally the bypass should operate under a pressure head of not less than 2.0 m (having allowed for friction losses), the nozzles should be of a diameter not less than 12 mm, and the total bypass flow should be about 25% of the circulating water flow. Fig. 5.7 shows typical bypass pipe locations.

It is helpful, when combating a tendency to form ice, to reduce the number of cells working, which will increase the water loading and the temperatures of the water on the remaining cells.

It should be noted that complete isolation has to be attained as the persistence of minor water leaks on to the packing can lead to a major ice build-up. Precautions should also be taken to avoid any static water in sections of pipework.

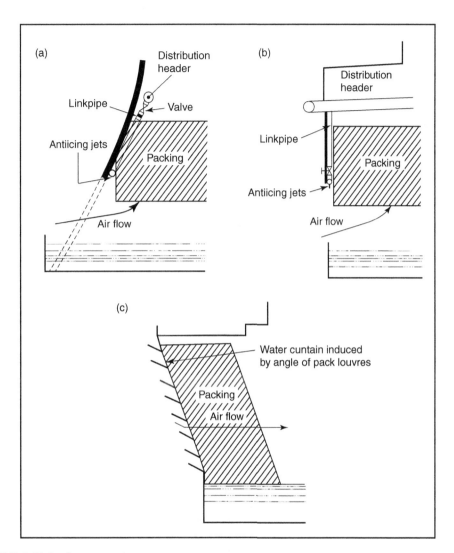

**FIGURE 5.7  Antiicing Arrangements**

## 5.5.2 **AVOIDANCE OF ICING BY AIR CONTROL**

As the air temperature drops, so does the recooled water temperature. It is often the case that below a certain temperature no further economic benefit occurs.

In such circumstances, in a mechanical draught tower, some fans may be stopped allowing the water temperature to rise sufficiently to keep the tower free of ice. A further refinement would be reducing fan speed in suitable circumstances, particularly where the saving in power has significant economic effect. In extreme conditions (about −20°C) it may be desirable to provide for reversed airflow as an effective means of clearing air inlets partially blocked by ice. In crossflow towers, antiicing may be brought about by reducing the fan speed so that a curtain of warm water is formed in front of the packing, falling from louvre to louvre down the face of the tower.

### 5.5.3 EFFECTS OF ANTIICING FLOW ON TOWER CAPACITY

When the antiicing pipe is in use, the effect on tower capacity is the equivalent of a bypass equal to the antiicing flow as the cooling effect on the bypass flow will not be substantial. At the time of the year when such a system is in use, unless the towers serve generating plant, the maximum capacity of the tower is not usually required and no disadvantage is to be expected from the use of an antiicing facility.

## 5.6 SITING, SPACING, AND ENVIRONMENTAL CONSIDERATIONS

The siting and spacing of a cooling tower installation should be considered from economic, thermal and environmental aspects.

### 5.6.1 SITING

#### 5.6.1.1 Tower levels
The cooling tower should be located at a suitable site and due consideration should be given to the question of drainback from the system resulting in loss of water and flooding.

#### 5.6.1.2 Air restrictions
On small industrial tower installations, due to aesthetic reasons or sound attenuation requirements, enclosures or barriers are sometimes built to shield the towers. These barriers or enclosures should be spaced and designed to achieve the minimum of air restriction with the maximum maintenance working area.

The exclusion of birds and bird droppings may also necessitate the provision of barriers, which should be subject to the same considerations. The total flow area in the barrier or enclosure should be a minimum of twice the area of the tower inlet openings on that side.

#### 5.6.1.3 Recirculation
The extent of recirculation depends mainly upon wind direction and its velocity, tower length and atmospheric conditions. Further factors that may exert some influence are spacing, topography or geographical situations with respect to downdraught, exit air speed, tower height and the density difference between exit air and ambient temperatures.

#### 5.6.1.4 Orientation of cooling towers
The orientation of cooling towers should be as follows:

1. Towers with air inlets on one side should be oriented so that the air inlets face the prevailing wind.
2. Towers with air inlets on opposite faces of the cooling tower should be oriented so that the air inlets face at 90 degree to the prevailing wind.

3. Large mechanical draught towers should preferably be divided into banks, each of which should have a length-to-width ratio of about 5 to 1.
4. The wind loading on any tower within a group will be affected by the grouping and spacing and should be considered in their structural design (see BS 4485: Part 4).

**Note:** The prevailing wind direction, determined by the local topography, should be taken as that obtained during periods of maximum duty.

## 5.6.2 ENVIRONMENTAL CONSIDERATIONS

The effects of drift, blow-out, fogging, and noise are further contributing factors that may need consideration when siting a tower installation.

### 5.6.2.1 Drift

When towers are sited adjacent to high-voltage electrical equipment, drift may cause flashover and icing problems. Drift can also constitute a hazard, particularly under icing conditions, on public footpaths and roadways and may also create a nuisance in adjacent residential areas.

Drift may also create a health hazard by virtue of its bacterial population and towers should be sited so as to avoid drift into open windows and re- entrainment of cooling water droplets in the intake air to ventilation equipment. Effective eliminators at the tower discharge should be capable of reducing the drift to an acceptable level (see BS 4485: Part 2: 1988).

### 5.6.2.2 Blow-out

Blow-out is water blown out from the air inlet and occurs to a greater extent on natural draught towers of counterflow design than on mechanical induced draught towers. It can produce a nuisance factor with similar detrimental effects to drift, although the radius of area affected would be smaller. Where blow-out creates a nuisance, it may be reduced by the following means:

1. diagonal partitions or a central division situated so that the prevailing winds are prevented from blowing across the tower basin;
2. inclined louvre boards positioned around the air opening at the base of the tower, sections of which may be removable to permit access.

## 5.6.3 SPACING

When the long axis of one bank is perpendicular to the prevailing wind direction (Fig. 5.8a), the influence on another bank will be minimized when the distance, X, between the banks is greater than their average length. The long axes of the tower banks should be in line.

When the long axis of the existing tower is parallel to the wind direction (Fig. 5.8b), the influence of the existing tower on the new will be minimized if the distance, X, is greater than their average length.

When the long axis of the existing tower is at 45 degree to the wind direction (Fig. 5.8c), the influence of the existing tower on the new will be minimized if the distance, X, between the towers measured normal to the wind direction is greater than their average length.

Spacing of large natural draught cooling towers is considered to be adequate if adjacent cooling towers are spaced so that the distance between the towers is equal to or greater than half the base diameter of the large tower. Should a tower be sited in close proximity to large buildings such as

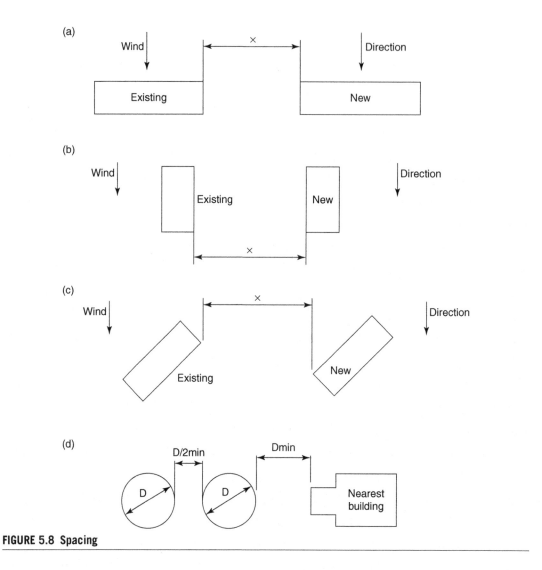

**FIGURE 5.8 Spacing**

the turbine and boiler houses, the nearest point of the tower relative to the buildings should be at least one tower diameter away (Fig. 5.8d).

### 5.6.3.1 Discharge of cooling water

Discharge of cooling tower purge (blow down) directly to the environment will be subject to Regulations of Environment Protection Organization. Discharge temperature restrictions and the elimination of navigation hazards may necessitate auxiliary cooling or predilution of the discharge by further abstracted water. The said Regulations will almost certainly place limitations on the type of water treatment, which may be used. Chromates are now almost universally banned and restrictions on the use of a range of other inhibitors and biocides are becoming increasingly widespread. Discharge to sewers may also entail consent subject to conditions. These conditions should also be observed when draining the system for maintenance.

### 5.6.3.2 Fogging

Fogging arises from the mixing of the warm moist air discharged from the tower with cooler ambient air, which lacks the capacity for absorbing all the moisture as vapor. This mixing results in the excess moisture condensing as fog.

When fogging exists, it is a nuisance factor that could create visibility and icing hazards. It is an intrinsic feature of any evaporative cooling tower and is worst during periods of low ambient temperatures and high relative humidity. The dissipation of fog, where it occurs, depends mostly on the characteristics of the prevailing atmosphere.

Measures may be taken to reduce fogging, but these add substantially to the capital and operating costs of the tower. For example, heating the moist air discharge reduces the fog, as does an increase in the volume of air through the tower.

Fan stacks discharging warm vapors at a high elevation can be a partial solution, although this may be expensive depending on the stack diameter and the height required to derive some benefit. This problem is solved far more easily if the tower installation is sited where the least possible nuisance may be caused. The high-level discharge of the vapor plume from a natural draught tower generally prevents this from being a hazard.

### 5.6.3.3 Noise

The likely noise level from fans and falling water in a cooling tower installation should be considered in the context of the existing noise level at the proposed site.

It is necessary that the noise aspect be given early prominence in that to effect a significant reduction in noise may need a design of tower quite different from and possibly less economical than that which may have been suitable in a position of no noise restriction. For data on noise level requirements reference should be made to the Company's project specification.

### 5.6.3.4 Noise abatement

If noise is to be a minimum at any point of sensitivity, the following recommendations are made:

1. The basic noise level should be as low as possible. Fan power should be low and the water noise reduced to a minimum.
2. The towers should be sited the maximum distance from the point of sensitivity.
3. If a multicell mechanical draught unit is to be installed, the fans should be in line with the point of sensitivity and the air inlets in the broadside on position.
4. Motors should be located behind the fan flares when viewed from the sensitive point.
5. In forced draught towers the fan axis should be away from the sensitive point.
6. It may be possible to analyze the sound spectrum, locate and abate discrete frequencies, and thereby reduce the general noise level.
7. Antivibration mounting may be necessary to minimize the transfer of vibration from a tower to a supporting building. In certain cases the fan and driver may be independently supported on antivibration mountings and flexibly connected to the tower, which may then be rigidly mounted. This has the advantage of eliminating the need for flexible circulating water connections, make-up lines, and drains.

8. Silencers on fan intakes or discharge stacks are possible aids in the reduction of noise, but may introduce prohibitive additional air resistance.
9. The use of multispeed drives for the fan provides for a reduction in noise level when operated at reduced speed.
10. Employment of centrifugal fans in the case of forced draught cooling towers may contribute to a reduction in noise level.
11. As fan noise is roughly proportional to fan driver power, the airflow directly influences the degree of noise. Towers that operate at high L/G ratio may therefore be less noisy than their most economical equivalent where there is a noise problem to be solved.
12. If fan noise is to be reduced on an existing tower, it may be possible to reduce the fan capacity and therefore fan power by operating at an increased water flow and a different heat balance. If the tower is small and of the forced draught kind, then a change from axial to centrifugal fans may be possible.

## 5.7 GUARANTEES

The Vendor should give the predicted performance of the tower over a range of atmospheric conditions.

The Vendor should guarantee that the cooling tower and its appurtenances and accessories should perform successfully and satisfactorily in continuous operation over the entire range of duty, without undue noise or injurious vibration, or sagging of the Redwood wood members due to the temperature involved.

The tower should be designed for a wet-bulb temperature that will not be exceeded more than 2.5% of the time in five hot spring and summer months (eg, May to Sep.).

A 5.5°C approach between cold-water temperature and wet-bulb temperatures should be used as stated in 8.3 unless otherwise specified.

There should be no leakage of water from sides or ends of the tower.

The manufacturer should furnish curves showing performance of cooling tower over a considerable range of atmospheric wet-bulb temperatures, for various heat loads and water qualities.

The field performance of the tower should be determined by CTI standard testing procedures and as specified by 6.5.2. If the tower does not meet the performance specifications, the Vendor should be obligated to make the corrections without charge until performance specifications are met.

Drift losses should not exceed 0.01% of design flow rate as per 8.8.

Water loading should not exceed 407 L/min per m (407 L/min per m) of tower cross-section area in the horizontal plane.

### 5.7.1 PROCESS REQUIREMENTS

The Vendor's proposal should include three separate performance curve sheets, one each for 90%, 100%, and 110% design water quantity, showing tower temperatures versus ambient air wet-bulb temperature. Curves should cover cooling ranges corresponding to 60%, 80%, 100%, and 120% of design heat duty, at wet-bulb temperatures ranging from 5.5°C (10°F) below to 1.6°C (3°F) above the specified ambient wet-bulb temperature. Performance curves should be based on constant fan power equal to the design power.

### 5.7.2 **INQUIRY AND BID FORM**

The Vendor's proposal should include a data sheet containing all applicable information listed in the A(2) and the following:

1. Percent recirculation (design).
2. Air temperature leaving tower, °C.
3. Grade, dimensions, and other applicable specification for materials of construction.
4. Detailed description of deicing facilities, if required.
5. Cold-water temperature when one cell is shutdown, and design duty and flow is distributed between remaining cells.
6. Description of Vendors standard practice for incising of lumber (if incising is proposed).
7. Recommended spare parts based on 2 years' continuous operation.

### 5.7.3 **OPERATING AND MAINTENANCE MANUALS**

All manuals should be written specifically for the equipment being furnished and should contain (but not be limited to) the following information:

1. Mechanical data on fan, hub, coupling, and fan drivers, including shafts bore and keyway dimensions, design pitch for fan blades, type of lube oil recommended.
2. Sectional drawings of fan, hub coupling, fan drivers (including gears) showing location of parts, part numbers, and materials.
3. Instructions for installation, operation, and maintenance of the tower and mechanical equipment.
4. Instructions for fan hub and blade assembly, and recommended procedures for field dynamic balancing of the installed fan.

### 5.7.4 **SUPPLEMENTAL PROPOSAL DATA**

A supplemental data sheet should be submitted with proposals to include the following additional information for fan data:

1. State whether the fans are multispeed and, if so, give air delivery per fan at all speeds, at inlet conditions.
2. State whether the fans are reversible and, if so, give qualifications for operating in the reverse direction.
3. Fan efficiency (including gear loss).

Polypropylene, polyethylene, or polyvinyl chloride fill should be quoted as a base bid or as an alternative.

## 5.8 **PROCESS DESIGN OF COOLING WATER CIRCUITS**

This section specification covers:

1. Cooling water circuits for internal combustion engines.
2. Cooling water circuits for reciprocating compressors.
3. Cooling water circuits for inter cooling and after cooling facilities.

## 5.8.1 INTERNAL COMBUSTION ENGINES, COOLING SYSTEMS

When the fuel is burnt in the cylinder, a part of the heat developed during combustion, and flows to cylinder walls. If the temperature of cylinder walls is allowed to rise above a certain limit (about 150°C) then the oil lubricating the piston starts evaporating. This action damages both piston and cylinder. The high temperature developed may sometimes cause excess thermal stresses and hence cracking of the cylinder head and piston. The hot spots may also cause preignition in the combustion space. In order to avoid any damage, the heat flowing to the cylinder walls must be carried away.

All heat carried away from an engine should finally be conveyed to the atmosphere. However, the methods of cooling may be divided into two main groups of direct or air-cooling and indirect or liquid-cooling.

In cooling of engine cylinders, all three means of heat transfer, that is, conduction, convection, and radiation will be utilized. But, conduction will play the most important part in carrying the heat through the thin layers of hot gases and water in contact with cylinder walls and will be the sole object of process design in the Standard.

The quantity of heat lost per second to the heating surface, that is, inside surface of cylinder wall, head and exhaust valve cages by combustion gases should be considered.

### 5.8.1.1 Requirements of cooling system

Unless otherwise specified, the required cooling water system should include the following features:

1. The closed cooling water system should either use distilled or treated soft water, which is passed through a heat exchanger where it is cooled and then passed through the cylinder jacket.
2. The heat exchanger used may usually be of shell and tube type. Using of air-cooled heat exchanger should be based or Company's agreement.
3. Within the cylinder jackets, only liquid phase cooling should be permitted.
4. The system should be capable of providing required quantity of water for cooling of cylinder jackets, cylinder heads, exhaust valve cages, and circulating oil.
5. The following operating condition should be considered in the design of a cooling system:
   a. An uninterrupted flow of cooling water will always be maintained through the cylinder jackets.
   b. The water used for cooling of a cylinder jacket should be free from scale and impurities and should not be of a corrosive nature.
   c. The inlet water temperature to cylinder jackets should be maintained at 63°C to 68°C.
   d. The maximum water temperature rise within the cylinder jackets including the heat absorbed from cylinder heads and exhaust valves should not exceed 10°C.
   e. The system should be designed to meet the working pressure of not less than 520 kPa and testing pressure of 800 kPa.
   f. An automatic control system should be considered for controlling of inlet water temperature.
   g. Thermometers, complete with thermowells should be fixed at cooling system outlets.
   h. A protection device should be established at the cooling system outlet to monitor and act, if the temperature rise exceeds a critical value specified by the manufacturer.
   i. A cooling water high temperature alarm should be provided on cylinder outlet. The alarm should actuate and the compressor should shut down when the discharge temperature of any cylinder exceeds the rated discharge temperature by 22°C.

j.  The quantity of circulating water by each pump should meet the temperature rise across each and all of the cylinder, cylinder head and the exhaust valve cage and circulating oil.

k.  The system should be provided with an appropriate draining connection. The connection should provide facilities for perfect washing, cleaning and draining of the system.

l.  Low inlet water temperature to cylinder jackets will increase the viscosity of the lubricating oil and consequently the piston frictions. Vendor should make necessary provisions to control the inlet water temperature at a specified range.

6.  Unless specified otherwise, Vendor should furnish a detailed drawing for his proposed closed water-cooling system. Company's recommended drawing should typically be as per Fig. A.1, in Appendix "A".

7.  Unless otherwise specified, the Vendor should supply closed water-cooling piping with a single inlet and a single outlet connection on each cylinder.

### 5.8.1.2 Equipment/devices and process design

Unless otherwise specified, the process design of the following equipment/devices should constitute a combined, self-contained closed water-cooling system for internal combustion engines.

1.  Soft water circulating pumps.
2.  Soft water circulating piping.
3.  Reservoir (or surge tank) for soft water.
4.  Soft water circulating, heat exchanger (or cooler).
5.  Thermometers for measuring inlet and outlet temperatures.
6.  Temperature regulator to control the outlet temperature.
7.  A soft water high temperature protective device to control the excessive cylinder jacket temperature.

Fig. 5.9a, illustrates the required equipment/devices of an internal combustion engine's closed water-cooling system.

## 5.8.2 RECIPROCATING COMPRESSORS' COOLING SYSTEM

When air/gas is compressed, its temperature and pressure will rise and a considerable heat will be generated due to a rise in temperature. Part of the heat so generated, will be transferred to the cylinder wall increasing the wall temperature which will reduce the lubricating efficiency in cylinder and might result in an overheated and warped rod. The heat of compression will also results in a loss by boosting of pressure.

It has been found desirable to remove part of this heat, which has traveled to cylinder wall in order to get rid of any damage to cylinder barrel and heads. Any heat removed also results in a slight reduction in the compression brake kilowatt.

Unless otherwise specified, the following standards, codes, and specifications to the extent specified herein, form the Company's minimum requirement for the process design of a complete closed cooling water system for reciprocating compressor cylinder jackets, it's engine cylinder jackets, it's lubricating oil cooling circuit and the compressor packing box cooling in part or in integral.

Throughout this section, references are mainly made to the API Standard 618 and API Specification 11P along with other internationally acceptable codes, standards and engineering practices and many important applied design book and resources.

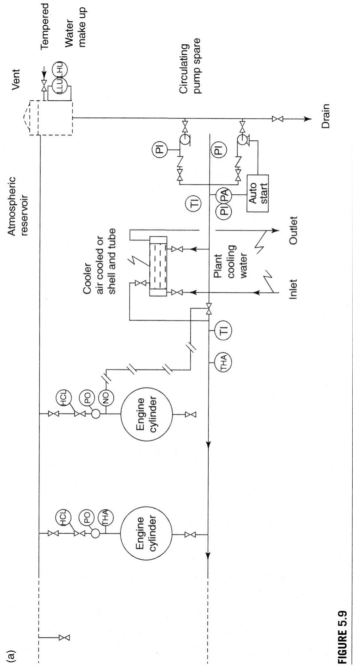

**FIGURE 5.9**

(a) Typical closed cooling water system for multiple engine installations,

(b)

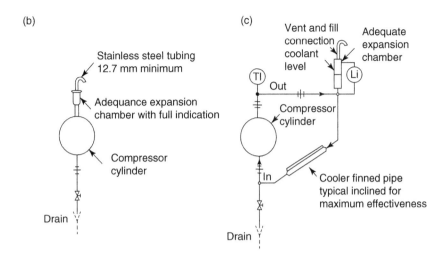

Stainless steel tubing
12.7 mm minimum

Adequance expansion
chamber with full indication

Compressor
cylinder

Drain

(c)

Vent and fill
connection
coolant
level

Adequate
expansion
chamber

TI

Out

Li

Compressor
cylinder

In

Cooler finned pipe
typical inclined for
maximum effectiveness

Drain

When jacket water temperature is
to be controlled by steam spraying
the following precautions should
be observed:

A silent (water-hammer-cushion type)
steam sprayer should be placed in the
water inlet line to the jacket system.

The water flow rate must remain
constant in accordance with the
manufacturer's requirements.

The steam flow into the water should
be regulated automatically to maintain
the water jacket temperatures.

(d)

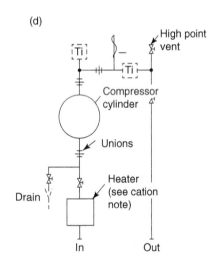

High point
vent

Ti

Ti

Compressor
cylinder

Unions

Drain

Heater
(see cation
note)

In          Out

**FIGURE 5.9 (cont.)**

(b) Static (standpipe) system, (c) thermosyphon system, (d) forced liquid coolant system typical cooling provisions for reciprocating compressor cylinder. Notes: The console shown in the plan is typical; more or less equipment may be furnished. Heaters used to preheat the cylinder cooling water (if needed to meet the requirements) may be electric, hot water, or steam. They must be sized to take into account heat losses of surface areas of the cylinder, pipe, and fittings. Good judgment must be exercised so that heaters will not be undersized.

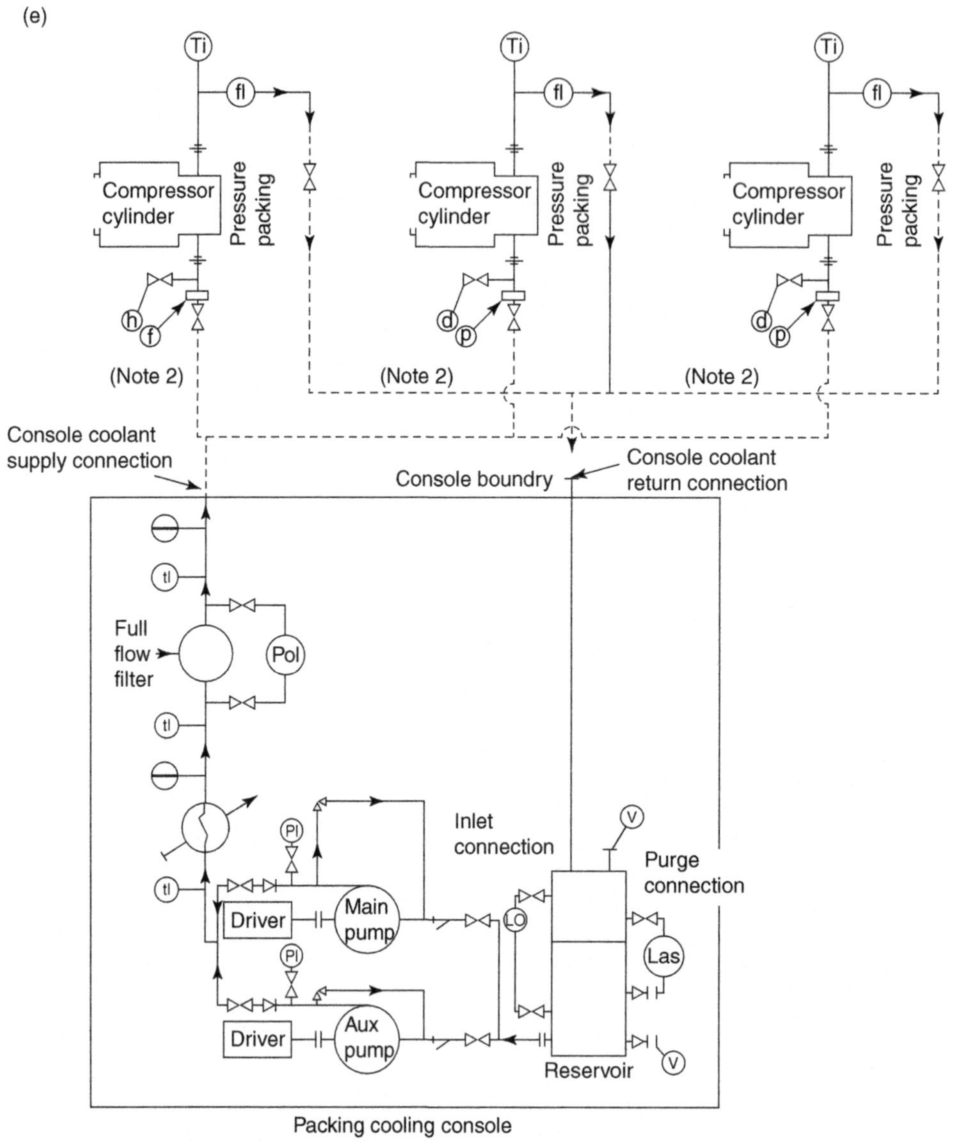

**FIGURE 5.9** *(cont.)*

(e) Typical self-contained cooling system for piston rod pressure packing. Notes: The system shown is typical; more or less equipment may be furnished. If a packing cooling console is not supplied, individual filters are required.

### 5.8.2.1 Methods

The heat traveled to the cylinder wall can be carried off either by direct air-cooling or by indirect liquid cooling process. Most cylinders have water jackets to remove the heat and maintain required cylinder and/or liner temperature. Usually small kilowatt power units may use air-cooling system.

According to API Standard 618, three following methods of cooling may be used, depending on the extended period of time where, cylinders will or will not be required to operate fully unloaded.

1. Thermosyphon coolant systems may be used where cylinders will not be required to operate fully unloaded for extended period of time and either (1) expected maximum discharge temperature is between 88°C and 99°C or (2) the rise in adiabatic gas temperature is less than 66°C (Fig. 5.9c).
2. Static field coolant system may be used when the cylinders will not be required to operate fully unloaded for extended period of time. The expected maximum discharge temperature is less than 88°C and the rise in adiabatic gas temperature (difference between suction and discharge temperatures based on the isentropic compression) is less than 66°C (Fig. 5.9b).
3. Forced liquid coolant system should be provided, where cylinders will operate fully unloaded for extended period of time and either (1) the expected maximum discharge temperature is above 99°C or (2) the rise in adiabatic gas temperature is 66°C or greater (Fig. 5.9d).

Unless otherwise specified, forced closed cooling water system should be used for taking away the heat which has traveled to the cylinder wall. The water should be pumped through the secondary cooler and then back to cylinder jacket for reuse.

The Vendor is required to evaluate the Company's proposed standard coolant system against his own standard coolant system or any other standard coolant system and should recommend the use of the most efficient, effective, and techno-commercially feasible other coolant system together with strong convincing proves. However, the employment of any other coolant system will solely be upon the Company's written approval.

### 5.8.2.2 Cylinder jackets cooling

Unless otherwise specified, the following requirements should be considered when the closed cooling water system is used only for cylinder jacket cooling:

The cylinder jackets when designed, all protective measures must be taken to prevent the process gas flow into the cooling water circuit.

A liberal supply of cooling water for cylinder jacket and cylinder head must be maintained.

The cylinder cooling system provided should be designed to prevent gas condensation in the cylinder, that may dilute or remove lubricant or may cause knocking.

The use of untreated or scale depositing water that will cause fouling and plugging of the water passage, reducing cooling efficiency should strongly be avoided.

### 5.8.2.3 Calculation of heat rejected to circulating cooling water

The Vendor/Manufacturer will furnish complete design data on quantity of compression heat to be removed from cylinder jacket and the head in joules per kilowatt per hour.

The Vendor/Manufacturer if deemed necessary may furnish an integral closed cooling water system for compressor cylinder jackets, engine cylinder jackets, lubricating oil circuit and compressor packing boxes, he should provide separate design data on the quantity of heat rejected to the cooling system from each section separately and as a whole along with quantity of water circulating and pressure drops.

A closed cooling water system for packaged reciprocating compressor should be furnished either in separate cylinder jacket cooling or integral with engine cylinder, lubricating oil, and cooling of compressor packing boxes, within the temperature limit recommended by the manufacturer for the specified compression services.

### 5.8.2.4 Integral cooling system

Not withstanding the requirements set forth, the following requirement should be considered when the closed cooling water system is used for an integral compressor and engine cylinder jackets, lubricating oil, and packing box cooling:

1. The cooling circuit should include engine lubricating oil, engine cylinder jacket, compressor cylinder jacket, and packing boxes.
2. Elevated deaerating type reservoir with gage glass, vent line, cooling water level switches, overflow, filling connections, and drains.
3. Cooling water temperature control should be provided.
4. Plugged manual drain connection(s) for complete draining of the system.

### 5.8.2.5 Lubricating oil-cooling system

Since cooling of the lubricating oil should be considered as an integral part of the closed cooling water system, the following recommendations apply:

1. Since liquid coolant is used, the design should minimize the chance of the lube oil being contaminated.
2. The coolant pressure should be less than the lube oil pressure at all times.
3. Adequate cooling water circulating rate for removal of total heat rejection to lube oil, should be maintained.

### 5.8.2.6 Packing box cooling system

If a separate closed cooling circuit is specified for the piston rod pressure packing the criteria should be followed.

When packing is cooled by forced circulation, the Vendor should supply a suitable filter of appropriate mesh rating.

Where cooling of packing is required, the Vendor should be responsible for determining and informing the Company on the minimum requirements such as flow, pressure, pressure drop, and temperature as well as filtration and corrosion protection criteria.

### 5.8.2.7 Design features

Unless otherwise specified, the provisions set forth when applicable, should be considered as the design feature for a closed cooling water system for reciprocating compressors.

The cooling water supply to each cylinder jacket should be at temperature of at least 6°C above gas inlet temperature as per API Standard 618.

The quantity of cooling water circulation by each pump should be regulated to maintain a rise in cooling water temperature across only of the individual cylinder and cylinder head between 6°C to 11°C (as per API Standard 618).

An oil detection device should be provided in the water supplied to pumps suction head.

In case of compressor cylinder cooling, the following should be observed:

1. Coolant inlet temperature less than 6°C greater than gas inlet temperature may cause gas constituent condensation.
2. Cooling water rate and velocity should be provided by the Vendor to prevent fouling of cylinder jacket system.
3. Cooling water exit temperature more than 17°C above gas inlet temperature may cause compressor capacity reduction. (As per API Specification 11P).

Installations for reciprocating compressor integral with engine cylinder, lubricating, and packing box cooling system should be as indicated in Fig. E.1 of Appendix "E". The system should be capable of providing the following:

1. Cooling water for reciprocating compressor cylinder jackets.
2. Cooling water for engine cylinder jackets.
3. Cooling water for lubricating oil coolers.
4. Cooling water for cooling of compressor packing boxes (or in the case of refrigeration services, water for warming of packing).

Typical installation for cylinder jackets and cylinder head cooling system of the reciprocating compressor should be as indicated in Fig. C.1 of Appendix "C".

Typical installation for packing box cooling system of the reciprocating compressor should be as per Fig. D.1 of Appendix "D".

### 5.8.2.8 Working pressure

The system should be designed for not less than 520 kPa working pressure and a hydraulic test pressure of 800 kPa.

### 5.8.3 COOLING WATER RESERVOIR

The cooling water reservoir should be located above the highest point of the closed cooling water system.

Working capacity of the reservoir should be at least equal to the normal capacity of the pump per 5 smin.

The reservoir should be furnished with appropriate level indicator, with sufficient length covering working range of the reservoir and normal expansion and contraction of the system.

Level control system should be provided for automatic control and maintaining of desired level and pumping suction head.

The reservoir should be furnished with level switches and alarm, vent/overflow and filling connection and drain.

The reservoir vent/overflow line size should have a diameter not less than ½ the diameter of the pump suction line.

The design of the suction line from the reservoir to the pump should not provide any air pocket.

Reservoir should be furnished with necessary chemical injection facilities to control the more corrosive nature of circulating water.

Continuous steam injection line should be provided for the required rate of steam flow to the reservoir's top position for blanketing and spilling out the air.

The Company should specify whether the installation is to be indoor or outdoor and the climatic conditions, including maximum and minimum temperature. The Vendor should take all necessary protective measures in design by proper winterizing of the reservoir and other auxiliaries in cooling water system.

## 5.8.4 COOLERS

Unless otherwise specified by the Company, shell and tube heat exchanger should be used for forced closed cooling water system on reciprocating compressors and internal combustion engine cylinder cooling.

The Vendor should advise, when a cooler other than shell and tube exchanger is preferred. However, the use of any cooler other than shell and tube should only be made upon the Company's written approval.

Heat exchanger's tube-bundle should be designed removable with clean and noncorrosive fluid flowing through the tube side.

Unless otherwise specified, process design of shell and tube exchanger should be in accordance with the following conditions:

| | |
|---|---|
| Velocity in exchanger tubes | 1.5–2.5 m/s |
| Maximum allowable working pressure | $\geq$690 kPa (ga) or 6.9 bar (ga) |
| Test pressure | 1.5 × MAWP |
| Maximum pressure drop | 100 kPa or (1 bar) |
| Maximum inlet temperature | 32°C |
| Maximum outlet temperature | 49°C |
| Maximum temperature rise | 17°C |
| Minimum temperature rise | 11°C |
| Shell corrosion allowance | 3.2 mm |
| Fouling factor on water side | 0.35 m². K/kW |

**Note:**

The Vendor should notify experts if the criteria for minimum temperature rise and velocity in exchanger tube result in a conflict. The criterion for velocity in exchanger tubes is intended to minimize waterside fouling; the criterion for minimum temperature rise is intended to minimize the use of cooling water. The Company will approve the final selection. (Mod. To API Std. 618,).

The following recommendations should be applied for the coolers used in closed cooling water system in general and for the lubricating oil cooler in particular:

The heat exchange surface should be located and arranged so that it can be removed for maintenance or replacement.

Provide means for draining both sides of the cooler during shutdown.

- Vent connections should be provided on the cooler to permit air removal.
- When dual coolers are used, the three-way change over valve should be designed so that oil flow will not be interrupted when transferring from one cooler to the other.

- Since liquid coolant is the main coolant to be used, the design should minimize the chance of lube oil being contaminated.
- Separate seals or gaskets should be provided for the coolant and the lube oil sealing. The space between the seals should be open to the ambient.

### 5.8.5 PIPING AND APPURTENANCES

Unless otherwise specified by the Company, the Vendor should supply a closed cooling water piping system for all equipment mounted on the compressor package. The piping should be arranged to provide single flanged inlet and outlet connections at the edge of the skid. Necessary valves and bypasses should be provided for temperature control.

The piping of the cooling system should be prepiped, factory skid mounted and complete with various pressure and temperature indicators, alarm, and other specific instrumentation required.

The inlet water connections should be located at the lowest point of cylinder, so that water can easily be drained from the cylinder when compressor is shut down.

The discharge connection should be at the highest point to ensure complete filling of water jackets with no air pockets.

The water piping should be provided with a valve controlling the flow of water.

If the Company does not specify the extent of closed cooling water piping, the Vendor should supply piping with single inlet and a single outlet connection on each cylinder requiring cooling.

Coolant piping should be arranged so that air cannot be trapped. Where air trap cannot be avoided, vent equipment should be provided. Low points should have drains.

Both, cooling water inlet line and cooling water outlet to each compressor cylinder should be provided with a gate valve. A globe valve with union should be provided on the main outlet line from each cylinder. A sight flow and temperature indicator should be installed in the outlet line from each cylinder.

Note: Where more than one cooling water inlet and outlet point exist on a cylinder, one sight flow indicator and regulating globe valve should be provided for each outlet point on each cylinder.

For the packaged reciprocating compressor, Vendor should supply all necessary piping, valves, and fittings for all instruments and instrument panels.

The cylinder cooling system piping should be equipped with vents and low point drains. Manual block valves to permit working on the compressor unit or auxiliary equipment without draining the engine cooler should be furnished.

Internals of piping and appurtenances should be accessible through openings or by dismantling for complete visual inspection and cleaning.

External drain and vent piping should be of Schedule 80 carbon steel and of not less than DN 25 (Diameter Nominal 25 mm) size. However, vent connection in the packing case and inter connecting tubing, should be of 300 series stainless steel and at least 6.35 mm outside diameter with a minimum wall thickness of 1.24 mm.

### 5.8.6 CIRCULATING PUMPS

The equipment and auxiliaries, should be designed for a minimum service life of 20 years and at least 3 years of uninterrupted operation. This should be considered as a design criterion.

The equipment rated operating point should be specified on the data sheets along with any other anticipated operating conditions.

The Vendor should specify on the data sheets the NPSHR when pump is operated on water at the rated capacity and rated speed, when water temperature is less than 66°C.

Pumps should have mechanical seals with flushing line to maintain a seal chamber pressure greater than the maximum suction pressure and, to ensure that the temperature and pressure in the seal chamber prevent vaporization while providing continuous flow through the seal chamber.

Pumps should be provided with constant-speed motor drivers or steam turbine drive (if required by the Company). The motor driver of the pumps (main and spare) should be on the secondary selective electric system (emergency power).

Each pump should be designed for the capacity required to maintain the complete cooling requirement of the system.

Each pump should be operative as the spare of the other and should automatically be started upon loss of pressure in the discharge of the main pump.

Fig. 5.9 shows a typical self-contained cooling system for piston rod pressure packing.

Fig. 5.10 shows a typical closed cooling water system for reciprocating compressors integral with engine, lubricating oil and compressor packing box cooling installations.

## 5.9 DESIGN REQUIREMENTS AND FEATURES FOR INTERCOOLING AND AFTER-COOLING FACILITIES

Vendor should supply water-cooled intercoolers and after coolers for skid mounted reciprocating compressors.

1. Unless otherwise specified, the process design of the intercoolers and after coolers should be based on "Process Design of Compressors" and the Appendix "C" thereof.
2. Not withstanding, shell and tube heat exchangers should be used for intercooling and aftercooling and should be in conformity with API 619, and API 680, on the following:
   a. Water-cooled intercoolers and after coolers for nonflammable, nontoxic services (air, inert gas, and so forth) should be designed and constructed in accordance with the ASME Boiler and Pressure Vessel Code. For flammable or toxic gas services, TEMA Class R heat exchangers should be furnished and should be in accordance with API Standard 660 and Section VIII, Division 1 of the ASME Code.
   b. Water should be on the tube side of the heat exchanger.
   c. Intercoolers should be mounted separately or on the machine, as specified by the Company.
   d. Relief valves should be provided on the process side of both intercoolers and aftercoolers.
   e. Rupture discs on the shell side should be furnished only when specified by the Company.

Intercoolers and aftercoolers should be provided by the Vendor with facilities to separate, collect, and discharge condensate through a continuous drainer. Condensate collection pots should be per the ASME Code, Section VIII or other pressure vessel code specified by the Company.

Air-cooled heat exchanger should only be used for skid-mounted compressor's intercoolers and aftercoolers upon the Company's requirements and approval.

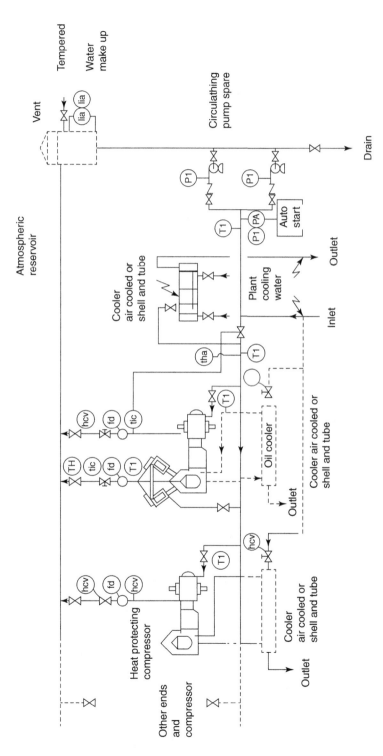

**FIGURE 5.10 Typical Closed Cooling Water System for Reciprocating Compressors Integral with Engine, Lubricating Oil, and Compressor Packing Box Cooling Installations**

When air-coolers are specified by the Company, they should conform either to API Standard 661, or should be of Vendor's standard. The Vendor should inform the Company on the advantages of Vendors's standard and should acquire Company's written approval.

Unless otherwise specified, air-cooled heat exchangers used for intercoolers and aftercoolers should have automatic temperature control. This control may be accomplished by louvers, variable pitch fans, bypass valve or by any combination thereof. The proposed control systems should be approved by the Company.

Caution should be exercised because of the susceptibility of heat exchangers and their supporting structures to pulsation-induced vibration.

Thermometer wells should be located at the inlet and outlet of the intercoolers and aftercooler and temperature reading should be taken at these points.

# HOT OIL AND TEMPERED WATER CIRCUITS

## 6.1 INTRODUCTION

Thermal liquids are used for process heating and cooling in the form of liquid, vapor, or a combination of both. In addition to steam, thermal liquid includes: hot and tempered water, mercury, Na, K, molten salt mixtures, hot oils, and many others, each of which can be used for a specified field of application and can operate in different temperature ranges.

This chapter is intended to cover the minimum requirements and recommendations deemed necessary in process design of hot oil and tempered water systems.

### 6.1.1 HOT OIL HEATER APPLICATION

These heaters furnish a heating bath 300°C or higher, which is hot enough for process applications, such as dry desiccant or hydrocarbon recovery regeneration gas. Another less severe application is heavier hydrocarbon vaporization prior to injection into a gas pipeline to raise the heating value.

Manufactured heat transfer oils are blended to about a 90–95°C operating range. For example, Fig. 6.2, gives typical heat transfer properties for a 150–300°C polyphenyl ether.

### 6.1.2 PROCESS DESIGN OF HOT OIL SYSTEM

A simplified schematic of major components of a hot oil system is given in Fig. 6.1. The heat transfer medium is pumped through a fired heater to the heat exchanger, and returns to the pump suction surge tank. In some cases a fired heater may be replaced by a waste heat source, such as the exhaust stack of a gas turbine.

While the system is ordered and designed as a packaged system, all necessary equipment such as ladders, platforms, guards for moving parts, etc., should be supplied as part of the package.

Indoor equipment should be suitably protected against damage by infiltration of moisture and dust during plant operation, shutdown, wash down, and the use of fire protection equipment.

Outdoor equipment should be similarly protected, and in addition, it should be suitable for continuous operation when exposed to rain, snow or ice, high wind, humidity, dust, temperature extremes, and other severe weather conditions.

The system should be laid out such that it makes all equipment readily accessible for cleaning, removal of burners, replacement of filters, controls and other working parts, for adjustment, and lubrication of parts requiring such attention. For similar reasons, the heater front and rear doors should be hinged or deviated.

Maintenance tools specially designed for the equipment should be furnished with the system.

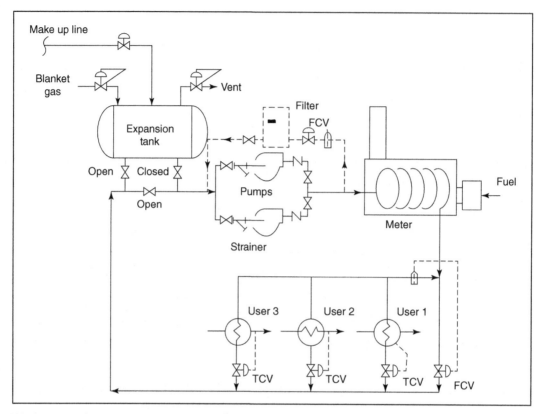

**FIGURE 6.1 Typical Hot Oil System**

Spare parts must be readily available. If a stock of parts is not maintained by the manufacturer, critical spare items should be furnished with the system.

### 6.1.3 ADVANTAGES AND DISADVANTAGES OF HOT OILS

#### 6.1.3.1 Advantages

The advantages of hot oils are:

- low vapor pressure at ambient temperature
- always liquid and easy to handle
- blended for a specific temperature range
- higher specific heat than normally occurring hydrocarbons

#### 6.1.3.2 Disadvantages

The disadvantages of hot oils include:

- Escaping vapors are environmentally undesirable.
- When overheated, the oils will oxidize and coke on the fire tube. Also, they can be ignited.

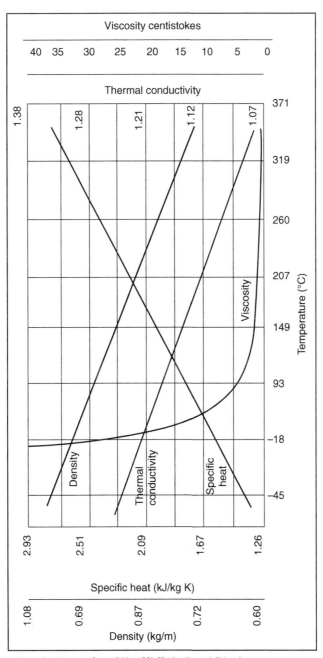

**FIGURE 6.2 Typical Heat Transfer Properties of Hot Oil (Polyphenyl Ether)**

- Ethers if used, are expensive.
- Ethers are hydroscopic and must be kept dry.

## 6.1.4 **DESIGN**

The following features and criteria should be considered in process design of each component of the hot oil system.

### 6.1.4.1 *Heater design*

Process design of the heater is critical for satisfactory operation. The heat transfer fluid must have sufficient velocity, generally 1.2–3 m/s, to avoid excessive film temperatures on the heater tubes.

Design and capacity of the heater should be limited so that the maximum film temperature does not exceed the maximum recommended operating temperature of the fluid.

Hot spot occurrence should be avoided, since it can lead to tube failure and fluid degradation.

The heater should be rated for the specified output. Multiple identical units may also be employed for the designed total heat load, but care should be taken in system design to ensure adequate and proportional flow through the heaters.

The preferred thermal efficiency of the heater is supposed to be 80%, based on LHV of fuel. The contractor should specify the expected and guaranteed values for the thermal efficiency and the basis for their estimation.

Based on total outside surface area of the fire-tube(s) and the return flue(s), the average heat flux should not exceed 17.35 kJ/sm$^2$. The flame characteristics and combustion chamber design should ensure that the maximum heat flux at any point is limited to 23.66 kJ/s m$^2$.

Heating medium (hot oil) should clearly be specified and its discharge temperature from the heater should be limited to a specified value in the data sheet.

Under normal operating conditions, the rise in heating medium temperature across the heater should not exceed the allowable DT specified in the data sheet.

The heater should be designed to give an efficient heater operation over the complete operating load range.

Each heater (in case the use of multi-heater units) should have a self-supporting stack designed to carry the total exhaust under the maximum firing conditions.

### 6.1.4.2 *Firing system design*

The hot oil heater should be designed for a continuous and reliable operation.

The burner(s) should be designed for a minimum of 120% of normal full load firing and be suitable for firing the specified fuels (oil, gas, or both) without undue maintenance or adjustment.

In case of a forced draft type heater, the burner design should incorporate air/fuel ratio system(s) to ensure complete combustion with minimum amount of excess air. The air/fuel ratio system should be effective throughout the burner firing range ie, from low to high fire positions.

The burner nozzles and other parts exposed to the radiant heat of the combustion chamber should be made from heat resisting alloy steel.

The burner fuel and air openings should be arranged to provide suitable velocities for complete mixing, resulting in efficient combustion of the fuel.

Each burner should have observation ports to permit sighting and inspection of the flame.

Suitable igniter(s) should be provided for firing fuel oil or gas, and should be of adequate output to permit safe ignition of the fuel.

### 6.1.4.3 Combustion air fan design
In the case of a forced-draft type heater, the following should be considered:

- The fan should be designed for maximum ambient temperature.
- The fan(s) performance should be stable over the complete firing range ie, maximum firing down to shut off.
- Inlet screens should be provided at the fan(s) inlet.
- Combustion air fan(s) should be sized to handle a minimum of 120% of the normal full quantity of combustion air.
- Outlet ducts of air fan (s) should have some equipment like damper(s) for adjustment of the amount of intake air to the heater.

### 6.1.4.4 Heater control and instrumentation
The heater package should be provided with the following control and instrumentation as a minimum:

1. fuel gas/oil control system complete with:
   a. pressure gages in important locations
   b. pressure regulators
   c. pressure relief valves
   d. flow control valves(s) on heater inlet line
   e. strainers
   f. isolating valves
2. fuel gas/oil emergency shut-off valve
3. heater TIC (modulating type) for hot oil temperature control
4. heater manual control for fire regulation
5. temperature switches should be supplied for the following alarm and shut-down functions:
   a. hot oil high temperature alarm
   b. hot oil very high temperature shut-down
   c. hot oil low temperature alarm
   d. stack high temperature alarm
   e. fuel oil low temperature alarm
6. Pressure switches should be supplied for the following alarm and shutdown functions:
   a. hot oil high pressure alarm
   b. hot oil low pressure alarm
   c. fuel gas/oil high supply pressure alarm
   d. fuel gas/oil high supply pressure shut-down
   e. fuel gas/oil low supply pressure alarm
   f. fuel gas very low supply pressure shutdown
7. Flow switches installed on the flow transmitter output should be supplied for the following functions:
   a. turning the unit down to low flame position in a low flow condition
   b. shutting the unit down (including circulation pumps) in very low flow condition

8. Additionally, the following instrumentation should be provided on the hot oil heater:
   a. pressure gage complete with isolation and bleed valve
   b. temperature indicators complete with thermowells for hot oil inlet and outlet streams
   c. stack exhaust temperature indicator complete with thermowell
   d. ASME rated relief valve(s), factory set, and sealed and located suitably. Relief valves should have stainless steel trim
9. Automatic start-up/shut-down sequence control is normally not recommended, but if specified by the company should consist of:
   a. pre-ignition purge of the combustion chamber
   b. ignition
   c. pilot proving
   d. firing rate modulation between low flame position and the maximum output
   e. post purge after shutdown
10. The burner management system, if specified by the company, should be housed in a locally mounted panel, suitable for the area classification in which it is installed.
11. The burner management system should incorporate a remote shutdown facility so that the main burner(s) and pilot(s) can be extinguished by a push-button located in central control room.
12. All other instrument, and controlling systems as per vendor specification, such as flame failure detector, pilot flame monitoring etc. should also be considered.

### 6.1.4.5 Hot oil surge tank design

A surge tank should be provided, suitably sized to handle expansion of inventory in the whole system, and should be designed as a pressure vessel.

The surge tank should be arranged on the pump suction side and should be blanketed with fuel gas or inert gas.

The surge tank should be located inside the heater building.

The surge tank should be provided complete with the following:

1. level gage(s), spanning the entire operating range
2. pressure gage
3. pressure relief valve
4. blanket gas pressure make-up regulator
5. pressure regulator to vent tank over pressure due to expansion or filling
6. make-up connection complete with isolation valve and non-return valve
7. level alarm high and low

### 6.1.4.6 Circulating pump design

At least two centrifugal pumps (one operating and one spare) with 10% over design capacity at the design head should be provided. However, the following requirement should also be considered in the design.

A strainer should be provided at the suction of each pump.

Valving around each pump should include the following:

1. suction isolation gate valve of the same size as the suction line
2. non-return valve in the discharge line

**3.** isolation gate valve in the discharge line
**4.** casing vent and drain valves

In the design of pump suction and discharge lines, the differential thermal expansion of the lines due to temperature variation should be taken into account.

### 6.1.4.7  Hot oil filter

A hot oil filter should be provided to handle a slipstream equal to 10% of the design flow rate.

The filter should be of the disposable cartridge type, capable of removing all solid particles above 5 μm.

The hot oil filter should be provided complete with the following features:

**1.** DP gage.
**2.** isolation valves at inlet and outlet
**3.** restriction orifice, sized for the required flow rate
**4.** vent and drain valves
**5.** relief valve
**6.** filter by-pass line

### 6.1.4.8  Carbon filter

A bulk-pack type carbon absorber that is consistent with the heating medium, should be provided downstream of the heating medium filter to remove any products of degradation.

The design flow rate of the carbon filter should be the same as the flow rate of the hot oil filter.

The carbon filter should be provided complete with the following features:

**1.** DP gage.
**2.** isolation valves at inlet and outlet
**3.** vent and drain valves
**4.** relief valve
**5.** filter by-pass line

## 6.1.5  PROCESS DESIGN OF TEMPERED WATER SYSTEM

Using tempered water as a cooling medium for solutions that would freeze or crystallize at usual cooling water temperatures is common practice.

A special tempered water circulating system should be designed to minimize the chance of fouling by deposition of these types of materials on heat exchanger surfaces.

Condensate or treated process water should be used as circulating tempered water and the water should be treated efficiently for corrosion inhibition according to material specification and design procedures. Provisions for corrosion inhibitor facilities should be made. A tempered water system typically consists of a surge drum, circulating pumps, air cooler and/or shell-tube heat exchanger and associated piping and measuring devices.

Single or multiple user(s) may be incorporated in a single tempered water system. All components of the system should accordingly be designed to maintain required cooling load capacity when all or parts of the user(s) are in operation.

The water return from the various users should be cooled to the specified temperature depending on the local climatic conditions.

All applicable parts of general requirements set forth in the standard should be considered for this part unless it is contrary to the vendor specification.

### 6.1.5.1 General layout and operational facilities

For layout and operation of the system, the vendor should take into account an adequate space being allowed for operational access, cleaning, and maintenance.

The arrangement of installations should be such that, instruments and indicators can readily be seen from the appropriate working position. Valves and controls should be closely arranged and easily accessible.

In case the system is admitted for indoor installation, access facilities should be arranged such that major items of the system can be brought in and taken out or removed as necessary.

### 6.1.5.2 Process design requirements

Design of the system and its associated controls should take into account the following:

1. The nature of the application.
2. The type of installation ie, indoor or outdoor installation.
3. Cooling load patterns.
4. Tempered water supply requirements.
5. Economic factor and minimizing the use of primary energy.

### 6.1.5.3 Surge drum

Atmospheric surge tanks should be designed as per API-650 with a steam blanketing system for oxygen removal.

Make up water to surge drum will be taken from the condensate system, moderately heated to the required temperature by hot condensate circulation and chemical treatment should be practiced to minimize corrosion.

The drum should be blanketed with steam. Provisions should be considered for automatic controlling of the water level and adding make-up water as required.

The pressurized-type surge drum operating above 7.0 kPa should have a minimum design pressure of 110 kPa (abs). The drum should be designed for full vacuum.

When considering the design, it needs to be taken into account that all applicable loads, including wind load, earthquake, and hydrostatic testing load act simultaneously.

All outline drawings should be furnished and should contain the data indicated in Table 6.1. Location of the drum marking or nameplate should be indicated on this drawing.

A manufacturer's data report should be furnished and should contain the same information as required by form U-1 of ASME Code, Section VIII, Division 1.

Provisions for entering, cleaning, venting, and draining of the vessel should be considered on the basis of basic practice as specified by the vendor/manufacturer.

The surge drum should be furnished with the following auxiliaries:

1. pressure gage
2. temperature gage
3. level gage
4. steam blanketing regulator
5. all inlet, outlet piping nozzles

**Table 6.1  Data in Drawing**

| Description | Unit | Data |
| --- | --- | --- |
| Design pressure | kPa | — |
| Design temperature | °C | — |
| Operating pressure | kPa | — |
| Operating temperature | °C | — |
| Maximum allowable stress at design temperature | kPa | — |
| Hydrostatic test pressure at upper-most part of drum | kPa (ga) | — |
| Hydrostatic test temperature | °C | — |

### 6.1.5.4 Piping design

All applicable portions should be considered in pipe sizing and design of the tempered water system. It should apply to auxiliary piping, connecting the equipment of the system and the piping between the system and the consuming units. However, the following considerations should be taken into account in order to accomplish the whole requirements of piping process design.

1. In a piping layout, the location of operating and control points such as valves, flanges, instruments, vents, and drains should enable operation of the system with minimum difficulties.
2. The piping system should be laid out to allow easy repair or replacement of any portion of the system.
3. Basic design data for each line should be given in the line designation table as per the company's project specification.
4. The actual minimum corrosion allowance should be listed for each line in the line designation table.
5. Flanges or other removable connections should be provided throughout the piping system to permit complete removal of the piping.
6. A gate valve should be installed in each instrument take-off connection except thermowells and should be located close to the pipe.
7. All piping connections to equipment should be suitable for the equipment design and the hydraulic test pressure.

### 6.1.5.5 Circulating/recirculating centrifugal pumps

The circulating and recirculating pumps should be centrifugal with 100% spare and should be designed in accordance with the following requirements:

1. Unless otherwise specified, the pumps and auxiliaries should be suitable for unsheltered outdoor installation in the climatic zone specified.
2. Flanged suction and discharge nozzles should be integral to the casing.
3. Gate valves should be used for vents and drains.
4. All required vents should be valved.
5. Nameplate data should be in SI units.

6. Vendor's piping should terminate with a flanged or threaded connection, of a line rating at least equal to the design pressure and design temperature rating of the equipment.
7. Vendor should specify the spare parts and should include his proposed method of protection from corrosion during shipment and subsequent storage.
8. Vendor's proposal should state the minimum flow rate recommended for sustained operation on the specified tempered water.
9. Isolation gate valves should be used at the suction and discharge line with the same respective line size.
10. Non-return valves should be installed in discharge lines.
11. Globe valve should be sized for blanketing line.

The pumps should be of proven modern design and should have operating characteristics as specified in the pump data sheet.

The arrangement of equipment, including piping and auxiliaries should provide adequate clearance area and safe access for operation and maintenance.

The equipment and component parts should be warranted against defective materials, design, and workmanship for a specified guaranteed period.

### 6.1.5.6 Other requirements

The whole system should be arranged and sized so that the design-cooling load can be met by an appropriate flow of tempered water within the applicable system temperature limits.

Circulating piping should be thermally insulated and traced where the local climatic condition implies.

Isolating valves are normally fully open or fully shut and should be provided to facilitate isolation of individual items of equipment.

An appropriate and compatible automatic control should be arranged for the system elements.

# HEAT TRACING AND WINTERIZING

This chapter covers the requirements for protection of process and utilities and all associated equipment, flow lines, and instruments against the temperature which would cause congealing or freezing of contents, interfere with operation or cause damage to equipment or pipe lines and for heat conservation requirement as would be determined by process conditions. The heat conservation system should be designed for continuous operations while, winterizing should be for seasonal operation. The two systems should be separate from each other.

## 7.1 APPLICATION AND METHODS

To avoid operating difficulties in process and utility units in colder climates and the hazard of freezing which may cause damage to equipment or blockage of lines, different methods, depending on the extent of protection required, will be used. The extent of protection may vary considerably between one location and another. However, the requirements for other ambient temperature conditions are also covered herein.

## 7.2 METHODS OF WINTERIZATION

The following methods of winterization are required to be used to provide adequate protection and this will be determined by a combination of climatic and process conditions depending on seasonal operation and heat conservation:

1. Heating. Winterizing with heating should not be used where other methods can be used. Any of the following heating systems may be used:
   a. internal heat tracing
   b. external heat tracing
   c. jacketing
   d. electrical tracing
   e. routing along and/or insulating together with a hot line.

   The heating medium for nonelectric tracing and jacketing can be steam, hot oil, and hot water.
2. Insulation. Thermal insulation is the reduction of heat transfer (the transfer of thermal energy between objects of differing temperature) between objects in thermal contact or in the range of radiative influence. Thermal insulation can be achieved with specially engineered methods or processes, as well as with suitable object shapes and materials.

Heat flow is an inevitable consequence of contact between objects of differing temperature. Thermal insulation provides a region of insulation in which thermal conduction is reduced or thermal radiation is reflected rather than absorbed by the lower-temperature body.

The insulating capability of a material is measured with thermal conductivity ($k$). Low thermal conductivity is equivalent to high insulating capability ($R$-value). In thermal engineering, other important properties of insulating materials are product density ($\rho$) and specific heat capacity ($c$).

The insulation alone may be used to prevent solidification or an increase in viscosity where the liquid in the equipment has sufficient sensible heat for normal operating flow rates and will be of value only for a short time exposure; for long-term exposure, however, it is only successful when heat is continuously added by the process. Although heat input at normal flow rates may be adequate, it may fall to an unsatisfactory level at low throughputs, or during startup or shutdown, or when a line is blocked off in error, and as a result, freezing of the system may take place.

3. Vent/drain on lines and equipment. Winterizing by draining requires particular attention which must be given to vents and drains on utility lines and equipment for eliminating low spots or dead ends in which water and other liquids can collect and freeze. During shutdown and any nonoperating period these lines may be completely drained.
4. Hot air circulation. Hot air circulation may be used in aerial exchangers, instrument and equipment housing, and flushing connections to displace viscous-material.
5. Bypasses around equipment to provide continuous flow. Bypass line connection in appropriate points as specified herein, and in accordance to project specification should be provided for use in maintaining flow.

## 7.3 CONDITIONS REQUIRING WINTERIZATION

Process, utilities, equipment, and pipe lines and other equipment/pipe lines should be winterized when any of the following conditions apply to the fluids contained:

1. Pour point or freezing point is above the lowest ambient temperature.
2. Undesirable phase separation, deposition of crystals, or hydrate formation will occur at any ambient temperature.
3. Ice or hydrate formation occurs due to pressure reduction of moisture-bearing gases.
4. Viscosity at any ambient temperature is so high that an inadequate flow rate is obtained with the pressure available for starting circulation.
5. Corrosive compounds form if condensation occurs.
6. Lines which are normally dry, eg, flare lines which may carry moisture during an operating upset may require some protection.

## 7.4 REQUIREMENTS

The requirements for protection should be based on the winterizing temperature specified in the project specification, and should consist of two parts:

1. Lines and equipment that appear on P&I diagrams. The extent and degree of protection should be specified by the project engineer and shown on the P&I diagrams, by *relevant* standard nomenclature.

Protection described by standard nomenclature on P&I diagrams may be inadequate for projects involving low winterizing temperatures in combination with unusual process fluid properties. On such projects consideration should be given to the use of high thermal conduction cement bonding of tracers, steam jacketing, electric heating, shelters, and other special designs.

2. Lines and equipment not shown on the P&I diagrams. Protection should be provided by the design contractor/consultant to the extent and in the manner provided herein.

Layout, design, and details that are to be followed by the design contractor in winterizing all equipment should be as specified herein.

Protective heating of piping and instruments should be indicated on the process engineering flow schemes and on the piping data sheets.

The extreme case of the lowest minimum temperature should not be selected, but in general, equipment should be designed for protection against the minimum temperature prevailing after rejection of the lowest 1% of the hourly temperature readings in the coldest month, or in 1% of the daily minimum temperatures for the year; the readings should as far as possible be based on the average of records obtained over a period of years and not those of a single year.

The amount of winterizing protection should be based on minimum atmospheric temperature as shown on the site data sheet. For heat conservation during operation, fluids with a pour point of 10°C and higher should be traced to maintain a temperature at least 22°C above their pour point. Molten sulfur lines should be maintained between 118°C (245°F) and 158°C (316°F).

### 7.4.1 PIPING

A list of all piping that requires tracing should be prepared for each unit in the project specification.

Sections of gas systems in which ice or condensate would otherwise be produced, due to atmospheric cooling or auto refrigeration, should be traced. Protection will also be required where there is hydrate formation at temperatures above 0°C.

Careful consideration should be given to the design and protection of lines which are dry during normal operations, but which may contain sufficient moisture to be troublesome during an operational upset, for example, flare lines, etc.

### 7.4.2 PROCESS PIPING

Compressor suction lines between the knockout drum and the compressor should be heat traced and insulated if the ambient temperature is below the dew point of the gas at compressor suction or when handling hydrocarbon gas components heavier than ethane.

Intermittently used process piping containing liquids such as tars or chemicals, which will congeal during nonflowing conditions, should be provided with valves for venting and draining, blowing out with air or flushing with light stock in preference to heat tracing.

Tank car and tank truck loading lines should be heat traced and insulated and provided with valved flushing or blow-out connections.

Blowing out piping with air should be confined to lines containing stocks of low volatility, which are well below their flash points. Piping to be blown discharges into tankage. Venting capacity should be provided to prevent pressurizing the tankage.

### 7.4.2.1 Piping for water services

When daily mean temperature is below −1°C, underground water systems (including sewers) should be installed at a minimum of 300 mm below the frost line. An above ground portion of water systems should be winterized by such means as heat tracing or draining component of the system after each use. All piping in saltwater service, if heat traced, should be cement lined. A typical pipe line external tracing detail is shown here in Appendix B.

Where branch single service lines rise from below ground, block valves should be provided in the risers just above the ground. The following arrangement will provide protection against freezing:

1. A bypass should be provided just under the block valves, from the supply back to the return for use in maintaining circulation. This bypass should be DN 20 (¾″) for lines DN 80 (3″) and smaller, DN 25 (1″) for lines DN 100 (4″) to DN 200 (8″) and DN 40 (1½″) for lines larger than DN 200 (8″). Bypasses should be covered with 25 mm of insulation.
2. A drain should be provided in the line at a minimum distance above the block valve, except that for 150 mm and larger size valves a drain should be provided in the valve body above the seat.
3. 25 mm of insulation should be provided around the piping, from the ground up to and including the block valves in the water risers.

Where a header for multiple services rises from below ground, protection should be provided in the same manner as overhead headers.

## 7.5  INSTRUMENTS

Winterization of instrumentation systems should be in accordance with Chap. 8Section 8.5 of API RP 550. However, when electronic instruments are heat traced the type of heat tracing should be to the instrument manufacturer's recommendation. Consideration should be given to the use of electrical heat tracing and also to thermostatic control to ensure the manufacturer's specified operating temperatures are not exceeded.

Proposals for winterization are to be discussed and agreed with the company.

Where practical, instruments should be installed in heated buildings to simplify protection requirements and facilitate maintenance.

Electronic instruments which may be damaged by freezing should either be installed in heated housing or located in buildings to maintain the temperature within the manufacturer's recommended temperature rating.

When installation of instruments in a heated building is impractical, protection is required for instruments on water, steam, hydrocarbons, or other liquid services which are subject to freezing or congealing. A safe temperature should be maintained for hydrocarbons with pour points −12°C and above.

Protection is also required on gas or air service where condensate may render instruments an inaccurate operation or make inoperative and on liquid services where moisture is likely to enter lead lines and instruments.

Precaution should be taken to prevent excessive heating of mechanical and electrical instrument components.

Enclosed analyzer cabinets should be heated.

Preferred practices are insulation, heat tracing, and heated instrument housing to maintain the manufacturer's recommended temperature rating.

Instrument piping should be winterized by sealing with an antifreeze solution where possible. Protective heating of lead lines should be installed in a manner which will prevent the liquid from overheating and boiling away.

Locally mounted pressure gages and instruments, and seals required for corrosion protection, should be winterized.

Unless otherwise specified by the manufacturer the following locally mounted instruments should not be housed in cabinets. They are winterized by sealing, heat tracing, or by a combination of both:

1. Alarm pressure switches.
2. Control pilots for control valves.
3. Displacer type level instruments.
4. Float-type alarm units.
5. Pressure gages.

Protection houses are used for local instruments where the process fluid is in the instrument body. Houses are also used for instruments that require regular access that are outdoors and are otherwise unprotected. Such instruments include controllers and recorders. All housings are of galvanized sheet metal construction with hinged doors and a steam or electric heating coil. Blank metal doors are furnished for blind transmitters. A lucite window is furnished in the door on houses for indicating transmitters and meters. The housings for analyzer transmitters are furnished as part of the instrument assembly.

### 7.5.1 FLOW INSTRUMENTS

Differential pressure instruments having factory-filled bellows or diaphragm assemblies are specified with a fill material that does not require winterizing. Care should be taken not to overheat diaphragms above their design temperature.

### 7.5.2 LEVEL INSTRUMENTS

1. Protective provisions for differential type level instruments conform to that described for differential type flow instruments.
2. External float instruments are heat-traced for the following services:
   a. Steam
   b. Water
   c. Caustic
   d. Viscous hydrocarbon with a pour-point −12°C and above
   e. Light hydrocarbon where hydrate formation is possible.

### 7.5.3 CONTROL VALVES

1. Control valves are not traced with the associated process piping, except that valves are steam traced on gas or vapor services with high pressure drops, where hydrates may be formed, or where freezing or congealing may occur.
2. When control valves that are used as direct connected regulators require winterizing, the pressure control line and valve diaphragm chamber containing the process fluid should be heat traced and

insulated. When the diaphragm chamber is sealed, the pressure control line should be heat traced and insulated from the point of seal to the process line connection.

3. In light hydrocarbon-vapor services, where hydrate formation or frosting due to low temperature is likely, only the control valves body should be heat-traced.

4. Steam atomizing control valves with diaphragm and in contact with heavy process fluid should be protected with a seal pot filled with glycerin.

## 7.5.4 TEMPERATURE INSTRUMENTS

Bulb type temperature instruments should be specified with fill material that does not require winterizing for the particular zone in which they are to be installed.

## 7.5.5 LEAD LINES AND INSTRUMENTS

1. Lead lines and instruments containing fluids subject to winterizing should be protected by seals, tracing, or heated housings.

2. Seals are a nonfreezing solution compatible with the process fluids, and piping and instrument materials. A 60% ethylene glycol and water solution is utilized for most hydrocarbon process fluids. (Refer to API RP- 550 for other sealing fluids).

3. Those liquids of seal pots that are in contact with process fluids are protected for the zone in which they are installed.

4. In areas where steam or electricity is not readily available for heat tracing, consideration should be given to use of instruments equipped with a mechanical diaphragm type seal at the process connection, in lieu of heat tracing.

5. Instruments with dry gas or dry air purging do not require protection for the lead line.

6. Instrument lines and gage glasses which are steam traced and contain liquids that boil at tracing steam temperature should be separated from the tracer lines by insulation of 25 mm.

## 7.5.6 ROTAMETERS

Rotameters are winterized in accordance with the requirements of the process line in which they are installed. They are housed when they are recording and when they are outdoors.

## 7.5.7 ANALYZING INSTRUMENTS

1. A sample system for analyzers which require a liquid stream should be protected in the same manner as the pipeline from which the sample is obtained, using caution to insure the sample is not damaged by overheating or vaporizing.

2. Gas samples to analyzers which contain anything condensable should be provided with heat tracing to prevent condensation.

3. Heated housing should be provided for the analyzer and have temperature control and sample conditioning systems. Each analyzer installation should be investigated for winterizing requirements.

## 7.5.8 VALVES

Where necessary, relief valves and adjoining piping should be suitably protected.

The vent line relief valves discharging to the atmosphere should have a suitable drain hole at the lowest point.

Flanged shutoff valves and check valves in vertical lines should have bodies trapped and valved above the disk or seat if the commodity will freeze or congeal during shutdown.

Water seals and traps should be steam traced.

Low points of flare lines should drain into vessels that are suitably protected against freezing.

---

## 7.6 EQUIPMENT

### 7.6.1 DRUM, VESSEL, STORAGE TANKS

A drum or vessel containing hydrocarbon and water which operate normally at 52°C or above, should be protected by insulating the nozzles, block valves, and drain piping in contact with water.

A drum or vessel containing hydrocarbon and water which operates normally below 52°C should be protected by steam tracing and insulating the nozzles, block valves, and drain piping in contact with water.

All other process vessels containing fluids that may congeal during dormant periods should be insulated and if necessary should be heat traced.

Bottoms of fuel gas drums and low points in above ground gas lines should be insulated and steam traced.

Tanks containing liquids difficult to pump or flow when cold should be equipped with heaters.

Steam coils in tanks should consist of a number of sections arranged in parallel flow, thus avoiding the total loss of tank heating in the event of a coil section leaking.

Roofed or open water tanks (except for potable water tanks) should have steam connections to heat and agitate water at intervals to prevent freezing.

Storage tanks should be equipped with freeze proof type water draw valves.

Consideration may have to be given in some extreme cases to insulating the roofs of some cone roof tanks to prevent internal corrosion due to condensation of sulfurous vapors.

### 7.6.2 EXCHANGERS

Heat exchangers and coolers containing liquids which may congeal or freeze at ambient temperature should have sufficient valved drain points to insure complete drainage upon shutdown.

Evaporators in the chlorine service should be housed in a heated, forced ventilated building if the ambient temperature may drop below 13°C.

### 7.6.3 PUMPS AND COMPRESSORS

Although compressors may be housed in enclosed or partly sheeted buildings, consideration should be given to the protection of any exposed parts such as water lines, lubricating and seal oil lines, air and oil filters, suction lines, and knock-out drums. Tracing of the oil sumps of compressors may be required to assist startup.

Pumps should have plugged drains on all water-cooled jackets and pedestals.

Compressors and auxiliaries enclosed in buildings should have winterizing protection when the system is shut down.

Heating should be provided for every reciprocating pumps lubricator.

Pumps and associated piping should be protected as required by the nature of fluids handled and duration of anticipated nonoperating periods. Circulating fluid from an active pump through a nonoperating pump is a preferred practice for pumps handling viscous fluids.

The pumps in intermittent services which cannot conveniently be drained whenever not in use, should be traced or jacketed or located in a heated enclosure if the liquid handled would otherwise freeze or become too viscose to pump.

Seal and flushing oil piping should be steam traced whenever cooling to the ambient temperature would diminish the quantity or pressure of seal oil at the point of consumption below the equipment manufacturer's recommended minimum.

Suitable provision should be made in the water systems of equipment which is on intermittent duty or immediate standby duty and which cannot therefore be drained to ensure that it will not freeze during idle periods. A particular case to be checked is that of machines that have thermostatic valves for controlling water flow which may cut off the water completely during idle periods.

Pumps handling fluids with a pour point above 10°C should meet the following heat conservation requirements: Pumps are not to be traced (except asphalt and high viscosity fluids pumps which should be jacketed) All piping including dead legs should be heat traced and insulated to maintain the fluid temperature at least 22°C above pour point. Back-flow circulation should be used through nonoperating pumps. Flushing oil should be used to clear the systems, and the pumps should be drained when taken out of service.

### 7.6.4 PUMP WINTERIZATION

Winterizing is required for services containing water and other fluids with pour points above minus 18°C but less than 10°C. Pumps are not to be heat traced:

1. Hydrocarbon pour point 0–10°C. Heat trace dead legs. Insulate but do not heat trace the lines to and from the pumps. Use backflow circulation to keep nonoperating pump lines above pour point. It is important to drain pumps when taken out of service during cold weather.
2. Aqueous fluids. Trace and insulate all lines to maintain 24°C if required by climatic or process conditions. Use back-flow circulation through nonoperating pumps. It is important to drain pumps when taken out of service during cold weather.

## 7.7  MISCELLANEOUS ITEMS
### 7.7.1  HYDRANTS, MONITORS, SPRAY, AND DELUGE SYSTEMS

Self-draining provisions should be incorporated for fire hydrants used on underground piping. Monitors and hose reels should be installed with self-draining or manually operated valves for aboveground piping and underground piping above the frost line.

Spray and deluge systems should be drained through manually operated valves located at the main operating valve.

### 7.7.2  EMERGENCY (SAFETY) SHOWERS AND EYEWASH

Self-drain yard hydrants should be used for safety shower and eyewash. For extreme climatic conditions provisions should be taken to maintain the water temperature tolerable for users.

Heat tracing for winterizing piping to safety showers and eyewash should be thermostatically controlled electric heating.

Steam tracing should not be used for the piping of safety showers.

### 7.7.3  DRAINAGE SEPARATORS, SUMPS, AND LINES

Separators and sumps should be provided with steam injection points or steam coils where necessary to keep the fluid in a pumpable condition.

Drain lines from equipment should be suitably sloped and traced where necessary.

Chemicals and supplies should be stored at a temperature above their freezing point unless other storage temperatures are specified.

## 7.8  DESIGN

### 7.8.1  INTERNAL STEAM TRACING

Internal steam tracing is limited to lines of DN 150 and larger, for long outside plot lines below 100°C in noncorrosive service. Internal tracing is acceptable only if steam leakage into the product can be tolerated.

For DN 25 inner pipe tracers, the decreased flow area and pressure drop in process line should be checked.

Tracers should not exceed 50 m length for 350 kPa (ga) or 3.5 bar (ga) steam, thus the maximum line length between two expansion loops can be 100 m.

### 7.8.2  EXTERNAL STEAM TRACING

The basic concepts and requirements for steam tracing systems of each project should be described in a project engineering specification. The company's requirements specified herein should adhere to such specification.

The terminology used for steam tracing systems is:

- tracing steam header, self-explanatory;
- distribution manifold (DM), station to serve tracers;
- leads, from DM to tracer;
- tracer, heating pipe along process line;
- tail, from tracer to collection manifold (CM);
- CM station to transmit condensate;
- tracing condensate header, self-explanatory.

The level of pressure to be supplied for steam tracing supply and condensate return systems should be specified for each project.

Steam traced lines should have an insulation, layer (spacer) between fluids flowing lines, tracer and fasteners to prevent high temperatures at contact points causing stress corrosion cracking of the pipe. Some products may be sensitive to the temperature of hot spots. The product properties and the heating limitations should be checked. For any of the following fluids are being handled, a protective layer or spacer should be placed between the tracers and piping or equipment:

1. Acid or caustic fluids.
2. Heat sensitive fluids.
3. Fluids having electrolytic properties where hot spots will accelerate the corrosion rate.

Additionally, spacers should be provided between tracers and piping, or equipment which is lined with glass, rubber, plastic, or other heat degradable materials. For spacer detail refer to Appendix D.

Piping in all sizes should be individually traced.

Each heat tracing circuit should have its own block valves at the supply and CM. These valves should be tagged with a metal tag designating the line or instrument served by a number or description and should indicate supply or return.

All heat tracing circuits should be manifolded wherever possible at take-offs from the main supply and return headers with block valves at the headers.

When approved by the company, external steam tracing may be replaced by other heating methods, ie, electrical heating, internal steam, or steam jacketed piping. Such alternates should be shown and identified in specific job requirements.

The type of tracing, the purpose of the tracing, and the size of tracer should be indicated on the flow diagrams using the following symbols:

1. Piping (list the following symbols after piping number and specification):

| | |
|---|---|
| ET | Electric traced and insulation |
| ETT | Electric traced with heat transfer cement and insulation |
| SJ | Steam jacketed pipe and insulation |
| ST | Steam traced and insulation |
| STS | Steam traced with spacers and insulation |
| STT | Steam traced with heat transfer cement and insulated |

2. Valves (list the following symbols next to the valve):

| | |
|---|---|
| TB | Trace body and insulate |
| TBB | Trace body and bonnet and insulate |

3. Equipment (list symbol the following equipment title):

| | |
|---|---|
| ST 25 mm | Insulation-steam traced |
| ET 25 mm | Insulation-electric traced |

4. Instruments (list the symbol next to the instrument number circle):

| | |
|---|---|
| ST | Steam traced and insulated |
| ET | Electric traced and insulated |
| WS | Winter seal |

5. WISI on P & IDS should have the tracer insulate from the line using 15 mm thick × 50 mm long × 25 mm wide blocks of insulating material at approximately 300 mm intervals. Blocks and tracer should be banded to the pipe and the entire assembly insulated with an outer layer. A typical cross section of WISI is given in Appendix D-IV.

Steam tracer lines should generally follow those listed below:

| Line Traced, DN (in.) | Traced, DN (in.) |
|---|---|
| 25–40 (1–1½) and smaller | 10 (3/8) tubing |
| 50, 80, and 100 (2, 3, and 4) | 15 (½) tubing or pipe |
| 150–300 (6–12) | 15 (½) or 20 (¾) pipe |
| 350 (14) and over | 20 (½) or 25 (1) pipes |

Contractors' methods of determining heat transfer requirements should be approved by the company. The following figures are minimum requirements as dictated by experience. However, the figures may be adjusted by the company for individual job specifications:

1. The lengths specified are the maximum permissible in all climates subject to freezing, and using 4.1 bar (ga) steam. Maximum lengths may be increased proportionally for higher pressure steam and should be decreased proportionally for lower pressure steam.

| Tracing Line (Tracer) | Maximum Length (m) |
|---|---|
| DN 15 (½ in.) pipe or tubing and less | 50 |
| Over 15 (½ in.) | 100 |

2. Maximum depth of pocket should be as below:

| Steam Pressure | | Pocket Depth (m) |
|---|---|---|
| kPa (ga) | bar (ga) m | |
| 280 | 2.8 | 3 |
| 410 | 4.1 | 4 |
| 520 | 5.2 | 6 |
| 690 | 6.9 | 8 |
| 1030 | 10.3 | 10 |

## 7.8.3 STEAM SUPPLY

Steam tracing headers should be supplied from the top of a main steam line, to avoid intake of condensate. The steam-tracing header should have a block valve as close as possible to the main header; so as to isolate the tracing system.

Block valves should be provided at the high points of the branch lines in a horizontal position near the headers. The section of line between the header and the block valve should be of minimum length (Fig. 7.1).

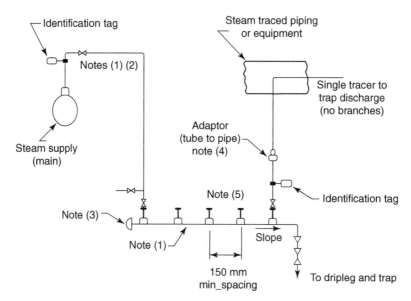

**FIGURE 7.1 Typical Arrangement of Piping and Tubing Components for Steam Supply and Distribution**

Notes: (1) All take off connections located at the top of headers; (2) block valves at the main connection for each distribution header; (3) preferential location of distribution header based on layout considerations to be; at accessible locations in elevated pipeways; at platforms; near grade; (4) tube to pipe adaptor normally located at start of equipment or piping to be traced. At a change of material (material spec. break), the carbon steel piping should be braced; (5) tracer tubing should be grouped together, whenever practicable, to permit insulation as a unit.

Steam supply to individual tracers should be taken from a DM and have a block valve for each tracer. This steam DM with all valves, drains, and connections is the steam supply station.

For isolation during maintenance there should be at least one flanged connection next to an isolating block valve between each manifold and its header. Tracer lines with a direct connection to the steam main line should have a normally open block valve placed at the branch and a second valve near the line to be traced.

A separate steam distribution header should be provided for tracing instruments and attendant piping, connected in such a manner that the instrument tracing will not be shut off when steam is shut off to other users. An exception is allowed for local pressure indicators, PD meters, gage glasses, and control valves, which may be protected by the pipeline or equipment tracers.

The steam supply for tracers required continuously for equipment protection against ambient temperatures should be independent of the steam supply required intermittently for winterization.

The number of individual tracers taken from a steam supply line or manifold header should be as in Table 7.1.

### 7.8.4 STEAM TRAPPING

Each tracer should have its own steam supply valve and trap.

Steam traps should be grouped together on a condensate CM. A maximum of 12 steam traps should be connected to one condensate-collecting manifold.

| Table 7.1 Steam Supply Station | | | | |
|---|---|---|---|---|
| | | Number of Leads | | |
| Branch From Header to DM Line Size DN (in.) | (DM) Steam Supply Manifold Size DN (in.) | No. of Size DN 15 (½ in.) | No. of Size DN 20 (¾ in.) | Recommended No. of Spare Lead Connections |
| 20 (¾) | 25 (1) | 1–2 | 1 | — |
| 25 (1) | 40 (1½) | 3–5 | 3 | 1 |
| 40 (1½) | 50 (2) | 6–15 | 4–6 | 1 |
| 50 (2) | 80 (3) | 16–30 | 7–12 | 2 |

Valves and piping at the trap should be the same size as the trap size.

All steam traps should have strainers upstream of the trap or should have integral strainers. All strainers should be equipped with blowdown valves.

The back pressure on the steam traps should not be higher than that recommended by the trap manufacturer.

All steam condensate piping including trap discharge to the header should be sized for two-phase; ie, they should be sufficiently large to handle the condensate and any flashed steam.

Condensate recovery should be 100%, however in exceptional condition when approved by the company, the cases where it is not practicable to recover, discharge piping should be short, without elbows and discharged into sewage or a properly designed soak away sump. However, alternative means of tracing (electrical) should also be considered.

Inverted bucket type traps should be acceptable where there is no danger of condensate freezing. Such applications should be subject to the company's approval.

In a severe climate if requested by the company, steam traps on process units should be protected by enclosing them in cabinets of steel or other suitable materials. Each cabinet should contain at least six traps and allow easy access for maintenance.

Isolating block valves for steam traps should be provided for winterizing services.

Traps should be preferably installed with the flow down. If the trap is in a horizontal run, it should be installed on its side to prevent freezing.

## 7.8.5 CONDENSATE REMOVAL

Tracer condensate header connections on condensate mains should be shown on drawings.

A typical arrangement of piping and tubing components is illustrated in Fig. 7.2.

Condensate collecting piping for grouped tracer traps should be such as to avoid excessive back pressure on traps and trap discharge lines, and should be based on the lowest expected steam supply pressure. Minimum size of condensate collecting piping for grouped tracer traps should normally be as follows:

- 1–2 traps DN 20 (¾ in.)
- 3–5 traps DN 25 (1 in.)
- 6–15 traps DN 40 (1½ in.).

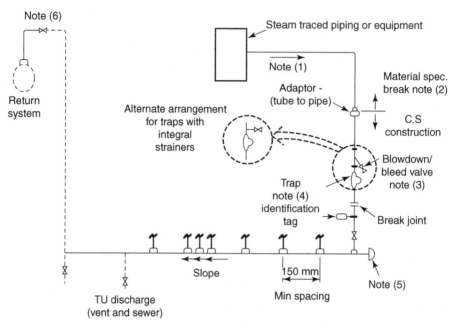

**FIGURE 7.2 Condensate Collection Stations**

Notes: (1) Tracer configuration to permit gravity flow of condensate to traps. If a tracer must rise vertically 1 m or more, it must be trapped before the vertical rise also; (2) tube-to-pipe adaptor normally located at end of "effective tracer" of equipment or piping being traced; (3) Y-type strainer required unless trap is furnished with an integral strainer; (4) orientation of traps and block valves to be vertical whenever practicable. Traps should always be installed to insure self-draining; (5) collection header and steam traps should be located near the grade or at the platform; (6) connection to the return system should be located at the top of the header.

## 7.8.6 STEAM TRACING LINES (TRACERS)

The tracer should be held in place with steel bands or 1.5 mm soft galvanized wire loops spaced 1 m apart. On tracers DN 20 (¾ in.) and larger spacing may be increased to 1.5 m.

Single or multiple tracers should be used, depending on pipe size and heat transfer requirements. Tracers should be DN 15 (½ in.) and DN 20 (¾ in.) seamless steel pipe schedule 80, for pipes DN 40 (1½ in.) up to and including DN 600 (24 in.).

Tracers should be in contact with piping or equipment except where spacers are specified.

Tracer material should be specified as herein below:

1. Steam tracing lines should be carbon steel pipe to ASTM A-53, Grade B, API 5L Gr.B, or soft annealed copper tubing to ASTM B-68 or equal or exceptionally steel tubing to ASTM A-179 as specified herein. Tubing is only acceptable in locations where piping is not practical, such as burner manifolds and instrumentation.

2. Copper tubing should not be used where steam or process temperature exceeds 200°C.

**Table 7.2  Minimum Tube Wall Thickness**

| Size | | Steel Tubing | | Copper Tubing | |
|------|------|------|------|------|------|
| DN | in. | mm | in. | mm | in. |
| 10 | 3/8 | 0.86 | 0.034 | 0.81 | 0.032 |
| 15 | ½ | 1.24 | 0.049 | 0.81 | 0.032 |
| 20 | ¾ | 1.24 | 0.049 | | |
| 25 | 1 | 1.65 | 0.065 | | |

3. The use of type 304 stainless steel seamless tubing to (BS 3605, Grade 801 or ASTM A-269) with stainless steel compression fittings is acceptable.
4. Minimum tube wall thickness should be in accordance with the data as specified in Table 7.2 above.

### 7.8.7  INSTALLATION OF TRACERS

Piping tracers should be installed as follows:

1. Horizontal pipe. Along the bottom half of the pipe, for pipe sizes DN 40 (1½ in.) and smaller the tracer may be helically wound.
2. Vertical pipe. Multiple tracers equally spaced around the circumference. Single tracers should be helically wound.
3. Pipelines supported by shoes or similar devices should have the bottom tracer located as close as possible to the support.
4. Tracing of the run pipe should extend to the first block valve of any branch connection.
5. For service operations sensitive to cold spots (waxes, asphalts, sulfur, etc.), the points where tracers leave the insulation should not be coincident, to avoid leaving any pipe length totally untraced.
6. Flanges, valve bonnets, and packing glands should be traced only when specified. When such components are not to be traced, the run tracer should be bent to follow the contour of the main pipe.
7. Bends should be used wherever practical and fittings kept to a minimum. Unions should be used when an item is traced and its removal is required for frequent maintenance.
8. Joints in tracing lines should be located at pipe flanges. Expansion loops should as far as possible be installed in horizontal plane and pockets should be avoided.
9. Expansion loops should be provided for all straight runs of tracers longer than 7.5 m unless otherwise specified as follows:
   a. Spacing should not exceed 30 m.
   b. The sum of the effective legs of the loops should be at least 0.6 m for 7.5 m runs, and 1 m for 30 m runs.
   c. Loops should be oriented to be self-drained.

**10.** The number of tracing line required and the arrangement of tracing line(s) around the pipe should be specified on the basis of the pipe size and severity of the flowing fluid and climatic condition. An X-section of a typically single and multiple arrangement of tracer.

Where control valves and bypasses are traced, the tracing should be arranged so that the control valve can be removed without interfering with the tracing of the bypass.

Each individual tracing line should be provided with its own trap and, in parallel systems, to each leg. Groups of tracers that are self-draining may be arranged to drain to a level controlled condensate pot or a collection header with an integrated type of steam trap.

Each tracing line should have its own steam supply valve, and steam trap. Piping connections to steam and condensate headers will be shown on the piping arrangement plan drawings and isometric drawings.

No provision should be made for expansion movement of 13 mm or less on DN 15 and smaller tracers, since the sag or offset will take care of this amount of expansion.

For tubing or piping tracers larger than DN 15, anchoring should generally be made at the midway point, and the piping arrangement at the ends of the tracers should be sufficiently flexible to allow for expansion of tracers.

Where it is impossible to allow for end movements, or in cases where for special reasons the unanchored length of pipe tracer exceeds 40 m, expansion loops should be provided. Minimum radius of expansion loop should be 6 times the outside diameter of the tracers at the bends of the loop.

Insulation should be slotted at expansion loops, and at anchored tracer ends where the tracers leave the pipe.

Anchors or guide clips should be installed on tracers near valves, flanges, expansion loops, and turns to avoid damage to insulation due to tracer expansion.

Tracers, connections, and fittings should be provided under the following conditions:

**1.** The use of threaded fittings should be minimized.
**2.** Threaded pipe fittings should be seal welded except at steam traps and piping downstream of traps.
**3.** Break joints in tracer system should be provided in equipment that must be removed for maintenance. Tracer-fitting break joints and all other mechanical joints should be located outside of the equipment insulation.

## 7.8.8 INSTRUMENTATION TRACER INSTALLATION

Tracer arrangements for instrumentation should be per API RP 550, section 8 except as modified below:

**1.** Tubing tracer size should be as shown in Table 7.2.
**2.** Lead lines for differential pressure instruments should have common heating and insulation.
**3.** The selection of "light" or "heavy" tracing methods should be based on the following criteria assuming that heating does not:
   **a.** boil away the process fluid in the lead lines, or,
   **b.** heat the instrument above its recommended maximum operating temperature.
**4.** Instrument houses should be large enough to accommodate the valve manifold so that separate heating and tracing of the valve manifold is not required.

### 7.8.9 **INSULATION**

Insulation for traced piping should be in accordance with insulation specification. In addition, the following design criteria should be considered.

1. The thickness of insulation used should be in accordance with the project specification.
2. The insulation cover should be applied tightly around the line and tracer. All tubing unions are to be made outside of the insulation and separately wrapped with nonasbestos ropes.
3. The following components of the tracer system should not be insulated:
   a. Traps.
   b. The portion of trap inlet piping that is required to be kept uninsulated for proper operation of the trap.
   c. Traced pipelines in sleeves under road crossings or similar underground routing. In such cases the insulation should be terminated at about 300 mm within the sleeve.
4. Traced lines should be covered with oversized or "extended leg" insulation.
5. When the plant is installed in a location where prolonged freezing temperatures are likely to occur, the condensate recovery piping should be insulated and where practical, insulated together with the associated steam piping.

#### 7.8.9.1 *Steam jacketing*

Where heat input into process piping is required (such as asphalt and liquid sulfur services), and when steam or electric tracing using heat transfer cement is impractical, steam jacketing of piping should be used. All steam jackets should be provided with valved drains at the low points.

#### 7.8.9.2 *Identification*

Identification tags should be permanently installed at each end of the tracer. Tags should identify the equipment or pipe line being traced. Weather tracing is for winterizing or process protection and the location of the inlet valve (supply point) and trap.

## 7.9 **INSPECTION AND TESTING**

For a steam tracing system the following visual inspections should be made prior to insulating:

a. Tubing and pipe bends should be visually inspected for kinked or flattened sections. All such sections should be cut out and replaced.
b. The tracer attachment should be checked to ensure freedom of movement towards expansion loops.
c. A tracer attachment should be inspected at expansion loops, at equipment break points, and at other changes of direction where movement of the tracer could damage the insulation.

The tracer system should be pressure tested, prior to insulation. Hydrostatic pressure should be at least 700 kPa or 1½ times design pressure whichever is greater.

## 7.10 **ELECTRICAL TRACING**

Electric tracing should be used in preference to steam tracing under the following conditions:

1. Where the temperature must be accurately controlled or limited. Temperature control by throttling or on-off control of steam is usually not practical because of water logging of sections of the tracer, freezing problems, and temperature gradients along the length of the tracer. Some examples of lines that require temperature control are as follows:
   **a.** Water lines to safety showers and eyewash fountains because of safety hazards if the line is overheated.
   **b.** HF, $H_2SO_4$, and NaOH lines in certain concentrations because of corrosion or stress-corrosion cracking at elevated temperatures.
   **c.** Fuel oil lines where high steam temperatures would cause coke formation and fouling of the line.
   **d.** Boric acid lines in −7°C to −1°C. Temperature must be maintained within a −7°C to −1°C band to prevent precipitation both above and below a certain temperature range.
2. Where instrument lines are monitoring and controlling important processes.
3. When lined pipes are used to avoid corrosion and/or abrasion. Most rubber and polymer linings should not exceed 93°C.
4. Where plastic pipes and tanks are used.
5. Where the minimum ambient is less than −12°C for instrument lines because of condensate freezing problems that occur with steam tracing.

   **Note:**
   When minimum ambient is below −12°C, strong consideration should be given to electric heat tracing of traps on steam tracing systems.

   Electrical tracing should be used as specified in the job specification. For detailed requirements of engineering and construction refer to the following general consideration:

1. Due consideration should be given to the classification of a dangerous area and all equipment specified accordingly.
2. Consideration should be given but not limited to: environmental conditions, pipe material, pipe size and length, fittings, type and thickness of insulation, lowest ambient design temperature, fluid flow conditions, type of control required such as thermistors, thermostats, etc., and area classification.
3. Overlimit thermostats should be used with constant wattage heaters.
4. Electric tracing used underground should be designed to allow maintenance without any requirement for excavation.
5. Installation involving a major amount of heat tracing should include a central indication panel, whereby the status of each loop should be monitored with a current transformer.
6. All thermostatic devices should be enclosed in an approve enclosures by the company.
7. All installation should be in conformity with the manufacturer's recommendation. The manufacturer's data should be utilized along with published charts to determine the size and amount of cable to be installed.
8. Thermostats should be installed to maintain desired temperatures.
9. All mineral insulation (MI) cable installation should be a continuous run. Each run should be supplied with an individual thermostat.

## 7.11 USE OF HEAT TRANSFER CEMENT

Heat transfer cement should be utilized on tubing when a process line requires a high heat input and more than three tracers would be required when using 4 bar (400 kPa) steam. All tubing joints must be outside the insulation in heat transfer applications.

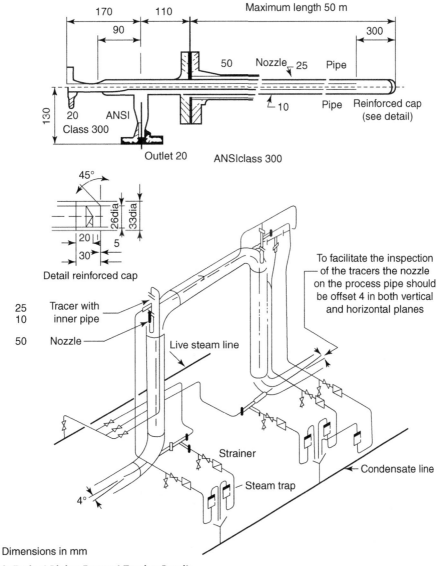

Dimensions in mm

**FIGURE 7.3 Typical Piping External Tracing Details**

One tracer

Two tracer

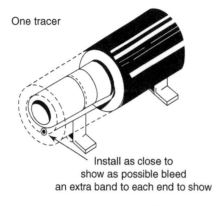
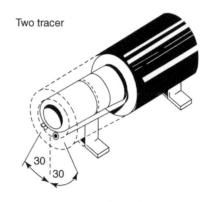

Install as close to
show as possible bleed
an extra band to each end to show

30
30

Tracer installation with cement

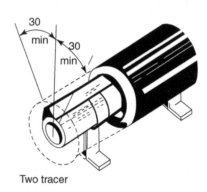

30
min
30
min

One tracer

Two tracer

**FIGURE 7.4 Tracer Installation**

When the heat transfer cement isused it should have the following specification and characteristics unless otherwise specified.

(a) Thermal conductivity                 13.02 W/m.K (11.2 k cal/m$^2$/h/°C/cm).
(b) Electrical resistance                     0.635 ohms/cm$^2$/cm.
(c) Specific heat                               0.209 J/kg/°C(0.50 k cal/kg/°C).
(d) Linear shrinkage                         1% maximum.
(e) The linear coefficient of thermal expansion should be approximately the same as for materials of the pipeline and tracer.
(f) Compressive and tensile strengths should be high enough to withstand stresses due to differential expansion and contraction.
(g) The heat transfer cement should be nonreactive with the materials of the pipeline, tracer, and insulation.
(h) The heat transfer cement should be air curing, weather resistant, and capable of application by unskilled labor without the use of special equipment (Fig. 7.3).

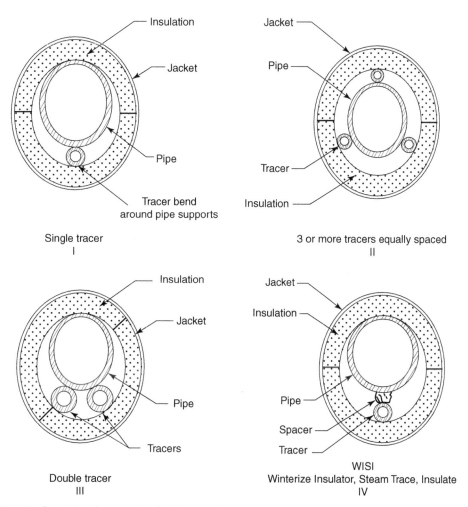

**FIGURE 7.5 Typical X-Sections for Required External Tracing**

### 7.11.1 LINE TRACING DETAILS

**1.** Tracers on vertical pipes are to be equally spaced around the pipe.
**2.** Guide clips and anchors are to be installed every 300 mm.
**3.** Tracers are to be banded to the pipe with 12.7 × 0.38 mm stainless steel bands installed tight enough to make tracers contact the pipe without crimping or deforming the tracer tubing (Figs. 7.4 and 7.5).

# WATER SUPPLY AND DISTRIBUTION SYSTEMS

## 8.1 QUALITY OF WATER

Water is the most vital of all life-sustaining substances and is irreplaceable. All of the waters found in nature have some impurities. The raindrops, as they fall, absorb dust, dissolve some oxygen, carbon dioxide, and other gasses. At the ground level they take up silt and other inorganic matter.

Surface water retains all these impurities for an indefinite period, but that part of the rainfall which percolates into the soil will lose the suspended silt and bacteria through natural filtration.

The total dissolved solid is the numerical sum of all dissolved solids determined by chemical tests. In general, the total concentration of dissolved salts (TDS) is an indication of the overall suitability of water. The quality of water for drinking and irrigation diminishes as the value of TDS increases.

### 8.1.1 QUALITY STANDARD FOR DRINKING WATER

Water intended for human consumption must be free from organisms and hazardous concentrations of chemical substances. The situation, construction, operation, and supervision of a water supply system, including its storage and distribution, must be such as to exclude any possible pollution of the water. The standards for drinking water prescribed by the World Health Organization (WHO) 1971 is the most accepted international standard on the potability of water. In addition, many countries have established national standards for drinking water supplies. Table 8.1 presents the substances and characteristics affecting the acceptability of water for domestic use.

### 8.1.2 BACTERIAL POLLUTION

The greatest danger to drinking water is that it may have been contaminated by sewage, human excrement, or animal pollution. Drinking of such water infected by living pathogens of diseases such as dysentery may result in epidemics.

The organisms most commonly used as indicators of pollution are *E. coil* and the coliform group as whole. Both are considered to be of faecal origin. Water circulating in the distribution system, whether treated or not should not contain any organism that may be of faecal origin. Frequent bacteriological examinations are essential for hygienic control. When repairs or extensions to water supply installations are carried out, it is essential that a bacteriological examination of the water should be performed, after the part of the system concerned has been disinfected and before it is put into service. Efficient treatment, usually through chlorination, yields water free from any coliform organisms, however polluted the original raw water may have been.

**Table 8.1 Substances and Characteristics Influencing the Acceptability of Water for Domestic Use**

| Substance Characteristics | Undesirable Effect Produced | Highest Desirable Level | Maximum Permissible Level |
|---|---|---|---|
| Substances causing Discolouration | Discoloration | 5 units[a] | 50 units[a] |
| Substances causing Odors | Odors | Unobjectionable | Unobjectionable |
| Substances causing tastes | Tastes | Unobjectionable | Unobjectionable |
| Suspended matter | Turbidity | 5 units[b] | 25 units[b] |
| Total solids | Taste | 500 mg/L | 1500 mg/L |
| Ph range | Taste | 7.0–8.5 | 6.5–9.2 |
| Anionic detergents | Taste and foaming | 0.2 mg/L | 1.0 mg/L |
| Mineral oil | Taste and odor after chlorination | 0.01 mg/L | 0.30 mg/L |
| Phenolic compounds (As phenol) | Taste, particularly in chlorinated water | 0.001 mg/L | 0.002 mg/L |
| Total hardness | Excessive scale formation | 2 mEq/L (100 mg/L $CaCO_3$) | 10 mEq/L (500 mg/L $CaCO_3$) |
| Calcium (as Ca) | Excessive scale formation | 75 mg/L | 225 mg/L |
| Chloride (as Cl) | Taste; corrosion in hot-water systems | 200 mg/L | 600 mg/L |

[a]*Platinum cobalt standard.*
[b]*Turbidity standard.*

## 8.1.3 CHEMICAL SUBSTANCES

A number of chemical substances, if present in certain concentrations in supplies of drinking water, may constitute a health hazard. Table 8.2 presents the limits of toxic substances in drinking water, as prescribed tentatively by the WHO, assuming an average daily intake of 2.5 L of water by a man weighing 70 kg.

## 8.1.4 QUALITY STANDARD FOR IRRIGATION WATER

Water acquired from any one of the water resources, if used in farmlands then no limit of turbidity is set. It is enough to prevent entry of wastewaters contaminated with oil and other chemicals harmful to plants into the irrigation water supply system. But if irrigation water is to be supplied in a separate pipe distribution system (other than drinking water networks) for the needs of irrigation, courtyard washing etc. of residential houses, the recommended maximum turbidity level of such irrigation water is around 20–25 ppm. This limit is attainable with plain sedimentation.

**Table 8.2 Tentative Limits for Toxic Substances in Drinking Water**

| Substance | Upper Limit of Concentration (mg/L) |
|---|---|
| Arsenic (as As) | 0.05 |
| Cadmium (as Cd) | 0.01 |
| Cynanide (as Cn) | 0.05 |
| Lead (as Pb) | 0.1 |
| Mercury (as Hg) | 0.001 |
| Selenium (as Se) | 0.01 |

*Source: International Standards for Drinking Water, WHO, Geneva (1971).*

## 8.2 WATER RESOURCE SELECTION AND PROTECTION

The water supply should be obtained from the most desirable source feasible and effort should be made to prevent or control pollution of the source. If the source is not adequately protected against pollution by natural means, the supply shall be adequately protected by treatment. Sanitary surveys shall be made of the water supply system, from the source of supply to the connection of the customer's service piping to locate any health hazards that might exist.

## 8.3 WATERWORKS SYSTEM

A waterworks system is designed and executed either as a new project or expansion of an existing system to supply a sufficient volume of water at the required quality and at adequate pressure from the supply source to consumer points for domestic, irrigation, industrial, fire-fighting, and sanitary purposes. Hence, the design of a waterworks system needs adequately sized components. The water supply facilities consist of water intake facilities and water treatment facilities dependent on the need with regard to quality requirements, storage, transmission, pumping, distribution etc.

### 8.3.1 INTAKES

Water intakes consist of the opening, strainer, or grating through which the water enters into a conduit conveying the water, usually by gravity, to a sump or is pumped onshore to presedimentation channels or to the mains for direct use or treatment plants as required. In designing and locating intakes, the following aspects must be considered:

1. The source of supply and the fluctuation of water level or water table.
2. The navigation requirements, if any, and the scouring possibilities of river or lake bottom.
3. The location with respect to sources of pollution.
4. In order to minimize the possibility of interference with the supply, the intakes should be duplicated, wherever possible.
5. Permission should be obtained from the concerned official authorities.

### 8.3.2 INTAKES FROM IMPOUNDING RESERVOIRS AND LAKES

As the water of an impounding reservoir varies in quality at different levels it is recommended to take water from about 1 m below the lowest water surface. As the water level in reservoirs is expected to fluctuate, gates at various heights should be provided.

In the case of lake intakes, it is advisable to have the intake opening 2.5 m or more above the lake bottom to prevent entry of silts. To minimize the entry of floating matter it is recommended that the entry velocities are limited to 0.15–0.2 m/s.

### 8.3.3 RIVER INTAKES

The river intakes, dependent on the formation of the river bed and the amount of total suspended solids of raw water could be located onshore with the intake line preferably on piles to convey water into the onshore sump for pumping or it could be a simple offshore jetty structure to support low head borehole pumps or high pressure centrifugal pumps as required by the system.

In case of turbid rivers with clayey-sand stratum it is preferable to install low lift borehole pumps in pairs on simple jetty structures with piles complete with screen and delivery pipe conveying the water to presedimentation channels in order to reduce the turbidity to acceptable limits for irrigational needs. This settled raw water can be supplied directly (a) for irrigational and other nonpotable needs of the residential houses or plants (fire-fighting etc.) and (b) can be transmitted to the water treatment plant for potable needs, thus decreasing the initial and operational costs of required clarification and filtration. Generally the mouth of intake should be 1 m higher than the river bottom level.

### 8.3.4 WELL INTAKES

Ground water utilization is mainly through open wells, tube wells, springs, and "Ghanats." For an estimation of ground water flow Darcy's law indicates that flow in water-bearing sands varies directly with the slope of the hydraulic gradient.

### 8.3.5 SEAWATER INTAKE FOR DESALINATION PLANTS

Increasing water consumption and depletion of existing sweet water resources has led to considerable interest in conversion of saline or brackish waters. The saline water intake level preferably should be at least 3 m below seawater level as the quality of seawater at shallower depths is more saline. Consequently, this water is transformed into potable water by desalination or distillation processes.

Information on the characteristics of the water-bearing formations and the well should be obtained by conducting pumping tests performed in observation wells.

## 8.4 METHODS OF DISTRIBUTION: GENERAL

Water can be distributed to consumers through pipes in three ways, as local conditions and other considerations would permit.

### 8.4.1 **GRAVITY DISTRIBUTION**

Whenever the source of supply is a lake or impounding reservoir at some elevation above the consumers' so that sufficient pressure can be maintained in the mains for domestic, industrial, and fire fighting needs, this is the most reliable and cheapest method. However, the main pipeline leading from the source to the town should be well safeguarded against accidental breaks. Motor pumps might be needed for fire fighting purposes.

### 8.4.2 **DISTRIBUTION OF WATER SUPPLY BY MEANS OF PUMPS WITH MORE OR LESS STORAGE**

In this method of direct supply to consumers, the excess water pumped during periods of low consumption is stored in elevated tanks or reservoirs. During periods of high consumption, the stored water is drawn upon to augment that of the pumped supply. This method allows fairly uniform rates of pumping and usually is economical because of a two-directional flow which normally keeps the pipe sizes of the mains a little lower than one-directional flow systems and because of the fact that pumps can be operated at their rated capacity.

A variation to this method, still by means of pumping and storage is the method of direct supply of drinking water to high level tanks and from they're onto consumers.

In this method, drinking water is being pumped from pumping station through different feeders or trunk mains to each of the high-level storage tanks positioned at the center of groups of houses. From there on, through a network of distribution pipes, the drinking water of the tanks under their static pressure is being consumed by the community.

### 8.4.3 **DISTRIBUTION OF WATER BY MEANS OF DIRECT PUMPING WITHOUT STORAGE**

In this method the pumps force water directly into the mains with no other outlet other than the draw-off points where water is actually consumed. This method is the least desirable system for a drinking water supply. But it is the recommended system for an irrigation water supply that should have a completely separate distribution network.

Whenever the quality of water supplied to irrigational draw-off points and other nonpotable uses it is inferior to that of drinking water.

In this direct pumping method without storage, as consumption varies, the pressure in the mains would fluctuate. Hence to cope with the varying rate of consumption several pumps usually four to six in number should be provided in each pump house allowing a 50% safety factor so that in a pump house with six installed pumps, four of them in series can meet the peak demand. An advantage of direct pumping is that a high pressure fire service pump can be provided and operated in times of large fires to increase the water pressure to any desired level that the mains can withstand.

## 8.5 **PREFERRED PIPE DISTRIBUTION SYSTEM**

Whenever a reliable municipal water supply system exists, purchase of such water together with associated water supply services, whether metered at source of use (dwelling) or at the boundary of a private housing development is preferred. But, if water winning, treatment, and supply to plants and housing developments is the responsibility of the company, the following supply system of water distribution is recommended.

### 8.5.1 IRRIGATION WATER (NONPOTABLE WATER) SYSTEM

#### 8.5.1.1 Intake from turbid river

In such a case the requirements and recommendations, that is, distribution of settled raw water by means of direct pumping without storage (within distribution network) is preferred.

As regards the pattern of network of pipes within the network of streets, the gridiron pattern with central or looped feeder is preferred. The gridiron network of pipes normally with single mains of 100–150 mm at one side of the roads would suffice. The tree system or branching pattern with dead ends would be permitted temporarily on the outskirts of the community, in which ribbon development follows the primary arteries of roads and streets, but tees with plugs should be provided at dead ends with an aim to have a gridiron pattern at a later phase.

#### 8.5.1.2 Intake from well, lake or reservoir

In such a case, it is recommended that there be just one distribution system for both potable and nonpotable needs as set out in the gridiron pattern of network.

The minimum pipe size of distribution mains, ordinarily, should not be less than 150 mm with dual mains, one at each side of the roads.

### 8.5.2 DRINKING WATER (POTABLE) SYSTEM

#### 8.5.2.1 Intake from turbid river

Whenever the source of water is a turbid river, the river intake facilities, together with sedimentation channels should be sized for total water requirements, that is, potable and nonpotable needs of the water supply project so that part of the settled raw water that meets the quality requirements be conveyed to the water treatment plant.

The separate distribution system of drinking water meeting the quality requirements should be designed in accordance with guidelines adopting one of the alternative storage methods as appropriate.

#### 8.5.2.2 Intake from well, lake, or reservoir

In such a case as indicated in 10.1(b) only one distribution network should be provided for both potable and non-potable needs and the recommendations are applicable dependent on the local conditions. As regards the pattern of network of distribution mains, the dual mains not less than 150 mm with one at each side of the roads is preferred for both cases of (a) and (b).

## 8.6 DESIGN OF WATERWORKS DISTRIBUTION SYSTEM

A waterworks distribution system, dependent on whether it is a gravity distribution or pump distribution, includes pipe network, storage of some sort, pumping station, and components such as valves, fire hydrants etc. Once the best layout of water distribution system as required in a specific project based on general recommendations set out in the standard is decided, the design of its features should be made in accordance with applicable design criteria given hereunder.

### 8.6.1 **DESIGN OF PIPE DISTRIBUTION NETWORK**

The design principles of pipe distribution network supplying potable or nonpotable water to communities or plants as regards hydraulics are the same. The differences are in design criteria and design of some of the components that are needed in one system and either not or less needed in another.

### 8.6.2 **DESCRIPTION OF DISTRIBUTION SYSTEM**

Apart from water intake facilities, the water distribution systems generally comprise of:

1. The primary feeders or arterial mains that carry large quantities of water from the distribution pumphouse or natural artificial reservoirs (gravity supply system) to high level tanks simultaneously connected to consumers or not.
2. The secondary feeders may be included in the pipe network to carry considerable amounts of water from the primary feeders to the various areas with looped pattern for normal supply and fire fighting.
3. The small distribution mains form the bulk of the gridiron network supplying water to service pipes of residences and other draw-off points. Ordinarily the pipe sizes of distribution mains should not be less than 150 mm, with the cross pipes also 150 mm at intervals of not more than 200 m.

### 8.6.3 **FACTORS AFFECTING DESIGN OF WATER DISTRIBUTION NETWORK**

Topography of the land will affect pressure, while existing and expected population densities, and commercial and industrial needs, will affect both pipe size and the location and capacity of storage tanks.

For the preliminary layout of the distribution network, the city or town road layout of the housing project can be utilized.

The plan of distribution mains should allow for expected future expansions and population growth at execution phases.

## 8.7 **DESIGN CRITERIA**

### 8.7.1 **RATES OF WATER USE**

In the absence of recorded actual water consumption rates in identical projects, the average rates quoted hereunder can be used for design of distribution networks of residential communities:

1. Drinking water
   a. 250 L/person per day for moderate climate regions
   b. 350 L/person per day for hot climate regions
   Typical rates of water use for various establishments can be used for the design of building piping and reference can be made to Table 8.3.
2. Irrigation water
   a. Hot climate regions:
      – 100 m$^3$/day per hectare of planted areas such as lawns, cultivation etc. for hot climate regions. Irrigation draw-off water points (20 mm or ¾in. taps) shall be provided 15 m apart in planted areas of housing schemes so that with a 10 m long flexible hose pipe all planted areas may be reachable.

**Table 8.3 Typical Rates of Water Use for Various Establishments**

| User | Range of flow, L/(person or unit)d |
|---|---|
| Assembly hall, per seat | 6–10 |
| Automobile service station | |
|   Per set of pumps | 1800–2200 |
|   Per vehicle served | 40–60 |
| Country club | |
|   Resident type | 300–600 |
|   Transient type, serving meals | 60–100 |
| Dwelling unit, residential | |
|   Apartment house on individual well | 300–400 |
|   Apartment house on public water supply, unmetered | 300–500 |
|   Boarding house | 150–220 |
|   Hotel | 200–400 |
|   Lodging house | 120–200 |
|   Private dwelling on individual well on metered supply | 200–600 |
|   Private dwelling on public water supply, unmetered | 400–800 |
|   Factory, sanitary wastes, per shift | 40–100 |
|   Hospital | 700–1200 |
|   Office | 40–60 |
|   Recreation park, with flush toilets | 20–40 |
| Restaurant (including toilet) | |
|   Average | 25–40 |
|   Kitchen wastes only | 10–20 |
| School | |
|   Day, with cafeteria or lunchroom | 40–60 |
| Store | |
|   First 7.5 m of frontage | 1600–2000 |
|   Each additional 7.5 m of frontage | 1400–1600 |
|   Swimming pool and beach, toilet, and shower | 40–60 |

- 400L/day per dwelling or 20 L/m² of concrete pavement for washing a courtyard.
- 100L/min lasting for 15 min for car washing once a week.
- 12L/ton for cooling capacity when cooling the central air conditioning plants and/or A/C units.
- Firefighting water, in separate systems of potable and nonpotable needs, shall be provided from a settled raw water system through fire hydrants that are 100 m apart at a rate of

8 m³/min for a duration of a maximum of 10 h with fire hydrants each having 2 hose couplings. The maximum amount of water required for control of an individual fire is 4.5 m³/min in large communities. In residential districts the required flow ranges from 1.9 m³/min to 9.5 m³/min. A single hose stream at 1.4 kg/cm² pressure is considered to be 1 m³/min. Hence, at least eight hydrant hose connections taken from four fire hydrant stands should fight each individual heavy fire.

**b.** Moderate regions:
- 50 m³/day per hectare of grass plantation.
- 300 L/day per dwelling or 15 L/m² of concrete pavement for washing of the courtyard.
- For car washing the same as in hot climates.

For firefighting the same as in hot climates.

*Note:* In a water distribution system that provides for both potable and nonpotable needs, the rates of water use should be added up.

## 8.7.2 VELOCITIES

Velocities at peak flows in all feeders and distribution mains usually do not exceed 1 m/s with 2 m/s as the upper limit, which may be reached near large fires.

## 8.7.3 PRESSURE RATINGS

There are wide differences in the pressures maintained in distribution systems dependant on the limitation of building heights imposed by the municipality and firefighting factor.

Table 8.4 gives the required pressures for different height limitations and firefighting source and means.

**Table 8.4  Pressures**

| Building Heights and Fire Fighting Factor | Pressure (kPa) |
|---|---|
| Residential districts having houses max. Two stories in height for ordinary service (fire fighting from other source) | 150–300 |
| Res. districts with building heights limited to four stories but where direct hose streams are used for fire fighting | 400–500 |
| Same as above, but for commercial districts | 500–600[a] |
| In company owned and operated housing areas with max. three storey buildings and separate settled raw water system for irrigation and fire fighting | Min. 2 kg/cm = 220 kPa for drinking water mains[b] and 500 kPa for settled raw water network |
| Required pressure in separate settled raw water distribution for irrigation and fire fighting | 500 |

[a]*During heavy fire demands when a pumper is used a drop in pressure to not less than 200 kPa is permitted in the vicinity of fire in commercial districts.*
[b]*The practical available pressure at the highest and farthest tap of any building with design pressure of 2 kg/cm² would not be less than 0.5 kg/cm², that is, 5 m of water column pressure.*

## 8.7.4 **NECESSARY STORAGE**

Water is stored to equalize pumping rates over the day, to equalize supply and demand over a long period of high consumption, and to furnish water for such emergencies as firefighting or accidental breakdowns.

Elevated storage is furnished in earth or masonry reservoirs situated on high ground or in elevated tanks.

The capacity of the elevated tank or tanks will depend upon the load characteristics of the system, which should be carefully studied before any decision is made. To equalize the pumping rate, that is, allow a uniform rate throughout the day will ordinarily require storage of 15–30% of the maximum daily use. Future increases in demand must, of course, be considered.

In the absence of data the maximum daily consumption can be taken as 180–200% of the annual average. The main functions of elevated storage tanks in order of importance are:

- To discharge water into the mains during peak loads, thus raising the hydraulic grade line in nearby distribution mains, thus economizing on the size of mains.
- To provide adequate volume of water having at least a 20-m water pressure for firefighting purposes.
- Storage may also be needed to equalize demand over a long-continued period of high use, such as a cold period in winter or a dry period in summer. Such storage is particularly needed when wells of limited capacity are the sources of supply or when water must be filtered. In such cases, periodic testing of the quality of stored water is mandatory.

Wherever land at reasonable cost is available at the site of water winning, it is recommended to provide ample storage in the vicinity of source of supply at ground level such as artificial lakes etc. so that the needed disinfections (chlorination etc.) can be done at the water purification stage.

The tanks can be equipped with valves, which close the entrance pipe when the tank is full and reopen when the pressure in the mains drops during peak demand.

## 8.7.5 **DESIGN FLOWS**

The sizing of the pipes in drinking water distribution network shall be designed for a peak demand rate of four times average rate of consumption, whereas pipes in the settled raw water distribution network shall be designed for a peak demand of twice the average rate of predicted flow.

## 8.7.6 **FLOW IN PIPES**

Since flow is usually turbulent in pipes used for water supply the friction factors depend upon the roughness of the pipe and also upon the Reynolds number which, in turn, depends in part upon the velocity in the pipe and its diameter.

Various pipe-flow formulas are available to predict head losses as a function of velocity in pipes, and of these the Hazen-Williams formula (although empirical) is more often used in the design of water distribution systems, however, use of equivalent formulas are acceptable when approved by the authorized representative of the owner (AR).

Hazen–Williams formula states that: $v = (C)r^{0.63}S^{0.54}$

in which $v$ is the velocity in the pipe in distance per second, $r$ is the hydraulic radius of the pipe, $S$ is the hydraulic gradient, $C$ is a constant depending upon the relative roughness of the pipe, and $K$ is

| Table 8.5  Hazen-Williams Coefficients for Various Pipe Materials | |
|---|---|
| **Description of the Pipe** | **Values of C** |
| Extremely smooth and straight | 140 |
| Cast iron | |
| New | 130 |
| 5 years old | 120 |
| 10 years old | 110 |
| 20 years old | 90–100 |
| 30 years old | 75–90 |
| Concrete or cement lined welded steel, as for cast iron pipe, 5 years older riveted steel, as for cast iron pipe, 10 years older | 120–140 |
| Plastic | 150 |
| Asbestos cement | 120–140 |
| Note: In future the use of cast iron and asbestos cements pipes should be minimized. | |

an experimental coefficient and equal to 0.849. For values of $C$ coefficient in Hazen–Williams formula refer to Table 8.5.

To be on the safe side the designer can use a value of $C = 100$ for a cast iron pipe for designing water distribution systems considering this to represent its relative roughness at some future time. This may be too large where very corrosive or incrusting water is encountered. Where such waters are encountered the use of cement lined, plastic, or asbestos cement pipe or treatment of the water to reduce its corrosiveness may be advantageous.

For sizing of pipes, the nomogram shown in Fig. 8.1 can be used for velocity or flow when pipe size and slope of hydraulic grade line are known and vice versa.

## 8.8  DESIGN PROCEDURE

In the design of a new system the pipe sizes should be assumed and the system investigated for the pressure conditions that will result from various demand requirements.

On the preliminary layout of the distribution network all the required feeder mains, secondary mains, and small distribution mains with distinct line thickness indicating at the same time the location of storage tanks, valves, and hydrants should be marked. Next, the rates of demand for all purposes should be estimated and marked on the appropriate pipe.

Finally, the "Hardy Cross Method of Analysis," can be used which balances either the head losses or quantities. The balancing of head losses is preferred. The pipe flow formulas of Hazen–Williams, Manning, or Chezy can be safely used in the design of domestic water supply projects.

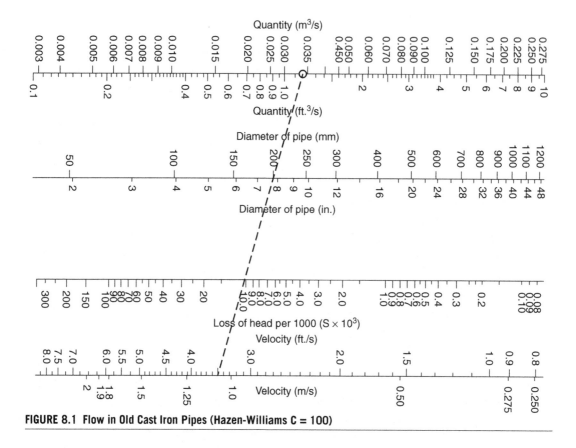

**FIGURE 8.1  Flow in Old Cast Iron Pipes (Hazen-Williams C = 100)**

## 8.9  DESIGN OF COMPONENTS

The components involved in the waterworks distribution system apart from water reservoirs and elevated tanks are the pumps and pumping stations, various types of valves used in the conveyance of water, fire hydrants, and the arrangement of service pipes, that is, the pipe extending from the main to the consumer's water meters if any, or extending up to dwelling units isolating the main valve just inside the curtilage.

### 8.9.1  PUMPS AND PUMPING STATIONS

Apart from supplies of water originating in mountainous areas that can be furnished to consumers entirely by gravity, usually, however, it is necessary to raise the water by means of pumps at one or more points in the system. Pumps are needed, therefore, to lift water from a lake, reservoir, or river to sedimentation channels or to a water treatment plant, and from thereon another lift will be needed to force the water into the mains and elevated storage tanks. Sometimes a booster pump may be needed at certain points of trunk or feeder mains to keep the pressure at desirable levels.

- In pumping stations supplying drinking water in single distribution systems with storage for potable and nonpotable needs, the installed total pumping capacity including stand-by should be 30 percent more than the peak demand of all seasons.
- Whereas the installed total pumping capacity of pumping stations supplying water (mostly settled river water) by direct pumping without storage, should be 50% more than the peak demand of dry weather seasons.
- It is good practice to provide one or two spare pumps and motors in each pumping station over and above the installed stand-by capacities to replace units under repair.
- The pumps used in water supply systems are divided into three general classes, reciprocating, rotary, and centrifugal. The reciprocating class typically consists of a piston or plunger which alternatively draws water into a cylinder on the intake stroke and then forces it out on the discharge stroke. The rotary type contains two rotating pistons or gears which interlock and draw water into the chamber and force it practically continuously into the discharge pipe. The centrifugal type has an impeller with radial vanes rotating swiftly to draw water into the center and discharge it by centrifugal force.
- The centrifugal or impeller-type pumps are the most used pumps in water engineering field. The duties covered by impeller-type pumps range from the pumping of small flows, minimum 450 L/min, against very high heads of maximum 300 m (generally performed by multistage diffuser pumps) to the pumping of large quantities, say 225 m³/min against very low heads of minimum 1.5 m. Fig. 8.2 shows schematic location of valves in distribution systems.

### 8.9.2 **VALVES**

Various types of valves are used in conveyance of water. The fields of usefulness of each type together with some design criteria are given in Table 8.6.

### 8.9.3 **FIRE HYDRANTS**

Hydrant spacing is dictated by the required firefighting water flow because the capacity of a single hydrant is limited. A single hose stream is considered to be approx. 1 m³/min that serves a minimum

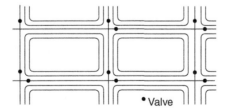

**FIGURE 8.2 Schematic Location of Valves in Distribution System**

*Notes:*
1. Recommended spacing of valves in a high value district is 150 m and 250 m in other districts.
2. Underground valves should be placed in valve chambers.
3. Sluice or gate valves should not be left partially open for long periods as the velocity of the water may erode the seating.
4. Valves larger than 300 mm (12 in.) require gearing unless the pressure is low.

**Table 8.6 Various Types of Valves**

| Item | Types of Valves | Field of Application and Some Technical Recommendations |
|------|-----------------|----------------------------------------------------------|
| 1 | Gate valves (USA) sluice valves (UK) with rising or nonrising stem as required | Commonest and most used valve in distribution networks. |
| 2 | Globe valves | Seldom used in water distribution network because of their high head loss characteristics. They can be used in household plumbing where their low cost outweighs their poor hydraulics. |
| 3 | Float valves | Used on the inlets to service reservoirs and tanks. |
| 4 | Check or reflux or nonreturn valves | One directional nonreturn flow valves closed automatically by fluid pressure. Used to prevent reversal of flow on gravitational supply mains (also in effluent disposal rising mains), or when pumps are shut down. They are installed also at the end of a suction line to prevent drainage of suction line and creation of vacuum. |
| 5 | Foot valves | Check valves installed at entry of suction lines. |
| 6 | Butterfly valves | Less costly than gate valves, used in both low-pressure applications and in filtration plants. In large pipe operations have numerous advantages over gate valves including lower cost, compactness, min. Friction wear and ease of operation. Should not be used for conveyance of sewage effluents of all kind. |
| 7 | Pressure regulating-reducing valves | These automatically reduce pressure on the down-stream side to any desired magnitude and are used on lines entering low altitude areas of a city where, without such reduction, pressure would be too high. Not applicable in oil industries' small housing communities. |
| 8 | Pressure-sustaining valves | Similar to pressure-reducing valves, except that, instead of tending to close when the downstream pressure rises, they tend to close if the upstream pressure falls, thus maintaining a more or less constant upstream pressure. |
| 9 | Altitude valves | Used to close automatically a supply line to an elevated tank when the tank is full. Flow from the tank is permitted when a selected and set low pressure below the valve indicates that water from the tank is required. This kind of valve can be used at the foot of inlet pipe of water tanks in distribution system. |
| 10 | Air and air-relief valves | Required to discharge air when a main is being filled and to admit air when it is being emptied. The inlet to the air valve should be provided with a sluice valve to isolate it for repairs.<br>    Air valves must be fixed at high points on the pipeline and close to main valves. |
| 11 | Wash-out valves | Proportional to the mains and located adjacent to large open drainage channel (major nullahs). In order to enable a particular length of the pipe to be emptied as required. |
| 12 | Sluice gates (USA) or penstocks (UK) | A barrier plate free to slide vertically across a water or sewage channel, or an opening in a lock gate |

of 3720 m² area. Thus the recommended spacing of fire hydrants is approximately 60 m when installed on drinking water networks and 100 m apart when installed on separate settled raw water networks. Ordinarily hydrants are located at street intersections where streams can be taken in any direction. In high value districts additional hydrants may be necessary in the middle of long blocks. The fire hydrant stands diameter in cold climate regions should be 150 mm and in hot climates 100 mm.

### 8.9.4 THE SERVICE PIPE

The pipe extending from the distribution mains (preferably small distribution mains) to the housing or dwelling's water meter, or stop valve with no meter is known as the service pipe. The service pipes must be durable to minimize unsightly breaks in the high-grade paving for repairs.

In order of preference, copper high density and high strength flexible plastic pipes and/or rigid PVC and galvanized iron pipes with screw joints can be used for service pipes. If flexible pipe materials are used it is recommended to provide steel sleeves for their protection just after the corporation cock or ferrule tapping. The ferrule tappings should be located on the top section of the mains preferably slightly inclined toward the curtilage of the housing unit. Thus, a flexible connection could be obtained more easily and the possibility of any sediment entering from the mains to the service pipe will be reduced. The size of service pipe should not be less than 20 mm (¾in.) to insure good pressure in the building. If, for any reason, its length be more than 10 m, a 30 mm ferrule is needed for a single dwelling unit, 40 mm for a two-family house and 50 mm for an apartment house having not more than 25 families.

Corporation cocks or ferrules requiring over a 50 mm hole are not ordinarily tapped into the mains of even adequate size.

For ferrule connections to the mains, the following limitations should be respected.

For 100 mm mains maximum ferrule size permitted should be less than 20 mm; for 150 mm, 25 mm; for 200 mm, 30 mm; for 250 mm, 40 mm and for 300 mm, 50 mm. Table 8.7 shows substances and characteristics influencing the acceptability of water.

## 8.10 PROCESS DESIGN OF WATER SYSTEMS

This section lists the minimum requirements for the process design and selection of various water supply systems, used in OGP Industries, and consists of the following systems:

- water treatment system
- Raw water and plant water System.

### 8.10.1 RYZNAR STABILITY INDEX

An empirical method for predicting scaling tendencies of water based on a study of operating results with water of various saturation indices.

$$\text{Stability Index} = 2\text{pH}_S - \text{pH} \qquad (8.1)$$

Where, $\text{pH}_S$ = Langelier's Saturation pH.

**Table 8.7 Substances and Characteristics Influencing the Acceptability of Water**

| Substance Characteristics | Highest Desirable Level (mg/L) | Maximum Permissible Level (mg/L) |
|---|---|---|
| **Toxic substances** | | |
| Arsenic, As | — | 0.05 |
| Barium, Ba | — | 1 |
| Cadmium, Cd | — | 0.01 |
| Chromium, Cr | — | 0.05 |
| Cyanida, Cn | — | 0.05 |
| Lead, Pb | — | 0.05 |
| Nitrate, $NO_2$ | — | 1 |
| Slenium, Se | — | 0.1 |
| Silver, Ag | — | 0.05 |
| Mercury, Hg | — | 0.01 |
| Boron, B | — | 1 |
| **Other substances** | | |
| Total dissolved solids (TDS) | 500 | 2000 |
| Total hardness as $CaCO_3$ | — | 500 |
| Magnesium, Mg | [a] | 150 |
| Zinc, Zn | 5 | 15 |
| Copper, Cu | 0.05 | 1 |
| Iron, Fe | 0.1 | 1 |
| Manganese, Mn | 0.05 | 0.5 |
| Sulfate, $SO_4$ | 250 | 400 |
| Chloride, Cl | 200 | 600 |
| Nitrate, $NO_3$ | — | 45 |
| Ammonium, $NH_4$ | 0.05 | 0.5 |

[a]*If the density of sulfate ion more than 250 mg/L the density of magnesium ion shouldn't be higher than 30 mg/L. if and only if the density less than 250 mg/L then the density of magnesium ion could go up to 150 mg/L.*

This index is often used in combination with the Langelier Index to improve the accuracy in predicting the scaling or corrosion tendencies of water. The following Table 8.8 illustrates how to use this index.

## 8.10.2 LANGELIER'S INDEX

This is a technique of predicting whether water will tend to dissolve or precipitate calcium carbonate. If the water precipitates calcium carbonate, scale formation may result. If the water dissolves calcium

| Table 8.8  Ryznar Stability Index | |
|---|---|
| **Ryznar Stability Index** | **Tendency of Water** |
| 4.0–5.0 | Heavy scale |
| 5.0–6.0 | Light scale |
| 6.0–7.0 | Little scale or corrosion |
| 7.0–7.5 | Corrosion significant |
| 7.5–9.0 | Heavy corrosion |
| 9.0 and higher | Corrosion intolerable |

carbonate, it has a corrosive tendency. To calculate Langelier's Index, the actual pH value of the water and Langelier's saturation pH value ($pH_S$) are needed. Langelier's saturation pH value is determined by the relationship between the calcium hardness, the total alkalinity, the total solids concentration, and the temperature of the water. Langelier's Index is then determined from the expression $pHpH_S$. Fig. 8.3 is a chart used for determining Langelier's Index. The interpretation of the results obtained are shown later in Table 8.9.

Also note that the presence of dissolved oxygen in the water may cause water with a "Zero" Langelier's Index to be corrosive rather than "neutral."

Caution must be observed in employing Langelier's Index for controlling corrosion or deposit formation, since there are factors that may make its application inappropriate. These include temperature differences within a system, changing operating conditions, or the presence of chemical treatment in the water.

## 8.10.3 **WATER TREATMENT SYSTEM**

Water treating requirements for refinery and/or plant services depend upon:

1. the quality of the source of make-up water
2. the manner in which the water is used
3. environmental regulations
4. site climatic conditions governing wastewater disposal.

These factors should be considered in selecting the overall plant process and utility systems.

### 8.10.3.1 Source water types

Source or make up water is normally either ground or surface water, neither of which is ever chemically pure.

Ground waters contain dissolved inorganic impurities, which come from the rock and sand strata through which the water passed.

Surface waters often contain silt particles in suspension (suspended solids) and dissolved organic impurities (dissolved solids).

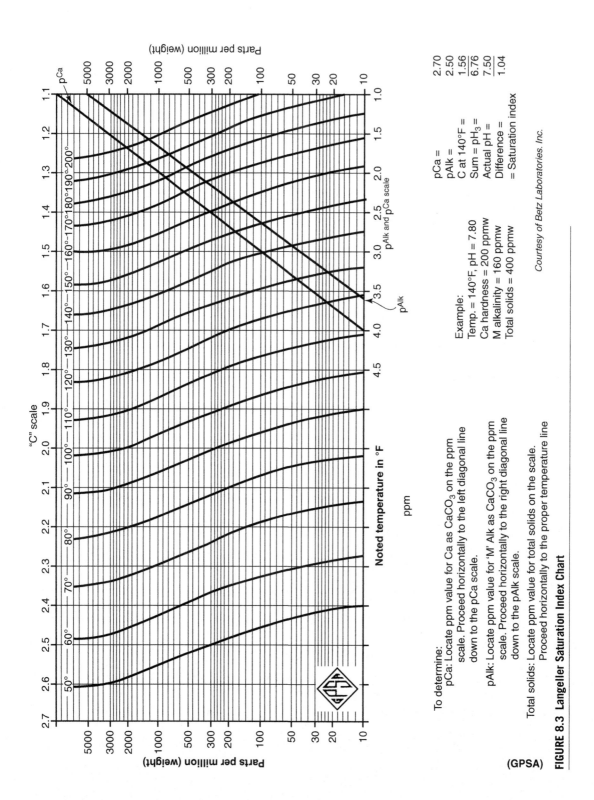

To determine:

pCa: Locate ppm value for Ca as $CaCO_3$ on the ppm scale. Proceed horizontally to the left diagonal line down to the pCa scale.

pAlk: Locate ppm value for 'M' Alk as $CaCO_3$ on the ppm scale. Proceed horizontally to the right diagonal line down to the pAlk scale.

Total solids: Locate ppm value for total solids on the scale. Proceed horizontally to the proper temperature line

Example:
Temp. = 140°F, pH = 7.80
Ca hardness = 200 ppmw
M alkalinity = 160 ppmw
Total solids = 400 ppmw

pCa =                     2.70
pAlk =                    2.50
C at 140°F =              1.56
Sum = $pH_3$ =           6.76
Actual pH =              7.50
Difference =             1.04
= Saturation index

Courtesy of Betz Laboratories. Inc.

(GPSA)   **FIGURE 8.3 Langelier Saturation Index Chart**

| Table 8.9 The Interpretation of the Results for Tendency of Water | |
|---|---|
| **pH–pH$_s$** | **Tendency of Water** |
| Positive value | Scale forming |
| Negative value | Corrosive |
| Zero | Neither scale forming nor corrosive |

Table 8.10 lists some of the common properties or characteristics and the normal constituents of water, together with corresponding associated operating difficulties and potential methods of water treatment.

## 8.10.4 QUALITY OF SOURCE WATERS

The type of water treatment depends on the quality of the source water and the quality desired in the finished water.

Adequate information on the source water is thus a prerequisite for design. This includes analysis of the water and where the supply is nonuniform, the ranges of the various characteristics.

The quality of many sources will change little over the lifetime of the treatment plant except for the seasonal changes that should be anticipated in advance. Other sources can be expected to deteriorate substantially as a result of an increase in wastes. A reasonably accurate prediction of such changes in quality is difficult to make.

In some instances, it is best arrived at by judgment based on past trends in quality, a survey of the source, and evaluation of future developments relating to the supply.

Ground water sources tend to be uniform in quality, to contain a greater amount of dissolved substances, to be free of turbidity, and to be low in color.

Surface water supplies receive greater exposure to wastes, including accidental spills of the variety of substances.

Generalizations like the earlier, although useful, are not a substitute for the definitive information required for plant design.

To provide adequate protection against pollution, special studies in the design of intakes should have to be made to indicate the most favorable locations for obtaining water.

In connection with deep reservoirs, multiple intakes offer flexibility in selecting water from various depths, thus overcoming poorer water quality resulting from seasonal changes.

For ground water sources, the location and depths of wells should be considered in order to avoid pollution and secure water of favorable quality.

As a typical guidance the raw water specifications of three refineries are presented in Tables 8.11, 8.12, and 8.13, for ground water, surface water, and seawater.

Water treating requirements for gas processing plants depend upon (1) the quality of the source or makeup water, (2) the manner in which the water is used, (3) environmental regulations, and (4) site climatic conditions governing wastewater disposal. These factors should be considered in selecting the overall plant process and utility systems.

Many gas processing plants, especially smaller plants, are designed to be "water free," utilizing air for all cooling services, a heating medium for process heat requirements, and electric motor drivers.

**Table 8.10 Common Characteristics and Impurities in Water**

| Constituent | Chemical Formula | Difficulties Caused | Means of Treatment |
|---|---|---|---|
| Turbidity | None, usually expressed in Jackson Turbidity Units | Imparts unsightly appearance to water; deposits in water lines, process equipment, boilers, etc.; interferes with most process uses. | Coagulation, settling, and filtration |
| Color | None | Decaying organic material and metallic ions causing color may cause foaming in boilers; hinders precipitation methods such as iron removal, hot phosphate softening; can stain product in process use | Coagulation, filtration, chlorination, adsorption by activated carbon. |
| Hardness | Calcium, magnesium, barium, and strontium salts expressed as $CaCO_3$ | Chief source of scale in heat exchange equipment, boilers, pipe lines, etc.; forms curds with soap; interferes with dyeing, etc. | Softening, distillation, internal boiler water treatment, surface active agents, reverse osmosis, electrodialysis |
| Alkalinity | Bicarbonate $(HCO_3^{-1})$, carbonate $(CO_3^{-2})$, and hydroxyl $(OH^{-1})$, expressed as $CaCO_3$ | Foaming and carryover of solids with steam; embrittlement of boiler steel; bicarbonate and carbonate produce $CO_2$ in steam, a source of corrosion | Lime and lime-soda softening, acid treatment, hydrogen zeolite softening, demineralization, dealkalization by anion exchange, distillation, degasifying |
| Free mineral acid | $H_2SO_4$, HCl, etc. expressed as $CaCO_3$, titrated to methyl orange end-point. | Corrosion | Neutralization with alkalies |
| Carbon dioxide | $CO_2$ | Corrosion in water lines and particularly steam and condensate lines | Aeration, deaeration, neutralization with alkalies, filming and neutralizing amines |
| pH | Hydrogen Ion concentration defined as $$pH = \log\frac{1}{(H^{+1})}$$ | pH varies according to acidic or alkaline solids in water; most natural waters have a pH of 6.0–8.0 | pH can be increased by alkalies and decreased by acids |
| Sulfate | $(SO_4)^{-2}$ | Adds to solids content of water, but, in itself, is not usually significant; combines with calcium to form calcium sulfate scale | Demineralization, distillation, reverse osmosis, electrodialysis |
| Chloride | $Cl^{-1}$ | Adds to solids content and increases corrosive character of water | Demineralization, distillation, reverse osmosis, electrodialysis |
| Nitrate | $(NO_3)^{-1}$ | Adds to solid content, but is not usually significant industrially; useful for control of boiler metal embrittlement | Demineralization, distillation, reverse osmosis, electrodialysis |
| Fluoride | $F^{-1}$ | Not usually significant industrially | Adsorption with magnesium hydroxide, calcium phosphate, or bone black; alum coagulation; reverse osmosis; electrodialysis |

**Table 8.10  Common Characteristics and Impurities in Water** *(cont.)*

| Constituent | Chemical Formula | Difficulties Caused | Means of Treatment |
|---|---|---|---|
| Silica | $SiO_2$ | Scale in boilers and cooling water systems; insoluble turbine blade deposits due to silica vaporization | Hot process removal with magnesium salts; adsorption by highly basic anion exchange resins, in conjunction with demineralization; distillation |
| Iron | $Fe^{+2}$ (ferrous) $Fe^{+3}$ (ferric) | Discolors water on precipitation; source of deposits in water lines, boilers, etc.; interferes with dyeing, tanning, paper mfr., etc. | Aeration, coagulation, and filtration, lime softening, cation exchange, contact filtration, surface active agents for iron retention |
| Manganese | $Mn^{+2}$ | same as iron | same as iron |
| Oil | Expressed as oil or chloroform extractable matter, PPMW | Scale, sludge and foaming in boilers; impedes heat exchange; undesirable in most processes | Baffle separators, strainers, coagulation and filtration, diatomaceous earth filtration |
| Oxygen | $O_2$ | Corrosion of water lines, heat exchange equipment, boilers, return lines, etc. | Deaeration, sodium sulfite, corrosion inhibitors, hydrazine or suitable substitutes |
| Hydrogen sulfide | $H_2S$ | Cause of "rotten egg" odor; corrosion | Aeration, chlorination, highly basic anion exchange |
| Ammonia | $NH_3$ | Corrosion of copper and zinc alloys by formation of complex soluble ion | Cation exchange with hydrogen zeolite, chlorination, deaeration, mixed-bed demineralization |
| Conductivity | Expressed as microhms, specific conductance | Conductivity is the result of ionizable solids in solution; high conductivity can increase the corrosive characteristics of a water | Any process which decreases dissolved solids content will decrease conductivity; examples are demineralization, lime softening |
| Dissolved solids | None | "Dissolved solids" is measure of total amount of dissolved matter, determined by evaporation; high concentrations of dissolved solids are objectionable because of process interference and as a cause of foaming in boilers | Various softening process, such as lime softening and cation exchange by hydrogen zeolite, will reduce dissolved solids; demineralization; distillation; reverse osmosis, electrodialysis |
| Suspended solids | None | "Suspended solids" is the measure of undissolved matter, determined gravimetrically; suspended solids plug lines, cause deposits in heat exchange equipment, boilers, etc. | Subsidence, filtration, usually preceded by coagulation and setting |
| Total solids | None | "Total solids" is the sum of dissolved and suspended solids, determined gravimetrically | See "Dissolved Solids" and "Suspended Solids" |

**Table 8.11  Raw Water Specification—Sample 1**

| Source | | Ground (Well) Water (mg/kg) mg/L |
|---|---|---|
| pH | | 8.1 |
| Total hardness | as $CaCO_3$ | 181 |
| Calcium | as $CaCO_3$ | 123 |
| Magnesium | as $CaCO_3$ | 58 |
| Totalalkalinity | as $CaCO_3$ | 165.6 |
| Sodium | as $CaCO_3$ | 55 |
| Sulfate | as $CaCO_3$ | 43 |
| Potassium | as $CaCO_3$ | 3 |
| Chloride | as $CaCO_3$ | 25.4 |
| Nitrate | as $CaCO_3$ | 5 |
| Silica | as $SiO_2$ | 16 |
| Total iron | | Traces |
| Suspended solids | | 5 |
| Dissolved solids | | 356.7 |
| Chemical oxygen demand (COD) | | 1 |

**Table 8.12  Raw Water Specification—Sample 2**

| Source Constituent | Surface (River) Water mg/L (mg/kg) as $CaCO_3$ |
|---|---|
| Calcium | 146 |
| Magnesium | 45 |
| Sodium | 73 |
| Total cations | 264 |
| Bicarbonate | 146 |
| Chloride | 50 |
| Sulfate | 68 |
| Nitrate | Traces |
| Total anions | 264 |
| Total suspended solids (as ions) | 0–10 |
| Iron as Fe | Traces |
| Silica as $SiO_2$ | 8.0 |
| Temperature range | 15.6–37.8°C (60–100°F) |
| pH | 7.9 |

**Table 8.13  Raw Water Characteristics—Sample 3**

| Source | | Ground (Well) Water (mg/kg) mg/L |
|---|---|---|
| pH at 35°C | | 8.3 |
| Sulfate | As $CaCO_3$ | 3,650 |
| Chloride | As $CaCO_3$ | 25,140 |
| Calcium | As $CaCO_3$ | 550 |
| Magnesium | As $CaCO_3$ | 1,750 |
| Bicarbonate | As $CaCO_3$ | 175 |
| Nitrate | As $CaCO_3$ | 0.9 |
| Nitrite | As $CaCO_3$ | 0.01 |
| Silica | As $SiO_2$ | 2.7 |
| Dissolved solids | | 47,000 |
| Relative density (sp. gr.) at 20°C | | 1.032 |
| Temperature range | | 20–35°C |

Such plants have essentially no makeup water requirements and wastewater treatment requirements are minimized.

### 8.10.4.1 Source waters

Source or makeup water is normally either ground water or surface water, neither of which is ever chemically pure. Ground waters contain dissolved inorganic impurities, which came from the rock and sand strata through which the water passed. Surface waters often contain silt particles in suspension (suspended solids) and dissolved organic matter in addition to dissolved inorganic impurities (dissolved solids).

## 8.10.5 PRELIMINARY WATER TREATMENT

Regardless of the final use of source water and any subsequent treatment, it is often advisable to carry out general treatment close to the intake or well. The purpose is to protect the distribution system itself and at the same time to provide initial or sufficient treatment for some of the main uses of water.

In the case of surface water, general protection should be provided against clogging and deposits.

The obstruction or clogging of apertures and pipes by foreign matter can be avoided by screening or straining through a suitable mesh. The protection used is either a bar screen, in which the gap between the bars can be as narrow as 2 mm, or a drum or belt filter, with a mesh of over 250 μm.

According to the requirements of the equipment and the amount of pollution (slime) in the water a 250 μm filter may be used on an open system, or microstraining down to 50 μm may be necessary in certain specific cases. In some cases, rapid filtration through silicous sand may be necessary after screening and will eliminate suspended matter down to a few micrometers. Where there is a large amount of suspended matter, grit removal and/or some degree of settling should be provided.

In the case of ground water the main risks are abrasion by sand or corrosion.

For abrasion the pumps should be suitably designed, and the protection, which concerns only the parts of the system downstream, the pumps will take the form of very rapid filtration through sand, straining under pressure of use of hydrocyclones, if the grit is of the right grain size.

Corrosion frequently occurs on systems carrying underground water and leads to the formation of tuberculiform concretions, which must not be confused with scale. This corrosivity is often caused by the lack of oxygen. The best method, therefore, of preventing corrosion is by oxygenation and filtration processes that have the dual advantage of removing the grit and any iron present, and of feeding into the water the minimum amount of oxygen needed for system to protect itself.

• Water analysis

Water analyses are conventionally expressed, for both cations and anions, in parts per million by weight (ppmw) except for hardness and alkalinity which are usually expressed in ppmw of calcium carbonate ($CaCO_3$). These ppmw values can be converted to a common basis (such as milliequivalents/L) by dividing by the equivalent weight of the ion and multiplying by the specific gravity of the water solution. This permits the summation of oppositely charged ions such that total cations will then equal total anions. Cation and anion concentrations in milliequivalents/L can be converted to ppmw $CaCO_3$ by multiplying by the equivalent weight of $CaCO_3$ ($100.08/2 = 50.04$) and dividing by the relative density of the water solution (Table 8.23).

## 8.10.6 TREATMENT PROCESS SELECTION

The quality of the source, giving due consideration to variations and possible future changes, the quality goals for the finished water, and cost, shall form the basis for selecting a treatment process.

Often various types and combination of treatment units would be used to achieve the performance desire. Determination of the most suitable plan should be on a comparative cost study, which includes an evaluation of the merits and liabilities of each proposal.

The experience acquired through treatment of the same or similar source shall provide an excellent guide in selecting a plan.

Where experience is lacking or where there is the desire for a different degree of performance, special studies should be indicated. Tests conducted in the laboratory, in existing plants, or in pilot plants should then be employed to obtain information for design purposes.

## 8.10.7 PLANT SIZING AND LAYOUT

For plant sizing and layout the following considerations should be followed:

1. coordination of plant elements to provide for efficient production of a high quality effluent. Major considerations in treatment plant design include, frequency of basin cleaning, length of filter runs, and effluent quality
2. economic but durable construction
3. simplification of equipment and operations
4. centralization of operations and control
5. chemical feed lines as short and direct as reasonably possible
6. simplified chemical handling and feeding
7. essential instrumentation only
8. appropriate use of automation.

## 8.10.8 **POTABLE WATER QUALITY**

In refineries and/or plants water systems the minimum specification required for potable water should be as per local authority of municipal water supply, and is recommended to be based on the establishment of water system performance goals and potable water standard specifications as required by relevant national standards.

## 8.10.9 **BOILER WATER QUALITY CRITERIA**

### 8.10.9.1 Boiler water chemistry

There are four types of impurities of concern in water to be used for the generation of steam:

1. Scale-forming solids, which are usually the salts of calcium and magnesium along with boiler corrosion products. Silica, manganese and iron, can also form scale.
2. The much more soluble sodium salts which do not normally form scale, but can concentrate under scale deposits to enhance corrosion or in the boiler water to increase carryover due to boiler water foaming.
3. Dissolved gases, such as oxygen and carbon dioxide, which can cause corrosion.
4. Silica, which can volatilize with the steam in sufficient concentrations to deposit in steam turbines.

### 8.10.9.2 Boiler blowdown

Blowdown should be employed to maintain boiler water dissolved solids at an appropriate level of concentration. At equilibrium the quantities of dissolved solids removed by blowdown exactly equals those introduced with the feedwater plus any injected chemicals.

### 8.10.9.3 ABMA and ASME standard

The American Boiler Manufacturers' Association (ABMA) and the American Society of Mechanical Engineers (ASME) have developed suggested limits for boiler water composition, which depend upon the type of boiler and the boiler operating pressure. These control limits for boiler water solids are based on one or more of the following factors.

### 8.10.9.4 Sludge and total suspended solids

These result from the precipitation in the boiler of feedwater hardness constituents due to heat and to interaction of treatment chemicals, and from corrosion products in the feedwater. They can contribute to boiler tube deposits and enhance foaming characteristics, leading to increased carryover.

### 8.10.9.5 Total dissolved solids

These consist of all salts naturally present in the feedwater, of soluble silica, and of any chemical treatment added. Dissolved solids do not normally contribute to scale formation, but excessively high concentrations can cause foaming and carryover or can enhance "underdeposit" boiler tube corrosion.

### 8.10.9.6 Silica

This may be the blowdown controlling factor in softened water containing high silica. High boiler water silica content can result in silica vaporization with the steam, and under certain circumstances, silicous scale. This is illustrated by silica solubility data in Fig. 8.4 where silica content of boiler water is not as critical for steam systems without steam turbines.

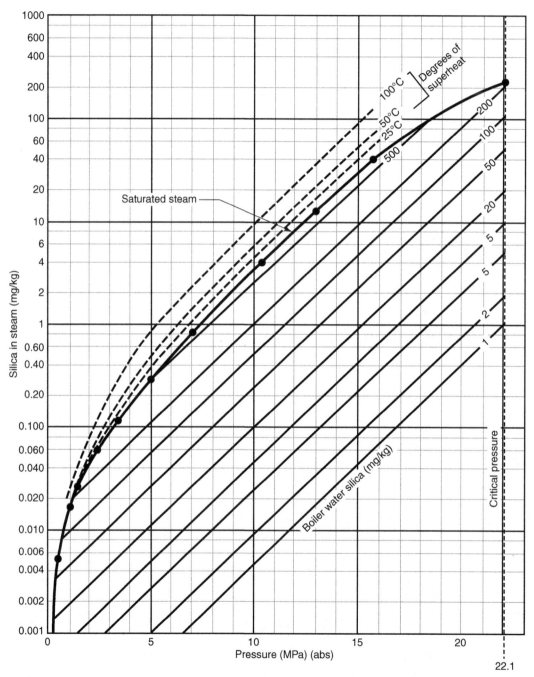

**FIGURE 8.4** Relationship Between Boiler Pressure, Boiler Water Silica Content, and Silica Solubility in Steam (GPSA 2004)

### 8.10.9.7 Iron

Occasionally in high pressure boilers where the iron content is high in relation to total solids, blowdown may be based upon controlling iron concentrations, high concentrations of suspended iron in boiler water can produce serious boiler deposit problems and are often indications of potentially serious corrosion in the steam/steam condensate systems.

- While there are other considerations (such as corrosive or deposit forming tendencies) in establishing limits for boiler water composition, the ABMA recommendations as per Table 2 clearly indicate that boiler feedwater purity becomes more important as operating pressures increase.

### 8.10.9.8 Common deposits formed in water systems

The deposits may be classified generally as scale, sludge, corrosion products, and biological deposits. The more common types of deposits are shown in Table 8.14. Table 8.15 shows recommended boiler water limits and associated steam purity at steady state full load operation drum type boilers.

## 8.10.10 PROCESSES DESIGN

### 8.10.10.1 Coagulation and flocculation

In water treatment, coagulation is defined as a process by which colloidal particles are destabilized, and is achieved mainly by neutralizing their electric charge. The product used for this neutralizing is called a coagulant.

Flocculation is the massing together of discharged particles as they are brought into contact with one another by stirring. This leads the formation of flakes or floc. Certain products, called flocculating agents, may promote the formation of floc.

Separation of the floc from the water can be achieved by filtration alone or by settling.

Coagulation and flocculation are frequently used in the treatment of potable water and preparation of process water used by industry.

Certain dissolved substances can also be adsorbed into the floc (organic matter, various pollutants, etc.).

### 8.10.10.2 Main coagulants

The most widely used coagulants are based on aluminum or iron salts. In certain cases, synthetic products, such as cation polyelectrolytes, can be used.

Cation polyelectrolytes are generally used in combination with metal salt, greatly reducing the salt dosage, which would have been necessary. Sometimes no salts at all are necessary, and this greatly reduces the volume of sludge produced.

### 8.10.10.3 pH value for coagulation and dosage

For any water, there is an optimum pH value, where good flocculation occurs in the shortest time with the least amount of chemical.

For actual application of coagulating agents, the dosage and optimum pH range should be determined by coagulation control or a jar test.

**Table 8.14 Common Deposits Formed in Water Systems**

| Name | Chemical Composition | Deposit Formation Conditions $T < 100°C$ With or Without Evaporation | $T > 100°C$ No Evaporation | $T > 100°C$ Evaporation | Water Vapor or Steam |
|---|---|---|---|---|---|
| Acmite | $Na_2O.Fe_2O_3.4SiO_2$ | | | × | |
| Analctive | $Na_2O.Al_2O_3.4SiO_2.2H_2O$ | | | × | × |
| Anhydrite | $CaSO_4$ | | × | × | |
| Aragonite | $CaCO_3$ | × | × | × | |
| Biological | | | | | |
| (a) Non-spore bacteria | | × | | | |
| (b) Spore bacteria | | × | | | |
| (c) Fungi | | × | | | |
| (d) Algae and diatoms | | × | | | |
| (e) Crustaceans | | × | | | |
| Brucite | $Mg(OH)_2$ | | × | × | |
| Burkeite | $Na_2CO_3.2Na_2SO_4$ | | | | × |
| Calcite | $CaCO_3$ | × | × | × | |
| Calcium hydroxide | $Ca(OH)_2$ | | | × | |
| Carbonaceous | | × | × | × | × |
| Copper | $Cu$ | | | × | |
| Cuprite | $Cu_2O$ | | × | | |
| Ferrous oxide | $FeO$ | | × | | |
| Goethite | $Fe_2O_3.H_2O$ | × | × | × | |
| Gypsum | $CaSO_4.2H_2O$ | × | × | × | |
| Halite | $NaCl$ | | | | × |
| Hematite | $Fe_2O_3$ | | | × | |
| Hydroxylapatire | $Ca_{10}(PO_4)_6(OH)_2$ | × | × | × | |
| Magnesium Phosphate (basic) | $Mg_3(PO_4)_2.Mg(OH)_2$ | | × | × | |
| Magnetite | $Fe_3O_4$ | | × | × | × |
| Oil (chloroform extractable) | | × | × | × | × |

**Table 8.14  Common Deposits Formed in Water Systems**

| Name | Chemical Composition | Deposit Formation Conditions | | | |
|---|---|---|---|---|---|
| | | $T < 100°C$ With or Without Evaporation | $T > 100°C$ No Evaporation | $T > 100°C$ Evaporation | Water Vapor or Steam |
| Quartz | $SiO_2$ | | | | × |
| Serpentine | $3MgO.2SiO_2.2H_2O$ | | × | × | |
| Siderite | $FeCO_3$ | | | | × |
| Silica (amorphous) | $SiO_2$ | | | | × |
| Sodium carbonate | $Na_2CO_3$ | | | | × |
| Sodium disilicate | $Na_2Si_2O_6$ | | | | |
| Sodium ferrous phosphate | $NaFePO_4$ | | | × | × |
| Sodium silicate | $Na_2SiO_3$ | | | × | × |
| Tenorite | $CuO$ | | | | |
| Thenardite | $Na_2SO_4$ | | | × | × |
| Xonotlite | $5CaO.5SiO_2.H_2O$ | | | × | |

Note: Symbol "×" means the formation of deposits in the relevant conditions.

**Table 8.15  Recommended Boiler Water Limits and Associated Steam Purity at Steady State Full Load Operation Drum Type Boilers**

| Drum Pressure Bar (ga) | Range Total Dissolved Solids Boiler Water, mg/kg (max.) | Range Total Alkalinity, mg/kg (max.) | Suspended Solids Boiler Water, mg/kg (max.) | Range Total Dissolved Solids Steam, mg/kg (max. Expected Value) |
|---|---|---|---|---|
| 0.20–69 | 700–3500 | 140–700 | 15 | 0.2–1.0 |
| 20.76–31.03 | 600–3000 | 120–600 | 10 | 0.2–1.0 |
| 31.10–41.38 | 500–2500 | 100–500 | 8 | 0.2–1.0 |
| 41.45–51.72 | 400–2000 | 80–400 | 6 | 0.2–1.0 |
| 51.79–62.07 | 300–1500 | 60–300 | 4 | 0.2–1.0 |
| 62.14–68.96 | 250–1250 | 50–250 | 2 | 0.2–1.0 |

### 8.10.10.4 Choice of coagulant

Coagulant should be chosen after the raw water examination in the laboratory by means of a flocculation test, while considering the following factors:

1. nature and quantity of the raw water
2. variations in the quality of the raw water (daily or seasonal especially with regard to temperature)

3. quality requirements and use of the treated water
4. nature of the treatment after coagulation (filter coagulation, settling)
5. degree of purity of reagents, particularly in the case of potable water.

### 8.10.10.5 Sedimentation

1. The process by which suspended or coagulated material separates from water by gravity is called sedimentation.
2. Sedimentation alone is an effective means of water treatment but is made more effective by coagulation.
3. Presedimentation basins or sand traps are sometimes used when waters to be treated contain large amounts of heavy suspended solids. This decreases the amount of sediment which accumulates in the sedimentation basin as a result of the coagulation and sedimentation process.
4. If water is to be filtered in the course of treatment, coagulation and sedimentation will reduce the load on filters.

### 8.10.10.6 Type of sedimentation tanks

1. The effectiveness of a sedimentation tank depends on the settling characteristics of the suspended solids that are to be removed and on the hydraulic characteristics of the settling tank.
2. The hydraulic characteristics of a settling tank depend on both the geometry of the tank and the flow through the tank.
3. Most sedimentation tanks used in water purification today are of the horizontal-flow type.
4. Horizontal-flow tanks may be either rectangular or circular in plan. Circular, horizontal-flow tanks may be either center feed with radial flow, peripheral feed with radial flow, or peripheral feed with spiral flow.
5. Fig. 8.5 shows the flow patterns in horizontal-flow type of sedimentation tanks.
6. In horizontal-flow tank design, the aim should be to achieve as nearly as possible the ideal condition of equal velocity for all points lying on each vertical line in the settling zone (The ideal basin condition). This, in effect would be complete separation of the four zones of the tank.
7. The sedimentation basins should be equipped with mechanical equipment for continuous removal of settled solids.

## 8.10.11 PRACTICAL SEDIMENTATION BASIN

1. The situation in practical sedimentation basins is modified because of the relative density (specific gravity) and shape of the particles, coagulation of particles, concentration of particles, and movement of water through the settling tank.
2. The relative density of suspended matter may vary from 2.65 for sand to 1.03 for flocculated particles or organic matter and mud containing 95% water.
3. Floc particles resulting from coagulation with aluminum compounds have a relative density of about 1.18, and those obtained using ferrous sulfate as a coagulant have a relative density of 1.34. These values can be increased by clay or silt or decreased by organic matter. However, most of the particles in a settling basin settle at velocities within Stock's law.
4. Because of the difference in shape, size, and relative density of particles, there is a wide range of settling velocities. This results in some subsiding particles overtaking others, thus increasing the natural tendency of suspended matter to flocculate.

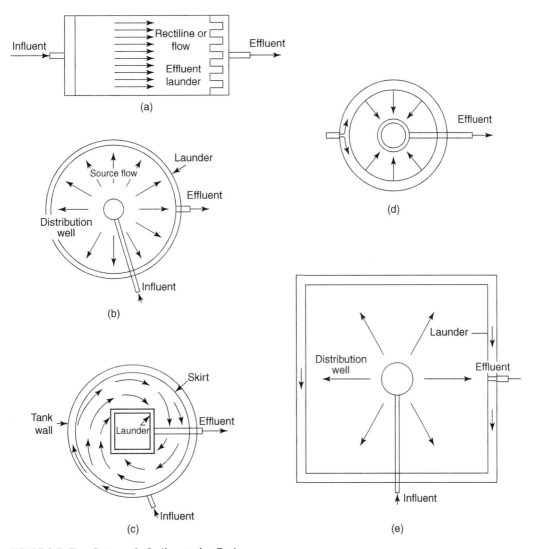

**FIGURE 8.5  Flow Patterns in Sedimentation Tanks**

(a) Rectangular settling tank, rectilinear flow, (b) center-feed settling tank, source flow, (c), (d) and (e) peripheral-feed settling tank.

## 8.10.12  FACTORS INFLUENCING THE DESIGN OF SEDIMENTATION BASINS

Sedimentation basins are often designed on the basis of existing installations, which are handling the same type of water. Experience and judgment of the engineer are also instrumental in the design. However, there are some important points, other than structure, which should be considered in the design of a basin.

The basin should be large enough to insure an adequate supply of treated water during periods of peak load.

The characteristics and type of water treatment also affect the design of the basin. Such things as the nature of the suspended material and the amount and type of coagulant needed, if any, should also be considered.

The influence of temperature is also important, since the viscosity of the water is less on a warm summer day than in cold weather.

The number of basins depends upon the amount of water and the effect of shutting a basin down. It is desirable to have more than one basin to provide for alternate shutdown of individual basins for cleaning or repairs.

Basins vary in shape—square, rectangular, and round. However, regardless of shape, most basins have slopping bottoms to facilitate the removal of deposited sludge.

Sedimentation basins are equipped with inlets in order to distribute the water uniformly among the basins and uniformly over the cross-section of each basin. Inlet and outlets should be designed to avoid short circuiting through the basin.

## 8.10.13 HYDRAULIC PROPERTIES OF SEDIMENTATION BASIN

### 8.10.13.1 Surface area or surface overflow rate

The surface area of the tank is one of the most important factors that influences sedimentation. For any particular rate of inflow the surface area provided determines the tank overflow rate, $v = Q/A$. If there were such a thing as an ideal tank, the tank overflow rate could be made equal to the settling velocity of the particles that the tank was designed to remove.

Because no ideal tank exists, it is customary to reduce the tank overflow rate and to increase the detention time over those indicated by theoretical analysis. It is recommended that a correction factor of 1–1.25 to both values when settling a discrete solid is applied.

For the sedimentation of flocculent particles from dilute suspensions, the settling velocity will generally be decreased by a factor of 1.25–1.75.

The net effect of these corrections is to provide the range of settling velocities for different applications summarized in Table 8.16.

The higher settling velocities or tank overflow rates should be used for warmer water; the lower settling velocities for colder waters.

The settling velocity used in the settling tank design overflow rate is one of the major factors determining tank efficiency.

### 8.10.13.2 Depth

1. The theoretical detention time is equal to the volume of the tank divided by the flow rate. Hence, if A and Q are constant, the theoretical detention time is directly proportional to the tank depth.
2. As the performance of the tank depends on the flocculation of the suspended solids, and the degree of flocculation depends on the detention time, the tank performance in removal of flocculent particles will depend on its depth.

**Table 8.16  Typical Sedimentation Tank Overflow Rates**

| Type of Water | Treatment | Overflow Rate ($m^3/m^2$ h) |
|---|---|---|
| Surface water | Alum floc | 0.61–0.93 |
| Surface or ground water | Lime softening | 0.93–1.54 |
| | Clarification in upflow-units | 2.44–4.52 |
| | | 1.83–2.44 (cold water) |
| | | 2.44–3.66 (warm water) |
| | Softening in upflow-units | 1.83–6.11 |
| | | To 3.06 (surface) |
| | | To 4.40 (well) |

3. The efficiency of removal, however, is not linearly related to the detention time. For example, if 80% of the suspended solids were removed with a detention time of 2 h, a detention time of 3 h might remove only 90%.
4. The raw water entering a sedimentation tank will have a greater density than the water in the tank, as it will contain more suspended solids. The heavier influent water will tend to form density currents and move toward the bottom of the tank, where it can interfere with the sedimentation process. Density currents are more apt to occur in deep tanks.
5. Sedimentation basins are commonly designed to remove solids resulting from chemical coagulation of surface water and lime soda as softening of surface and ground waters. In a properly designed basin, a detention time of between 2–4 h is usually sufficient to prepare the water for subsequent filtration. When the water is to be used without filtration, longer detention time (up to 12 h) may be provided.

### 8.10.13.3 Velocity through basin
The velocity of flow through a settling basin will not be uniform over the cross section perpendicular to the flow even though the inlets and outlets are designed for uniform distribution. The velocity will not be stable because of density currents, and the operation of the sludge removal mechanism. In order to minimize these disturbances, the velocity through a sedimentation tank should be kept between 0.0026 and 0.015 m/s.

### 8.10.13.4 Inlet and outlet conditions
1. The inlet to a sedimentation tank should be designed to distribute the water uniformly between basins and uniformly over the full cross section of the tank.
2. The inlet is more effective than the outlet in controlling density and internal currents, and tank performance is affected more by inlet than by outlet conditions.
3. The best inlet is one that allows the water to enter the settling tank without the use of pipelines or channels.

4. The head loss in preamble baffle ports or basin inlet ports should be relatively large compared to the kinetic energy of the water moving past the permeable ports. This is required to assure equal distribution of flow between tanks and between inlet ports.

5. As flocculent solids will frequently be involved, the velocities in the influent channels must be kept low, usually between 0.15 and 0.60 m/s, to prevent break up of the floc. Similar low velocities are required through the inlet ports to reduce the danger of inertial currents interfering with sedimentation.

6. It has been found that relatively minor changes in an inlet can completely change the hydraulic performance of a settling tank.

7. The main purpose of the inlet is to provide a smooth transition from the relatively high velocities in the influent pipe to the very low uniform velocity distribution desired in the settling zone, in such a way that interference with the settling process is minimal.

8. The purpose of the outlet is the same except that the transition is from the settling zone to the effluent pipe.

9. The water level in settling basins is usually controlled at the outlet. This control, however, may be set by means other than the outlet weir, as for example, by a succeeding unit.

10. It may be desirable to encourage deliberate fluctuation of water level in the settling basins to make use of the storage in them, or to break up ice.

11. Basin outlets are often of the V-notch weir type, and these are quite often provided with means for vertical adjustment to aid in control of the overflow. The V-notches help in keeping a uniform flow over the weir at low water levels.

12. The effect of weir rates, cubic meters per hour per meter of weir, on sedimentation, is not well known, but weir rates are usually limited to commonly accepted values (Table 8.17).

13. Circular basins with the inlet at one side and the outlet on the opposite side are not very efficient because of dead areas in the tank and short circuiting of water flow across the tank. The efficiency of circular tanks is much greater if the water is fed to the tank from an inverted siphon located in the center of the tank, and the effluent taken from a weir passing around the entire periphery.

14. A square basin may be operated in the same manner or may be fed from one side with effluent removed from the opposite side.

15. The use of baffles in sedimentation tanks should be limited to the inlets and outlets and as remedial measures in poorly designed tanks.

**Table 8.17 Typical Weir Overflow Rates**

| Type of Service | Weir Overflow Rate (m³/m h) |
| --- | --- |
| Water clarification | <26 |
| Water treatment | |
| Light alum floc (low-turbidity water) | 6–7.5 |
| Heavier alum flow (higher-turbidity water) | 7.5–11.2 |
| Heavy floc from lime softening | 11.2–13.4 |

### 8.10.13.5 Sludge handling

1. The bottom of a settling tank is normally sloped gently toward a sludge hopper where the sludge is collected.
2. The sludge usually moves hydraulically toward the hopper.
3. Sludge scraper mechanisms are used to prevent the sludge from sticking to the bottom and to help its flow.
4. The sloping bottom and the sludge hopper provide a certain amount of storage space for the sludge before it is removed.
5. The movement of the sludge scrapper mechanism should be quite slow so as not to disrupt the settling process or to resuspend the settled sludge. The velocity of the scrapers should be kept below 18.3 m/h for this reason.
6. Some scrapper mechanisms in circular tanks carry vacuum suction pipes instead of squeegees for removing relatively light, uniform solids.

## 8.10.14 EQUIPMENT USED IN CLARIFICATION

The equipment used for clarification can be of many types; however, the equipment used should provide the correct environment to carry out each step: coagulation, flocculation, and sedimentation.

Older design for clarification units provided separate chemical addition, flash mixing, flocculating, and settling facilities. Modern combined units provide all three steps in one unit, such as sludge recirculation (solids contact) type or sludge blanket unit and so on.

In theory, the sludge blanket unit provides better clarification than the sludge recirculation type as a result of filtering action provided by the sludge bed and the gentle handing of the flocs.

A basic clarification system consists of the clarifier and a chemical feeding system, which meters chemical additives in proportion to flow (as per Standard E-PR-492, "Process Requirements of Chemical Injection Systems").

The size of a standard clarification unit is based upon an upflow rate of approximately 2.5 m$^3$/m$^2$h (1.2–3.7 range), with a total retention time of 1.5–4 h. The clarified water will contain approximately 5–10 mg/kg of suspended matter.

Fig. 8.6 illustrates the addition of "tubes" to a clarifier. These tubes either substantially increase the allowable upflow rate or reduce carryover at the same upflow rate.

Greater throughput capacities can be achieved by the parallel plate of a "lammella" type separator shown in Figs. 8.7 and 8.8.

Design criteria of clarifiers should be based on steady operation at maximum load. However, it is expected that the actual load will fluctuate over the range from 10 to 100% of design flow rate and the clarifier should have the capacity to perform satisfactorily under these conditions.

Consideration should also be given to anticipation of flow rate limitations and chemical dosages during difficult treatment periods considering high turbidity, low temperature, and/or polluted conditions. The manufacturer shall either take such potential difficulties into consideration in design or state the limitations imposed by such conditions.

## 8.10.15 FILTERS FOR WATER TREATING SYSTEMS

### 8.10.15.1 Design criteria

Multiple units shall be provided to allow continuous operation at full system design capacity with two units out of service (eg, one unit shutdown for maintenance and one unit in backwash mode).

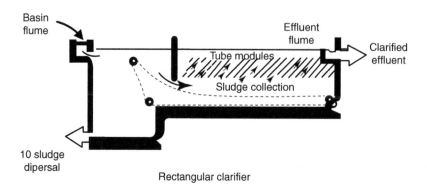

Rectangular clarifier

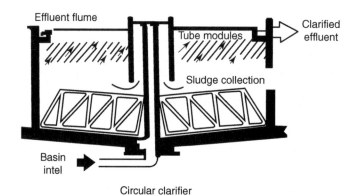

Circular clarifier

**FIGURE 8.6 Typical Tube Settler Arrangement**

Design service flow rates shall be as described in Table 8.18.

Air securing shall be used for units treating effluent at temperatures less than 93°C. The design air scour rate shall be 90 m³/m²h minimum.

If plant air is unavailable, a separate air compressor shall be included within the system.

Subsurface washers shall be furnished for units treating effluent at temperatures 93°C and greater. Subsurface wash rate shall be 12.3 m³/m²h minimum.

Design backwash rates can be per the following Table 8.19:

Bed depth in filters shall be 750 mm minimum. For "in-depth" filters at least two different density media, of different sizes, shall be furnished.

For hot pressure type units only washed anthracite coal should be used.

Freeboard shall be a minimum of two-thirds total bed depth, measured from the top of the filter media to the tangent line at the top of vessel.

Filter media traps shall be furnished on the outlet of each pressure filter unit to prevent filter media from entering downstream equipment in the event of underdrain failure. Maximum pressure drop through the trap shall not exceed 35 kPa when the unit is operating at maximum design flow rate.

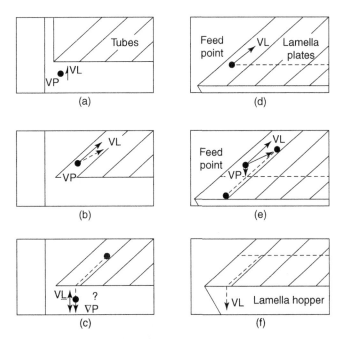

**FIGURE 8.7** Lamella Gravity Settler Versus Tube-Type Setter

(a) Particles are carried up to tubes because their downward velocity ($V_p$) is less than the rise rate ($V_L$) of the liquid; (b) particles enter tubes where the resultant vector force carries them to the tube walls; (c) particles slide down tube walls. However, as particles leave tubes $V_L$ is still greater than $V_p$. unless the particles have flocculated in the tubes. Thus, any increase in capacity achieved is limited lo this further flocculation in the tubes. (d) Feed containing particles is presented to Lamella plates from the side by means of a bottomless feed box; (e) particles enter Lamella plates where the resultant vector force carries them to the plate surfaces. However, unlike the tube settler, the particles slide down the plate past the point of entry into an area of "zero" velocity a feature found only in Lamella design; and (f) particles exit Lamella plates into a quiescent zone where there is no upward liquid velocity to hinder solids settling.

Characteristics of potable water filters should be as per Table 8.20.

Anthracite or marble should be used instead of quartz sand, when any trace of silica must be avoided in an industrial process or when they are easier to obtain.

## 8.10.16 QUANTITIES OF SUSPENDED SOLIDS WHICH CAN BE REMOVED BY BED FILTRATION

The following consideration should be made as a guiding principle:

**1.** The suspended solids lodge between the grains of the filter material.

Since sufficient space should always be left for the water to percolate, the sludge should not, on average, fill more than one quarter of the total volume of voids in the material.

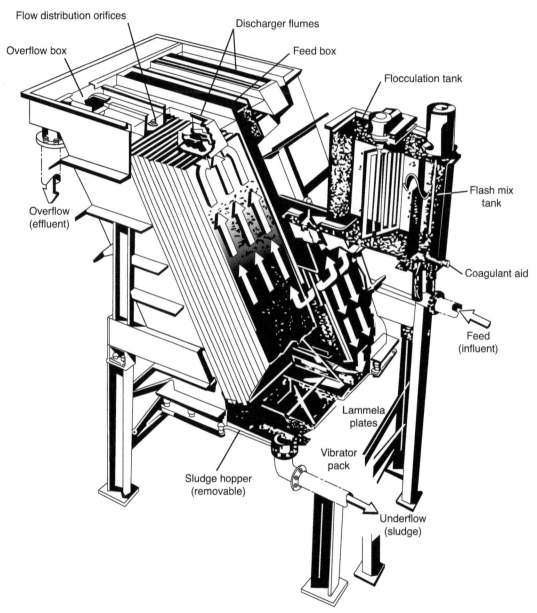

**FIGURE 8.8 Closely Spaced Inclined Plates (Lamella) Multiply the Available Settling Surface in a Small Volume and Reduce Installation Space**

**Table 8.18  Flow Rates in Different Types of Water Filters**

| S. No. | Unit | Flow Rate (Maximum With one Unit Backwashing, $m^3/m^2$ h) |
|--------|------|-------------------------------------------------|
| 1. | Downflow, cold pressure type ($<65°C$) | 9.7 |
| 2. | Downflow, hot pressure type ($>65°C$) | 10.8 |
| 3. | Downflow, gravity type | 9.7 |
| 4. | Upflow type | [a] |

[a]Vendor's recommendation if not specified in the project specification.

**Table 8.19  Minimum Backwash Rates in Different Types of Water Filters**

| Filter Media | Minimum Backwash Rate ($m^3/m^2$ h) |
|--------------|-------------------------------------|
| Sand | 36.7 |
| Anthracite coal | 29.2 |
| Activated carbon | 24.5 |

**Table 8.20  Characteristics of Potable Water Filters**

| Filter Type | Permissible Filtration Rates ($m^3/m^2$ h) | | Design Pretreatment to Reduce Turbidity in Applied Water to (mg/kg) | | Head Required (m) | | Length of Filter run (h) | | Min. Thickness (mm) | |
|-------------|----------|----------|-----|------|--------------|------|------|------|--------|------|
| | Max. day | Max. Rate | Avg. | Max. | Clean Filter | Max. | Avg. | Min. | Gravel | Sand |
| Rapid-sand gravity | 4.9 | 12.2 | 2 | 5 | 0.3 | 2.4 | 36 | 5 | 304.8 | 508 |
| Pressure | 4.9 | 12.2 | 2 | 5 | 0.3 | 7.6 | 48 | 5 | 304.8 | 609.6 |
| Slow-sand | 2.4 | 7.3 | 1 | 3 | 0.6 | 1.2 | 1000 | 250 | 304.8 | 1066.8 |
| Diatomite | 2.4 | 7.3 | 1 | 3 | 2.1 | 21.3 | 6 | 0.5 | — | — |

**2.** Irrespective of grain size, one cubic meter of filtering material contains about 0.45 $m^3$ of voids, the volume available for the retention of particles is about 0.11 $m^3$, provided that the effective grain size of the filtering medium is suitable to the nature of the particles.

**3.** When the suspended solids are based on colloidal floc, their dry matter content does not exceed 10 $kg/m^3$; the quantity that can be removed per $m^3$ of filter material is therefore no more than $0.11 \times 10 = 1.1$ kg.

## 8.10.17 PROCESS USED FOR BOILER FEED WATER TREATMENT

### 8.10.17.1 Hot process treater

Vessel should be of sludge blanket type employing a central downcomer. A separate (not integral) deaerator compartment is preferred.

The treater should be so sized that the rising rate of settled water is such that effluent-treated water has a turbidity of less than 10 mg/kg.

Separate clean and dirty backwash compartments shall be sized to meet normal filter backwash and sodium zeolite regeneration requirements without increasing flow rate through the unit to more than 10% of normal design.

The clean backwash water compartment should be replenished by filtered water at a much slower rate than the backwash rate.

The dirty backwash water should be returned at a set rate so that heat and water are recovered.

Treaters shall be insulated.

The treatment should be carried out at low pressure corresponding to vapor pressure for temperatures chosen between 102 and 115°C as required.

Units shall be designed for continuous service and uninterrupted operation for a period of 2 years.

All equipment shall be suitable for unsheltered outdoor installation for the climatic zone specified.

The total detention time of the vessel should not be less than 90 min at rated capacity of flow.

The maximum allowable upflow rate (rinse rate) through the unit shall be 3.7 $m^3/m^2h$ at water temperatures above 90°C. This rate shall not be exceeded when backwashing filters rinsing softeners. This rate shall be reduced to 1.0 $m^3/m^2h$ or less for waters containing appreciable organic matter turbidity or magnesium to meet guaranteed effluent turbidity of less than 10 nephelometric[1] units for a range of 10–100% of design raw water throughput.

Chemical mix tanks and pumps should be provided for hot-process treater.

Incoming water should be provided at a pressure sufficient to overcome the following losses:

1. pipe friction
2. static head to the top of the softener
3. vent condenser
4. spray nozzle
5. water flowmeter
6. water level control valve
7. vessel operating pressure (exhaust steam pressure).

### 8.10.17.2 Ion exchange

Classification of ion exchange resins: Ion exchange resins are classified according to their specific application as per Table 8.21 (see DIN 19633).

### 8.10.17.3 Design criteria

Design criteria for an ion exchange system should be based upon:

1. the required flow rate
2. influent water quality

---

[1]An instrument for determining the concentration of particle size of suspensions by means of transmitted or reflected light.

**Table 8.21  Classification of Ion Exchange Resins**

| Type | Application[a] | Ionic Form in the Ready-to-use Condition | Regenerating Agent. Aqueous Solution of |
|---|---|---|---|
| Cation exchange resins | | | |
| Strongly acidic | (1) | Na | NaCl |
| | (3) | H | HCl, $H_2SO_4$ |
| Weakly acidic | (2) | H | HCl, $H_2SO_4$, $CO_2$ |
| | (5) | H, Na | HCl, $H_2SO_4$, NaOH |
| Anion exchange resins | | | |
| Strongly basic | (3) | OH | NaOH |
| | (4) | Cl, $HCO_3$ | NaCl, $NaHCO_3$ |
| | (6)[b] | Cl, OH | NaCl, NaOH |
| Weakly basic | (3) | Free base | NaOH |
| | (5) | Free base | NaOH |
| | (6)[b] | Free base | NaOH |

[a]*(1) reduction of calcium ion concentration, (2) reduction of hydrogencarbonate concentration, (3) reduction of salt content, (4) reduction of the content of certain ions, for example, nitrate ions, sulfate ions, (5) reduction of heavy metal ion content, and (6) reduction of the organic substance content, for example, humic acids.*
[b]*Macroporous types shall be used for this purpose.*

3. desired effluent water quality
4. exchange capacity and hydraulic characteristics of the exchanger
5. period between regenerations
6. type of operation: manual or automatic
7. flexibility required, that is the number of softener units.

The ion exchangers are not economically suitable for demineralizing waters containing more than 1000–2000 mg/kg of dissolved solids, except in a few specialized industrial applications (see American Water Works Association, AWWA).

The process of ion exchange for softening waters is preferable to the precipitation process when one or more of the following conditions exist:

1. less than 100 mg/kg of hardness expressed as calcium carbonate is present in the water
2. an extremely low dissolved solids content is required
3. only a limited volume of treated water is required.

Relative exchange capacity of cation exchangers and regenerative salt dosage would be as per Table 8.22.

The anion exchangers have typically an exchange capacity calculated as $CaCO_3$ of 27.4–57.2 g/L at a sodium dosage of 1.05–7 kg/kg removed (see AWWA).

**Table 8.22 Relative Exchange Capacity of Cation Exchangers[a]**

| Cation Exchanger | Nominal Exchange Capacity (g/L) | Regenerative Salt Dosage | | Effective kg/kg Hardness Removed |
| --- | --- | --- | --- | --- |
| | | Volumetric (kg/m³) | | |
| Greensand | 6.4 | 20.2 | | 3.1 |
| Processed greensand | 12.6 | 39.5 | | 3.1 |
| Synthetic siliceous | | | | |
| Zeolite | 25.2 | 79.2 | | 3.1 |
| Resin, polystyrene | 73.2 | 201.7 | | 3.1 |
| Resin, polystyrene | 50.3 | 80.0 | | 1.7 |

[a]*See AWWA.*

Interstage degasification in demineralization systems should be considered at flows over 22.7 m³/h and alkalinity over 100 mg/kg.

When ultra pure water is required, using the mixed bed demineralizer is recommended.

The demineralized water storage(s) shall be designed in order to store the produced demineralized water and to cover the following users:

1. make-up to deaerators
2. process units
3. regeneration of condensate treatment.

## 8.10.18 MISCELLANEOUS PROCESSES

### 8.10.18.1 Evaporation

Vapor compression and multistage flash evaporators may be recommended for producing high purity water from brackish and seawaters.

The vendor(s) standard specifications for the required evaporation system may be considered upon the company's approval.

### 8.10.18.2 Reverse osmosis

The applied pressure for brackish water purification is typically in the range of 2760–4140 kPa (ga) [27.6–41.4 bar (ga)] and for seawater purification, in the range of 5520–6900 kPa (ga) [55.2–69.0 bar (ga)].

Recovery of product (desalted) water with reverse osmosis units ranges from 50 to 90% of the feedwater depending upon the feedwater composition, the product water quality requirement, and the number of stages utilized.

For water containing from about 250 to 1500 mg/kg dissolved solids, an economic comparison of ion exchange and reverse osmosis is recommended to select the more cost effective process.

Reverse osmosis may be considered for desalination of seawater.

In many cases, the reverse osmosis product water shall be treated by one of the ion exchanger processes, if high quality feedwater is required.

A pretreatment system shall be provided to avoid fouling or excessive degradation of the membrane. Typically pretreatment will include filtration to remove suspended particles and the addition of chemicals to prevent scaling and biological growth.

Heating feedwater to provide optimum operating temperature of 25°C for a reverse osmosis system, shall be considered.

The process design of a reverse osmosis system shall be based on feedwater and product water qualities and rates.

Different types of reverse osmosis module layouts, for example, parallel, series including reject staging and product staging shall be proposed by the vendor(s) and the final configuration will be selected upon the company's approval.

### 8.10.19 ELECTRODIALYSIS

Recovery of product (deionized) water with electrodialysis units ranges from 50 to 90% of the feedwater depending upon the number of stages and degree of recirculation utilized.

Operating costs consist mainly of power costs (typically 1.6–2.7 kWh/m$^3$ of product water) and membrane cleaning and replacement costs.

Based upon combined capital and operating costs, the electrodialysis process is most economical when used to desalt brackish water (1000 to 5000 mg/kg dissolved solids) to a product water concentration of about 500 mg/kg dissolved solids.

The process design of an electrodialysis unit is to be based on feedwater and product water qualities and rates.

## 8.11 RAW WATER AND PLANT WATER SYSTEMS
### 8.11.1 SURFACE WATER INTAKE

The entrance of large objects into the intake pipe should be prevented by the use of a coarse screen or by obstructions offered by a small opening in the cribwork or riprap placed around the intake pipe.

Fine screens for the exclusion of small fish and other small objects should be placed at an accessible point, for example, at the suction or wet well at the pumping station where the screens can be easily inspected and cleaned.

The area of the openings in the intake crib should be sufficient to prevent an entrance velocity greater than about 0.15 m/s, in order to avoid carrying matter that can settle into the intake pipe.

Intake ports may be placed at various elevations so that water of the best quality may be taken. They should also be placed so that if one or more of the ports is blocked another can be opened.

Submerged ports should be designed and controlled to prevent air from entering the suction pipe. The difficulty can be minimized by maintaining an entrance velocity not greater than 0.15 m/s, preferably much less than this and maintaining a depth of water over the port of at least three diameters of the port opening.

The capacity of the intake should be sufficient for future demands during the life of the structure.

Grids with parallel bars, preferably removable, may be placed over intake ports with openings between bars not less than 25–50 mm.

The grids have sometimes been electrically charged to keep fish away.

Self-cleaning screens of the moving-belt type over intake ports are in successful use.

Vigorous reversal of flow through the intake port and screen or grid is sometimes created as an expedient for cleaning. Provision should be made for such reversal in the design of the intake structure and conduit.

### 8.11.2 INTAKE CONDUIT AND INTAKE WELL

The conduit conveying water from the intake should lead to a suction well in or near the pumping station.

Either a pipe, lying on or buried in the bottom of the body of water, or deep tunnels may be used as intake conduits.

The capacity of the conduit and the depth of the suction well should be such that the intake ports to the pumps will not draw air.

A velocity of 0.6–0.9 m/s in the intake conduit, with a lower velocity through the ports, will give satisfactory performance.

The horizontal cross-sectional area of the suction well should be three to five times the vertical cross-sectional area of the intake conduit.

Pumps should be started gradually to avoid drawdown in the suction well, and they should be stopped gradually to prevent surge.

The intake well acts as a surge tank on the intake conduit, thus minimizing surge.

The intake conduit should be laid on a continuously rising or falling grade to avoid accumulation of air or gas, pockets of which would otherwise restrict the capacity of the conduit.

Where air traps are unavoidable, provision should be made to allow gas to be drawn off from them. Where pipes are used, they should be weighted down to avoid flotation.

It is recommended that the intake works be in duplicate because of the almost complete dependence of the waterworks on its intake, the intake conduit, and the suction pit. Two or more widely separated intakes are highly desirable.

### 8.11.3 AQUEDUCTS

An aqueduct is a conduit designed to convey water from a source to a point, usually a reservoir, where distribution begins.

An aqueduct may include canals, flumes, pipe lines, siphons, tunnels, or other channels, either open or covered, flowing at atmospheric pressure or otherwise.

The choice between available types of conduits in an aqueduct depends on topography, available head, quality of water, and possibly other conditions.

Water in aqueducts should be protected against pollution by infiltration of nonpotable ground water, the overflow into the aqueduct of polluted surface waters, and all other possible sources of pollution to which water flowing low or atmospheric pressure may be exposed.

### 8.11.4 **GROUND WATER INTAKE**

It is important to keep water entrance velocity through the screen openings between 0.03 and 0.06 m/s. Such velocities will minimize head losses and chemical precipitation or encrustation.

Care should be taken in estimating the effective screen area. It is not uncommon to allow for as much as 50% plugging of the screen slots by formation particles.

The total open area required should be obtained by adjusting either the length or diameter of the screen because the slot is not arbitrary.

It is sometimes advisable to construct an artificial gravel-pack well to permit an increase in screen slot size. When a well screen is surrounded by an artificial gravel wall the size of the openings is controlled by the size of gravel used and by the type of openings.

Actual screen design should not be final until samples of the aquifer front the actual well location, are available for proper sieve analysis. However, experience and samples from test wells in the area permit preliminary calculations of the openings.

The actual open area per meters of screen depends upon the type of construction and the manufacturer.

### 8.11.5 **WATER TREATMENT OF INTAKE WATER**

For intake water from surface water or ground water the preliminary water treatment should be considered.

For intake water from seawater, the water treatment should be done.

### 8.11.6 **PUMPING STATIONS**

#### 8.11.6.1 *Location*
Conditions to be considered in the location of water works pumping station include:

1. sanitary protection of the quality of water
2. the hydraulics of the distribution system
3. possibilities of interruption by fire, flood, or other disaster
4. availability of power or of fuel
5. growth and future expansion.

Floods offer a hazard, which may be minimized by favorable location and site protection.

To obtain the greatest hydraulic advantage, the station should be located near the middle of the distribution system.

The danger of interruption of service by fire should be considered and the station should be protected by firewalls, fireproofing, and a sprinkler system.

Pumping stations may sometimes be located underground when conditions are favorable.

Underground locations require care to avoid flooding and dampness.

Attention should be given to accessibility illumination, ventilation, and heating and providing adequate space for operation and maintenance.

#### 8.11.6.2 *Choice of power*
The type of primary or of auxiliary power selected should be the most reliable, the most available and the least expensive. If all three conditions cannot be fulfilled they should be rated in the order stated, with the greatest reliability as the most important.

Auxiliary sources of power include internal combustion engines, gas turbines, steam engines, or a secondary source of electric energy.

### 8.11.6.3 Standby equipment

Standby pumping equipment should be provided in pumping stations.

One or more horizontal shaft centrifugal pumps may be equipped with an electric motor on the pump axis on one side of the pump and an internal combustion engine on the other side; or two electric motors, driven from different circuits, may be placed on the same shaft.

### 8.11.6.4 Compressed air

Use of compressed air in pumping stations should be considered as an auxiliary power for starting engines, blowing boiler tubes, operating control systems, pumping wells, and for other purposes.

### 8.11.6.5 Power rating of pumping station

There should be sufficient power in a waterworks pumping station to supply the peak demand without dangerously overloading the power equipment.

The highest efficiency and the greatest economy of operation require that the total load on all the units in the pumping station should be divided among them in such a manner that each can operate at its rated capacity.

### 8.11.6.6 Piping in the pumping station

The layout of the piping in a pumping station may be as important as the efficiency of operation in affecting economy. Short, straight, well-supported pipelines, devoid of traps for sediment or for vapor, sloping in one direction to drains and with adequate cleanouts, should be the object of the designer.

A few details to be observed in making connections to centrifugal pumps are:

1. The elbows on the suction side of a double suction pump should be normal to the suction nozzle, or a special elbow with guide vane followed by a piece of straight pipe to the suction nozzle should be used, to equalize the flow on each side of the impeller;
2. On the pump discharge, where change of direction is necessary, a constant-diameter bend should be used, followed by a long, straight reducer or increaser. Such an installation will most effectively interchange velocity and pressure heads;
3. Eccentric reducers with the top of the pipe and the top of the reducer at the same level should be used in horizontal suction piping to avoid the creation of an air pocket;
4. An increaser whose length is ten times the difference in diameter may be considered "long."

## 8.11.7 REFINERY AND/OR PLANT, PLANT WATER

Raw water pumped from outside the refinery after treatment for surface water, or usually without any treatment for ground water would be stored as clarified water in storage tanks.

Raw water pumped from sea intake after treatment in desalination plant would be stored as desalinated water in storage tanks.

Clarified (or desalinated) water stored in the clarified (or desalinated) water tanks is pumped to the plant water header for refinery and/or plant use, make-up water for the cooling tower, make-up water for the boiler feedwater treatment plant, feed to the potable water sand filters, and make-up for fire water tanks.

Seawater after desalination shall be considered as make-up for firewater, machinery cooling water, and the demineralization plant.

Seawater shall be considered in an emergency as an alternate source for fire water.

After branching from the main header of plant water for fire water tanks and the potable water system, one back-flow preventer should be installed on the header to insure against back-flow of possibly contaminated water from process units into lines used for drinking and sanitary services.

### 8.11.8 **POTABLE WATER SYSTEM**

Potable water is clarified water that has been filtered in sand filters and chlorinated.

Two potable water systems should be provided, one for refinery and or plant use and the other for the employee housing with identical flow scheme, if required.

The scheme of each system with different capacities should consist of sand filtration, chlorination, and storage tanks.

A cross connection should be provided to furnish water in an emergency from the discharge of refinery and/or plant potable water pumps to the housing potable water tanks.

Each system should have a separate chlorination system.

### 8.12 **COOLING WATER DISTRIBUTION AND RETURN SYSTEM**

The water pumping station for a once-through cooling-water system should be located at points where reliable quantities of suitable water are obtainable.

The choice between a once-through and recirculated cooling-water system should be based on availability of sufficient water of satisfactory quality, on process temperatures, on atmospheric conditions, and on equipment maintenance and operating costs.

After effective treatment refinery wastewaters may be considered as part of the make-up for recirculated cooling towers.

Where fresh water supplies are limited and salt water is available, a once-through system should be selected.

At plants located on rivers or lakes or where abundant supplies of both salt and fresh water are present, a choice should be made between a recirculated or once-through system.

The choice of cooling water system should be based on comparisons of initial and operating costs.

### 8.12.1 **DESIGN CRITERIA**

A stand-by pump(s) shall be provided for cooling-water recirculating pumps.

It is recommended that two main recirculating pumps and one 50% capacity standby pump are installed for recirculating water in cooling water system.

A steam ejector should be provided for each pump pit to remove water from that as it may accumulate from rain, leakage, etc.

The cooling water blowdown stream should normally be discharged into the nonoily water sewer system. Provision should be provided to divert this stream into the oily water sewer if oil is discharged into the cooling water return header.

Corrosion probe connections shall be provided at each recirculating pump suction and in the cooling water return header to the cooling tower.

Flow meters should be considered to measure the flow rates of supply and return headers.

Where the supply of fresh water is limited, a warm line softener or other treating systems may be provided to reduce cooling water make-up and blowdown by recovering a portion of the boiler plant blowdown and treating it in conjunction with slipstream taken from the cooling water return to the cooling tower.

The surge tank shall be provided upstream of the cooling water warm line softener.

Effluent from a cooling-water warm line softener shall be pumped to the pressure filters for final treatment after a pH adjustment using acid.

In case seawater is used, the minimum pressure of the machinery cooling water at the outlet of the machinery cooling water/sea cooling water cooler shall be higher than the normal pressure of the sea cooling-water circuit.

## 8.13 WATER ANALYSIS

1. Water analysis is conventionally expressed, for both cations and anions in mg/kg except for hardness and alkalinity, which are usually expressed in mg/L (mg/kg) of calcium carbonate ($CaCO_3$).
2. These mg/kg values can be converted to a common basis (such as milliequivalents/L) by dividing by the equivalent mass of the ion and multiplying by the relative density (specific gravity) of the water solution. This permits the summation of oppositely charged ions such that total cations will then equal total anions.
3. Cation and anion concentrations in milliequivalents/L can be converted to mg/L (mg/kg) $CaCO_3$ by multiplying by the equivalent mass of $CaCO_3$ ($100.08/2 = 50.04$) and dividing by the mass density of the water solution (Tables 8.23–8.25).

**Table 8.23  Water Analysis Calculation**

| Water Analysis Ion | mg/kg | M[a] | Equivalent Mass |
|---|---|---|---|
| Calcium ($Ca^{+2}$) | 100.1 | 40.08 | $\dfrac{40.08}{2} = 20.04$ |
| Magnesium ($Mg^{+2}$) | 20.4 | 24.32 | $\dfrac{24.32}{2} = 12.16$ |
| Sodium ($Na^{+1}$) | 12.0 | 23.0 | $\dfrac{23.0}{1} = 23$ |
| Bicarbonate ($HCO_3^{-1}$) | 366.0 | 61.02 | $\dfrac{61.02}{1} = 61.02$ |
| Sulfate ($SO_4^{-2}$) | 48.1 | 96.06 | $\dfrac{96.06}{2} = 48.03$ |
| Chloride ($Cl^{-1}$) | 7.1 | 35.46 | $\dfrac{35.46}{1} = 35.46$ |

[a]*Relative molecular mass in kg/kmol or g/mol.*

**Table 8.24  Cations**

| Ion | Milliequivalents/L (Relative Density = 1) | mg/L (mg/kg) $CaCO_3$ |
|---|---|---|
| $Ca^{+2}$ | 100.1/20.04 = 5.00 | (5.0)(5.04) = 250 |
| $Mg^{+2}$ | 20.4/12.16 = 1.68 | (1.68)(50.04) = 84 |
| $Na^{+1}$ | 12.0/23 = 0.52 | (0.52)(50.04) = 26 |
| $HCO_3^{-1}$ | — | — |
| $SO_4^{-2}$ | — | — |
| $Cl^{-1}$ | — | — |
| Totals | 7.20 | 360 |

**Table 8.25  Anions**

| Ion | Milliequivalents/L (Relative Density = 1) | mg/L (mg/kg) $CaCO_3$ |
|---|---|---|
| $Ca^{+2}$ | — | — |
| $Mg^{+2}$ | — | — |
| $Na^{+1}$ | — | — |
| $HCO_3^{-1}$ | 366/61.02 = 6.00 | (6)(50.04) = 300 |
| $SO_4^{-2}$ | 48.1/48.03 = 1.00 | (1)(50.04) = 50 |
| $Cl^{-1}$ | 7.1/35.46 = 0.20 | (0.2)(50.04) = 10 |
| Totals | 7.20 | 360 |

Total hardness is the sum of calcium and magnesium and is therefore equal to 334 mg/L (mg/kg) as $CaCO_3$ (250 + 84).

Correspondingly, alkalinity is the sum of $CO_3^{-2}$, $HCO_3^{-1}$, and $OH^{-1}$ ions and is equal to 300 mg/L (mg/kg) as $CaCO_3$.

## 8.14 POTABLE WATER STANDARD SPECIFICATION
### 8.14.1 BACTERIOLOGIC REQUIREMENTS

As an operating goal, the AWWA committee recommends no coliform organisms on the basis that the water supplies of utilities with high standards of operation have shown a fraction of one coliform organism per liter over periods of many years.

The goal is considered practical in view of modern disinfection procedures and desirable for improved health conditions.

**Table 8.26 Comparison of Physical Characteristics of Water**

| Characteristic | Recommended Limits[a] | AWWA Goals[b] |
|---|---|---|
| Turbidity-units | 5 | <0.1 |
| Color-units | 15 | <3 |
| Odor (threshold odor number) | 3 | No odor |
| Taste | | None objectionable |

[a]Limits that should not be exceeded according to latest USPHS Drinking Water Standards.
[b]The figures in this column are taken from AWWA committee report.

## 8.14.2 PHYSICAL CHARACTERISTICS

Table 8.26 compares the levels for turbidity, color, taste, and odor as expressed in Public Health Service Drinking Water Standards (USPHS) and the report of AWWA Committee 2640 P.

The committee report presents those levels that should be approached by well designed and operated systems and which reflect a high degree of consumer acceptability

## 8.14.3 CHEMICAL CHARACTERISTICS

Table 8.27 gives the maximum concentration of various chemical substances allowed by the USPHS Standards and recommended by the AWWA report, and the recommended lower, optimum, and upper control limit for fluoride concentrations, are shown in Table 8.28

- The AWWA Committee report establishes a hardness of 80 mg/L (as $CaCO_3$) as the desirable objective for potable water supplies. Although this is not of concern to utilities with sources of supply that are naturally quite soft, it is significant to utilities in hard water areas because softening must be provided to achieve this goal.
- A number of limits expressed in Table 8.27 are based on aesthetic rather than health considerations. For example, the limits for iron and manganese are based on the staining and other objectional properties of these elements.
- Limits for the concentration of sodium have not been set by the Standards or Committee report.
- Fluoride is considered as an essential constituent of drinking water for the prevention of tooth decay in children. Conversely, excess fluoride may give rise to dental fluorosis (spotting of the teeth) in children. In the Drinking Water Standards it is also recommended that fluoride in average concentrations greater than twice the optimum values shall constitute grounds for rejection of the supply (Table 8.28).

## 8.14.4 RADIOACTIVITY

- In establishing reasonable, long-term limits for radioactivity in potable water, the population's total exposure to radiation must be considered. This requires the assessment of the intake from such sources as food and milk, as well as from potable water, together with an evaluation of the effects of specific radioactive substances.
- The USPHS standards specify that water supplies other sources of radioactivity intake of radium-226 and strontium-90 when the water contains these substances in amounts not exceeding 3 pico curie (pCi)/L and 10 pCi ($10 \times 10^{-12}$ Ci)/L, respectively.

**Table 8.27 Comparison of Chemical Characteristics**

| Substance | USPHS Drinking Water Standards | | AWWA Committee Goal (mg/L) |
| --- | --- | --- | --- |
| | Concentration That Should not be Exceeded (mg/L) | Concentration Which, if Exceeded, Constitutes Grounds for Rejection of Supply (mg/L) | |
| Alkyl benzene sulfonate (ABS) | 0.5 | | |
| Aluminum (Al) | | | <0.05 |
| Arsenic (As) | 0.01 | 0.05 | |
| Barium (Ba) | | 1.0 | |
| Chloride (Cl) | 250 | | |
| Cadmium (Cd) | | 0.01 | |
| Chromium ($Cr^{+6}$) | | 0.05 | |
| Copper (Cu) | 1 | | <0.20 |
| Carbon chloroform extract (CCE) | 0.2 | | <0.04 |
| Cyanide (CN) | 0.01 | 0.2 | |
| Iron (Fe) | 0.3 | | <0.05 |
| Lead (Pb) | | 0.05 | |
| Manganese (Mn) | 0.05 | | <0.01 |
| Nitrate ($NO_3$) | 45 | | |
| Phenols | 0.001 | | |
| Selenium (Se) | | 0.01 | |
| Silver (Ag) | | 0.05 | |
| Sulfate ($SO_4$) | 250 | | |
| Total dissolved solids | 500 | | 200 |
| Zinc (Zn) | 5 | | <1.0 |

- In the known absence of strontium-90 and alpha emitters, the water supply is acceptable when the known concentrations do not exceed, 1000 pCi/L.
- Considerable knowledge has been acquired on the treatment of water to remove radioactive substances and it is desirable to maintain the levels of radioactivity in raw water well below the established limits.

## 8.14.5 TYPICAL PLANT DESIGN OF DEMINERALIZED WATER*

Based on the following data the plant design of demineralized water should be done:

- TAC concentration of the raw water in French degrees;
- SAF concentration of the raw water in French degrees ($SO_4^{-2} + Cl^{-1} + NO_3^{-1}$);

**Table 8.28 Allowable Fluoride Concentration**

| Annual Average of Max. Daily Air Temperature[a] (°C) | Recommended Control Limits, Fluoride Concentrations, mg/L[b] | | | Maximum Contaminant Level (MCL) |
|---|---|---|---|---|
| | Lower | Optimum | Upper | |
| 12 or lower | 0.9 | 1.2 | 1.7 | 2.4 |
| 12.1–14.6 | 0.8 | 1.1 | 1.5 | 2.2 |
| 14.6–17.7 | 0.8 | 1.0 | 1.3 | 2.0 |
| 17.7–21.4 | 0.7 | 0.9 | 1.2 | 1.8 |
| 21.5–26.2 | 0.7 | 0.8 | 1.0 | 1.6 |
| 26.3–32.5 | 0.6 | 0.7 | 0.8 | 1.4 |

[a]*Based on temperature data obtained for a minimum of 5 years.*
[b]*From "Drinking Water Standards," US Public Health Service, No. 956, 1962.*

- Silica content as T $SiO_2$ (1 French degree = 12 mg/L $SiO_2$);
- T $CO_2$. Content of carbon dioxide (carbonic acid) in the water after passing through the cation exchanger and where appropriate after elimination of carbon dioxide (carbonic acid), in French degrees;
- Volume V of water to be supplied between regeneration processes, in m³, including service water if appropriate;
- Exchange capacity C of the resins expressed in French degrees/liter per liter of consolidated resins.

The anion exchanger is calculated first: the volume in m³ to be used is given by one of the following equations:

$$V_a = \frac{V \times SAF}{C} \qquad \text{(for a weak - basic exchanger)}$$

or

$$V_a = \frac{V \times (SAF + TCO_2 + TSiO_2)}{C} \qquad \text{(for a strong - basic exchanger)}$$

Then the volume in m³ of cation exchanger is calculated, allowing for the additional water $\alpha$. $V_a$ necessary to rinse the anion exchanger, where (may vary from 5 to 20 depending on the type of resin)

$$V_c = \frac{(V + \alpha V_a)(SAF + TAC)}{C}$$

The volumes calculated shall then be compared with the hourly output to be treated. There are upper limits to the flow rate or to the bed volume.

If $V_c$ or $V_a$ are too low, they should be adjusted, possibly by increasing the cycle volume V.

Table 8.29 shows the concentration of solution (values of the different degrees).

**Table 8.29  Table Showing the Concentration of Solution (Values of the Different Degrees)**

| | Formula | Molecular Mass | Value of the various units (mg/L) | | | |
|---|---|---|---|---|---|---|
| | | | mEq/L | Fr. | Germ. | Eng. |
| 1. Calcium and magnesium salts and oxides causing hardness in water (degree of hardness) | | | | | | |
| Calcium carbonate | $CaCO_3$ | 100 | 50 | 10.0 | 17.8 | 14.3 |
| Calcium bicarbonate | $a(HCO_3)_2$ | 162 | 81 | 16.2 | 28.9 | 23.1 |
| Calcium sulfate | $CaSO_4$ | 136 | 68 | 13.6 | 24.3 | 19.4 |
| Calcium chloride | $CaCl_2$ | 111 | 55.5 | 11.1 | 19.8 | 15.8 |
| Calcium nitrate | $Ca(NO_4)_2$ | 164 | 82 | 16.4 | 29.3 | 23.4 |
| Quicklime | $CaO$ | 56 | 28 | 5.6 | 10.0 | 8.0 |
| Hydrated lime | $Ca(OH)_2$ | 74 | 37 | 7.4 | 13.2 | 10.5 |
| Magnesium carbonate | $MgCO_3$ | 84 | 42 | 8.4 | 15.0 | 12.0 |
| Magnesium bicarbonate | $g(HCO_3)_2$ | 146 | 73 | 14.6 | 26.1 | 20.9 |
| Magnesium sulfate | $MgSO_4$ | 120 | 60 | 12.0 | 21.4 | 17.1 |
| Magnesium chloride | $MgCl_2$ | 95 | 47.5 | 9.5 | 17.0 | 13.5 |
| Magnesium nitrate | $Mg(NO_3)_2$ | 148 | 74 | 14.8 | 26.4 | 21.2 |
| Magnesia | $MgO$ | 40 | 20 | 4.0 | 7.1 | 5.7 |
| | $Mg(OH)_2$ | 58 | 29 | 5.8 | 10.3 | 8.2 |
| 2. Anions | | | | | | |
| Carbonate ion | $CO_3$ | 60 | 30 | 6.0 | 10.7 | 8.6 |
| Bicarbonate ion | $HCO_3$ | 61 | 61 | 12.2 | 21.8 | 17.4 |
| Sulfate ion | $SO_4$ | 96 | 48 | 9.6 | 17.3 | 13.7 |
| Sulfite ion | $SO_3$ | 80 | 40 | 8.0 | 14.3 | 11.4 |
| Chloride ion | $Cl$ | 35.5 | 35.5 | 7.1 | 12.7 | 10.2 |
| Nitrate ion | $NO_3$ | 62 | 62 | 12.4 | 22.1 | 17.7 |
| Nitrite ion | $NO_2$ | 46 | 46 | 9.7 | 16.4 | 13.1 |
| Phosphate ion | $PO_4$ | 95 | 31.66 | 6.32 | 11.25 | 9.03 |
| Silicate ion | $SiO_2$ | 60 | 60 | 12.0 | 21.4 | 17.1 |
| 3. Acids | | | | | | |
| Sulfuric acid | $H_2SO_4$ | 98 | 49 | 9.8 | 17.5 | 14 |
| Hydrochloric acid | $HCl$ | 36.5 | 36.5 | 7.3 | 12.8 | 10.3 |
| Nitric acid | $HNO_3$ | 63 | 63 | 12.6 | 22.5 | 18 |
| Phosphoric acid | $H_3PO_4$ | 98 | 32.66 | 6.52 | 11.64 | 9.31 |

*(Continued)*

**Table 8.29 Table Showing the Concentration of Solution (Values of the Different Degrees)** (*cont.*)

| | Formula | Molecular Mass | Value of the various units (mg/L) | | | |
|---|---|---|---|---|---|---|
| | | | mEq/L | Fr. | Germ. | Eng. |
| **4. Cations and oxides** | | | | | | |
| Calcium | Ca | 40 | 20 | 4.0 | 7.15 | 5.7 |
| Magnesium | Mg | 24.3 | 12.1 | 2.43 | 4.35 | 3.47 |
| Sodium | Na | 23 | 23 | 4.6 | 8.2 | 6.6 |
| | Na$_2$O | 62 | 31 | 6.2 | 11.1 | 8.8 |
| Potassium | K | 39 | 39 | 7.8 | 13.9 | 11.2 |
| | K$_2$O | 94 | 47.1 | 9.4 | 16.8 | 13.4 |
| Iron | Fe | 55.8 | 27.9 | 5.6 | 10.0 | 8.0 |
| Aluminum | Al | 27 | 9 | 1.8 | 3.2 | 2.6 |
| | Al$_2$O$_2$ | 102 | 17 | 3.4 | 6.1 | 4.85 |
| **5. Bases** | | | | | | |
| Sodium hydroxide | NaOH | 40 | 40 | 8.0 | 14.3 | 11.4 |
| Potassium hydroxide | KOH | 56 | 56 | 11.2 | 20.0 | 16.0 |
| Ammonia | NH$_4$OH | 35 | 35 | 7.0 | 12.5 | 10 |
| **6. Various salts** | | | | | | |
| Sodium bicarbonate | NaHCO$_3$ | 84 | 84 | 16.8 | 30 | 24 |
| Sodium carbonate | Na$_2$CO$_3$ | 106 | 53 | 10.6 | 18.9 | 15.1 |
| Sodium sulfate | Na$_2$SO$_4$ | 142 | 71 | 14.2 | 25.3 | 20.3 |
| Sodium chloride | NaCl | 58.5 | 58.5 | 11.7 | 20.9 | 16.7 |
| Sodium phosphate | Na$_3$PO$_4$ | 164 | 54.7 | 10.9 | 19.5 | 15.6 |
| Sodium silicate | Na$_2$SiO$_3$ | 122 | 61 | 12.2 | 21.8 | 17.4 |
| Potassium carbonate | K$_2$CO$_3$ | 138 | 69 | 13.8 | 24.6 | 19.7 |
| Potassium bicarbonate | KHCO$_3$ | 100 | 100 | 20 | 35.7 | 38.5 |
| Potassium sulfate | K$_2$SO$_4$ | 174 | 87 | 17.4 | 31.1 | 24.8 |
| Potassium chloride | KCl | 74.5 | 74.5 | 14.9 | 26.6 | 21.2 |
| Potassium phosphate | K$_3$PO$_4$ | 212.3 | 70.8 | 14.1 | 25.2 | 20.2 |
| Ferrous sulfate | FeSO$_4$ | 152 | 76 | 15.2 | 27.1 | 21.7 |
| Ferric sulfate | Fe$_2$(SO$_4$)$_3$ | 400 | 66.6 | 13.3 | 23.8 | 19 |
| Ferric chloride | FeCl$_3$ | 162.5 | 54.2 | 10.8 | 19.3 | 15.4 |
| Aluminum sulfate | Al$_2$(SO$_4$)$_3$ | 342 | 57 | 11.4 | 20.3 | 16.3 |

## 8.15 STANDARD SPECIFICATION OF DEMINERALIZING UNIT

This section covers the general requirements for the design, construction, and inspection of automatic regenerating type demineralizing units for the production of boiler feedwater.

For materials, design, fabrication, and inspection of the demineralizing units not specified in this specification, the specified code, or standard in each individual case shall be applied per the latest editions.

### 8.15.1 DESIGN

The demineralizing unit shall consist of, but not necessarily be limited to, the following equipment:

- cation resin bed for exchanging acidic hydrogen
- anion resin bed for exchanging basic hydroxide
- degasifier (decarbonator) removing carbon dioxide formed in the cation resin bed, if required
- mixed bed polisher containing strong cation and anion resin for exchanging acidic hydrogen and basic hydroxide respectively, if required
- regenerating equipment including chemical storage tanks, measuring tanks, pumps, blowers, instrumentation for control, neutralizing equipment for regeneration-effluent, interconnecting piping, and others.

### 8.15.2 CHEMICAL FOR RESIN REGENERATION

The following chemicals shall be used for the regeneration of resin:

- $H_2SO_4$ or HCl solution for cation exchanger
- NaOH solution of anion exchanger
- skid-mounted chemical storage tanks shall be provided and equipped with a chemical transfer pump for regeneration purposes
- the capacity of each chemical storage tank shall be designed to enable maximum operation within five cycles.

Two separate chemical transfer pumps shall be provided for each chemical. These chemical pumps, one for sulfuric acid transfer and the other for caustic solution transfer, shall be driven by individual motors.

The unit shall be designed to minimize consumption, and its guaranteed values shall be satisfied.

### 8.15.3 DEMINERALIZED WATER QUALITY

The following data on demineralized water quality shall be specified:

- electrical conductivity
- total hardness
- silica
- pH
- other requirements.

### 8.15.4 TYPE OF DEMINERALIZING UNIT

The type of demineralizing unit is to be decided, based on the raw water analysis. The use of a two-bed two-tower, two-bed three-tower unit or others suitable for the specified raw water quality and treated water quality shall be considered.

### 8.15.5 CAPACITY AND OPERATION

- The unit shall continuously produce a net flow to the service. However, where adequate demineralized water storage is available to meet standby and service requirements, a single unit may be permitted.
- Expected turn-down ratio of a demineralized water flow rate to a designed value shall be suitable for boiler feed operation, as specified.

Basically, operation of the demineralizing unit shall be controlled automatically from the local panel. Regeneration of the unit shall be started automatically by the following signals:

- signal when conductivity is exceeding the set value
- signal when scheduled water volume is obtained
- signal when automatic generation is manually initiated.

In addition to the earlier automatic regeneration, the regeneration system shall enable manual operation of each value.

The demineralizing unit shall be installed in a nonhazardous area. Therefore, the unit shall be designed for suitable outdoor installation.

## 8.16 EQUIPMENT DESIGN

### 8.16.1 RESIN VESSEL

- Cation/anion resin vessels shall be made of carbon steel with an inner rubber lining or equivalent to protect against corrosion.
- Where diluted sulfuric acid and caustic soda is used, vessel internals shall be made of rubber-lined carbon steel, Type 316 stainless steel or equivalent.

### 8.16.2 FEED PUMP

- The raw water feed pump design shall be based on the hydraulic calculations applicable between the raw water feed line and the rundown line to the demineralized water tank.
- To make the earliest possible determination of the pump specifications, the pressure drop at the demineralizing unit shall be assumed to be less than 140 kPa (1.4 bar) and it is recommended that the pressure drop be less than 100 kPa (1.0 bar) under normal operation.

### 8.16.3 REGENERATION WATER PUMP

Regeneration water pumps shall be provided.

The design of chemical pumps for the regeneration of the resin shall be based on the following considerations:

- capacity of demineralizing unit
- operating schedule of demineralizing unit
- concentrations of chemical solutions.

### 8.16.3.1 Piping
The demineralized water (treated water) outlet piping shall be made of rubber-lined carbon steel or stainless steel.

### 8.16.3.2 Miscellaneous
- resin traps shall be provided in the outlet of each unit
- resin charges for each resin vessel shall be included
- activated carbon removing system, where applicable, shall be provided.

### 8.16.3.3 Neutralization equipment
- Neutralization equipment shall consist of recirculation and/or discharge pumps, mixing blower or mixer and bulk waste water tank with adequate internals.
- Operation of this system shall be done automatically as a rule. However, manual regeneration may be employed for any unit whose regeneration interval is more than 7 days.

## 8.16.4 INSPECTION AND TESTING
### 8.16.4.1 Shop tests and inspections
The following tests and inspections on respective parts or equipment in the demineralizing unit shall be performed after completion at the fabrication shop:

- dimensional inspection
- material check against the mill test certificate
- mechanical running test of the equipment
- electrical test
- instrument checks
- pressure and leakage test on vessels.

### 8.16.4.2 Field test
The following tests shall be done after completion in the field:

- instrument test
- running test on overall system
- performance test
- leakage test on piping.

### 8.16.4.3 Performance characteristics
The following performance characteristics shall be guaranteed:

- treated water output capacity per hour and per cycle
- inlet water flow rate of both operation and regeneration
- treated water quality per items:
  - electrical conductivity
  - silica

- total hardness
- pH at 250°C.
- Operating/regeneration cycle time.
- Chemicals both for regeneration and neutralization per cycle (kg/cycle) and per each treated water $(kg/m^3)$.
- Waste water quantity per cycle $(m^3/cycle)$ and per each treated water $(m^3/m^3)$.
  Table 8.30 shows typical ion exchange flows and dimensions.

**Table 8.30  Typical Ion Exchange Flows and Dimensions**

| Diameter (mm) | Area (m²) | Min. Resin (m³ at 760 mm) | Back Wash Rate (m³/h) | | | Unit(s) (m³/h max.)ᵃ | | |
|---|---|---|---|---|---|---|---|---|
| | | | b | c | d | 1 | 2 | 3 |
| 610 | 0.25 | 0.23 | 2.04 | 4.31 | 8.63 | 4.31 | 5.68 | 11.36 |
| 762 | 0.46 | 0.34 | 3.41 | 6.81 | 13.40 | 6.81 | 9.08 | 18.17 |
| 914 | 0.66 | 0.51 | 4.77 | 9.54 | 19.30 | 9.54 | 12.72 | 25.43 |
| 1067 | 0.89 | 0.68 | 6.59 | 13.17 | 26.34 | 13.17 | 17.49 | 34.97 |
| 1219 | 1.17 | 1.91 | 8.63 | 17.03 | 34.06 | 17.03 | 22.71 | 45.42 |
| 1372 | 1.48 | 1.13 | 10.90 | 21.57 | 43.15 | 21.57 | 29.52 | 59.05 |
| 1524 | 1.82 | 1.39 | 13.40 | 26.80 | 53.37 | 26.80 | 36.34 | 72.67 |
| 1676 | 2.21 | 1.70 | 16.12 | 32.47 | 64.72 | 32.48 | 43.15 | 86.30 |
| 1829 | 2.63 | 2.01 | 19.30 | 38.61 | 77.21 | 38.61 | 52.23 | 104.47 |
| 1981 | 3.08 | 2.35 | 22.71 | 45.42 | 90.84 | 45.42 | 61.32 | 122.63 |
| 2134 | 3.58 | 2.72 | 26.12 | 52.23 | 104.46 | 52.23 | 70.40 | 140.80 |
| 2286 | 4.11 | 3.12 | 29.98 | 60.18 | 120.36 | 60.18 | 79.48 | 158.97 |
| 2438 | 4.67 | 3.57 | 34.06 | 68.13 | 136.26 | 68.13 | 90.84 | 181.68 |
| 2591 | 5.28 | 4.02 | 38.61 | 72.21 | 154.43 | 77.21 | 102.19 | 204.39 |
| 2743 | 5.91 | 4.53 | 43.15 | 86.30 | 172.60 | 86.30 | 113.55 | 227.10 |
| 2896 | 6.59 | 4.98 | 48.37 | 96.52 | 193.03 | 96.52 | 129.45 | 258.89 |
| 3048 | 7.29 | 5.55 | 53.37 | 106.74 | 213.47 | 106.74 | 143.07 | 286.15 |
| 3353 | 8.83 | 6.80 | 64.72 | 129.45 | 258.90 | 129.45 | 170.32 | 340.65 |

ᵃ*Service flow is based on rates of 14.7 m³/m² h, single unit maximum rate, and 19.6 m³/m² h short term rate, on multiple units when one unit is out of service for regeneration. To provide a continuous supply of treated water, most plants use two units, so that while one is being regenerated the other continues to provide finished water. Plants having variable demand store finished water to eliminate surges in flow, minimizing equipment size. Treated water storage permits smaller plants to install only a single unit, relying on stored water to provide requirements during regeneration. This is risky, since it provides no margin for error or for normal maintenance.*
ᵇ*Backwash rate of 7.3 m³/m².h for anion resins at 21°C.*
ᶜ*Backwash rate of 14.7 m³/m².h for cation resins at 21°C.*
ᵈ*Backwash rate of 29.3 m³/m².h for cation resins at 104°C.*

## 8.17 **DISINFECTION**

The purpose of disinfection is to render water safe for human consumption, free from pathogenic bacteria and therefore incapable of transmitting disease.

Chlorine, in its various forms (liquid, gas, or hypochlorite) is the major chemical widely used in disinfecting water. Other disinfectants are iodine, bromine, ozone, chlorine dioxide, ultraviolet light, and lime, which might be considered if chlorine gas is not readily available. The important applied disinfectants are as stated in later section.

### 8.17.1 **CHLORINATION**

While the principal use of chlorine is as a disinfectant, its application is practiced for prevention and destruction of odors, iron, and/or color removal simultaneously. Chlorination is classified according to its point of application and its end result.

### 8.17.2 **PLAIN CHLORINATION**

Whenever surface waters are used with no other treatment than chlorination, its role is extremely important as the principal if not the only safeguard against disease.

Such otherwise untreated waters are likely to be rather high in organic matter and require high dosages and long contact periods for maximum safety. The chlorine can be added to the water in the pipe leading from an impounding reservoir to the residential township. For disinfection alone a dose of 0.5 mg/L or more may be required to obtain a combined available residual in the township distribution system.

### 8.17.3 **PRECHLORINATION**

Apart from diatomaceous earth filters for which prechlorination is a must, in case of all kinds of rapid filters, the chlorine may be added in the suction pipes of raw-water pumps or to the water as it enters the mixing chamber. Its use in this manner may improve coagulation and reduce tastes and odors, may keep the filter sand cleaner, and increase the length of filter runs. Frequently the dosage is such that a combined available residual of 0.1–0.5 mg/L goes to the filters. The combination of prechlorination with postchlorination may be advisable or even necessary if the raw water is so highly polluted that the bacterial load on the filters must be reduced in order that a satisfactory coliform count or most probable number (MPN) be obtained in the final effluent.

### 8.17.4 **POSTCHLORINATION**

This usually refers to the addition of chlorine to the water after all other treatments. It is standard treatment at rapid sand filter plants, and when used without prechlorination and with low residuals it is sometimes called marginal chlorination. The chlorine may be added in the suction line of the service pump, but it is preferable to add it in the filter effluent pipe or in the clear well so that an adequate contact time will be assured. This should be at least 30 min before any of the water is consumed if only postchlorination is given. Dosage will depend upon the character of the water and may be 0.25–0.5 mg/L in order to obtain a combined available residual of 0.1–0.2 mg/L as the water leaves the plant.

Greater residuals will probably be needed if it is desired to hold a disinfecting effect throughout the distribution system.

### 8.17.5 BREAK-POINT CHLORINATION

Depending on the amount of free and saline ammonia existing in filtered water, the breakpoint curve varies.

The breakpoint indicates complete oxidation of the available ammonia and any other organic amines, and the residual above the breakpoint is mostly free available chlorine. Usually the chlorinous and other odors will disappear at or before the breakpoint. Dosage is likely to be −10 mg/L in order to obtain a free available residual of about 0.5 mg/L or more. The chlorine, when the breakpoint procedure is applied, usually, but not always, is added at the influent to the plant. In some cases ammonia has been added to water lacking in it in order to form a more pronounced breakpoint.

### 8.17.6 AMMONIA CHLORINE TREATMENT (CHLORAMINATION)

The chlorine combines with the ammonia and other organic amines to form chloramines. They are less active than hypochlorous acid (HOCl) and their disinfecting efficiency is considerably less than HOCl, but its bactericidal effects are maintained over a longer period. The beneficial aspect of ammonia chlorine treatment is that little or no combination of chlorine with organic matter occurs, to produce undesirable odors. Hence chloramines have been used as an odor preventive with satisfactory bactericidal effects, if longer contact periods are provided.

Ammonia is usually added in the ratio or 1 part of ammonia to 4 parts of chlorine, although experiment may indicate the desirability of a higher or lower proportion.

It is used as the gas, as a solution of the gas in water, or as ammonium sulfate or ammonium chloride. The ammonium sulfate used is the chemical fertilizer, which is sufficiently pure to be used for this purpose.

If taste and odor control is the main object of chloramination, the ammonia should be added in advance of the chlorine.

### 8.17.7 SUPERCHLORINATION AND DECHLORINATION

The process of superchlorination followed by dechlorination is best defined as the application of chlorine to water to produce free residual chlorination, in which the free available chlorine residual is so large that dechlorination is required before the water is used. Whenever highly polluted waters have to be disinfected for drinking purposes, adoption of this process is recommended.

Where superchlorination, or alternatively, breakpoint chlorination is practiced, it is desirable to provide a baffled contact tank of approximately 30 min retention and in no case less than 20 min, in order to ensure complete sterility. The most common dechlorinating agent is sulfur dioxide, which is added either at the end or in the last bay of the contact tank so as to ensure complete mixture and dechlorination to leave the desired residual. It is normally found that the dose of sulfur dioxide is approximately 20% greater than the theoretical amount calculated to combine with the chlorine removed. In certain small supplies sodium thiosulphate or sodium bisulphite have been employed as dechlorinating agents. If complete dechlorination is desired then passage of the highly chlorinated water through a bed of granular carbon is often used.

Very occasionally a small dose, 0.2–0.5 mg/kg (mass ppm$_m$ or ppm by mass), of potassium permanganate is also added to the contact tank prior to filtration in order to control tastes.

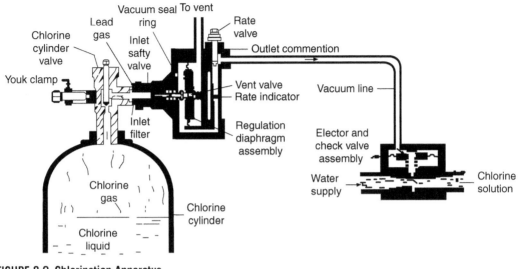

**FIGURE 8.9 Chlorination Apparatus**

## 8.17.8 **CHLORINATORS**

For injection of chlorine gas into water a variety of chlorinators have been designed and manufactured. The two recommended types are shown in Figs. 8.9 and 8.10.

The type to be chosen largely depends upon the capacity needed. Chlorine gas is obtained in pressurized cylinders ranging from 45 to 1000 kg capacity. Small plants commonly use the 45 kg cylinders, while large plants requiring 75–100 kg/day generally use 1000 kg containers as a matter of convenience and economy.

Satisfactory chlorinators must feed the gas into the water at an adjustable rate, and it must do this although the pressure in the gas container changes as the temperature changes. While some chlorinators apply the measured amount of gas to the water through a porous porcelain diffuser, most types dissolve the gas in water and feed the solution.

In the chlorinator a water-operated aspirator draws chlorine through a regulator and measuring device. The flow can be set to provide a predetermined dosage of chlorine. The quantity of water required and the pressure at which it must be delivered are a function of the size and design of the system. At the injector the chlorine is dissolved in the water and is conveyed to the point of application. In some instances the injector may be at the point of application. The inlet valve closes if the vacuum fails. If it should not close, the chlorine pressure forces the seal of the vent, and the chlorine is discharged outside the building.

The V-notch chlorinator also operates under a vacuum generated by an injector, with the chlorine flow controlled by a variable V-notch orifice. The amount of chlorine fed is shown by a feed rate indicator of the rotameter type. The differential valve and the chlorine pressure regulating valve maintain a constant vacuum differential across the V-notch. The feed rate is changed by changing the area of the V-notch orifice.

Certain precautions are needed in chlorination practice. Chlorinators and cylinders of chlorine should be housed separately and not in rooms used for other purposes. Ventilation should be provided to give a complete air change each minute and the air outlet should be near the floor since chlorine is heavier than air. Switches for fans and lights should be outside the room and near the entrance. The

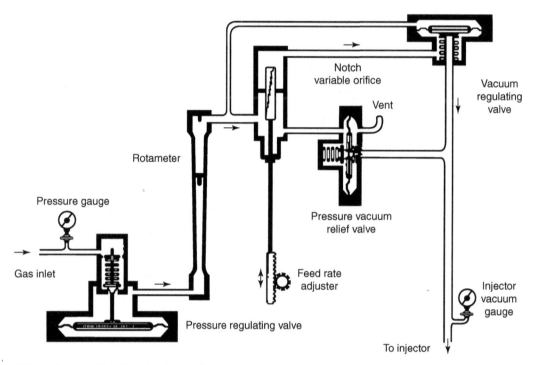

**FIGURE 8.10 Simplified Diagram of the V-Notch Chlorinator**

entrance door should have a clear glass window to allow observation from outside. Chlorinator rooms should be heated to 15°C with protection against excessive heat. Cylinders should be protected against temperatures greater than that of the equipment.

### 8.17.9 HYPOCHLORINATION

Chlorinated lime has been largely displaced, not only by chlorine gas, but also by improved commercial compounds of sodium and calcium hypochlorite.

Hypochlorination is especially applicable to emergency use where supplies are endangered and there would be considerable delay in obtaining chlorine gas and chlorinators.

### 8.17.10 CHLORINATION

Chlorination has been proved to be responsible for increases in the concentration of volatile halogenated organics, particularly chloroform, bromodichloromethane, dibromochloromethane and bromoform, which are commonly found in chlorinated water.

Ozone is a strong oxidizing agent and may be applied in any situation where chlorine has been used. Dosages range from 0.25 mg/L for high quality groundwaters to 5 mg/L following filtration for poor quality surface waters.

Effective ozone dosages for viruses range from 0.25 to 1.5 mg/L at contact times of 45 s–2 min. Ozone, unlike chlorine and the other halogenes, is not particularly sensitive to pH within the range of pH 5–8, but is significantly affected by temperature.

The disadvantages of ozonation, which have restricted its use worldwide, are its cost relative to chlorine, the need to generate it at the point of use, and its spontaneous decay which prevents maintenance of a residual in the distribution system. However, the fact of production of halogenated hydrocarbons by present popular chlorination practice would make the said disadvantages of lesser importance especially if production of ozone should be or is generated for other needs.

### 8.17.11 ACTIVATED CARBON

Activated carbon is used in water processing primarily as a short-term treatment to correct seasonal taste and odor problems.

Powdered activated carbon is generally less than 0.075 mm in size and thus has an extremely high ratio of area to volume.

It is applied as a slurry at points of raw water entry, at the mixing basin, split feed: with a portion in the mixing basin and the balance just ahead of the filters, either at constant rate or at a heavy rate immediately after the filter is washed, followed by a light rate (during a seasonal time period).

The dosages used vary from 0.25 to 8 mg/L with 7 to 2 mg/L being most common.

The carbon can also be used in impounding reservoirs to reduce algae or either odors. In this case it should be applied as a slurry and sprayed over the water surface at $1–10$ g/m$^2$.

The required dosage of activated carbon can be controlled by means of the threshold odor test.

## 8.18 WATER SUPPLY AND SEWERAGE SYSTEMS

This section provides general guidance on the construction of water and sewage networks and installation of internal pipes and fittings at buildings, including inspection of both services.

### 8.18.1 CONSTRUCTION OF WATER MAINS

#### 8.18.1.1 Transport and storage of pipeline components

Pipeline components must be protected against damage. Only suitable equipment may be used for loading and unloading. In the case of pipes made of unplasticized PVC, in particular, abrupt stresses at temperatures below +5°C must be avoided because of the influence of temperature on their resistance to impact.

In the case of pipes with external protection, wide slings or other devices, which do not damage the external protection, must be used. The use of chains or bare steel wire is not permitted.

Pipes with a bituminous coating must be handled in such a way that the effects and influences of high and low temperatures do not give rise to damage. If wire or chain hooks are used, they must be padded in order to prevent damage to the pipe ends.

#### 8.18.1.2 Transport to the construction site

During transport to the construction site, pipeline components must be kept apart by suitable intermediate layers and be secured against rolling, shifting, sagging, and vibration.

### 8.18.1.3 Storage

Pipeline components must be so stored that they do not come into contact with harmful substances. Earth, mud, sewage, or similar substances must not internally soil the pipeline components. If such soiling has been unavoidable, the pipeline components must be cleaned before being installed.

Pipes with bituminous external protection must not be stored directly on ground covered with vegetation because plant sprouts and shoots can grow into the external protection. Damage due to stony storage surfaces must also be avoided. Storage on stacking timber is recommended.

Stacking and stack heights must be so selected that damage to and permanent distortion of the pipes and damage to the external protection do not occur. The instructions of the pipe manufacturer determine the maximum stack height of pipes.

In the case of plastic pipes, stacks of pipes must not exceed the following heights:

| | |
|---|---|
| Pipes of PVC | 1.5 m |
| Pipes of PE | 1.0 m |

Stacks of pipes must be secured against rolling apart. If pipeline components must be stored in the open in frosty weather, care must be taken that they do not freeze to the ground. Where long periods of storage are involved, pipeline components made of materials sensitive to temperature and light (eg, plastics, rubber) or with external protection sensitive to light or temperature must be protected against sunlight (eg, by covering, coating with white paint, etc.).

### 8.18.1.4 Transport on the construction site

Where necessary, suitable transport equipment must be used for transport on the construction site. Dragging or prolonged rolling are not permissible.

### 8.18.1.5 Pipe trenches

The width of pipe trenches as a working space must be sufficient for proper installation of pipeline components. Before the pipes are laid, the pipe trench must be checked for correct depth and width and also for the condition of the trench bottom.

### 8.18.1.6 Dewatering

During pipe-laying operations, working spaces (pipe trenches and socket holes) must be kept free of water.

### 8.18.1.7 Depth of cover

The pipe trench must be formed and excavated in such a way that all pipes are finally laid at a frost-free depth (depth of cover usually 1.0–1.9 m, depending on climate, nominal diameter and soil conditions).

### 8.18.1.8 Trench bottom

The trench bottom must be so constructed that the pipeline rests on it throughout its length. If necessary, appropriate cavities must be excavated in the trench bottom at joint points. Unintended high and low points must be avoided.

The bottom of trench excavations shall be carefully prepared to a firm even surface so that the barrels of the pipes when laid are well bedded down for their whole length. Mud, rock projections, boulders, hard spots, and local soft spots shall be removed and replaced with selected fill material consolidated to the required level. The width of the excavations shall be sufficient to allow the pipes to be properly bedded, jointed, and backfilled. Joint holes or recesses, made as short as practicable, shall be formed in the trench bottoms so that joints can be made properly.

Where rock is encountered, the trench shall be cut at least 150 mm deeper than other ground and made up with well rammed material.

### 8.18.1.9 Bedding

The bedding shall ensure even distribution of pressure in the bedding zone. Pipeline components must be laid in such a way that neither linear nor point support occurs. Cavities of adequate size must, therefore, be provided in the trench bottom for sockets and couplings.

In the case of the usual type of bedding, a support angle of about 60° is suitable for pipes and fittings of small and medium nominal diameter (up to about DN500).

If pipes (or fittings) are designed for a different form of bedding or for a larger support angle, the bedding and embedding are to be constructed accordingly.

### 8.18.1.10 Sand and gravely sand bedding

In rocky and stony ground, the pipe trench must be excavated to a greater depth, depending on the material of the pipe and the external protection. The extra soil removed must be replaced by a layer containing no stones. For this purpose, depending on the material of the pipe, the external protection and the pipe diameter, compactible sand, gravely sand, screened soil—but no slag or other aggressive substances—are applied in a layer of suitable thickness and compacted. After compaction of the fill, the thickness of the stone—free layer of the bottom of the bedding at its lowest level must be 100 mm + 1/10 of the numerical value of the nominal diameter of the pipes in mm, with a minimum of 150 mm.

- **Bedding in nonload bearing soils**

   In the case of nonload bearing subsoil, such as soils with a high water content (boggy and tidal marsh soil), special embeddings are necessary such as, for example, mats, pile foundations, or reinforced concrete bearing plates.

### 8.18.1.11 Variations in bedding conditions

Where the bedding conditions in the direction of the pipeline axis is variant, unacceptable stressing of the pipeline components can arise as a result of differing movement in the bedding. Possible protective measures include, for example, a thicker sand embedding or flexible pipe joints coupled with appropriately short pipe lengths in the transition zone.

Also, in the case of crossings of other, already existing pipelines, provision must be made for bedding capable of bearing a load which can compensate for the different movements (Fig. 8.11). Protective measures must be coordinated with the operators of existing pipelines.

### 8.18.1.12 Special precautions in steeply sloping sections

In steeply sloping sections, suitable precautions must be taken to prevent the backfilled pipe trench from acting as a drain, with the result that the pipe embedding is washed away and the pipeline undermined. Appropriate measures are also necessary to prevent surface run-off along the backfilled pipe trench.

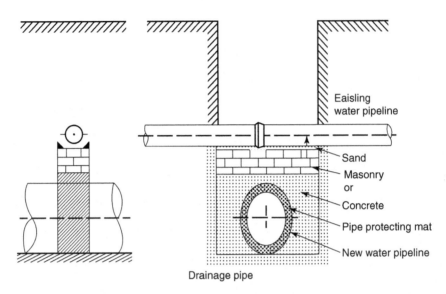

Eaisling
water pipeline

Sand

Masonry
or

Concrete

Pipe protecting mat

New water pipeline

Drainage pipe

**FIGURE 8.11 Example of Construction of Underpinning of a Crossing Pipeline of Small Nominal Diameter**

In hilly and steeply sloping sections, the pipeline must also be secured against slipping, for example, by means of force locking joints actuated by longitudinal forces (that transmit tensile forces in the direction of the pipeline axis), and or transverse arms.

### 8.18.1.13 Pipe laying
Pipes shall be laid in such a manner as will ensure even support throughout their length and shall not rest on their sockets or on bricks, tiles, or other makeshift supports.

*Note:* Pipes should be laid true to line to the general contours of the ground and at a sufficient depth for the pipe diameter to allow for the minimum cover below the finished ground level.

### 8.18.1.14 Ingress of dirt
Pipes should be kept clean and, immediately before laying each pipe and fitting, should be thoroughly cleansed internally and the open end temporarily capped until jointing takes place. Precautions shall be taken to prevent flotation of the capped pipes, in case the trench becomes flooded.

### 8.18.1.15 Protective coatings
Coatings, sheathings or wrappings shall be examined for damage, repaired where necessary, and made continuous before trench excavations are backfilled.

## 8.18.2 INSTALLATION OF PIPELINE COMPONENTS
### 8.18.2.1 Checking
Pipeline components must be checked for obvious damage, cleaned inside if necessary and any damage to the outer and inner protective coatings must be repaired before they are lowered into the pipe trench.

### 8.18.2.2 Placing in the pipe trench

If equipment is required for placing pipeline components, it must be capable of smooth and steady lowering without damage. Where stringing out is adopted, the bending radii must not be less than those permissible in each case.

### 8.18.2.3 Pipe cuts

Pipe cuts must be smooth. The start and end of the cut must not be staggered. Unevenness in the cut surface and burrs must be removed. Pipe ends must be properly machined as appropriate to the material and the type of joint.

### 8.18.2.4 Longitudinal gradient

Pipelines must be laid according to the structural drawings with the prescribed gradient.

### 8.18.2.5 Distances from underground installations

Distances from underground installations must be fixed in the light of the following safety considerations:

1. prevention of excessive transmission of forces
2. no undue temperature effects caused, for example, by district heating and cables
3. ensuring adequate working spaces for laying and repair work
   maintenance of a safety margin to avoid dangerous contacts with cables and proximity between pipelines and cables
4. electrically effective separation from all other metal conductors in the light of cathodic corrosion protection.

### 8.18.2.6 Distance from structures

The horizontal distance from foundations and similar underground installations shall not be less than 0.40 m.

### 8.18.2.7 Distance from pipelines and cables

Where there is (lateral) proximity or where the pipeline runs parallel to other pipelines and cables, the distance from them shall not be less than 0.40 m. At bottlenecks too, a distance of 0.20 m shall be maintained. If the spacing at bottlenecks has to be further reduced, suitable steps must be taken to prevent direct contact. Such steps must be agreed with the AR.

### 8.18.2.8 Crossings with pipelines and cables

In the case of crossings with pipelines and cables, a spacing of 0.2 m shall be adhered to. If this is not possible, contact must be prevented by, for example, interposition of nonconducting shells or plates. The possibility of transmission of forces must be excluded. This step must be agreed with the AR.

### 8.18.2.9 Protection of the pipeline against contamination

During laying, the pipeline must be protected against avoidable soiling. The pipeline must be cleaned. This can, for example, be done with a scraper, by pulling through a tightly fitting pipe brush and also, if entry into the pipeline is possible, by hand. When work is interrupted and on the conclusion of work, all openings must be closed so that they are watertight, by suitable means such as plugs, covers, and blank flanges.

### 8.18.2.10 Fitting of valves

Valves and their bypasses must be fitted in an unstressed condition. Any forces exerted must be harmlessly deflected. Where necessary, the weight loading must be taken up by foundations.

## 8.18.3 MAKING OF PIPE JOINTS

Pipeline components must be connected in such a way that the pipeline is tight and takes up static and dynamic stresses.

Bolts must be tightened crosswise around the entire circumference, in order to achieve even and adequate pressure in the sealing elements.

### 8.18.3.1 Force-locking pipe joints not actuated by longitudinal forces

Socket joints are usually force-locking pipe joints not actuated by longitudinal forces. When forming socket joints, care must be taken that the sealing rings seat accurately. In the case of bends, branches, and the like, forces must be deflected via abutments into the surrounding soil.

The nonmanipulative type of compression joint, as its name implies, does not require any working of the pipe end other than cutting square. The joint is made tight by means of a loose ring or sleeve that grips the outside wall of the pipe when the coupling nut is tightened. The manufacturer's recommendations should be followed.

### 8.18.3.2 Force-locking pipe joint actuated by longitudinal forces (manufacturer's instructions should be followed)

In the manipulative type of compression joint the end of the pipe is flared, cupped, or belled with special forming tools and is compressed by means of a coupling nut against a shaped end of the corresponding section on the fitting or a loose thimble.

### 8.18.3.3 Lubricants and sealing mediums

Only those lubricants and sealing mediums, which cannot adversely affect the quality of drinking water, may be used for making pipe joints in drinking water pipelines. In addition, lubricants and sealing mediums must not exert any harmful influence on pipeline components. There must also be no question of harmful interaction between pipeline components on the one hand and lubricants and sealing mediums on the other hand.

## 8.18.4 CORROSION PROTECTION

### 8.18.4.1 External protection by anticorrosive coatings (passive corrosion protection)

Water pipelines must be able to withstand the exposure to corrosion to be expected from outside. If the material is insufficiently resistant, external protection of adequate mechanical stress ability must be provided. The conditions of installation and operation are the decisive factors in the selection of pipe coatings, depending on the pipe materials selected.

No coating is necessary for plastic pipes.

### 8.18.4.2 Pipes of cement-bound materials

In normal cases, pipes of cement-bound materials do not require any external protection. DIN 4035 specifies the circumstances in which external protection is necessary for reinforced concrete pressure pipes.

### 8.18.4.3 Subsequent external protection

Repairs and additions to the pipe coating at faulty points and at pipe joints must be effected according to the manufacturer's instructions.

Flanges, after being cleaned, derusted, and dried, shall be protected by plastic strips or bitumen strips, by pouring round anticorrosive mediums or by shrunk-on formed parts.

In the case of bitumen and plastic coatings, care must be taken that no sharp-edged backfilling material is used, that no lasting heat effects, such as from district heating pipelines, are experienced and that no harmful substances such as oil and grease come into contact with the coating.

When laying pipeline components made of metallic materials with an electrically nonconducting coating to the pipes, the coatings must be tested with an electrical testing apparatus and, if necessary, properly repaired. The test voltage is at least 5 kV plus 5 kV per mm of thickness of the insulating layer, but with a maximum of 20 kV.

### 8.18.4.4 Internal protection

Water pipeline must be able to withstand the chemically corrosive stresses to be expected from inside. Should the material itself be insufficiently resistant, there must be appropriate internal protection.

### 8.18.4.5 Embedding the pipeline

To a large extent, the load and stress distribution at the circumference of the pipe determines the embedding. Bedding and embedding are often constructed together. For embedding, suitable soil, which does not harm the pipeline components or the coating, must be placed in layers on both sides of the pipeline and at least up to the middle of pipe diameter.

Pipelines liable to buoy upwards must be provided with safety precautions against uplift. In vegetation areas, placing and compacting in layers can, where appropriate, be dispensed with.

### 8.18.4.6 Backfilling of the pipe trench

Backfilling of the trench above the min. or max. height of embedding should be effected according to the requirements of the project. If trenches are in the roads, the degree of compaction of backfill should conform with specified compaction degree of the road subbase.

In vegetation areas, compacting and restoration of the surface of the ground must be done in such a way that damage due to plant growth is largely prevented.

## 8.18.5 SPECIAL STRUCTURAL MEASURES

### 8.18.5.1 Crossings with traffic routes

In the case of crossings of traffic routes, the specifications of the appropriate administrative bodies such as the Ministry of Roads, railway authorities, municipalities etc., must be observed and adhered to.

### 8.18.5.2 Working and safety strips

A working strip is necessary for the construction of pipelines. The width of the working strip depends on the nominal diameter of the pipeline and on local conditions.

Easements or "right of ways" serving the preservation and operating reliability of the pipeline must be procured by contracts or servitudes.

Other than in public traffic areas, a safety strip to ensure satisfactory maintenance and to exclude external influences, which could endanger the condition of the pipeline, shall protect water pipelines.

Structures having nothing to do with the pipeline must not be erected within the safety strip. The safety strip must be kept free of vegetation, which could adversely affect safety and maintenance of the pipelines.

The center of the safety strip shall correspond to the axis of the pipeline. The width of the safety strip shall be:

| Nominal size of pipeline | Width of safety strip (m) |
| --- | --- |
| Up to DN150 | 4 |
| Over DN150 up to DN400 | 6 |
| Over DN400 up to DN600 | 8 |
| Over DN600 | 10 |

Where there are compelling reasons, the figures listed may be reduced by up to 2 m over short stretches and at constrained points.

In the case of pipelines running parallel, the width of the safety strip increases by the distance between the outermost pipelines.

The final route of pipelines should permit ready and adequate access from public highways for the equipment and materials necessary to carry out planned inspections, maintenance, and emergency repairs.

## 8.18.6 PRESSURE TEST, DISINFECTION, FILLING OF THE PIPELINE

This section specifies the methods of testing the pipes, pipe joints, and fittings for leakage, and the safe positioning of a pressure pipeline before being put into operation.

### 8.18.6.1 Pressure test

Pipelines constructed with pressure pipes for the conveyance of drinking water, or water for industrial use must be subjected to an internal pressure test (preliminary test plus main test) according to requirements of DIN4279 Part 1–10, before backfilling.

The test pressure as per DIN4279 Part 1 is: $1.5 \times$ the nominal operating pressure for pipelines with a:

**10.1** permissible working pressure of up to 10 bar[2] and
**10.2** nominal operating pressure +5 bar for pipelines with a permissible working pressure of over 10 bar.

The duration of the test depends on pipe type and nominal width (see DIN4279 Part 1–8). In principle the test section should be chosen in such a way that the test pressure at the highest point of the pipeline is equivalent to at least 1.1 times the nominal pressure.

### 8.18.6.2 Preliminary test

The preliminary test comprises the preparatory steps leading up to the main test. It can be incorporated into the main test subject to local conditions and type of pipe.

---

[2]1 bar = 0.980665 atmospheric pressure ($kg/cm^2$)

### 8.18.6.3 Main test

The main test should be carried out at given test pressures, normally in sections of pipeline between 500 and 1500 m in length dependent upon local conditions, for example, geodetic variations in altitude.

The pressure test procedure comprising of bracing and anchoring of the pipeline, filling of the pipeline, measurement of pressure, and temperature and measurement of the water to be added should be executed as specified in DIN4279 Parts 1–8. Vents or other connections shall be opened to eliminate air from lines, which are to receive a hydrostatic test. Lines shall be thoroughly purged of air before hydrostatic test pressure is applied. Vents shall be open when systems are drained so as not to create buckling from a vacuum effect. Welded, flanged, or screwed connections must not be painted or otherwise covered before completion of pressure testing.

After completion of hydrostatic testing, all temporary blanks and blinds shall be removed and all lines completely drained. Any valves, orifice plates, expansion joints, and short pieces of piping, which might have been removed, shall be reinstalled with proper and undamaged gaskets in place. Valves, which were closed solely for hydrostatic testing, shall be opened. After lines have been drained with vents open, temporary piping supports, if any, shall be removed so that insulation and painting may be completed.

### 8.18.6.4 Execution of the test

If a preliminary test is conducted, the times for the duration of testing should be taken from Table 8.31.

If no preliminary test is conducted, the times for the duration of testing should be taken from Table 8.32.

If a preliminary test has taken place, the times for the duration of testing should be taken from Table 8.33.

### 8.18.6.5 Assessment of the test

The conditions of testing can be considered to be fulfilled if, at the end of the test, no drop in pressure can be established that exceeds the values given in Table 8.34.

**Table 8.31  The Duration of the Test of Pipe Type and Nominal Width**

| Nominal Width (NW) (m) | Duration of Test (h) |
| --- | --- |
| Up to 200 | 3 |
| 250 to 400 | 6 |
| Over 400 | 12 |

**Table 8.32  The Duration of the Test of Pipe Type and Nominal Width**

| Nominal Width (NW) (m) | Duration of Test (h) |
| --- | --- |
| Up to 200 | 3 |
| 250 to 400 | 6 |
| 500 to 700 | 18 |
| Over 700 | 24 |

| Table 8.33 The Duration of the Test of Pipe Type and Nominal Width | |
|---|---|
| **Nominal Width (NW) (m)** | **Duration of Test (h)** |
| Up to 400 | 3 |
| 500 to 700 | 12 |
| Over 700 | 24 |

| Table 8.34 The Drop in Pressure for Different Pressures | | |
|---|---|---|
| **Nominal Pressure Bar** | **Test Pressure Bar** | **Drop in Pressure Max.** |
| 10 | 15 | 0.1 |
| 16 | 21 | 0.15 |
| Over 16 | NP + 5 | 0.2 |

## 8.18.7 PERMISSIBLE LEAKAGE

Leakage allowed in a new mains is frequently specified in contracts, varying from 5.5 to 23 L/mm diameter per km per 24 h at the working pressure.

The specifications generally require that no pipe installation be accepted until the leakage is less than that indicated by the formula:

$$L = \frac{ND\sqrt{P}}{C}$$

in which $L$ is the allowable leakage, $N$ is the number of joints in the length of line tested, $D$ is the nominal diameter of the pipe, $P$ is the average test pressure during the leakage test, and $C$ is a constant depending on units and is equal to 326 (L/h, mm, kPa). For measurement of leakage a test pressure of 50% above the normal operating pressure for at least 30 min is recommended. It should be recognized that leakage can also occur from the service connections as well as the joints.

Care should be taken that no air has been retained in the pipe being tested.

## 8.18.8 DISINFECTION OF DRINKING WATER PIPELINES

Drinking water pipelines must be disinfected after successful fulfilment of the test. Before disinfection the main should be flushed at a velocity of at least 0.76 m/s. The use of a foam or rigid "pig," which is either driven through the line by the water pressure or pulled through by a cable is desirable. Mains have been satisfactorily disinfected with various chlorine compounds, potassium permanganate, and copper sulfate.

There is no satisfactory substitute for initial cleanliness of the mains. No disinfectant will kill bacteria, which are sheltered by debris. Filling should follow cleaning with "pig" and flushing at a velocity of at least 0.76 m/s with water containing a free residual chlorine concentration of at least 1.0 mg/L. A free residual of at least 0.5 mg/L must remain after 24 h. Following this procedure bacteriological analysis of the water should be conducted to insure its suitability. If total bacterial counts exceed 500/ mL or any coliform bacteria are found, the line should be filled with water containing 50 mg/L available chlorine, which should not decrease b low 25 mg/L in the 24-h holding period.

Water containing the disinfectant must be harmlessly disposed of. After being disinfected, the pipeline must be flushed until the water is of drinking water quality. In the case of pipes with a water-absorbent inner wall, it is advisable to carry out disinfection at the same time as the pressure test.

When it is necessary to repair or cut into an existing main, disinfection will also be necessary. No rule can be given as to methods, but by the use of fire hydrants for flushing and especially made taps the procedure given earlier may be followed.

### 8.18.8.1 Filling
Filling of the pipeline shall take place from the lowest point. For complete venting, adequate vents must be available at all high points. The inflow must be adjusted accordingly. The venting procedure must be controlled.

## 8.19 INSTALLATION OF WATER PIPES IN BUILDINGS
### 8.19.1 HANDLING OF MATERIALS

Pipes, fittings, and components shall be handled carefully to reduce damage. They should be stored so as to prevent contamination of the inside by dirt, mud, foul water, etc.

### 8.19.2 JOINTING OF PIPES

All proprietary joints shall be made in accordance with the manufacturer's instructions. Care shall be taken to establish satisfactory jointing techniques for all water service pipework. When making joints by welding, brazing, or soldering, precautions shall be taken to avoid the risk of fire. All burrs shall be removed from the ends of pipes and any jointing materials used shall be prevented from entering the waterways. All piping and fittings shall be cleaned internally and free from particles of sand, soil, metal filings, and chips, etc.

### 8.19.3 CAST IRON PIPES

Flexible mechanical joints shall be made in accordance with the manufacturer's instructions.

For molten lead joints, the spigot and socket shall be centered with rings of dry yarn caulked tightly into the bottom of the spigot to prevent the entry of lead into the bore of the pipe and to prevent contact of lead with the water.

Synthetic yarns that do not promote the growth of bacteria shall be used to prevent contamination of the water. The remainder of the joint space shall be filled with molten lead (taking care that no dross enters the joint), cold wire, strip, or spun lead (lead wool). The joint shall be caulked to a smooth finish with pneumatic tools or a hand hammer of mass not less than 1.5 kg. When working with spun lead, caulking tools shall be of a thickness to fill the joint space, ensuring thorough consolidation of the material to the full depth of the socket.

Lead joints shall be finished about 3 mm inside the face of the socket.

Flange joints shall be made with screwed or cast on flanges.

### 8.19.4 STEEL PIPES

Welded joints shall not be used where a protective lining would be damaged by heat, or where the pipework is employed as a primary circulation to an indirect hot water heating system.

Screwed joints in steel piping shall be made with screwed socket joints using wrought iron, steel, or malleable cast iron fittings. A thread filler shall be used. Exposed threads left after jointing shall be painted or, where installed underground, thickly coated with bituminous or other suitable corrosion preventative agent.

Flange joints shall be made with screwed or welded flanges of steel or cast iron using jointing rings and, if necessary, a suitable jointing paste. The nuts shall be carefully tightened, in opposite pairs, until the jointing ring is sufficiently compressed between the flanges for a watertight joint.

### 8.19.5 UNPLASTICIZED PVC PIPES

#### 8.19.5.1 Mechanical joints

Mechanical joints in unplasticized PVC piping of sizes 2 and upwards shall be made in accordance with BS4346: Part 2, by the use of push-fit integral elastomeric sealing rings which are compressed when the plain ended pipes are inserted into the adjoining sockets. The plain pipe ends shall be chamfered and the surfaces cleaned and lubricated.

The chamfered pipe end shall be inserted fully into the adjoining socket (except where provision is to be made for expansion), or as far as any locating mark put on the spigot end by the manufacturer. The sealing rings shall comply with BS2494.

#### 8.19.5.2 Compression joints

Compression joints shall only be used with unplasticized PVC piping of size 2 and smaller. The joints shall be of the nonmanipulative type. Care shall be taken to avoid overtightening.

#### 8.19.5.3 Solvent cement welded joints

Solvent cement welded joints in unplasticized PVC piping shall be made using solvent cement complying with BS4346: Part 3 recommended by the manufacturer of the pipe. The dimensions of the spigots and sockets shall comply with BSEN1452: Part 1–5.

Joints may also be made using integral sockets formed in the pipes and solvent cemented.

#### 8.19.5.4 Flanged joints

Flanged joints used for connections to valves and fittings shall use full-face flanges or stub flanges, both with corrosion resistant or immune backing rings and bolting.

#### 8.19.5.5 Polyethylene pipes

Mechanical joints shall be either plastics or metal proprietary compression fittings, for example, brass, gunmetal, or malleable iron. These shall include insert liners to support the bore of the pipe except where the manufacturer of the fitting instructs otherwise.

To ensure satisfactory jointing of the materials from which the pipe and fittings are made compatibility shall be established. The manufacturer's instructions shall be carefully followed.

No attempt shall be made to joint polyethylene piping by solvent cement welding.

## 8.20 JOINTING PIPES TO CISTERNS AND TANKS

Cisterns and tanks shall be properly supported to avoid undue stress on the pipe connections and deformation of the cistern or tank when filled, and holes shall be correctly positioned for the connection of pipes to cisterns and tanks. All debris, filings, borings, and blanks shall be removed from the inside of the cistern or tank.

Holes shall not be cut with flame cutters.

### 8.20.1 STEEL PIPES TO CISTERNS AND TANKS OF DIFFERENT MATERIALS (STEEL, GALVANIZED STEEL, OR GLASS REINFORCED PLASTICS)

The threaded end of the pipe shall be secured in the hole in the cistern or tank either by back nuts and washers both inside and outside (soft washers being used additionally with glass reinforced plastics or where there are irregular surfaces) or by using bolted or welded flanged connections.

### 8.20.2 VALVE CHAMBERS AND SURFACE BOXES

Surface boxes shall be provided to give access to operate valves and hydrants, and shall be supported on concrete or brickwork which shall not be allowed to rest on the pipes and transmit loads to them, allowance being made for settlement.

Alternatively, vertical guard pipes or precast concrete sections shall be provided to enclose the spindles of valves. Brick or concrete hydrant chambers shall be constructed of sufficient dimensions to permit repairs to be carried out to the fittings.

### 8.20.3 BRANCH PIPES OF RIGID MATERIALS

Any rigid branch pipe shall be connected to the ferrule on the pipe by a short length of suitable flexible pipe.

*Note:* This is to permit differential movement of the pipe and branch pipes without liability to fracture.

### 8.20.4 POSITION OF FERRULE IN A PIPE

The ferrule shall be set with the branch pipe leading off parallel to the pipe before turning into its proper course. Unless a swivel ferrule is used, the turn shall be arranged in a clockwise direction to ensure that if settlement of the pipe takes place, the joint between the ferrule and the pipe is being tightened.

#### 8.20.4.1 Allowance for movement

Underground piping of plastics shall be laid with slight deviations to allow for minor subsidence or temperature changes.

### 8.20.5 BRANCH PIPE CONNECTIONS

Branch pipes shall be connected to a main pipe by using one of two methods, depending upon the size of the pipes to be jointed namely ferrule, tee, or leadless collar.

The method used shall be as given in Table 8.35.

**Table 8.35  Method of Branch Pipe Connection**

| Nominal Size of Branch Pipes | | Nominal Diameter of Main Pipe | | | | |
|---|---|---|---|---|---|---|
| mm | in. | 80 mm | 100 mm | 150 mm | 200 mm | 250 mm and Over |
| 15 | ½ | F | F | F | F | F |
| 22 | ¾ | T | F | F | F | F |
| 25 | 1 | T | T | F | F | F |
| 35 | 1¼ | T | T | T | F | F |
| 42 | 1½ | T | T | T | F | F |
| 54 | 2 | T | T | T | T | F |

*F, Ferrule; T, tee or leadless collar.*

## 8.21  PIPEWORK IN BUILDINGS

### 8.21.1  ALLOWANCE FOR THERMAL MOVEMENT

In installations that do not have limited straight runs and many bends and offsets, allowance for expansion and contraction of the pipes shall be made by forming expansion loops, by introducing changes of direction to avoid long straight runs or by fitting proprietary expansion joints.

This is particularly important where temperature changes are considerable (eg, hot water distribution pipework) and where the pipe material has a relatively large coefficient of thermal expansion (eg, unplasticized PVC). In installations with limited straight runs and many bends and offsets, thermal movement is accommodated automatically.

### 8.21.2  FIXINGS FOR IRON PIPE

Iron pipe shall be secured by heavy weight holder bats of iron or low carbon steel either built in or bolted to the structure.

Steel, copper alloy, suitable plastics clips or brackets shall secure steel piping. Copper clips or brackets shall not be used for fixing steel piping.

Piping that is insulated shall be secured on clips or brackets that allow sufficient space behind the back of the pipe and the batten or wall to which the pipe is fixed for the insulation to be properly installed.

#### *8.21.2.1 Concealed piping*

Piping shall be housed in properly constructed builders' work ducts (riser shafts), or wall chases and have access for maintenance and inspection.

*Note:* Ducts and chases should be constructed as the building structure is erected and should be finished smooth to receive pipe fixings.

### *8.21.2.2 Piping passing through structural timbers*

Structural timbers shall not be notched or bored in such a way that the integrity of the structure is compromised. For recommended positions of notches and holes in timber beams and joists refer to BS6700.

### *8.21.2.3 Clearance of structural members*

Piping laid through notches, holes, cut-outs, or chases shall not be subjected to external pressure, and shall be free to expand or contract. Piping through walls and floors shall be sleeved. Where pipes are located in unheated surroundings such as open passages, they shall be laid at a depth sufficient to afford protection against freezing, if frost protection cannot be provided by other means. Pipe ducts shall be capable of being vented and drained.

## 8.22 DISINFECTION OF AN INSTALLATION

Flushing: Every new domestic water service directly connected to pressure main shall be thoroughly flushed with fresh water drawn direct from the water supplier's mains, immediately before being taken into use.

### 8.22.1 DISINFECTION: GENERAL

Where chlorinated water that has been used to disinfect an installation is to be discharged into a sewer, the authority responsible for that sewer shall be informed.

Where this water is to be discharged into a natural water course or into a drain leading to the same, the authority responsible for land drainage and pollution control shall be informed.

Where any pipework under mains pressure or upstream of any back-siphonage device within the installation is to be disinfected, the water supplier shall be informed.

### 8.22.2 DISINFECTION OF INSTALLATIONS WITHIN BUILDINGS THAT HAVE CENTRAL STORAGE CISTERNS

All visible dirt and debris shall be removed from the cistern. The cistern and distributing pipes shall be filled with clean water and then drained until empty of all water. The cistern shall then be filled with water again and the supply closed.

A measured quantity of sodium hypochlorite solution of known strength shall be added to the water in the cistern to give a free residual chlorine concentration of 50 mg/L (50 ppm) in the water. The cistern shall be left to stand for 1 h. Then each draw-off fitting shall be successively opened working progressively away from the cistern. Each tap and draw-off fitting shall be closed when the water discharged begins to smell of chlorine. The cistern shall not be allowed to become empty during this operation; if necessary it shall be refilled and chlorinated as previously mentioned. The cistern and pipes shall then remain charged for a further 1 h.

The tap farthest from the cistern shall be opened and the level of free residual chlorine in the water discharged from the tap shall be measured. If the concentration of free residual chlorine is less than 30 mg/L (30 ppm) the disinfecting process shall be repeated.

Finally, the cistern and pipes shall remain charged with chlorinated water for at least 16 h, for example, overnight, and then thoroughly flushed out with clean water until the free residual chlorine concentration at the taps is not greater than that present in the clean water from the water supplier's mains.

*Note:* Proprietary solutions of sodium hypochlorite should be used in accordance with the manufacturer's instructions having due regard for health and safety. A graduated container should be used to measure out the volume of solution required for disinfection. This can be calculated from the manufacturer's literature.

## 8.23 IDENTIFYING AND RECORDING PIPING LOCATIONS
### 8.23.1 LOCATION OF PIPE AND VALVES

Consideration shall be given to the need to locate the position of pipes and valves. Surface boxes shall be lettered to indicate what service is below them. Where possible, durable markers with stamped or set-in indexes shall be set up to indicate the pipe service, the size, the position, and depth below the surface.

### 8.23.2 IDENTIFICATION OF ABOVE GROUND PIPING

Where aesthetically acceptable, water piping shall be color banded.

In any building other than a single dwelling, every supply pipe and every pipe for supplying water solely for fire fighting purposes shall be clearly and indelibly marked to distinguish them from each other and from every other pipe in the building.

*Note:* Fire fighting water is not necessarily potable water. Settled raw water with 40–50 ppm turbidity is considered adequate in oil industries.

Record drawings: During the installation of a water supply system, records of all pipe runs, cisterns, valves, outlets, etc. shall be kept. On completion of the works, drawings shall be prepared on durable material of the "as fixed" installation.

### 8.23.3 IDENTIFICATION OF VALVES INSTALLED ABOVE GROUND

Every valve in hot and cold water service pipework installed above ground shall be provided with an identification label, either secured by noncorrodible, incombustible means to the valve or fixed to a permanent structure near the valve.

Labels secured to valves shall be of noncorrodible and incombustible material permanently and clearly marked, for example, by stamping or engraving, with a description of the service concerned and the function of the valve.

Alternatively, the label shall be marked with a reference number for the valve, instead of or in addition to the marking, and a durable diagram of the service, showing the valve reference numbers, shall be fixed in a readily visible position to a permanent part of the building or structure. Labels fixed near valves shall comply with the requirements for labels secured to valves except that they need not be incombustible.

## 8.24 CONSTRUCTION OF GRAVITY SEWERS
### 8.24.1 SETTING OUT SEWERS IN TRENCH

The center line and top width of trench should be accurately set out, marked, and referenced. Temporary benchmarks should be established in stable positions where they are unlikely to be disturbed. The transference of levels by straightedge and spirit level is not recommended and should be avoided where possible. On flat gradients the work should be set out and frequently checked by instrument.

### 8.24.2 EXCAVATION

The nature of the ground as revealed by the site investigation, the depth of the trench, and the avoidance of damage to existing (or proposed) structures or underground works, will determine the choice of the method of excavation and the type and strength of support required to the trench sides during construction.

The interruption of buried services is undesirable and may be dangerous. Every effort should be made to locate these accurately through the local authority before commencing excavation. Trial pits to confirm locations should be hand dug.

All pipes, ducts, cables, mains, or other services exposed in the trench should be effectively supported, or diverted, if necessary. Attention needs to be paid to the temporary and permanent support of these services to avoid subsequent damage. Care should also be taken to minimize ground movement, which may damage services alongside the trench.

### 8.24.3 DEWATERING

Where the proposed excavation formation level is below the groundwater table and the grading of the surrounding ground is suitable, it may be desirable to lower the groundwater table locally by pumping during the period of the excavation.

The pumping out of water can also carry with it fine material in suspension. This may cause subsidence of existing structures by loss of fines or water or both and it should be carried out under expert advice. Dewatering is usually only possible in soils coarser than silt (ie, 0.06 mm) with less than 10% passing a 75 $\mu$m sieve. Exceptionally, some coarse silts (0.02 mm) may dewater.

Pumping from a series of well points sunk adjacent to the line of the excavation may carry out dewatering.

### 8.24.4 PREPARATION OF TRENCH

Uniformity of support for a sewer is essential and the trench bottom should be carefully trimmed to the required depth and gradient to provide the proper formation level. A soft or uneven formation should be removed to an economical depth and the resulting cavity refilled with a material, which will give uniform support. If rock is encountered at formation level, this may also have to be removed and similarly replaced, except where concrete bedding is to be provided.

### 8.24.5 PIPE LAYING AND JOINTING

#### 8.24.5.1 Inspection of pipes preparatory to laying
All pipes, fittings, rubber joint rings, and preformed mastic seals should be carefully inspected prior to use, special attention being paid to joint surfaces, grade of mastic seal, and protective coatings and linings.

#### 8.24.5.2 Pipe laying
Pipe laying should start at the downstream end, the pipes are normally laid with the sockets upstream.

### 8.24.5.3 Laying pipes on trench as formation

Where the design permits and if the nature of the ground is such as to allow it to be trimmed to provide a uniform bearing, rigid pipes may be laid on the trench formation. Socket holes should be as short as practicable and should be scraped or cut in the formation, deep enough to give a minimum clearance of 50 mm between the socket and the formation.

If the formation has been overexcavated and does not provide continuous support, low areas should be brought up to the correct level by placing and compacting suitable material.

After the formation has been prepared, the pipes should be laid upon it true to line and level within the specified tolerances.

This should be checked for each pipe, and raising or lowering the formation, always ensuring that the pipes finally rest evenly on the adjusted formation throughout the length of the barrels, should make any necessary adjustments to level. Adjustment should never be made by local packing.

### 8.24.5.4 Laying pipes on granular bed

The trench should be excavated to a depth such as to allow the specified thickness of granular bedding material to be placed beneath the units. Any mud should be removed and soft spots either removed or hardened by tamping in gravel or broken stones. Rock projections, boulders, or other hard spots should also be removed.

In soft clay, disturbance of the trench bottom should be minimized by placing a layer of blinding material about 100 mm thick. Such precautions may also be necessary in bad weather and for wet ground conditions.

Granular bedding material should be placed to the correct level and should extend to the full width of the trench.

Pipes should be laid directly on the granular bed and should then be adjusted to correct line and level within the specified tolerances, that is, the pipe invert levels should be constructed to a tolerance of ±20 mm subject to the provision that the pipe gradient should not be less than 90% of that required in the design. Sidefill of either granular material or selected backfill material, depending upon the bedding specified, should be placed and compacted evenly on either side of the pipe taking care not to disturb the line and level.

Measures to prevent migration of fine material from pipe bedding should be undertaken where the pipeline is below groundwater level.

## 8.24.6 LAYING PIPES WITH CONCRETE BED, BED AND HAUNCH, OR SURROUND

Where in situ concrete bedding is required, the trench bottom should be prepared to give a firm foundation as described in 7.4, using a blinding layer if necessary. The level of this formation should allow for a depth of concrete under the pipe barrel of 0.25 D, where D is the nominal pipe bore, or 100 mm, whichever is the greater. The pipes should be supported clear of the trench bottom by means of blocks or cradles placed under the pipes immediately behind each socket and just clear of each spigot, or at both sides of the sleeve, where sleeve joints are used.

The blocks should extend the full depth of the bedding, which should be cast monolithically. Free standing blocks may be used if they are of a suitable size, which will not tilt, or rock when pipes are added to the line. A minimum layer of compressible material should be placed between the support and the pipe to permit the barrel of the pipe to rest uniformly on its bed after normal setting shrinkage of the concrete has occurred. Expanded polystyrene or impregnated fiber building board are suitable for this purpose.

The concrete bed should extend equally on each side of the pipe to a width of 1.25 Bc or Bc + 200 mm, whichever is the greater, where Bc is the outside diameter of pipe barrel, and should not be placed until the pipework has been inspected and deemed satisfactory.

Where flexible joints are employed, the overall flexibility of the pipeline should be maintained by the provision of flexible joints in the concrete. These should be formed through the full cross-section of the concrete by providing compressible materials at least 20 mm thick (or that required to ensure the flexibility of the pipe joint), such as expanded polystyrene or impregnated fiber building board, at the face of each pipe socket or at one face of each sleeve. This is to ensure that any subsequent flexing occurs only at a pipe joint.

Care should be taken as follows:

1. In the placing of concrete, that the pipes or lateral construction joints are not displaced and flexibility of the joint is not impaired; and
2. to avoid excessive shear loads developing at joints, especially immediately below road surfaces.

### 8.24.6.1 Pipes laid in made ground
Ground, which has been formed by loose tipping or has received inadequate compaction during placing may be subject to continuing consolidation and uneven settlement. Pipework laid in such ground should be of flexible material or should have flexible joints to allow uneven settlement to be accommodated without damage or loss of performance.

Pipe gradients should be as steep as conditions permit to ensure that a backfall is not induced as a result of settlement.

### 8.24.6.2 Pipes laid in fluid ground
Where pipework is to be laid in ground, which is fluid, or water logged expert advice should be sought.

## 8.25 JOINTING PIPES
### 8.25.1 FLEXIBLE JOINTS
The pipe manufacturer's instructions regarding the making of the joints should be followed closely. Only sliding ring joints should be lubricated, using the lubricants recommended by the manufacturers. The correct sealing rings should be used in jointing; if the rings are supplied separately from the pipes, care should be taken not to mix up different sizes.

Most types of flexible joint can be made in wet conditions, but it is preferable not to attempt jointing when the pipes are under water. The jointing faces and sealing rings should be clean and free from oil, grease, tar, mud, or sand particles, prior to placing the joint ring on the spigot or in the socket or collar as specified.

When joints cannot be made manually, mechanical pulling devices should be used. Any disturbance of the pipe bed should be minimized, and made good. The specified gap should be left between the end of the spigot and the socket of the next pipe to permit movement. With some rolling-ring types of joint there is a tendency for the ring to unroll with small pipes, unless the pipe is temporarily held in the trench, during the pulling-in of the joint.

## 8.25.2 RIGID JOINTS

While flexible joints are quicker to make and preferred, the older traditional rigid joints are still available with some pipe materials and for joining dissimilar pipes. Care should be taken to ensure that where they are used this is consistent with the overall design of the pipeline. If it is necessary to make the occasional rigid joint in a flexibly jointed pipeline, the rigid jointed section should be kept as short as possible.

Cement mortar for joints should be in a ratio of 1:3. The ends of the pipes should be wetted immediately before jointing and the joints kept damp and protected from the sun and wind until covered by the initial backfill. The interior of the pipe should be examined as each joint is made and any intrusions of mortar or gasket removed before further pipes are laid.

Concrete pipes with ogee or rebated joints are normally jointed with mortar, but they cannot usually be made completely watertight by this means alone.

## 8.26 ANCILLARY ITEMS

### 8.26.1 JUNCTIONS

Junctions should be inserted at intervals as required for present or future connections, during the construction of the sewers. All junctions, which are not immediately connected to laterals, should be closed with durable purpose made watertight caps. The position of each junction should be carefully measured and recorded for as built drawings.

### 8.26.2 GULLIES

Gullies should be set vertically and to the correct level; when necessary they should be surrounded in concrete, care being taken to prevent flotation.

### 8.26.3 MANHOLES

Manholes are either built by brickwork or concrete. Manholes are used to accept several secondary sewer lines to lead them into the main sewer line, or where change of direction in sewer line is required.

## 8.27 TRENCHLESS CONSTRUCTION

Trenchless construction is any means of constructing new, or rehabilitating existing pipes underground without excavating an open trench.

For guidance refer to BS6164: 1990 "Code of practice for safety in tunneling in the construction industry."

### 8.27.1 BACKFILLING

Generally care and attention should always be given to the placing and compaction of backfill, particularly where it forms part of a load-supporting system, for example, under roads.

### 8.27.2 SIDEFILL AND INITIAL BACKFILL

#### 8.27.2.1 Rigid pipes

As soon as possible after completion of the bedding or surround, selected fill should be placed by hand and carefully compacted between the pipes and trench sides and brought up in 150 mm to 250 mm layers to at least 150 mm of compacted material above the pipe crown, taking care to avoid uneven loading and damage to the pipe.

Where mortar joints or concrete beddings are used, sufficient time should be allowed for them to gain strength to avoid damage during the backfilling operation.

#### 8.27.2.2 Flexible pipes

The sidefill for flexible pipes other than ductile iron should be of the same granular material as that used for bedding. It should be taken to at least the level of the pipe crown and be carefully compacted. (The load carrying capacity of the pipes depends very largely on the compaction of the sidefill to provide the resistance to lateral deformation.) Selected backfill should then be placed and carefully compacted in layers to give at least one 150 mm layer above the pipe crown.

In most cases, for ductile iron pipes, tamped excavated natural material from the trench will be suitable for backfill. In instances of excessive depths, high vehicular loading or super loading from buildings, etc., or very poor soil properties it may be necessary to bring in a graded granular backfill.

Backfill should be built up in even layers not exceeding 300 mm, each layer being thoroughly compacted before further fill is added. Guidance on the selection of compacting equipment, with techniques and appropriate layer thickness for various types of soil is given in the standards. The aim is to compact the fill as nearly as possible to the same density and moisture content as that of the undisturbed soil in the trench sides, but this is rarely achieved in practice and some subsequent settlement usually occurs. The compacting equipment should be such as to ensure that the pipes are not overloaded during the filling.

### 8.27.3 BACKFILLING AROUND MANHOLES

The method of backfilling and compacting around manholes should be generally as for trenches. Care should be taken to raise the fill equally all round the manhole shaft to avoid unbalanced lateral loading. Care is also necessary in placing the fill around freestanding manhole shafts which have been constructed in advance of an embankment (eg, in a valley which is to be filled in subsequently); end-tipping of the fill in the vicinity of the manhole should be avoided.

Where there is a structural risk, precast concrete manholes should be surrounded with concrete 150 mm or more thick, possibly reinforced, depending on the loading conditions.

### 8.27.4 BACKFILLING AROUND SEWAGE PUMPING STATIONS

The method of backfilling and compacting around sewage pumping stations that are constructed using the open-cut procedure.

## 8.28 TESTING OF SEWERS

Sewers should be tested and inspected for infiltration and exfiltration to acceptable limits. Initial testing should be applied before any sidefill is placed. This will facilitate replacement of any faulty pipes or joints revealed by the test. Testing after placing backfill will reveal any leakage due to subsequent damage or the displacement of joints.

For the final acceptance test by AR the line should be water tested from manhole to manhole. Any short branches may be tested concurrently with the main line, but branches over approximately 10 m long should be tested separately.

## 8.28.1 CHOICE OF TEST METHOD

There are two test methods, which are relatively simple and technically acceptable; these are based on the loss of either water or air.

As sewers are designed to carry liquids, the water test is to be preferred, but under site conditions an air test is usually quicker and more economical. Hence, it is recommended that contractors carry out air tests also at their own cost to avoid replacing of faulty pipes and remaking of joints.

## 8.28.2 AIR TEST

The air test is easier to carry out than the water test and does not have the problem of providing and disposing of large quantities of water. It provides a rapid test, which can be carried out after every third, or fourth pipe is laid. This should prevent a faulty pipe or a badly made joint passing unnoticed until it is revealed by a test on a completed length. To replace a faulty pipe or remake a joint in the middle of a pipe run is a time consuming and costly operation. A smoke test can indicate the location of a failure. An air test will be performed only by the jurisdiction of AR.

## 8.28.3 WATER TEST

Gravity sewers up to and including DN 750 should be tested to an internal pressure represented by 1.2 m head of water above the crown of the pipe at the high end of the line. The test pressure should not exceed 6 m head of water at the lower end and if necessary the test on a pipeline can be carried out in two or more stages. The test pressure should be related to the possible maximum level of groundwater above the sewer.

When pipes larger than DN750 are to be tested, expert advice and special equipment may be needed.

Gravity drains and private sewers within curtilage up to and including 300 mm diameter should be tested to an internal pressure of 1.5 m head above the invert of the pipe at the high end of the line and not more than 4 m head at the lower end.

Testing should be carried out between inspection chambers, manholes, or other suitable points of access and through any accessible branch drains. Where the test head of water is in excess of 4 m at the lowest point of the pipeline under test (including the minimum test head of 1.5 m), solvent welded uPVC pipelines should be allowed to stand for 1–2 h before applying the test and should be suitably anchored to prevent flotation when the test is applied before backfilling the trench.

### 8.28.3.1 Test procedure

The following test procedures with the indicated limits of exfiltration can be adopted as acceptable choices.

1. Fit an expanding plug, suitably strutted to resist the full hydrostatic head, at the lower end of the pipe and in any branches if necessary. The pipes may need strutting to prevent movement.
2. Fit a similar plug and strutting at the higher end but with access for hose and standpipe.
3. Fill the system with water ensuring that there are no pockets of entrapped air.

4. Fill the standpipe to requisite level.
5. Leave for at least 2 h to enable the pipe to become saturated, topping up as necessary.
6. After the absorption period measure the loss of water from the system by noting the amount of water needed to maintain the level in the standpipe over a further period of 30 min the standpipe being topped up at regular intervals of 5 min.

The rate of loss of water should not be greater than 1 L/h per meter diameter per meter of pipe run. The water level in the manholes should not drop more than 100 mm.

### 8.28.3.2 Factors affecting the test
Excessive leakage may be due to the following:

1. porous or cracked pipes
2. damaged, faulty, or improperly assembled pipe joints
3. trapped air being dissolved
4. defective plugs
5. pipes or plugs moving.

### 8.28.3.3 Freedom from obstruction
As the work progresses the sewer should be checked for obstructions by visual inspection or by inserting a mandrel or "pig" into the line.

### 8.28.3.4 Straightness
A sewer should be checked for line and level at all stages of construction. Methods of checking include the following:

1. surveyor's level and staff
2. sight rails, boring rods, and travelers
3. laser beams with sighting targets
4. lamp and mirrors.

## 8.28.4 SOUNDNESS TESTS FOR ANCILLARY WORKS
Recommendations given in the standards for the materials, design, and construction of manholes, and similar underground chambers should ensure a high level of resistance to water penetration, both inwards and outwards.

Manholes should be so constructed that no appreciable flow of water penetrates the permanent works.

Where construction work has been effectively carried out, visual inspection of ancillary works may be sufficient for acceptance without specific testing. Inspection should always be made to reveal any possible weaknesses in the structure and particular attention should be paid to the following:

1. step iron and ladder housings
2. benching
3. pipes entering or leaving the structure
4. joints in brickwork or block work
5. joints between sections of the structure.

If required, the inspection should be followed by specific testing.

### 8.28.4.1 Specific testing

There may be a need for testing to be carried out in any of the following cases:

1. for petrol interceptors, suction wells, and similar structures
2. where unsatisfactory features have been revealed by inspection, for example, where there is reason to believe that materials or workmanship have been inadequate
3. in locations where there is fissured chalk or rock, or pervious subsoil
4. where frequent surcharging of the manhole is likely.

### 8.28.4.2 Test head

Manholes less than 1.5 m in depth to invert should be filled with clean water to the underside of the cover and the frame located at ground or surface level. Where the depth to the channel invert is 1.5 m or greater, the test head should be not less than 1.5 m. The test head for petrol interceptors, suction wells, and similar underground chambers should not be less than 0.5 m above the invert of the highest connection to the chamber.

Where the chamber is located in the ground subject to pore pressure, the test head should be the mean water table level based on seasonal variations or test heads previously specified, whichever is the greater.

### 8.28.4.3 Test procedures

Tests should not be carried out until structures have reached sufficient strength to sustain the pressure from testing.

The external faces of a structure should not normally be backfilled or concrete surrounded before the chamber is filled with water to the specified test level. Adequate stability should be ensured during the period of test and subsequent concrete placement and backfilling.

For the tests fit a bag stopper in the outlet of the manhole and expanding plugs or bag stoppers in all other connections.

Secure all plugs and stoppers to resist the full hydrostatic head and provide a means of safely removing the outlet bag stopper from the surface.

Fill the manhole with clean water to the required test level and allow to stand for at least 8 h for absorption, topping up the level as necessary. Carry out the tests as rapidly as possible.

*Note:* These tests should show that the construction is substantially watertight.

### 8.28.4.4 Acceptance criteria

The rate of water loss should not exceed 1 L/h per meter diameter per linear meter run of pipe. For various pipe diameters this rate of loss over a 30 min period may be expressed as follows:

DN100 pipe 0.05 L/m run
DN150 pipe 0.08 L/m run
DN225 pipe 0.12 L/m run
DN300 pipe 0.15 L/m run.

### 8.28.4.5 Testing of watertight structures

Sumps, suction wells, mud and oil interceptors, petrol interceptors, oil separators, septic tanks, and cesspools should be tested for water tightness but over the full height to surface level and without measurable loss of water after 30 min.

## 8.29 INSPECTION OF PIPELINES, SERVICES, AND INSTALLATIONS

### 8.29.1 PROCEDURE

Inspections and tests shall be undertaken as work proceeds. Prior notice shall be given to the water supplier before any statutory inspections or tests are undertaken. The installer shall keep records of all tests required by the specification.

*Note:* Testing should normally take the form of both interim or preliminary and final tests.

### 8.29.2 TIMING

The timing of tests shall be arranged as follows:

1. Interim tests. As soon as practicable after completion of the particular section, with particular attention to all work which will be concealed.
2. Final tests. To be carried out on completion of all work on the water services and prior to handing over.

### 8.29.3 INSPECTION

Visual inspections shall be carried out at both interim and final testing in order to detect faults in construction or material not shown up under test, but which could lead to failure at a later date, possibly after expiry of the contractual maintenance period.

This is particularly important in the case of an interim test where the installation will be covered up as work proceeds.

A careful record shall be kept of such inspections, and notes taken to facilitate the preparation of "as installed" drawings.

In the case of visual inspection of pipelines, particular attention shall be paid to the pipe bed, the line and level of the pipe, irregularities at joint, the correct fitting of air valves, washout valves, sluice valves, and other valves together with any other mains equipment specified, including the correct installation of thrust blocks where required, to ensure that protective coatings are undamaged. Trenches shall be inspected to ensure that excavation is to the correct depth to guard against frost and mechanical damage due to traffic, ploughing, or agricultural activities. No part of the pipe trench shall be backfilled until these conditions have been satisfied and the installation seen to comply with the drawings and specifications and the appropriate laws and regulations.

All internal pipework shall be inspected to ensure that it has been securely fixed.

### 8.29.4 TESTING OF INSTALLATIONS WITHIN BUILDINGS

When the installation is complete it shall be slowly filled with water, with the highest draw-off point open to allow air to be expelled from the system.

If the water is obtained from the water supplier's mains, it shall be taken in accordance with the supplier's requirements.

The installation, including all cisterns, tanks, cylinders, and water heaters, shall then be inspected for leaks.

It is desirable that the installation then be tested hydraulically in the following way. Subject the pipes, pipe fittings, and connected appliances to a test pressure of at least 1.5 times the maximum working pressure, with the pressure applied and maintained for at least 1 h and check that there is no loss of water or visual evidence of leakage. Water byelaws require that any pipe or fitting that is not readily accessible have to be capable of withstanding twice the maximum working pressure.

Each draw-off tap, shower fitting, and float-operated valve shall be checked for rate of flow against the specified requirements.

Performance tests shall also be carried out on any connected specialist items to show that they meet the requirements detailed in the specification.

Defects revealed by any of the foregoing tests shall be remedied and the tests repeated until a satisfactory result is obtained.

## 8.30 WATER DISTRIBUTION MAINS

It is essential that procedures for the operation, modification, repair, and inspection of pipelines are formulated and adhered to so that a pipeline continues to function safely while in use. The operating body should develop procedures for operating, modifying, repairing, and inspecting pipelines based upon the recommendations given here and upon best industry practice.

### 8.30.1 ROUTINE INSPECTION

Routine visual inspection of land pipelines should be made to check on the condition of the pipeline easement. Any third party activity on, or adjacent to the pipeline easement and which could affect the integrity of the pipeline should be investigated. The frequency of such inspection may vary dependent upon local conditions. Any excavation or development occurring near buried pipelines should be monitored.

Arrangements should be made with the company and occupiers to permit a routine programme of inspection of the route. In the absence of any such arrangement, except in cases of emergency, prior written notice of all pipeline inspections involving entry on land should be given to the occupiers.

All persons carrying out inspections should carry and produce on request adequate means of identification.

### 8.30.2 WATER ANALYSIS

Regular analyses of water samples at intervals not exceeding 6 months shall be carried out wherever bulk drinking water storage exceeds 1000 L.

Periodic chemical and bacteriological analysis of water samples is a useful guide to the condition of an installation. The collection and analysis of water samples is particularly recommended for new installations in large buildings or complexes and where extensive repairs or alterations have been carried out to such installations.

### 8.30.3 PIPEWORK

#### 8.30.3.1 Fixings and supports

Any loose or missing fixings or supports shall be made good. Provision for expansion and contraction shall be checked, particularly in the case of plastics pipework.

### 8.30.3.2 Joints

Leaking joints shall be tightened or remade, or where necessary the pipework shall be renewed, to stop all leakage.

### 8.30.3.3 Corrosion and scaling

If inspection reveals corrosion of the pipework or reduced flow rates indicate the possibility of corrosion products or scale obstructing waterways, the cause shall be investigated and appropriate remedial action taken.

*Notes:* (1) Pipes showing signs of serious external corrosion should be replaced. The replacement pipe should have suitable protection (eg, factory plastics coated, spirally wrapped, or sleeved with an impervious material) or should be of a corrosion resistant material compatible with the remaining pipework and (2) Internal corrosion of galvanized steel pipe is usually localized, requiring replacement of the affected section only. The whole pipe length between joints should be replaced to retain continuity of galvanizing or other protection.

### 8.30.3.4 Thermal insulation and fire stopping

Any damage to thermal insulation or fire stopping revealed during inspection shall be made good.

*Note:* Thermal insulation used for frost protection should be checked at the beginning of each winter.

## 8.30.4 TERMINAL FITTINGS AND VALVES

### 8.30.4.1 Terminal fittings

As soon as any sign of leakage from a float-operated valve (eg, dripping from a warning pipe) or tap is noticed the fitting shall be rewashered, reseated, or replaced as necessary to stop the leakage. The action of self-closing taps shall be checked at regular intervals and any necessary repairs or adjustments carried out.

In addition to preventing leakage, the free movement of infrequently used float-operated valves; particularly those fitted to the feed and expansion cisterns of hot water or space-heating systems should be checked at intervals not exceeding 1 year.

Spray heads on taps and showers should be cleaned at such intervals as experience indicates.

Gland packings on taps should be tightened or renewed as necessary to prevent any leakage while not impeding the normal operation of the fitting.

### 8.30.4.2 Stop valves

Stop valves shall be operated at least once per year to ensure free movement of working parts.

Any stiffness or leakage through the gland should be dealt with by lubrication, adjustment, or replacement of gland packings or seals. If there is any indication of leakage past the seating the valve should be rewashered, reseated, or replaced as necessary. If there is any indication that the waterway is blocked, the valve should be dismantled, cleared, and restored to good working order or replaced.

### 8.30.4.3 Relief valves

Easing gear fitted to relief valves shall be operated at least once per year to check that the valve has not stuck or become blocked. Any fault revealed shall be corrected immediately.

Since relief valves are explosion prevention devices it is important to check their operation at regular intervals. Operation of easing gear may sometimes cause the valve to leak and so involve additional attention; this is infinitely preferable to an explosion.

### 8.30.5 CISTERNS

Cisterns shall be inspected from time to time to ensure that overflow and warning pipes are clear, that covers are adequate and securely fixed, and that there are no signs of leakage or deterioration likely to result in leakage. Cisterns storing more than 1000 L drinking water shall be inspected at least once every 6 months, those storing less than 1000 L drinking water at least once per year.

## 8.31 INSPECTION OF SEWERAGE WORKS

Surveys of both sewers and manholes are undertaken for several reasons. The nature and detail of the data to be collected will depend on the objective, and may influence the method of recording findings.
Common objectives are as follows:

1. to investigate causes of poor performance, or of suspected deterioration (eg, frequent blockage, smell complaints, pipe fragments in manholes)
2. to locate and inspect sewers prior to change in their use, or in the use of the surface above them (with the objective of executing necessary repairs)
3. to record the condition of a sewer immediately prior to adjacent workings (with the objective of attributing liability for any associated damage)
4. to measure the flows and periodically update sewer records.

Internal sewer surveys may be classified as either operational (or service condition) surveys, or as structural surveys.

### 8.31.1 OPERATIONAL SURVEYS

Operational surveys are carried out to provide information on the following:

1. the necessity for, and nature and frequency of, cleaning to verify the recommendations given under 9.5
2. possible structural weaknesses which may be indicated by the type of sediment (inflow from voids, rubble, bricks, etc.)
3. tree root systems which might cause blockages
4. sediment which may indicate bad or flat construction even though the sewer may be sound
5. the necessity for flushing and cleansing operations
6. heavy concentrations of debris indicating the possibility of collapse
7. the presence of noxious gases indicating sediment and septicity problems which could lead to fabric deterioration.

### 8.31.2 STRUCTURAL SURVEYS

Structural surveys provide information on the following:

1. possible renewal or renovation costs;

**2.** major defects which can lead to collapse;
**3.** the condition and life of sewer fabric.

## 8.31.3 SURVEY METHOD

The surveying method for the condition of manholes, sewers, sewage pumping stations, outfall structures etc. is influenced by survey objective, costs, and size of sewer network.

## 8.31.4 PLANNED INSPECTIONS AND REPORTS

In normal circumstances, regular planned inspections should be made on the existing sewerage works as follows:

### 8.31.4.1 Manhole and sewer line surveys

Manholes and sewer lines should be surveyed at 12-month intervals over the first 3 years of its life and every 3 years thereafter.

The position of all manholes along a length of sewer can be determined by referencing the manhole covers to permanent features of detail on the ground (ie, buildings, fences, kerb lines, etc.).

In rural areas the manhole positions can be plotted on to 1/2500 scale maps giving a positional accuracy of ±1 m, and for urban areas on to 1/1250 scale maps giving positional accuracy of ±0.15 m.

The sewer center line should be determined in relation to each manhole center and plotted as an approximate line between manholes. The resulting information should be plotted on to the appropriate map.

All information recorded in the field should be properly indexed and cross-referenced for future use.

Sewer invert and cover levels should be related to Owner's Bench Marks. An accuracy of ±25 mm should be aimed at for all sewer leveling.

Cut marks can be established on manhole frames directly over the sewer invert if possible, and the distance from cut mark to sewer invert is measured very accurately down the manhole shaft. In shallow sewers the invert level may be read directly while surface leveling is being carried out.

It is convenient to store collected information on a standard data sheet. The level of the center of each manhole cover should be recorded in the field and booked to the nearest 10 mm on the sewer map.

The survey information thus carried out should be produced in the form of at least a written report.

### 8.31.4.2 Survey of sewage pumping stations, sewage treatment plants and sewage rising mains and outfalls

In normal circumstances, well-designed pumphouses, treatment plants, and outfalls should give satisfactory service without routine attention. Evidence of a build-up in the wet well of crude sewage pumphouse and continuous operation of the pumps in full capacity is a sign of a serious problem demanding emergency inspection.

The sewage pumping stations, sewage treatment plants, rising mains and outfalls should be physically checked and inspected regularly at 6-month intervals over the first 2 years of their life and every 2 years thereafter.

A build up of slime or sediment within the rising main and outfall may be observed by increases in the head required to maintain discharge at a given tidal state. If there is a marked decrease in the retention period in the outfall at a given rate of flow, the installation should be monitored regularly as regards hydraulic performance.

Systematic recording of the structural defects such as cracks, settlements, wear and tear at the intervals mentioned above, together with structural failure risk assessment could result in substantial savings by timely preventive measures.

As regards the ultimate outfall into the sea or river, engineering divers are briefed to note any signs of damage or other unsatisfactory conditions and should inspect the full route of the pipe. Particular attention should be given to the following:

1. abrasion of the outer protective coating
2. the physical condition of the outfall supporting structure
3. excessive growths of marine organisms at the mouth of outfall.

## 8.31.5  INSPECTION AND REPAIR OF ACCESSORIES

In addition to a detailed inspection of the sewers to ensure that they have been constructed or repaired to the specification and are clear of silt and other debris, all step irons, ladders, landings, guard rails, and chains should be checked to ensure that they are securely fixed and set in the correct position. Manhole covers should be correctly set and their frames properly bedded. A check should be made to ensure that any previously live connections have been remade.

The performance of a new or repaired sewer should be monitored for a period.

All statutory undertakers and those directly connecting to the sewer should be advised when the sewer has been put into use. Those responsible for sewer operation and maintenance, pumping stations and treatment works should be notified beforehand of the date when the sewer will be commissioned and advised as soon as it has been put into use.

Final inspection of sewers and manholes constructed under a contract should be made before the completion of the maintenance period in order that any defects can be remedied before the conclusion of the contract.

All drawings of works as executed should be kept. All inspection records should be kept for at least 12 months after completion of the works. Sewer maps should be updated in respect of new sewers, removed sewers, and closed sewers.

For operational, maintenance, and repair guidance refer to BS EN 752–7, Part 1: 1998.

## 8.31.6  BUILDING DRAINAGE

Drainage systems within curtilage should be inspected at regular intervals of 6 to 12 months and, where necessary, thoroughly cleaned out at the same time. Any defects discovered should be made good.

## 8.31.7  DRAINAGE OF ROOFS AND PAVED AREAS

### 8.31.7.1  Periodic inspection and cleaning

Gutters, rainwater pipes, outlets, and gratings should be inspected and thoroughly cleaned once a year, or more often if the building is in or near an industrial area, or is near to trees, or may be subjected to extremes of temperature.

Gullies and channels of paved areas (eg, car parks) should be inspected, and cleaned out, regularly. The frequency of inspection and cleaning will need to be based on local experience.

Defects should be remedied as soon as possible after being noted.

### 8.31.7.2 Jointing of potable water pipework

Table 8.36 below lists the jointing methods and materials that should be used for jointing potable water pipework.

| \multicolumn{5}{l}{**Table 8.36 Jointing of Potable Water Pipework**} | | | | |
|---|---|---|---|---|
| **S. No.** | **Type of Joint** | **Method of Connection** | **Jointing Material** | **Precautions/Limitations** |
| 1 | Lead to brass, gun-metal, pipe or fitting | Plumber's wiped soldered joint | Tallow flux | Of very limited application, see BS6700[a] |
| 2 | Galvanized steel (pipe to pipe or fitting) | Screwed joint, where seal is made on the threads | PTFE tape or proprietary sealants | PTFE tape only up to 40 mm (1½) diameter[a] |
| 3 | Galvanized (pipe to pipe or fittings) | Flanges | Elastomeric joint rings complying with BS2494, or corrugated metal | [a] |
| 4 | Long screw connector | Screwed pipework with BS2779 thread | Grummet made of linseed oil based paste and hemp | [a] |
| 5 | Shouldered screw connector | Seal made on shoulder with BS2779 thread | Elastomeric joint rings complying with BS2494 and plastics materials | — |
| 6 | Unplasticized PVC (pipe to fitting) | Solvent welded in sockets | Solvent cement complying with BS4346: Part 3 | — |
| 7 | Unplasticized PVC (pipe to fittings) | Spigot and socket with ring seal, flanges, union connectors | Elastomeric seal complying with BS2494 lubricants | Lubricant should be compatible with the unplasticized PVC and elastomeric seal |
| 8 | Cast iron (pipe to fitting) | Caulked lead | Sterilized gasket yarn/blue lead | [a] |
| | | Bolted or screwed gland joint | Elastomeric ring complying with BS2494 | — |
| | | Spigot and socket with ring seal | Elastomeric seal and lubricant | — |
| 9 | Plastic (pipe to tap or float-operated valve) | Union connector | Elastomeric or fiber washer | — |
| 10 | Pipework connections to storage cisterns (galvanized steel, reinforced plastics, polypropylene, polyethylene) | Tank connector/union with flange backnut | Washers: elastomeric, polyethylene, fiber | — |
| 11 | Polyethylene (pipe to fitting) | Nonmanipulative fittings | — | Do not use lubricant |
| | | Thermal fusion fittings | — | — |

*(Continued)*

**Table 8.36 Jointing of Potable Water Pipework** (*cont.*)

| S. No. | Type of Joint | Method of Connection | Jointing Material | Precautions/Limitations |
|--------|---------------|---------------------|-------------------|-------------------------|
| 12 | Polybutylene (pipe to pipe or fitting) | Nonmanipulative fittings | Lubricant on pipe end when Required | Lubricant if used should be listed and compatible with plastics |
| | | Thermal fusion fittings | — | — |
| 13 | Polypropylene (pipe to pipe) | Nonmanipulative fittings | Lubricant on pipe end when required | Lubricant if used should be listed and compatible with plastics |
| | | Thermal fusion fittings | — | — |
| 14 | Cross-linked polyethylene (pipe to fitting) | Nonmanipulative fittings | Lubricant on pipe end when required | Lubricant if used should be listed and compatible with plastics |
| 15 | Chlorinated PVC (pipe to fitting) | Solvent welded in sockets | Solvent cement | — |

*[a]Where nonlisted materials are to be used, due to there being no alternative, the procedure used should be consistent with the manufacturer's instructions taking particular note of the following precautions:*
- *use least quantity of material to produce good quality joints*
- *keep jointing materials clean and free from contamination*
- *remove cutting oils and protective coatings, and clean surfaces*
- *prevent entry of surplus materials to waterways*
- *remove excess materials on completion of the joint.*

## 8.32 WATER SUPPLY AND SEWERAGE EQUIPMENT

This section is the minimum requirement and deals with the material specification of pipes and fittings that are used for water supply and sewerage systems in residential areas of industrial projects including such auxiliary items as manhole covers and frames, step irons, ladders, and other components, with due consideration to the fact that commonly used pipes and fittings in water supply and sewerage projects of municipalities and water distribution authorities or organizations are mostly produced by local manufacturing firms based on BSI or DIN standards and in compliance with the recommendations given in Publication No. 128, dated 1993 Plan and Budget Organization.

### 8.32.1 NOMINAL SIZE (DN)

A numerical designation of size that is common to all components in a piping system other than components designated by outside diameters or by thread size (Table 8.37). It is a convenient round number for reference purposes and is only loosely related to manufacturing dimensions.

**Table 8.37 Nominal Size (DN)**

| Designation of thread size | 1/8 | 1/4 | 3/8 | 1/2 | 3/4 | 1 | 11/4 | 11/2 | 2 | 2 1/2 | 3 | 4 | 5 | 6 |
|----------------------------|-----|-----|-----|-----|-----|---|------|------|---|-------|---|---|---|---|
| Nominal size (DN) | 6 | 8 | 10 | 15 | 20 | 25 | 32 | 40 | 50 | 65 | 80 | 100 | 125 | 150 |

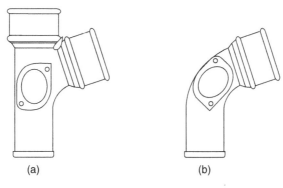

(a)                                                  (b)

**FIGURE 8.12  Right Hand Fittings**

(a) Right hand branch and (b) right hand bend.

*Notes:* (1) Nominal size is designated by the letters DN followed by the appropriate number, (2) This definition is identical to ISO 6708, and (3) The relationship between fitting size and nominal size is given hereunder for reference purposes.

### 8.32.2  RIGHT HAND FITTING

A bend or branch which is so constructed that, when it is viewed with the spigot downwards and with the access door facing the observer, the socket of the bend or the arm of the branch projects to the right (Fig. 8.12).

### 8.32.3  TERMS RELATING TO PRESSURES AND DIMENSIONS OF PIPES

For definition of technical words such as nominal pressure, nominal diameter, or nominal size etc., used in DIN, BS, and ISO Standards.

1. Equal fittings. Equal fittings where all outlets are the same size shall be specified by that one size, irrespective of the number of outlets (Fig. 8.13).
2. Unequal fittings. Unequal fittings shall be specified by the sizes of each outlet, the sequence being dependent upon the number of outlets, as follows:
    10.3  For fittings having two outlets. the larger outlet shall be specified first;
    10.4  For fittings having more than two outlets. these shall be specified in accordance with the sequence given in Fig. 2, for example a female reducing tee, having thread sizes of 1½ for outlet (1), 1 for outlet (2) and 1½ for outlet (3) shall have the fitting size designation of 1½ × 1 × 1½;
    10.5  tees Type B1 and pitcher tees Type E1 with equal outlets on the run and an increasing or reducing outlet on the branch shall be specified by stating the size of the run followed by the size of the branch, for example, a female reducing tee Type B1 having thread sizes of 1½ for outlet (1), 1½ for outlet (2) and 1 for outlet (3) shall have the fitting size designation of 1½ × 1.

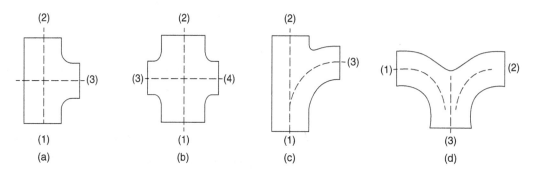

**FIGURE 8.13 Method of Specifying Outlets of Fittings Having More Than Two Outlets.**

*Note:* The method of specifying outlets is in accordance with method (b) in ISO 49.

## 8.32.4 NONPRESSURE PIPELINE COMPONENTS

General specification for nonpressure pipeline components including pipes, bends, "P" or "S" traps, branches, and ancillaries such as manhole covers and frames, gullies, ladders, step irons, etc. that are used in sewerage or drainage systems should comply with the requirements of the standard.

### 8.32.4.1 Rigid and semiflexible pipes

Whatever selection is made, pipes should have adequate strength to meet loading requirements, be sufficiently robust to withstand site handling and be sufficiently durable to remain watertight for the anticipated life of the system.

The various pipes and their application as recommended in BS standards are quoted hereunder as well as in Tables 8.1 and 8.2, moreover, the compatible standards in DIN, ISO, and ISIRI are also given in Table 8.1.

### 8.32.4.2 Fiber-cement pipes

The asbestos-cement pipes and fittings with flexible joints for use in gravity sewers or drains should comply with BS EN 588-1 which specifies the strength classification for pipe diameters from DN100–DN2500, in pipe lengths of 3, 4, and 5 m.

The asbestos cement pipes and flexible joints used in force mains (rising mains) should comply with BS EN 512, which specifies the hydrostatic classification for pipe diameters from DN50–DN2500 in pipe lengths of 3, 4, and 5 m. Asbestos-cement bends are available for diameters up to DN225. Cast iron fittings are also available.

### 8.32.4.3 Spigot and socket clay pipes

Vitrified or salt glazed clay pipes and fittings for use with gravity flow under atmospheric pressure should comply with BS65. They are available in nominal diameters from DN75–DN1000, and in lengths up to 3.0 m.

The following classifications of pipes are available:

1. normal, that is, suitable for all drains and sewers
2. surface water

**3.** perforated, suitable for soakaway land drains
**4.** extra chemically resistant.

Use of vitrified clay pipes for drainage of sanitary wastewater within curtilage of buildings and in sewers in the size range of DN150–300 is highly recommended when the residential town is not within the earthquake zone and moreover when they are manufactured locally.

### 8.32.4.4 Concrete pipes

Precast concrete pipes and fittings of circular cross section for the conveyance of sewage or surface water under gravity, should comply with the appropriate part of BS5911.

BS5911: Part 1 specifies requirements for flexibly jointed pipes in nominal diameters DN150–DN3000 in standard lengths 0.45 m to 5 m (3 m for pipes DN–600), in three strength classes for sewage or surface water.

### 8.32.4.5 Spigot and socket gray cast iron pipes and fittings

For nonpressure pipelines cast iron pipes and fittings should comply with BS437 and BS4622 respectively (Tables 8.38 and 8.39 for compatible acceptable standards). These pipes are manufactured in the size range DN50–DN225; BS437 makes provision for centrifugally cast pipes with flexible joints in lengths up to 5.5 m. Flexible joints for these pipes should comply with BS6087.

Pressure pipes and fittings with flexible or flanged joints should comply with BS4622.

Pipes with flexible joints are normally available in 5.5 m lengths and those with flanged joints in 4 m lengths.

Fittings complying with BS437 can be jointed directly to BS4622 pipes with lead caulked joints in the smaller diameters (DN100 and DN150 only).

*Note:* These pipes are suitable for below ground sewers and drains.

### 8.32.4.6 Cast iron spigot and socket soil, waste, and ventilating pipes and fittings

Cast iron spigot and socket soil, waste, and ventilating pipes (sand cast and spun) and fittings (for above ground use) should comply with BS416: 1973 (inclusive of latest amendments) manufactured in the size range of DN50–DN150 with type A sockets only.

*Note:* These pipes may be used as rainwater pipes when a heavier grade of pipe than that specified in BS460 is required.

### 8.32.4.7 Rainwater pipes and goods

Cast iron rainwater pipes and goods shall be manufactured by the sand cast or spun process with type A sockets only DN50–DN150 complying with BS460.

### 8.32.4.8 Corrugated metal pipes (semiflexible)

For helically corrugated pipes, in sizes from DN150–DN1500 made from galvanized steel sheet and in lengths from 6 to 9 m having bituminous coating, refer to BS EN 10142.

### 8.32.4.9 Ductile iron pipes (semiflexible)

Ductile iron pipes and fittings should comply with BS4772 and are manufactured in a range from DN80–DN1600.

Ductile iron pipes with flexible joints are manufactured in nominal lengths of 5.5 m for DN80–DN800 inclusive and nominal lengths of 8 m for DN900–DN1600 inclusive. Ductile iron pipes are suitable for both pressure and nonpressure applications.

**Table 8.38 Pipes Used in Nonpressure Pipelines**

| BS Standard are Chosen as Base / Nominal size | Rigid Pipes — Fiber BS EN 588-1 (ISO 881) | Fiber BS EN 512 (DIN19 moq pts 1.2.4.3) | Clay BS65 (DIN1230) | Concrete BS5911 Part + 100 (DIN4032) | Gray Cast Iron BS416 BS437 BS460 (DIN19572) | Flexible Pipes — Gray Cast Iron BS4622 | Ductile Iron BS4772 (ISO 7186) | Steel BS534 (DIN1530) | Unplaseleised PVC BS4660 (130&160DN) (ISO 8283 ISO 4435) (DIN19534 100–600DN) | BS5481 (230–630DN) (ISO 161.1) | BS3506 | Remarks |
|---|---|---|---|---|---|---|---|---|---|---|---|---|
| Nominal size | Nominal bore | Nominal bore | Nominal bore | Nominal bore | Nominal bore | Nominal bore | Nominal bote | Outside diameter | Outside diameter | Outside diameter | Min. Outside diameter | |
| DN75 | mm — | mm 75 | mm — | mm — | mm 75 | mm 80 | mm 80 | mm 76.1 / 88.9 | mm — | mm — | mm — | |
| 100 | 100 | 100 | 100 | 100 | 100 | 100 | 100 | 114.3 | 110 | — | 114.1 | BS486 ram be used in pressure mains also. |
| 125 | 125 | — | — | — | — | — | — | 139.7 | — | — | 140.0 | |
| 150 | 150 | 150 | 150 | 150 | 150 | 150 | 150 | 168.3 | 160 | — | 168.0 | The diameter Nominal (DN) of this table retets to BS std. only. |
| 175 | 175 | — | — | — | — | — | — | 193.7 | — | — | 193.5 | |
| 200 | 200 | 200 | — | — | — | 200 | 200 | 219.1 | — | 200 | 218.0 | The smallest Mombral size accepted on industries pipes is DN100 |
| 235 | 225 | 225 | 125 | 225 | 225 | — | — | 244.5 | — | — | 244.1 | |
| 250 | 250 | 250 | — | — | — | 250 | 250 | 273.0 | — | 250 | 273.6 | |
| 300 | 300 | 300 | 300 | 300 | — | 300 | 300 | 323.9 | — | 315 | 323.4 | |

Notes:
1. Pipes in diameters greater than 300 are available in many of these materials.
2. Most of the pipes listed are manufactured locally.

**Table 8.39  Chemical Resistance of Materials (for General Guidance)**

| Group | BS No. | Material. and Applications | Normal Domestic Sewage | Trade Effluent | | | | | Soil Environment Containing | |
|---|---|---|---|---|---|---|---|---|---|---|
| | | | | At Normal Temperature | | Organic Solvents | Containing Oil and Fat | | Sulfates | Acids |
| | | | | Acids | Alkalis | | Vegetable | Mineral | | |
| Ceramics | 65,1169 | Clayware pipes and fittings | A | S | S | S | S | S | S | S |
| | 771-1 | Bricks and blocks of fired | A | S | S | S | S | S | S | S |
| | | Brick- earth, clay or shale. | | | | | | | | |
| Concrete | 5911 | Concrete | | | | | | | | |
| | | Ordinary Portland cement | | E | A | A | E | A | E | E |
| | | Sulfate-resisting Portland cement. | | E | A | A | E | A | A | A |
| Fiber cement | BS EN 588-1 | Fiber cement pipes, joints and fittings (gravity) for sewage and drainage | A | E | A | A | E | A | A | E |
| Metals | 534 437 | Steel pipes and fittings | A | E | A | A | A | A | E | E |
| | 4772 | Gray iron pipes and fittings (gravity) ductile iron pipes and fittings | A | E | A | A | A | A | E | E |
| Plastics | 4660 5461 3506 | U PVC Gravity drain and Sewer pressure | A | A | S | E | A | A | S | S |

Jointing materials as specified by pipe supplier to meet the commentary requirements

Commentary:
Pipes and joints should remains sufficiently watertight to prevent ingress of ground water and egress of effluent when subjected to ground movement and settlement.
Flexible joints are generally available for the range of materials used for drainage and they can accommodate angular deflections, sail.
Displacement and draw within the joint. They are designed to resist loads without loss of water tightness where rigid joints are required.
They can be made by caulking metal or core pound or by working a cement mortar into the joints, by bolted flanges or by welding, the pipes together. Care should be taken on such jointing operations not to disturb the gradient of the line or the caulking of the bore.

Coatings (refer to ips E – TP – 270)

Commentary:
The coating most commonly applied to metal pipes does not necessarily provide adequate protection against all types of corrosion especially. Designed coatings which may be required for protection against concentrations or acids. Alkalis, sulfates, and other aggressive chemicals likely to be encountered in the ground and in liquid carbon.

*Notes: It is important to take account of quantities and concentrations of all types of chemical likely to be encounter. A, normally suitable; E, need expert advice, each case to be considered on its own merit; S, specially suitable.*

**Table 8.40 Index of Fitting Types, Symbols and Index to Tables**

| Type | Description | Symbol | | | |
|------|-------------|--------|--|--|--|
| A | Elbows | A1 | | A4 | |
| | Pipe ends | Female—equal | Female—reducing | Male and female—equal | Male and female—reducing |
| A | Elbows | A1/45 | | A4/45 | |
| | Pipe ends | Female—equal | | | |
| B | Tees | B1 | | | |
| | Pipe ends | Female—equal | Female—reducing | Female—increasing | Female—reducing |
| C | Crosses | C1 | | | |
| | Pipe ends | Female—equal | | | |
| D | Bends | D1 | D4 | D6 | |
| | Pipe ends | Female—equal | Male and female—equal | Male—equal | |

**Table 8.40 Index of Fitting Types, Symbols and Index to Tables (*cont.*)**

| Type | Description | Symbol |
|------|-------------|--------|
| D | 45 Degrees bends | D4/45 degrees |
| | Pipe ends | Male and female—equal |
| E | Pitcher tees | E1 |
| | Pipe ends | Female—equal / Female—reducing |
| E | Twin elbows | E2 |
| | Pipe ends | Female—equal |
| G | Long sweep bends | G1    G4 |
| | Pipe ends | Female—equal / Male and female—equal |
| kb | Return bends | kb1 |
| | Pipe ends | Female—equal |

(*Continued*)

**Table 8.40 Index of Fitting Types, Symbols and Index to Tables (*cont.*)**

| Type | Description | Symbol | | | |
|---|---|---|---|---|---|
| M | Sockets | M2 | | M3 | M4 |
| | Pipe ends | Female—equal | Female—reducing | Female—reducing | Male and female—reducing |
| N | Bushes | N4 | | | |
| | Pattern | I | II | | III |
| N | Hexagon nipples | N8 | | | |
| | Pipe ends | Male—equal | | Male—reducing | |
| P | Backnuts | P4 | | | |
| T | Gape | T1 | | T2 | |
| | Pattern | Hexagon | | Round | |
| T | Plugs | T8 | | T9 | T11 |
| | Pattern | Plain | | Beaded | Countersunk |

| Type | Description | Symbol | | |
|---|---|---|---|---|
| U | Unions | U1 | U11 | U12 |
| | Ends—pattern | Female—flat seat | Female—taper seat | Male and female—taper seat |
| UA | Elbow unions | UA11 | | |
| | Ends—pattern | Female—taper seat | | |

Table 8.40  Index of Fitting Types, Symbols and Index to Tables (*cont.*)

### 8.32.4.10  Commentary

Pipework within pumping stations is usually of ductile gray iron and the pipe joints are mostly flanged.

Pipe joints for use below ground should preferably be of the flexible type. If flanges are used on buried pipes the fastenings should be specially protected.

### 8.32.4.11  Step irons or ladder rungs

Step irons fixed in deep brick and concrete manholes should be fabricated from $\varphi$ 20 round bars in accordance with the standard drawing and be galvanized before installation. For more information refer to BS1247 (Tables 8.38 and 8.39).

### 8.32.4.12  Manhole covers and frames

Gray cast iron manhole covers and frames should be cast in accordance with standard drawings for heavy weight and for light weight cover and frame, applicable in both brick or concrete manholes, as chosen and required.

### 8.32.4.13  Safety chains (for sewage pumphouse)

Safety chains should be made of low carbon steel or of stainless steel, 10 mm nominal size, short-link, smooth welded chain to BS4942: Part 2. When made of low carbon steel they should be protected by hot dip galvanizing in accordance with BS729.

### 8.32.4.14 Ladders (for sewage pumphouse)

Fixed ladders should meet the dimensional requirements of BS4211 except that stringers should be not less than 65 mm × 20 mm in section and rungs 25 mm in diameter. When made of low carbon steel they should be protected by hot dip galvanizing in accordance with BS729.

For rise and tread dimensions, stringer and landing details refer to the specific engineering.

### 8.32.4.15 Handrails and handholds (in sewage pumphouse)

Handrails and handholds should be at least 25 mm in diameter. Low carbon steel tubes can be used for fabrication at the shop. They should be protected by hot dip galvanizing in accordance with BS729, before assembly.

### 8.32.4.16 Gullies

Gullies should comply with BS65, BS437 or BS5911: Part 2 as appropriate. A gully usually incorporates trap, or a sump, or both, to retain detritus. The top should be fitted with either a grating or sealed cover. Connections should be made below the grating or cover. Gullies may be specially designed to suit selected locations and the volume and nature of the flow.

Table 8.40 shows index of pipe fittings.

# FIRE-FIGHTING PUMP AND WATER SYSTEMS

The chapter provides a range of standard principal for specified fire pumps. A fire pump is a part of a fire sprinkler system's water supply and can be powered by electric, diesel or steam. The pump intake is either connected to the public underground water supply piping, or a static water source (eg, tank, reservoir, lake). The pump provides water flow at a higher pressure to the sprinkler system risers and hose standpipes. A fire pump is tested and listed for its use specifically for fire service by a third-party testing and listing agency, such as UL or FM Global. The main code that governs fire pump installations in North America is the National Fire Protection Association's NFPA 20 Standard for the Installation of Stationary Fire Pumps for Fire Protection.

The pumps covered are intended to include material, fabrication, design, and engineering features concerned with installation and use in water supply systems and mobile equipment in accordance with NFC Standards for centrifugal electrical and diesel engines fire pumps.

Included in the chapter are trailers used for mounting "foam" and other fire extinguishing agents and equipment such as premix Foam Liquid Concentrate (FLC), dry chemical, welding, cutting, and miscellaneous tools for special tasks.

Fire Pump Systems are engine drive pumps where electric power cannot be assumed or is not available and the pump is connected to an engine that runs on a 2-stroke, diesel, or unleaded fuel motor. Engine driven pumps are also ideal for water transfer, pumping water for stock watering, portable spray units, irrigation, and boom spraying.

Fire pumps may be powered either by an electric motor or a diesel engine, or, occasionally a steam turbine. If the local building code requires power independent of the local electric power grid, a pump using an electric motor may utilize, when connected via a listed transfer switch, the installation of an emergency generator.

The fire pump starts when the pressure in the fire sprinkler system drops below a threshold. The sprinkler system pressure drops significantly when one or more fire sprinklers are exposed to heat above their design temperature, and open, releasing water. Alternately, other fire hoses reels or other fire-fighting connections are opened, causing a pressure drop in the fire-fighting main.

Fire pumps are needed when the local municipal water system cannot provide sufficient pressure to meet the hydraulic design requirements of the fire sprinkler system. This usually occurs if the building is very tall, such as in high-rise buildings, or in systems that require a relatively high terminal pressure at the fire sprinkler in order to provide a large volume of water, such as in storage warehouses. Fire pumps are also needed if a fire protection water supply is provided from a ground level water storage tank.

Types of pumps used for fire service include: horizontal split case, vertical split case, vertical inline, vertical turbine, and end suction.

This engineering and material chapter specification covers the minimum requirements for fixed fire pumps and trailers carrying various fire extinguishing agents.

The adequacy and dependability of the water source are of primary importance and should be fully determined prior to the purchase of pumping equipment, with due allowance for its reliability in the future.

In this standard the water supply system, selection of pumps and their component parts as well as trailers used to take fire extinguishers and emergency tools to the scene of fire such as, foam equipment, premix FLC, dry chemical unit and foam/water monitor is discussed.

1. *Head.* The unit for measuring head should be the in meters. The relation between a pressure expressed in bars and a pressure expressed in meters of head is:

$$\text{Head in Meters} = \frac{\text{Pressure in bars}}{0.098}$$

2. *Velocity head.* The velocity head should be figured from the average velocity ($v$) obtained by dividing the flow in (m³/s) by the actual area of pipe cross-section in ($m^2$) and determined at the point of the gage connection.

Velocity head is expressed by the formula:

$$(m)h_v = \frac{V^2}{2g}$$

Where, g = The acceleration due to gravity and is ($9.807$ m/s²) at sea level and 45 degrees latitude; $V$ = Velocity in the pipe in (m/s).

## 9.1 METHOD OF TAKING WATER
### 9.1.1 PUMP UNITS TAKING SUCTION FROM OPEN WATER

At least two identical submerged pumps taking suction from open water should be installed; one electric motor driven, one diesel engine-driven, and if specified steam turbine.

The power of the drives, both the electric motor and the spare unit should be rated such that it is possible to start these units against an open discharge system which can be pressurized to 3 bar.

The electric motor should be provided with an automatic starting device, which should act after putting the fire alarm system into operation.

The spare unit should be provided with automatic starting facilities, which should act as present time if the electric motor or the pump does not function.

Manual starting and stopping of each unit should be possible from a control centre or from the fire station, and also should be possible at the pump site. Manual starting should be possible without the fire alarm coming into operation.

*Note:* In this section a diesel engine has been taken as a typical example of an independent power source for driving the spare pump.

### 9.1.2 PUMP UNITS TAKING SUCTION FROM STORED WATER

If water for fire fighting cannot be supplied direct from available open water under all conditions and at the required rate, or if owing to the excessively great distance it is not economically justified to install the fire-fighting pumps at that source, water storage facilities are required, for example, an open tank or pond having an adequate replenishment rate. This replenishment rate is of vital importance and the aim should be to obtain a rate equal to the installed capacity of one fire-fighting pump P-1 or P-2.

Apart from using plant-cooling water for this purpose, other sources, if available below ground level close to the premises can be utilized using electric motor-driven deep-well pumps.

If available from open water at an acceptable (not too great a distance) replenishment can be handled by a centrifugal pump driven by a diesel engine.

The water storage should be adequate to cover a period required to start the replenishment facilities. The reserve for fire fighting is to be as specified.

If the ambient temperature can fall below 0°C, provisions should be made to prevent the stored water from freezing, for example, by circulating.

Adequate care should be taken to keep the water in good condition, for example, to prevent algae growth.

If a tank has been selected for storing fire-fighting water, horizontal centrifugal pumps should be applied. In this case a pressurizing pump is not required, as the static head in the tank will keep the system filled with water.

The drivers should be diesel engine, electric motor, or as specified steam turbine as indicated on the individual standard of pump specification.

Each pump with driver should be mounted on the manufacturer's standard support plate. Horizontal drivers should be mounted on a separate drain-rim type baseplate. The baseplate should be of sturdy cast iron or fabricated steel construction. The baseplate should be provided with adequate accurately drilled holes for anchor bolts.

## 9.2 CONTROL SYSTEM

### 9.2.1 SUITABILITY

The control system should be suitable for:

Automatic and manual starting of the alarm siren, local and remote manual operation of the electric motor-driven pump, and the diesel engine-driven pump.

Starting the electric motor-driven pump automatically upon an alarm received from the fire alarm system. Starting the diesel engine-driven pump automatically if the electric motor-driven pump does not function properly within a preset time.

The siren should be started automatically in the case of a fire alarm, or manually by means of a "Fire" push button on the control panel.

### 9.2.2 PUMP CONTROL

The electric motor-driven fire-fighting pump should be started automatically in the case of a fire alarm, or manually either local or remote from the control panel and should have facilities for local and remote manual stopping.

Auxiliary contacts in the motor control circuit should be connected to a lamp to indicate the electric motor is running on the control panel.

If the electric pump fails to build up pressure within a time period of 0–30 s (adjustable), the diesel engine driven pump should be started automatically. This latter pump should also have facilities for local and remote manual starting and stopping.

### 9.2.3 DIESEL ENGINE CONTROL

All controls necessary for safe starting and operation of the diesel engine should be located on a local panel adjacent to the engine with duplication of the most important functions on the control panel.

For a diesel engine, the following additional requirements should be adhered to:

- The capacity of the fuel tank should be such that the engine can operate at full power for at least 10 h.
- This tank should be installed at such a level that the bottom is at least 0.2 m above the suction valve of the diesel injection pump.
- The tank should be provided with a level gage and facilities for refilling directly from drums.
- No clutch should be installed between diesel engine and pump.

## 9.3 ELECTRICAL SYSTEM

A 220-V AC/24-V DC rectifier to supply 24-V DC to the control system, and to keep the batteries fully charged (the batteries to be located near the engine starter motor).

Stationary batteries should be heavy duty and furnished in dual sets complete with necessary cables and connectors of proper size and length. An AC-powered battery charger should be furnished as specified by NFPA No. 20.

Battery trays (or boxes) should be provided with a plywood-lined metal bottom and legs to facilitate cleaning of surface beneath trays (or boxes).

Boxes should be constructed of 2 mm (14 ga) sheet steel of ventilated design and with a steel lid. The interior of the boxes should be painted with acid resistant paint.

### 9.3.1 BATTERY LOCATION

Storage batteries should be rack supported above the floor, secured against displacement, and located where they will not be subject to excessive temperature, vibration, mechanical damage, or flooding with water. They should be readily accessible for servicing.

For the control system (and for the alarm system when mounted in the control center) an electricity supply should be provided with rectifier and batteries, minimum voltage 24-V DC, capacity 24 h minimum, and suitable for variations in electricity supply voltage of $\pm10\%$. This electricity supply should also operate lamps, claxons, etc. and should be completely separate from other systems (including those for process safeguarding, telecommunication, etc.) but should be used for analogue signal transmission for fire-fighting water pressure control, if the distance would be too long for pneumatic signal transmission.

The siren should be connected to the AC power supply of the emergency generator.

The diesel engine-driven pump should have its own battery set for starting and control.

## 9.4 CENTRIFUGAL FIRE PUMPS
### 9.4.1 WATER SUPPLIES

The adequacy and dependability of the water source are of primary importance and should be fully determined prior to the purchase of pumping equipment, with due allowance for its reliability in the future.

Any source of water, that is, adequate in quality and quantity should provide the supply for fire pumps. Where the water supply is from a public service main, pump operation should not reduce the suction head below the pressure allowed by the local regulatory authority.

The minimum water level of a well or wet pit should be determined by pumping at not less than 150% of the fire pump rated capacity.

A stored supply should be sufficient to meet the demand, placed upon it for the expected duration, and a reliable method of replenishing the supply should be provided.

The head available from a water supply should be figured on the basis of a flow of 150% of rated capacity of the fire pump. This head should be as indicated by a low test.

### 9.4.2 SPECIAL REQUIREMENTS

- Suitable means should be provided for maintaining the temperature of a pump room or pump house, where required, above 5°C (40°F).
- Temperature of the pump room, pump house, or area where engines are installed should never be less than the minimum recommended by the engine manufacturer.
- The engine manufacturer's recommendations for water heaters and oil heaters should be followed.
- Artificial light should be provided in a pump room or pump house.
- Individual emergency lighting should be provided by fixed or portable battery operated lights, including flashlights.
- Provision should be made for ventilation of a pump room or pump house.
- Floors should be pitched for adequate drainage of escaping water or fuel away from critical equipment such as the pump, driver, controller, fuel tank, etc. The pump room or pump house should be provided with a floor drain, which will discharge to a frost-free location.

### 9.4.3 LISTED PUMPS

Centrifugal fire pumps should be listed for fire protection service.

#### 9.4.3.1 Rated pump capacities

Fire pumps should have the following rated capacities in liters per minute (LPM) and rated at net pressures of 2.7 bars or more.

A horizontal split case or vertical turbine pump should have a rated capacity equal to or greater than a value specified in Table 9.1.

An end suction or in-line pump should have a rated capacity of less than 1892 LPM' (500 GPM).

**Table 9.1 Pump Capacities**

| LPM | LPM | LPM |
|-----|-----|-----|
| 95 | 1514 | 7570 |
| 189 | 1703 | 9462 |
| 379 | 1892 | 11355 |
| 568 | 2839 | 13247 |
| 757 | 3785 | 15140 |
| 946 | 4731 | 17032 |
| 1136 | 5677 | 18925 |

## 9.5 OUTDOOR INSTALLATION

If specified the pumps should also be suitable for outdoor installation under local ambient conditions.

Extraordinary local ambient conditions, such as high or low temperatures, corrosive environment, sand storms etc. for which the pump must be suitable and should be specified by the purchaser.

## 9.6 PRIME MOVERS

The following have to be considered when determining the rated performance of the drive.

### 9.6.1 APPLICATION AND METHOD OF OPERATION OF THE PUMP

For instance in the case of parallel operation, the possible performance range with only one pump in operation taking into account the system characteristic should be considered.

- Position of the operating point on the pump characteristic curve.
- Shaft seal friction loss.
- Circulation flow for the mechanical seal (especially for pumps with low rate of flow).
- Properties of pumped liquid (viscosity, solids content, density).
- Power and slip loss through transmission.
- Atmospheric conditions at pump site.

Prime movers required as drivers for any pumps covered by this standard should have power output ratings at least equal to the percentage of rated pump power input given in Fig. 9.1, this value being never less than 1 kW. Where it appears that this will lead to unnecessary oversizing of the driver, an alternative proposal should be submitted for the purchaser's approval.

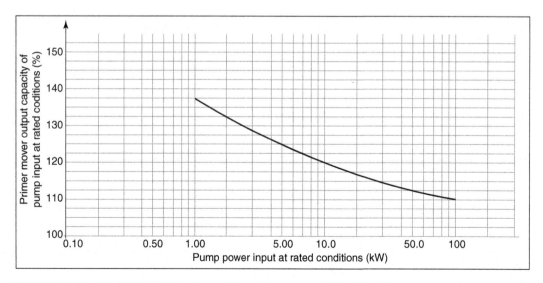

**FIGURE 9.1  Prime Mover Output, Percentage of Pump Power Input at Rated Conditions**

Various sections of the standard which are in NFPA Section 20 specify alarms to call attention to improper conditions that should exist in the complete fire pump equipment.

## 9.7 PRESSURE MAINTENANCE (JOCKEY OR MAKE-UP) PUMPS

Pressure maintenance pumps should have rated capacities not less than any normal leakage rate. They should have discharge pressure sufficient to maintain the desired fire protection system pressure.

A fire pump should not be used as a pressure maintenance pump.

## 9.8 HORIZONTAL PUMPS

### 9.8.1 TYPES

Horizontal pumps should be of the split-case, end-suction, or in-line design. Single stage end suction and in-line pumps should be limited to capacities of 1703 LPM or less.

### 9.8.2 FOUNDATION AND SETTING

The pump and driver should be mounted on a common baseplate and connected by a flexible coupling.

The baseplate should be securely attached to a solid foundation in such a way that proper pump and driver shaft alignment will be assured.

The foundation should be sufficiently substantial to form a permanent and rigid support for the baseplate.

The baseplate, with pump and driver mounted on it, should be set level on the foundation.

The deep-well, turbine-type pump is a vertical shaft centrifugal pump with rotating impellers suspended from the pump head by a column pipe that also serves as a support for the shaft and bearings.

It is particularly suitable for the fire pump service when the water source is located below ground and where it would be difficult to install any other type of pump below the minimum water level. It was originally designed for installation in drilled wells, but should also be used to lift water from lakes, streams, open swamps, and other subsurface sources. Both oil-lubricated enclosed-line-shaft and water-lubricated open-line shaft pumps should be used. Some health departments object to the use of oil-lubricated pumps; such authorities should be consulted before proceeding with oil-lubricated design.

## 9.9 VERTICAL SHAFT TURBINE-TYPE PUMPS

### 9.9.1 MAXIMUM DEPTH

Fire pumps should not be installed in a well where the pumping water level exceeds 61 m from the surface of the ground when pumping at 150% of rated capacity. In all applications the user should be supplied with data on the draw down characteristics of the well and the pump performance. The available discharge pressure at the discharge flange of the vertical pump can be determined from this data.

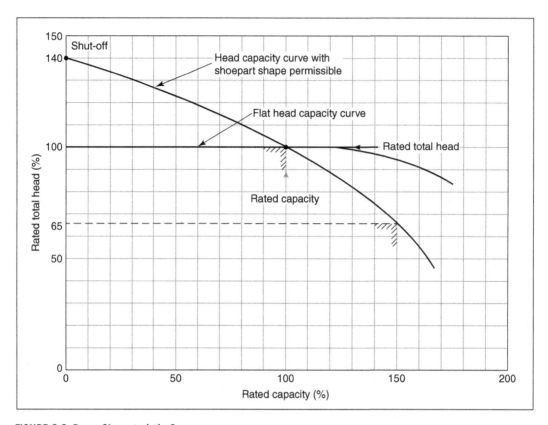

**FIGURE 9.2  Pump Characteristic Curves**

## 9.9.2 CHARACTERISTICS

Pumps should furnish not less than 150% of rated capacity at a total head of not less than 65% of the total rated head. The total shut off head should not exceed 140% of total rated head on vertical turbine pumps (Fig. 9.2).

## 9.9.3 WATER SUPPLY

The water supply should be adequate, dependable, and acceptable to the responsible authorities.

   The acceptance of a well as a water supply source should be dependent upon satisfactory development of the well and establishment of satisfactory water sources.

## 9.10 PUMPS SUBMERGENCE
### 9.10.1 WELL INSTALLATION

Proper submergence of the pump bowls should be provided for reliable operation of the fire pump unit. Submergence of the second impeller from the bottom of the pump bowl assembly should not be less than 3 m below the pumping water level at 150% of rated capacity.

   The submergence should be increased by 0.3 m for each 305 m of elevation above sea level.

### 9.10.2 WELL CONSTRUCTION

The vertical turbine-type pump is designed to operate in a vertical position with all parts in correct alignment. The well therefore should be of ample diameter and sufficiently plumb to receive the pump.

Wells for fire pumps not exceeding 1703 LPM developed in unconsolidated formations without an artificial gravel pack (tubular wells) will be acceptable sources of water supply for fire pumps not exceeding 1703 LPM.

## 9.11 PUMP

### 9.11.1 HEAD

The pump head should be either the aboveground or belowground discharge type. It should be designed to support the driver, pump column, and the oil tube tension nut or packing container.

### 9.11.2 PUMP HOUSE

The pump house should be of such design as should offer the least obstruction to the convenient handling and hoisting of vertical pump parts.

### 9.11.3 OUTDOOR SETTING

If in special cases it is considered that the pump does not require a pump room and the unit is installed outdoors, the driver should be screened or enclosed, and adequately protected against tampering. The screen or enclosure should be easily removable and have provision for ample ventilation.

## 9.12 POWER SUPPLY DEPENDABILITY

Careful consideration should be given in each case to the dependability of the electric supply system and the wiring system. This should include the possible effect of fire on transmission lines either in the property or in adjoining buildings that might threaten the property.

### 9.12.1 STEAM SUPPLY

Careful consideration should be given in each case to the dependability of the steam supply and the steam supply system. This should include the possible effect of fire on transmission piping either in the property or in adjoining buildings that might threaten the property.

## 9.13 ELECTRICAL DRIVE FOR PUMPS

This section outlines the minimum requirements for the source(s) and transmission of electric power to motors driving fire pumps and the minimum performance requirements of all intermediate equipment between the source(s) and the pump, including the motor(s), excepting the fire pump controller and its accessories. All electrical equipment should, as a minimum, comply with the provisions of material standards.

Power should be supplied to the fire pump by main and emergency power.

## 9.13.1 UTILITY SERVICE

Where power is supplied by a public utility service connection, the service should be located and arranged to minimize the probability of damage by fire from within the premises and exposing hazards.

## 9.13.2 SINGLE POWER STATION

Where power is supplied from a single private power station, the station should be of noncombustible construction, located and protected to minimize the probability of damage by fire.

### 9.13.2.1 Other sources

1. Where reliable power cannot be obtained from a private power station or utility service, it should be from two or more of either of the above or in combination, or one or more of the above in combination with an emergency generator, all as approved by the relevant authorities. The power sources should be arranged so that a fire at one source will not cause an interruption at the other source(s).
2. Emergency generator. Where power is supplied by an emergency generator, the generator should be located and protected in accordance with standards.

## 9.13.3 POWER SUPPLY LINES

### 9.13.3.1 Circuit conductors

1. The fire pump feeder circuit conductors should be physically routed outside of the building(s), excluding the electrical switchgear room and the pump room. When the fire pump feeder conductors must be routed through building(s), they should be buried or enclosed by 51 mm of concrete (or equivalent 1-h fire resistance) in order to be judged "outside of the building."
2. All pump room wiring should be in rigid, intermediate, or liquid-tight flexible metal conduit.
3. The voltage at the motor should not drop more than 5% below the voltage rating of the motor(s) when the pumps are being driven at rated output, pressure, and speed, and the lines between the power source and motors are carrying their peak load.

### 9.13.3.2 Capacity of lines

Each line between the power source and the fire pump motor should be sized at 125% of the sum of the full load currents of the fire pump(s), jockey pump, and fire pump auxiliary loads.

## 9.13.4 TRANSFORMERS

### 9.13.4.1 On-site power generator systems

Where on-site generator systems are used to supply power to fire pump motors to meet the requirements of 9.16.2.3, they should be of sufficient capacity to allow normal starting and running of the motor(s) driving the fire pump(s) while supplying all other loads connected to the generator.

Automatic shedding of loads not required for fire protection is permitted prior to starting of the fire pump(s).

Automatic sequencing of the fire pumps is permitted in accordance with 7-5-2-4 of NFPA Section 20 transfer of power should take place within the pump room.

Protective devices in the on-site power source circuits at the generator should allow instantaneous pick-up of the full pump room load.

## 9.13.5 ELECTRIC DRIVE CONTROLLERS AND ACCESSORIES

### 9.13.5.1 Location

1. Controllers should be located as close as is practical to the motors they control and should be within sight of the motors.
2. Controllers should be so located or so protected that they will not be damaged by water escaping from pumps or pump connections. Current-carrying parts of controllers should not be less than 305 mm above the floor level.
3. For controllers, which require rear access for servicing, a clearance of not less than 1.1 m should be provided at the rear of the controller and not less than 0.61 m on at least one side of the controller.

### 9.13.5.2 Alarm and signal devices remote from controller

When the pump room is not constantly attended, audible or visual alarms powered by a source, not exceeding 125 V, should be provided at a point of constant attendance. These alarms should indicate the followings:

1. Controller has operated into a motor running condition.
   This alarm circuit should be energized by a separate reliable supervised power source, or from the pump motor power, reduced to not more than 125 V.
2. Loss of line power on line side of motor starter, in any phase.
   This alarm circuit should be energized by a separate reliable supervised power source. The phase voltage providing starting coil excitation should be monitored to indicate loss of availability of such excitation.
3. Phase reversal on line side of motor starter.
   This alarm circuit should be energized by a separate reliable supervised power source, or from the pump motor power, reduced to not more than 125 V.

## 9.14 TRAILERS CARRYING FIRE EXTINGUISHING AGENTS

The capacity of trailer units should not be more than 4 tons. Towing chassis should be of four wheels with towing hook. The trailer should be suitable for use in the area as specified. The width should not exceed 2.40 m and steering should be possible by turning the front axle with an extended towing connection. The suspension of each axle should be designed for a reserve of at least 10% when the trailer is fully loaded.

The trailer of 4 tons should be provided with gravity brakes or air brakes for connection to the air-braking system of the towing vehicle.

Mudguards should be provided with mud flaps.

The lighting should be in accordance with traffic regulations. The lighting should be of 12 V or as specified.

The equipment compartment and the total superstructure should be fastened to the chassis beam by a method which will prevent harmful influences and ensure flexibility of the body work and super-structure. All fire fighting and emergency equipment housed should be stored properly, in such a way that they will not shift during driving, braking, and acceleration of the towing vehicle. The equipment should be so arranged that they can be quickly removed.

## 9.15 FOAM EQUIPMENT MOUNTED ON TRAILER UNIT

The purpose of the above trailer is to carry foam liquid to the scene of fire in order to replenish the major or general purpose fire trucks. It can also assist in direct fire fighting with foam by means of hoses connected between the water main and inductors installed on the trailer.

### 9.15.1 EQUIPMENT

The following equipment should be installed on the trailer:

- 4000 L tank for FLC,
- two inductors, each with a rating of approximately 400 LPM,
- inlet and outlet manifolds for foam liquid,
- lockers for materials.

The specific requirements given by the purchaser and details proposed by the supplier should be given on a purchase order.

### 9.15.2 TRAILER CHASSIS

The chassis of the trailer should be suitable for carrying the equipment. The towing pole eye should be suitable to fit the hook of towing vehicle.

The width of the trailer should not exceed 2.40 m.

The brake system should be of the positive air-brake type (air pressure to release brakes), actuated by the foot brake of the truck. A mechanical hand brake should also be provided.

Electrical accessories should comply with the traffic regulations.

### 9.15.3 FOAM SYSTEM

Two inlet valves A, one at each side of the trailer, fitted with hose coupling valves and chained caps of the same type and size as mounted on the fire-fighting truck, should be provided. Via each inlet the FLC can be pumped from the main FLC storage tank via hoses and check valve B into the tank mounted on the trailer. The trailer can also be filled by gravity from storage via the manhole.

Two outlet valves C, one at each side of the trailer, fitted with 65 mm hand suction round thread hose coupling of the same type and size as mounted on the truck should be provided. Via each outlet and strainer D the foam liquid can flow from the trailer tank to the foam liquid pump of the truck.

The purpose of the inductors H is to make solution for direct fire fighting by mixing FLC from the tank via D and check valve E, with water via a water hose connected to coupling F. Regulation of the water/FLC ratio should be done manually. For this purpose each inductor should be fitted with a simple regulator.

Where instantaneous couplings are required, inlets should be of the male type and outlets of the female type.

### 9.15.4 THE FOAM LIQUID TANK

The FLC tank on the trailer should be provided with internal transverse baffles and a 500 mm diameter expansion dome with a capacity of 3% of the tank volume. The manhole opening in the dome should be 500 mm in diameter and be provided with a quick-release lock.

The tank should be provided with:

- an inlet with an internal filling tube
- an outlet which prevents sludge from flowing to the inductors
- a sump with a bolted cover plate incorporating a sludge drain
- a gage tube K to measure the tank contents
- a breather valve L to prevent excessive underpressure or overpressure.

The tank should be made of stainless steel to be protected against corrosion.

### 9.15.5 ADDITIONAL REQUIREMENTS

The trailer should be provided with lockers and rolling shutters for storing small equipment.
The trailer should be capable of carrying the following materials:

- two 65-mm suction hoses, each 4 m long complete with coupling of the same type as mounted on the fire truck,
- 12 hoses 70 mm (2¾in.) × 25 m long; synthetic fabric and lined, complete with instantaneous couplings,
- two air foam-making branch pipes, approximate capacity 500 LPM at 10 bar each.

The trailer locker and tank should be painted externally with fire brigade red and the mud wings with black enamel. Steelwork should be treated against corrosion.
Grab handles should be chromium-plated or stainless steel.
The interior of the lockers should be covered with aluminum plate.
If the top of the tank is curved, a platform should be provided on the tank for access to the manhole. For this purpose a fixed ladder is also necessary.
A spare wheel, not mounted on the trailer, should be provided.

#### 9.15.5.1 Testing and quality assurance

Manufacturers should certify that tests have been carried out and give documentation for quality assurance before shipment from the factory.
Manufacturers should also supply minimum of two operating manual and spare parts list.

## 9.16 PREMIX FLC AND DRY CHEMICAL TRAILER UNIT

The purpose of this trailer is to have a mobile unit to be in hand at the site of construction work, maintenance or any emergency job in satellite areas where there is less possibility of quick reaching of fire service equipment in those areas.
The trailer width should not be more than 2 m and the total weight should not exceed 2 tons.

## 9.16.1 **FIRE FIGHTING SYSTEMS**

The trailer should be mounted with a premix foam solution of the type AFFF or FFFP with the capacity of 500 L, and two 200 kg dry chemical extinguishers.

Both systems should be actuated by dry air or nitrogen cylinder, each tank with separate actuating system.

Two hose reels with the combined outlet discharge nozzles and each hose reel fitted with 40 mm × 25–40 m hose and should be mounted on rear side of the trailer.

Dry chemical system should be of twin tanks, each tank provided with one pressurized air/$N_2$ cylinder.

Dry chemical should be of potassium bicarbonate (monex) compatible with the type of premix-foam.

The following portable fire extinguishers should also be provided:

2 × 10 kg dry chemical
2 × 5 kg $CO_2$.

Premix and dry chemical tanks should be provided with pressure gages on discharge part of the tanks.

### 9.16.1.1 *Trailer chassis*

The chassis of the trailer should be suitable for carrying fire-fighting system.

The trailer and premix/powder tanks and accessories should be painted externally with fire service red enamel and wings with black enamel.

### 9.16.1.2 *Tests and quality assurance*

Pressurized air/$N_2$ cylinders should be tested with specified test methods in accordance with BS or UL standards. Premixed and dry chemical tanks should be hydrostatically tested at twice of the precalculated pressure at ambient temperature of 50°C when discharge valve is closed.

Manufacturer should certify the quality assurance with documents before shipment.

### 9.16.1.3 *Marking*

Trailer unit should bear the following marking engraved on a brass plate attached to the chassis:

1. name of manufacturer or identifying symbol
2. catalog designation
3. date of manufacture
4. discharge rate.

## 9.17 **FOAM/WATER MONITOR TRAILER UNIT**

The purpose of this trailer is to carry a large capacity of foam/water monitor to an outbreak of major fire, such as a liquid hydrocarbon tank or major oil spill fire. By utilization of a high capacity of water fog, a water shield can be set to cool the area, preventing heat radiation.

The lightweight trailer is to be of three wheels that can be pulled by two men.

### 9.17.1 FOAM/WATER MONITOR

The monitor should have three hose inlet manifolds with instantaneous male couplings and ball valves. The piping should be laid under the monitor base plate and the monitor should be of duplex branch type, foam, and water.

Water branch to be of straight pattern but the nozzle tip can be changed to fog pattern type. The water discharge should be of 2000 LPM at 10 bar.

The foam monitor is to be provided with pick-up tube to take suction from foam drums. The pick-up tube to be of 40 mm diameter.

The water and foam solution capacity should be of 2000–3000 LPM.

Vertical movement of the monitor to be of gear type, from 20° to 70° and horizontal movement 180° by hand.

### 9.17.2 CONSTRUCTION AND MATERIAL

The monitor should be made of corrosion resistant materials preferably of brass or brass alloy. The foam-making barrel to be made of anodized aluminum alloy. The water discharge nozzle should also be made of brass. The inlet manifold should be made of material resistant to corrosion.

Monitors main valve to be of quick opening ball or gate valve and a pressure gage should be provided and fixed to the water inlet adjacent to the inlet valve.

The monitor and piping should be securely fixed on trailer chassis in such a way that during driving, braking, and acceleration the monitor remains effectively secured.

#### 9.17.2.1 Finishing

The trailer and parts made of brass to be painted fire service red. Trailer base plate and monitor barrel to be of aluminum.

#### 9.17.2.2 Tests and quality assurance

The manufacturer should certify by documents that the trailer has been road tested and the monitor has been hydrostatically tested and assure that the quality of materials used have met BS. AU. 24–30b and UL. 711.

#### 9.17.2.3 Marking

The unit should have the following marking cast or engraved on a brass plate fixed on chassis.

1. name of manufacturer or identifying symbol
2. catalog designation
3. manufacturing date
4. water and foam discharge rate.

#### 9.17.2.4 Equipment carried by trailer

1. 40 mm diameter 2.5 m length pick up tube,
2. minimum four lengths of standard fire hoses.

Figs. 9.3–9.7 show different types of line-up of fire-fighting pumps.

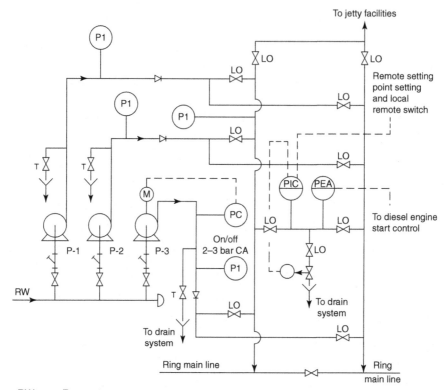

RW = Raw water

P-1 = Electric motor driven submerged fire-fighting pump (or steam turbine)

P-2 = Diesel engine-driven submerged fire-fighting pump

P-3 = Electric motor-driven pressurizing pump

LO = Locked open gate valve

T = Test valve

* = Operating range 6–20 bar g

** = Set point range 2–6 bar g

**FIGURE 9.3 Line-Up of Fire-Fighting Pumps (Close to Ring Main Line)**

## 9.18 FIRE-FIGHTING TRUCKS AND PUMPS

This section has been compiled to specify various fire trucks and pumping units used in the oil refineries, chemical plants, gas plants, and wherever applicable such as in production units, exploration, oil terminals, distributions, and affiliated industries. This standard covers a number of basic fire trucks equipped with selection of fire-fighting systems.

Depending upon the risk of the plants, the size of the area and fixed fire-fighting installations or facilities, the fire trucks and fire equipment should be so designed or selected to give satisfactory performance and to act quickly, and thus reducing loss of lives, injuries, and damages.

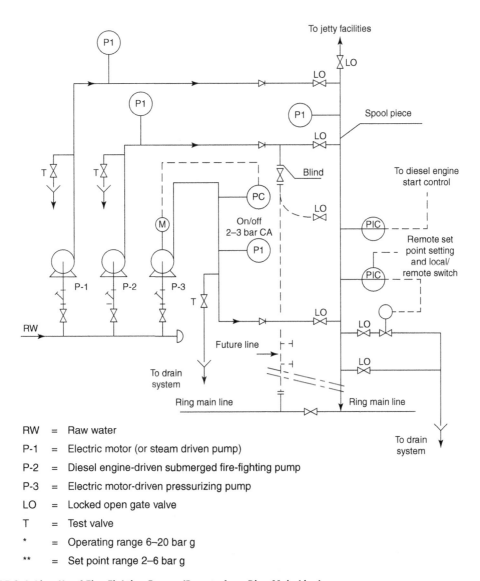

To jetty facilities

LO

LO

Spool piece

P1

P1

P1

Blind

To diesel engine
start control

PC

On/off
2–3 bar CA

LO

PIC

Remote set
point setting
and local/
remote switch

P1

PIC

M

P-1    P-2    P-3

T

LO

LO

LO

RW

Future line

To drain
system

Ring main line

LO

Ring main line

To drain
system

RW = Raw water
P-1 = Electric motor (or steam driven pump)
P-2 = Diesel engine-driven submerged fire-fighting pump
P-3 = Electric motor-driven pressurizing pump
LO = Locked open gate valve
T = Test valve
* = Operating range 6–20 bar g
** = Set point range 2–6 bar g

**FIGURE 9.4 Line-Up of Fire-Fighting Pumps (Remote from Ring Main Line)**

This standard will eliminate the use of similar types of trucks and equipment, which have different operational and maintenance procedures and is divided into the following sections:

- Engineering specification of major fire-fighting trucks
- Various fire-fighting systems for installation on major fire-fighting trucks
  - Proposed specification of general purpose and major foam
  - Tender fire-fighting trucks

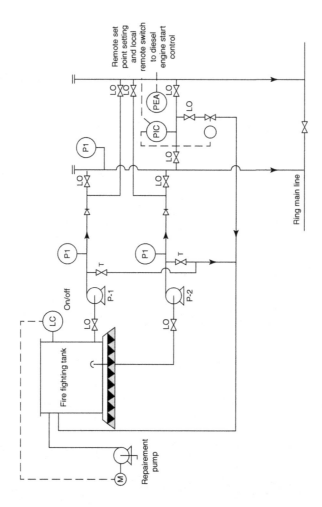

| P-1 | = | Electric motor-driven fire-fighting pump |
|-----|---|------|
| P-2 | = | Diesel engine-driven fire-fighting pump |
| LO | = | Locked open gate valve |
| T | = | Test valve |
| * | = | Operating range 6–20 bar g |
| ** | = | Set point range 2–6 bar g |

**FIGURE 9.5 Line-Up of Fire-Fighting Pumps (With Water Storage)**

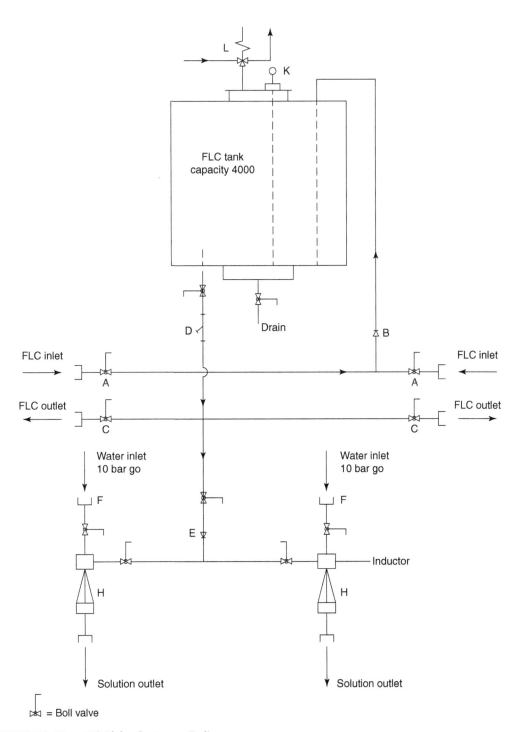

**FIGURE 9.6 Water FLC Piping System on Trailer**

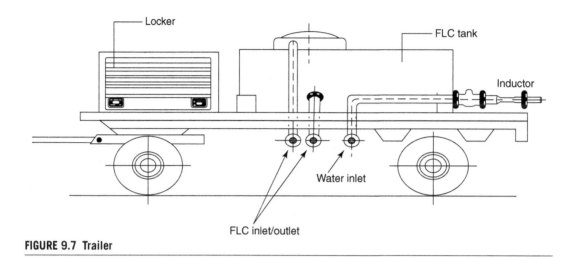

**FIGURE 9.7 Trailer**

- Specification of auxiliary fire and emergency vehicles comprising of the following:
  - Foam liquid dispensing truck
  - Dry chemical powder fire extinguishing truck
  - Twin agent fire extinguishing truck
  - Water tender
  - Emergency service and rescue vehicles
  - Hydraulic boom
- Brief description and list of proposed types of fire-fighting trucks
- Portable, trailer mounted, and fixed fire-fighting pumps
- Material procurement standard

This section describes the minimum engineering and material requirements for all types of fire-fighting trucks, emergency vehicles, and fire pumps utilized by petroleum industries and gives general concepts for the manufacturing design of vehicles and relevant material and equipment which have been installed on them.

The application of the chapter would make uniformity in the design of equipment, their operation and maintenance, and also will facilitate the training of fire-fighting personnel.

## 9.18.1 CATEGORIES

The fire-fighting vehicles used in petroleum industries are categorized by its load, liquid pumping capacity and its pressure therefore the following categories can be used.

A major fire-fighting truck with water and foam tanks in excess of 5000 L and pumping capacity of over 4000 LPM at 7 bar.

General purpose or medium size fire truck with water and foam liquid tank capacity of 3000–5000 L and pumping capacity of 2000–4000 LPM at 7 bar.

Auxiliary fire trucks such as fire-fighting boom, foam-liquid or water tenders, dry powder and twin agents trucks.

Light vehicles include three types as follows:

light fire-fighting vehicles with foam-water capacity of 1000 L and pumping capacity of 800 LPM at 7 bar
emergency combined rescue vehicles
emergency equipment carrier.

*Notes:*
1. For major fire-fighting and general purpose trucks at least one ton of weight of personnel and other equipment should be added.
2. For light vehicles at least 500 kg of weight of other equipment and personnel should be added.
3. Fire-fighting trucks used in oil, gas and petrochemical industries are generally designed and manufactured in accordance with the nature of services required, therefore all parts should be strong enough to withstand the expected general requirements with the minimum maintenance when under full load.

## 9.18.2 FIRE-FIGHTING TRUCKS, DESIGN SPECIFICATIONS

The purpose of the fire-fighting trucks for petroleum industries is to carry fire fighters, foam-liquid, fire-fighting chemicals and equipment to the scene of fire and inject FLC into the water stream, generating foam and utilizing chemical for fire fighting. The water required can be taken from the firewater main, open water, or other sources such as water tanks.

Design specification in this section includes major and general purpose fire trucks and the following main equipment should be installed.

- a water pump for boosting the pressure of water taken from water main or from other sources
- FLC tank
- FLC pump
- water inlet and outlet manifold
- orifices and instruments for water flow measurement and FLC lines and instrumentation to enable ratio control of these flows for making water/foam solution
- dry chemical fire extinguishing system
- premix water/foam system and hose reel equipment used for first aid fire-fighting operation or twin agent foam/dry chemical extinguishing system.

The specific requirements which are stated in this section gives general concepts for the design of fire-fighting trucks together with a fire-fighting system that can be installed on the vehicles.

## 9.18.3 THE VEHICLE

The fire-fighting vehicle should be designed for industrial purposes and in compliance with modern engineering practices. It is also essential to meet traffic regulations for overall weight, axle weight, power/weight ratio, lighting, etc. and this should be stated in the exchange of information with the manufacturer. The vehicle should function primarily on a firewater circuit (from hydrants) with a pressure of 6–16 bar. Certain vehicles should, however, also be provided with facilities and equipment for suction from open water.

The vehicle should be so constructed to assist visual inspection, maintenance, and repair. All equipment should be located so that it will be readily accessible. Fire-fighting systems should be simple and easy to operate, to facilitate training of personnel and use in an emergency.

The installation of mechanical, electrical, pneumatic, and hydraulic components should be located in such a way that dismounting or repair is not obstructed by the chassis structure or any other component, and electrical wiring and pneumatic tubing is not damaged during operating of the vehicle. The electrical system should be dustproof and waterproof to at least a minimum of IEC publication 529 (IP code) all instruments should be of marine-type construction.

The vehicle should also be driven on a steering pad around a circle of 30 m radius. The steering wheel rotation should increase with increasing speed to ensure the vehicle does not exhibit oversteer characteristics.

### 9.18.4 DIMENSIONS

Underchassis clearance of the vehicle should permit mobility in soft soils and rough terrain. The following should be the minimum dimensions:

| | |
|---|---|
| Angle of approach | 20° |
| Angle of departure | 20° |

Interaxle clearance angle 20° with 458 mm minimum clearance at midwheel base. Underaxle clearance 330 mm underaxle differential housing bowl.

Overall height, length, and width of the vehicle should be held to a minimum consistent with the best operational performance of the vehicle and the design concepts needed to achieve this performance and to provide optimum maneuverability and facilitate movement on the roads.

The vehicle should be constructed such that a seated driver, having an eye height of 800 mm, should be able to see the ground 6 m ahead of the vehicle and have vision up to 15° above the horizontal plane without leaving the driver's seat. The vision in the horizontal plane should be at least 90° on each side from the straight-ahead position.

### 9.18.5 WEIGHTS

The actual gross vehicle weight of the fully staffed, loaded, and equipped vehicle ready for service should not exceed the manufacturer's gross vehicle weight rating.

The weight should be distributed as equally as practical over the axles and tires of the fully laden vehicle. The difference in weight between tires on any axle should not exceed 5% of the average tire weight for that axle, and the difference in weight between axles should not exceed 10% of the weight of the heaviest axle. The front axle should not be the heaviest axle. Under no circumstances should axle and tire manufacturer's ratings be exceeded.

The center of gravity of the vehicle should be kept as low as possible under all conditions of loading. Single rear axle is generally recommended, but a double axle may be specified.

### 9.18.6 **CHASSIS**

The vehicle should be able to carry all of the equipment specified and to tow at least a trailer. The chassis should be provided with towing connections at the rear and two pull/push connections at the front, the additional forces when using these connections should be specified.

The vehicle should be suitable for use in areas as specified by the IPI authorities. It should have a wheelbase of approximately 4–4.50 m and be between 2.40 and 2.50 m wide.

Fenders and guards should be braced and firmly secured. Proper clearance should be provided for chains.

The steering mechanism for nondriving front axles should be capable of turning the front wheels to an angle of at least 30 degrees to either right or left. Power or power assist steering should be provided on all fire trucks.

The transmission and clutch should be of such a type as to operate smoothly and effectively under all conditions of service.

The chassis should be equipped with:

* Wheels fitted with radial tires suitable for wet roads.
* Fuel tank, 200 L minimum capacity with the possibility of refilling during operation (fitted on the inside of the chassis).
* Air pressure vessel if necessary fitted on the inside of the chassis.
* Stabilizers on front and rear axles.

### 9.18.7 **ENGINE**

#### *9.18.7.1 Diesel*

Unless otherwise specified the engine should be diesel and should have horsepower, torque, and speed characteristics to meet and maintain all specified vehicular performance characteristics in this standard. The engine manufacturer should certify that the installed engine is approved for this application.

The fully laden vehicle should consistently be able to accelerate from 0 to 80 km/h on dry level concrete pavement within the time specified in Table 9.2. Maximum speed should not be less than 100 km/h.

The above acceleration times should be achieved with the engine and transmission at their normal operating temperatures at any ambient temperature varying from −18°C to 58°C and at elevations of 600 m above sea level unless a higher elevation or lower minimum temperature is specified.

| Table 9.2 Acceleration on Dry Level Concrete Pavement | | |
|---|---|---|
| **Class** | **Minimum FLC or Water Capacity (L)** | **Acceleration Time (0–80 km/h) (s)** |
| 1 | 4,000 | 25 |
| 2 | 6,000 | 30 |
| 3 | 10,000 | 40 |
| 4 | 12,000 and over | 45 |

### 9.18.7.2 Fuel systems

The fuel system supplied by the engine manufacturer should be of sufficient size to develop the rated power. The manufacturer should supply fuel lines and fuel filters in accordance with the engine manufacturer's recommendations. To prevent engine shutdown due to fuel contamination, dual filters in parallel, with proper valving so that each filter can be used separately, may be desired.

Fuel tanks should not be installed in a manner that permits gravity feed.

A dry type air filter should be provided. Air inlet restrictions should meet the engine manufacturer's recommendations. Air inlet should be protected to prevent water and burning embers from entering the air intake system. The manufacturer should provide an air restriction, indicator, mounted in the cab, visible to the driver.

An engine governor should be installed which will limit the speed of the engine under all conditions of operation to that speed established by the engine manufacturer; this should be the maximum no-load governed speed. A tachometer should be provided on the instrument panel in the driving compartment for indicating engine speed.

### 9.18.7.3 Gasoline engines

Fuel lines and filters and/or strainers of an accessible and serviceable type, as recommended by the engine manufacturer, should be provided. The filters or strainers should be of a type, which can be serviced without disconnecting the fuel line. Where two or more fuel lines are installed, separate fuel pumps operating in parallel with suitable check valves and filtering devices should be provided. The fuel line(s) should be so located or protected as not to be subjected to excessive heating from any portion of a vehicle exhaust system. The line(s) should be protected from mechanical injury. Suitable valves and drains should be installed. The carburetor(s) of a gasoline engine should be nonadjustable, with the exception of the idle setting, of sufficient size to develop the rated power, and so located as not to be subjected to pocketing of vapor or excessive heating. An automatic choke should be provided. The gasoline feed system should include an electrically operated fuel pump located within or adjacent to the fuel tank.

### 9.18.7.4 Fuel tank

For light vehicles, the fuel tank should not be less than 100 L capacity. The capacity for apparatus with pumping equipment should be of a size, which should permit the operation of the pumping for not less than 3 h when operating at rated pump capacity. A suitable method of venting and means for draining directly from the tank should be provided. The tank fill opening should be conspicuously labeled as to the type of fuel used.

When a large capacity fuel tank is desired the capacity specified by purchaser in special provisions should be supplied.

Only one fuel tank is to be furnished where rated tank capacity is 150 L or less. The fuel gage should indicate the proportionate amount of fuel in the tank system at any time.

Tank and fill piping should be so placed as to be protected from mechanical injury, and not be exposed to heat from exhaust or other source of ignition. Tank should be so placed as to be easily removable for repairs.

Automatic engine shutdown systems should not be provided, but auto overspeed protection by means of a shut down valve in the air intake system in case of ignition of flammable gases should be provided.

### 9.18.7.5 Exhaust system

The exhaust piping and discharge outlet should be so located as not to expose any portion of the vehicle to excessive heating. Exhaust pipe discharge should not be directed toward the pump operator's position. Silencing devices should be provided. Exhaust backpressure should not exceed the limits specified by the engine manufacturer. Where parts of the exhaust system are exposed so that they are likely to cause injury to operating personnel, suitable protective guards should be provided. Spark arrestors should be also provided and the exhaust system should be of high-grade rust resistant materials.

### 9.18.7.6 Engine cooling systems

An adequate cooling system of sufficient capacity should be provided such that overheating will not occur during stationary use in tropical areas and during prolonged fire fighting under full operational conditions of both water and foam pumps (max. ambient temperature should be specified).

The cooling fluid should be a high-efficient cooling medium with an anti corrosion additive.

Radiator shutters, when furnished for cold climates, should be of the fail–safe automatic type, and should be designed to open automatically upon failure.

Adequate and readily accessible drain cocks should be installed at the lowest point of the cooling system, and at other such points as are necessary to permit complete removal of the coolant from the system. Drain cocks should not open accidentally due to vibration.

The radiator should be so mounted as not to develop leaks due to ordinary running and operating nor be twisted or strained when the apparatus operates over uneven ground. Radiator cores should be compatible with commercial antifreeze solutions and of straight tube construction for easy mechanical maintenance.

The cooling system should be provided with an automatic thermostat for rapid engine warming.

### 9.18.7.7 Brakes

Fire trucks' brakes should be of the most efficient and fail safe system. Brake performance should comply with applicable regulations at the date of manufacture.

The braking should feature service, emergency, and parking brake systems. Service brakes should be power actuation air, hydraulic, and air over hydraulic. Expanding shoe and drum brakes or caliper disc brakes, or the most reliable type should be furnished. A brake chamber should be provided for each wheel and should be mounted so that no part of the brake chamber projects below the axle.

Service brakes should be of the all-wheel type with split circuits so that failure of one circuit should not cause total service brake failure.

The service brakes should be capable of holding the fully loaded vehicle on a 50% grade.

As a minimum requirement the service brakes should be capable of bringing the fully laden vehicle to complete stop within 10.7 m from 32 km/h, and within 40 m from 64 km/h by actual measurement on substantially hard surface road, that is, free from loose material, oil, or grease.

The parking brake should be capable of holding the fully loaded vehicle on a 20% grade without air or hydraulic assistance.

The service brakes should provide one power-assisted stop with the vehicle engine inoperative, for the stopping distances specified above for the vehicle.

An emergency brake system should be provided which is applied and released by the driver from the cab and is capable of modulation by means of the service brake control. With a single failure in the service brake system of a part designed to contain compressed air or brake fluid, other than failure of a common valve, manifold, brake fluid housing, or brake chamber housing, the vehicle should stop in no more than 88 m from 64 km/h without any part of the vehicle leaving a dry, hard, approximately level roadway with a width equal to the vehicle width plus 110 cm.

*Note:* A newly developed service braking system should be considered.

### 9.18.7.8 Brakes-air system

When the vehicle is supplied with air brakes, the air compressor should meet the following criteria:

1. The compressor should be engine driven.
2. The compressor should have capacity sufficient to increase air pressure in the supply and service reservoirs from 5–7 bars when the engine is operating at the vehicle manufacturer's maximum recommended revolutions per minute (rpm) in a maximum of 25 s.
3. The compressor should have the capacity for quick build-up of tank pressure from 0.35 bars to the pressure required to release the spring brakes, and this build-up in pressure should be accomplished within 12 s.
4. The compressor should incorporate an automatic air-drying system immediately downstream from the compressor to prevent condensation build-up in all pneumatic lines.

Visual and audible low air pressure warning devices should be provided. The low-pressure-warning device should be visual and audible from the inside of the vehicle, and audible outside of the vehicle.

Service air reservoirs should be provided. The total of the service air reservoir volume should be at least 12 times the total combined brake chamber volume at full stroke. If the reservoir volume is greater than the minimum required, proportionately longer build-up time should be allowed.

Air reservoirs should be equipped with drain and safety valves.

If specified provision should be made for charging of air tanks by a pull away electrical connection used to power a vehicle-mounted auxiliary compressor.

When specified by the purchaser, a pull away air connection for charging of air tanks from an external air source should be provided.

### 9.18.7.9 Gears and power take-off

- A synchro-mesh gearbox should be fitted with a switch to operate the reversing lights and, when specified, an on/off buzzer. If an oil cooler is to be supplied, a connection for a temperature indicator/alarm will also be required.
- PTO (Power Take-Off) transmission for the booster pump should be selected to transmit the torque and power required by the booster (and foam) pumps when rotating at the required engine speed with all discharge branches completely open.
- PTOs should preferably operate electrically/pneumatically from the driver's cabin and, when specified, from the operating panel. A manually operated PTO may be specified when required.

### 9.18.7.10 Steering

The chassis should be equipped with power-assisted steering with direct mechanical linkage from the steering wheel to the steered axle(s) to permit the possibility of manual control in the event of power assist failure.

The power steering should have sufficient capacity to allow turning the tires stop-to-stop with the vehicle stationary on a dry, level, paved surface and fully loaded.

The wall-to-wall turning diameter of the fully laden vehicle should be less than three times the vehicle length.

### 9.18.7.11 Cabin

The cabin should provide seating for a minimum of driver plus one crew member including individually adjustable, suspension-type driver's seat and space for all instrument controls and equipment specified without hindering the crew. Additional crew of three to four will be seated in a separate crew compartment.

Wide opening doors should be provided on each side of the cabin with necessary steps and handrails to permit rapid and safe entrance and exit from the cabin. Cabin design should take into consideration the provision of ample space for the crew to enter and exit the cabin and carry out normal operations while wearing full protective equipment.

The cabin should meet the visibility requirements of clause 6.2.1. Interior cabin reflections from exterior and interior lighting should be minimized. The windshield should be of laminated or shatterproof safety glass with upper 20–30% tinted green. The windshield should be fitted with at least two wide arc wipers having two speeds and electrically operated washing system with at least two nozzles and all other windows should be constructed of approved safety glass. The cabin should be provided with wide gutters to prevent foam and water dripping on the windshield and side windows. There should be a quick opening passage providing access to the roof monitor.

The driver's cabin should be fitted with a sunscreen fitted above the windscreen on the outside.

Two adjustable sunshades having a minimum length of 380 mm and a minimum width of 130 mm are also required on the inside.

The crew cabin should be weatherproof, and should be fully insulated thermally and acoustically with a fire resistant material. The cabin may be of the unitized rigid body and frame structure type or it may be a separate unit flexibly mounted on the main vehicle frame. The cabin should be constructed from materials of adequate strength to ensure a high degree of safety for the crew under all operating conditions including excess heat exposure, and in the event of a vehicle roll-over accident.

The framework of the cabin should be built up from sections mounted on shock absorbers and be of such construction that harmful stresses will not occur during normal use and the crew will be offered maximum protection in the case of accidents.

The cabin should be equipped with driving mirrors on the right and left hand side of the vehicle, which should also be suitable as parking mirrors. The mirrors should be adjustable, free from vibration and fastened in such a way that they cannot move out of position under normal driving conditions.

Provision should be made for mounting radio and telephone. Operation of radiotelephone should be from the cabin and mounted in order to permit quick servicing and replacement. Suitable shielding should be provided to permit radio operation without undue interference.

Adequate measures should be incorporated for:

- protection of the crew during a frontal collision
- protection of head and neck of the crew in the event of a rear collision
- strength of the doors, the door frame, locks and hinges during a sidelong collision
- solidity of the seat attachments
- strength of the seat mounting and security of the seat safety belts.

The maximum noise level (under full load conditions) should be 85 dB.

### *9.18.7.12 Floor*

The floor of the cabin should be treated with an antiresonance material. The floor of the driver's compartment should be covered with a loose wear-resistant profiled rubber sheet fitted with a foam plastic underlayer, protected with a waterproof layer.

The floor of the crews' compartment should be covered with a minimum ripped sheet or other equivalent nonslip material.

The lower part of the doors should be covered on the inside up to a height of about 150 mm above the floor with aluminum sheet kicking plates. The doors should be treated internally with an antiresonance material and be protected from corrosion. The windows should be weatherproofed with rubber strips.

Drain holes should be fitted in the lower side of the doors. The door handles should be made of noncorrosive material; they should not protrude or have openings facing forward.

## 9.18.8 INSTRUMENTS, WARNING LIGHTS, AND CONTROLS

The minimum number of instruments, warning lights, and controls consistent with the safe and efficient operation of the vehicle, chassis, and fire fighting system should be provided.

All chassis instruments and warning lights should be grouped together on a panel in front of the driver. All fire fighting system instruments, warning lights, and controls should be grouped together by function so that accessibility is maintained.

All instruments and controls should be illuminated, with backlighting to be used where practical.

Groupings of both the chassis and fire-fighting system instruments, warning lights, and controls should be easily removable and be on a panel hinged for back access by the use of quick disconnect fittings for all electrical, air, and hydraulic circuits.

The following instruments or warning lights or both should be provided as a minimum:

1. speedometer/odometer
2. engine tachometer
3. fuel level
4. air pressure
5. engine temperature
6. engine oil pressure
7. voltmeter
8. water tank level
9. foam tank level
10. low air pressure warning
11. headlight beam indicator
12. clock.

The cabin should have all the necessary controls within easy reach of the driver for the full operation of the vehicle. The fire pump instruments should be provided at the rear over the pump.

## 9.18.9 ELECTRICAL SYSTEM AND DEVICES

Overall covering of conductors should be of moisture-resistant type. All connections should be made with lugs or terminals mechanically secured to the conductors. Wiring should be thoroughly secured in place and suitably protected against heat, oil, and physical damage where required. Wiring should be colored or otherwise coded.

Lighting equipment should be installed in conformity with relevant transport standards and should include the following:

- Headlights with upper and lower driving beam. A control switch, which is readily accessible to the driver, should be provided for beam selection. Fog lights with protectors against flying stones should also be fitted at the front of the vehicle.
- Dual tail lights and stoplights.
- Turn signals, front and rear, with a steering column mounted control and a visual and audible indicator.
- A four-way flasher switch should be provided.
- Spotlight, 152 mm minimum on both left and right sides of the windshield, hand adjustable type, with controls for beam adjustment inside the truck cabin.
- Adequate reflectors, and marker and clearance light, should be furnished to describe the overall length and width of the vehicle.
- Engine compartment lights, nonglare type, arranged to illuminate both sides of the engine with individual switches located in the engine compartment.
- Lighting should be provided for all top deck working areas.
- At least one back-up light and an audible alarm installed in the rear of the body.
- A flashing red beacon or alternate red and white flashing lights should be mounted on the top deck and visible 360° in horizontal plane. Mounting of beacon should also provide good visibility. A control switch should be provided on the instrument panel in the cabin for control of the beacon.

A warning siren should be provided having a sound output of not less than 100 dB. The siren should be mounted to permit maximum forward sound projection, but should be protected from foam dripping from the monitor or water splashed up by the tires.

A horn should be provided and should be mounted at the front part of the vehicle with the control positioned such that it is readily accessible to the driver.

Two searchlights 100 W (HALOGEN) 24 V with 30 m reeled cable and tripod should be mounted over the crew compartment. The searchlights can be removed and fixed on a tripod.

There should be two 12 V batteries connected in parallel 200 amp h capacity each at 20 h rate. Idle minimum charging rate of the alternator should be 30 amp. The electrical system should have negative ground including transistorized alternator and a fully transistorized voltage regulator. The alternator should be rated at 100% of anticipated load at 50% engine-governed speed, and if belt driven should be driven by dual belts.

Batteries should be securely mounted and adequately protected against physical injury and vibration, water spray, and engine and exhaust heat. When an enclosed battery compartment is provided, it should be adequately ventilated and the batteries should be readily accessible for examination, test, and maintenance.

Battery capacity and wiring circuits provided, including the starter switch and circuit and the starter to battery connections, should meet or exceed the manufacturer's recommendations. A master battery disconnect switch should be provided.

A built-in battery charger should be provided on the vehicle to maintain full charge on all batteries.

Grounded AC receptacle should be provided to permit a pull away connection from local electric power supply to battery charger.

An engine coolant-preheating device should be provided as an aid to rapid starting and high initial engine performance.

The electrical system should be insulated, waterproofed, and protected against exposure from ground fires.

Dashboard should contain the following switches:

- red revolving beacons with halogen lamps and a two-tone siren
- driving lamps and fog lamps
- floodlights
- compartment cabinets lighting (as a master switch)
- map reading-light
- heating element in the cooling system, when applicable
- heating and air conditioning
- electrical main switch (inside the cabin with a second switch outside for emergency use).

Lighting, installed over the water booster pump and elsewhere so that all gages, operating handles, operating panels, and their surroundings are properly illuminated.

This lighting should be switched on and off by a control switch in the driver's cabin and also from the control panel for operations at the rear, the lighting in the storage cabinets should operate automatically on opening and closing of doors and shutters. Light fittings in compartments should be to marine standard. All lighting should be protected against mechanical damage.

Weather-protected sockets for a portable floodlight at each side of the driver's cabin and at each side of the chassis. The sockets should be 2-pole, screwed connections, in accordance with DIN 14690. Material should be aluminum, brass, or stainless steel, plugs should also be supplied for:

- A red, or other color when specified revolving beacon at left and right-hand side.
- An adjustable from passenger position installed on top of the driving cabin searchlight.
- A double-tone siren of noise level 100 dB min. at 5 m.
- A red, or other color, when specified revolving beacon at the rear of the vehicle, normally positioned on the left.
- Two adjustable floodlights.
- A connection for trailer lighting, 24 V DC.
- A connection with 24-core, or as otherwise specified, screened cable for operating the cab-mounted mobile phone from the operating panel.
- All electric cables and wiring installed on the chassis should be run in metal conduit.

## 9.18.10 VENTILATION

Air vents should be distributed across the overall width of the dashboard (in order to "demist" the wind-screen), with adjustable outflow openings on the left-hand and right-hand side of the dashboard. The vehicle should be fitted with a blower having at least two speeds. A high-capacity ventilation system and defroster with defroster fan should also be provided.

## 9.18.11 HEATING AND AIR CONDITIONING

The cabin should be equipped with an adjustable heating system capable of achieving and maintaining a temperature of 15°C ±2°C inside the cabin within 20 min, when the outside temperature is −15°C, unless otherwise specified. When specified the cabin should also be equipped with an air conditioning system which will require a battery of increased capacity.

### 9.18.12  VEHICLE DRIVE

A range of gears providing the specified top speed should be provided with sufficient intermediate gears to achieve the specified acceleration.

All-wheel drive on these vehicles should incorporate a drive to the front and rear axles, which are engaged at all times during the intended service.

### 9.18.13  SUPERSTRUCTURE

The equipment compartments' water and foam tanks, pumps, and powder units should be attached to the chassis beams by a method which will prevent harmful influence, ensure flexibility of the body work superstructure and help to provide a better road grip for the wheels.

The fabrication of the superstructure should conform to the specification of the chassis.

### 9.18.14  SUSPENSION

The suspension system should be designed to permit the loaded vehicle to:

1. travel at the specified speeds over improved surface
2. travel at moderate speeds over unimproved surface
3. provide diagonally opposite wheel motion 25 cm above ground obstacles without raising the remaining wheels from the ground
4. provide at least 5 cm of axle motion before bottoming of the suspension on level ground
5. prevent damage to the vehicle caused by wheel movement
6. provide a good environment for the crew when traveling over all surfaces.

### 9.18.15  WHEELS, TIRES, AND RIMS

Vehicles should be required to have off-highway mobility while meeting the specified paved surface performance.

Tires should be selected to maximize the acceleration, speed, braking, and maneuvering capabilities of the vehicle on paved surfaces without sacrificing performance on all reasonable terrains.

The client should provide a tire description that reflects the off-road performance requirements necessitated by the soil conditions encountered. Soil conditions that may vary from an extremely fine grain soil, or clay to an extremely coarse grain soil, sand, or gravel in a dry, saturated, or frozen condition should be considered.

To optimize floatation under soft ground conditions, tires of larger diameter or width, or both, than are needed for weight carrying alone should be specified. Similarly, the lowest tire pressure compatible with the high-speed performance requirements should also be specified.

Front wheels should be single, and rear wheels dual. All tires should be of pneumatic truck type. For light vehicles rear wheels may be of single.

Each load-bearing tire and rim of the apparatus should carry a weight not in excess of the recommended load for intermittent operation for truck tires or the size used.

### 9.18.15.1 Body

The body should be constructed of materials that provide the lightest weight consistent with the strength necessary for off-pavement operation over rough terrain and when exposed to excess heat. The body may be of the unitized with-chassis-rigid-structure type or it may be flexibly mounted on the vehicle chassis. It should also include front and rear fenders parts of body, panels should be removable where necessary to provide access to the interior of the vehicle.

Access doors should be provided for those areas of the interior of the vehicle, which must be frequently inspected. In particular, access doors of sufficient size and number should be provided for access to:

- engine
- pump
- foam proportioning system
- battery storage
- fluid reservoirs.

Other areas requiring access for inspection or maintenance should be either open, or have removable panels.

### 9.18.15.2 Miscellaneous

Suitable, lighted compartments should be provided for convenient storage of equipment and tools to be carried on the vehicle. Compartments should be weather tight and self-draining.

A working deck should be provided and should be adequately reinforced to permit the crew to perform their duties in the roof, foam monitor area, cabin hatch area, water tank top fill area, foam liquid top fill area, and in other areas where access to auxiliary or installed equipment is necessary.

Handrails or bulwarks should be provided where necessary for the safety and convenience of the crew. Rails and stanchions should be strongly braced and constructed of a material, which is durable and resists corrosion.

Steps or ladders should be provided for access to the top fill area. The lowermost step(s) may extend below the angle of approach or departure or ground clearance limits if it is (they are) designed to swing clear. All other steps should be rigidly constructed. All steps should have a nonskid surface. Lower most step(s) should be no more than 558 mm above level ground when the vehicle is fully laden. Adequate lighting should be provided to illuminate steps and walkways.

A heavy-duty front bumper should be mounted on the vehicle and secured to the frame structure.

The entire vehicle and components except for chrome-plated, stainless steel and aluminum material, should receive a full anticorrosion treatment including an internally injected anticorrosion fluid or wax oil. The entire underside of the vehicle including the inside of the mudguards should be protected.

Finishing paint should be as follows:

- exterior of body: fire brigade red
- front and rear bumper: white
- mudguards and wheel hubs black
- both cabin doors should be provided with emblems and the unit name, sign-writing details should be supplied
- suction and discharge water lines: olive green
- suction and discharge FLC lines: yellow
- powder pipe: white.

Each coat of paint should have a thickness of 50–75 µm and the total thickness should at least be 120 µm.

All irregularities in painted surfaces should be rubbed down before the application of the finishing. Aluminum door shutters of all compartments should not be painted.

The battery compartment should be coated with an acid-resistant 2-component paint. The manufacturer should specify in his proposal the full paint procedure.

## 9.19 FIRE-FIGHTING SYSTEMS FOR INSTALLATION ON THE FIRE-FIGHTING VEHICLE

Depending on the application, the vehicle should be provided with a water foam system, a dry powder system or both. The minimum requirements for each system are given below.

When the vehicle is fully loaded with a full crew and the major items of equipment, chemicals, and water, it should be possible to add at least 500 kg of portable equipment, without exceeding 95% of the permissible load on the chassis. The manufacturer should provide the detailed load calculations for the vehicle and for each axle, so that compliance with the above requirement can be checked.

The overall load should be equally distributed over the front and rear axles and symmetrically distributed over the right and left-hand side wheels.

Under all circumstances the rear axle should never be subjected to more than 75% of the total load.

Under full load conditions the chassis should be in the horizontal position, any expected deviation from horizontal should be specified with the load calculations. This should be indicated as a difference in weight at the axle positions in mm for both the fully loaded and the unloaded conditions.

The manufacturer should also indicate the expected deviation in the loaded condition, but without the weight of water and foam.

- The type of water hose couplings should be instantaneous in accordance with BS 336.

However, unless otherwise specified, the couplings for suction from open water and for powder hose connections should be Storz or round thread.

| Hose coupling size for normal duties | 65 mm |
| For suction from open water | 75-100-125-150 mm |

### 9.19.1 FIRE-FIGHTING WATER AND FOAM SYSTEM

The water/foam system should include at least the following:

- Water supplied from hydrants through the vehicle manifold, by-passing the booster pump, with the addition of foam agent.
- Water supplied from hydrants via the booster pump to the discharge connections with the facility to add foam agent at each individual discharge connection.
- Water taken by suction from open water via the booster pump to the discharge connections, with the facility to add foam agent at each individual discharge connection.
- Each discharge connection should be suitable for water and for foam solution.

- Foam agent should be added via proportioners in each individual discharge, the foam percentage to be manually adjustable between 0 and 6%.
- Delivery of foam agent under pressure to fixed-installed systems.
- Foam supply source from the tank to fixed-installed foam pumps or inductors.
- When a water and foam tank is installed on the vehicle the water, with or without adding foam agent should be passed via the booster pump to the discharge connections.
- When specified it should be possible to drive the vehicle with a speed of 5 km/h on a road of given slope, when both the water and foam pumps are operating.

The above design criteria should be met by using approved equipment and components in an efficient manifold arrangement.

### 9.19.2 WATER PUMP PERFORMANCE REQUIREMENTS

#### 9.19.2.1 Capacity

1. The rated capacity of the fire pump used should be 1000, 2000, 3000, 4000, 5000, and 6000 LPM.
2. Unless otherwise specified the pump should deliver as the minimum requirement the percentage of the rated capacity shown below at the pressures indicated:
   a. 100% of rated capacity at 7 bar net pump pressure
   b. 70% of rated capacity at 10 bar net pump pressure
   c. 50% of rated capacity at 15 bar net pump pressure
3. For higher pressure up to 50 bar with lower output may be specified for H.P. hose reels with mist/spray nozzle.

#### 9.19.2.2 Suction capability

- When dry, the pump should be capable of taking suction and discharging water with a lift of 3 m in not less than 30 s through 6 m of suction hose of appropriate size, and not over 45 s for pumps of 6000 LPM or larger capacity.
- The pump vendor should certify that the fire pump is capable of pumping rated capacity at 7 and 10 bar net pump pressure, from draft, through 6 m of suction hose with strainer attached, under conditions as stipulated below.
- An altitude specified above sea level.
- Atmospheric pressure (corrected to sea level).
- Water temperature of 15.6°C.
- The suction system should be designed for efficient flow at the pumping rates. The pump suction line(s) should be of large diameter and shortest length consistent with the most suitable pump location.
- There should be a drain at the lowest point with a valve for draining all of the liquid from the pumping system when desired. Suction lines and valves should be constructed of corrosion-resistant materials.

#### 9.19.2.3 The type and other requirements

The pump should be of the centrifugal type, fitted at the rear of the chassis except when otherwise required, and be installed in such a way that there will be no axial force on the driving shaft, when in operation. Cross couplings should be considered for this purpose.

The water booster pump should be driven by a PTO, have a separate automatic priming system and be able to fulfill the characteristics given in clause 7.3.1.

The material of the pump casing and casing wear rings should preferably be of copper alloy to ASTM B 584-No. C 90500 with impeller and wear rings of copper alloy to ASTM B 148-No. C 95800.

The shaft material should be Monel K-500, with a sleeve of ANSI 316, Colmonoy 6 coated. Proposed equivalent materials should be subject to approval by the client authority.

The manufacturer should advise on the type of glands, bearings, and the material used as standard.

To safeguard the pump casing, a thermal relief valve, should be installed in the line-up discharging to atmosphere. The capacity should be such that when all discharge connections are closed, the water temperature will not exceed 60°C under full load.

The manufacturer should provide a copy of the pump test curves as certified by an independent institute.

### 9.19.2.4 Pump controls

Provision should be made for quickly and easily placing the pump in operation. The lever or other devices should be marked to indicate when in pumping position.

Any control device used in power train between the engine and pump should be arranged so that it cannot be unintentionally knocked out of the desired position.

Where the pump is driven with chassis transmission in neutral, in that propelling power can be applied to the wheels, while pumping, a device should be provided by which the chassis transmission can be positively held in neutral.

A nameplate indicating the chassis transmission control lever position to be used for pumping should be provided in the cabin and located so that it can be easily read from drivers' position.

Means should be provided for controlling the speed of the pump.

A priming device should be provided, it should function at an engine speed not exceeding the maximum no load governed speed and developing a vacuum of 50 cm Hg at an altitude of 600 m. The priming device should be controllable at the pump-operating position.

All pumping controls and devices should be installed so as to be protected against mechanical injury or the effect of adverse weather condition upon their operation.

### 9.19.2.5 Water tank

A water tank should have a minimum capacity of 2500 L (700 Gal) for general purpose fire truck and to be independent of the body or compartment and should be equipped with suitable mechanical method for lifting tank out of the body.

### 9.19.2.6 Construction

The tank should be constructed of stainless steel or fiberglass. The tank should have longitudinal and transverse baffles. The construction and connections should be made to prevent the possibility of galvanic corrosion of dissimilar metals.

The tank should be equipped with easily removable manhole covers over the tank discharge. Tanks should be designed to permit access within each baffled compartment of the tank for internal and external inspection and service. The tank should have drain valves.

Provisions should be made for necessary overflow and venting. Venting should be sized to permit agent discharge at the maximum design flow rate without danger of tank collapse, and should be sized to permit rapid and complete filling without pressure build-up. Overflows should be designed to prevent pressure build-up within the tank from overfilling and to prevent the loss of water from the tank during normal maneuvering, and to direct the discharge of overflow water directly to the ground.

The water tank should have a sufficient number of swash partitions.

The water tank should be mounted in a manner that limits the transfer of the torsional strains from the chassis frame to the tank during off-pavement driving. The tank should be separate and distinct from the crew compartment, engine compartment and chassis, and easily removable as a unit.

The water tank should be equipped with at least one top fill opening of not less than 13 cm internal diameter. The top fill should be equipped with an easily removable strainer of 6 mm mesh construction. The top fill opening should be equipped with a cap designed to prevent spillage.

### 9.19.2.7 Tank fill connection(s)

Tank fill connection(s) should be provided in a position where they can be easily reached from the ground.

The connection(s) should be provided with strainers of 6 mm mesh and should have check valves or be so constructed that water will not be lost from the tank when connection or disconnection is made.

The tank fill connection(s) should be sized to permit filling of the water tank in 2 min at a pressure of 5.5 bar at the tank intake connection.

### 9.19.2.8 Hose reels

Hose reels should have a minimum internal diameter of 2 cm and should have a minimum acceptance test pressure of 50 bar.

Hose reels should be equipped with a shut-off type nozzle designed to discharge both foam and water at a minimum discharge rate of 150 LPM each nozzle should have minimum foam discharge patterns from dispersed stream of 4.5 m width and 6 m range, to a straight foam stream with 10 m range. High-pressure fog/mist-spray nozzle to be provided and be changeable with foam branch.

Each reel should be designed and positioned to permit removal by a single person from any position in a 170° horizontal sector. Each reel should be equipped with a friction brake to prevent hose from unreeling when not desired. The nozzle holder, friction brake, rewind controls, and manual valve control should be accessible from the ground.

Flow to each reel should be controlled by a manually operated quarter turn ball type valve. Two hose reels of 50 m each equipped with rewined mechanism should be provided and fixed on the either side at the rear of the vehicle.

## 9.19.3 FOAM SYSTEM

### 9.19.3.1 Materials

All components of the foam system including the foam-liquid tank, piping, fill troughs, screens, etc., should be made of materials resistant to corrosion by the foam-liquid concentrate, foam-water solution, and water.

### 9.19.3.2 Foam-liquid concentrate tank

1. Foam-liquid concentrate tanks should be of the rigid type. The tank should be designed for compatibility with the foam concentrate being used and resist all forms of deterioration, which could be caused by the foam concentrate.

2. Tanks should be designed to permit access within each baffled compartment of the tank for internal and external inspection and service. Drain connection should be installed to flush out the bottom of the sump.

3. The tank outlets should be located above the bottom of the sump and should provide continuous foam-liquid concentrate to the foam proportioning system.

4. If separate from the water tank, the foam-liquid tank should be mounted in a manner that limits the transfer of the torsional strains from the chassis frame to the tank, during off-pavement driving. The tank should be separate and distinct from the crew compartment, engine compartment, and chassis, and should be easily removable as a unit.

5. A top fill trough should be equipped with a stainless steel 6 mm mesh screen and container openers to permit emptying 20 L foam-liquid concentrate containers into the storage tank at a rapid rate regardless of water tank level. The trough should be connected to the foam liquid storage tanks with a fill line designed to introduce foam-liquid concentrate near the bottom of the tank so as to minimize foaming within the storage tank.

6. Tank fill connection should be provided in a position where it can be easily reached from the ground to permit the pumping of foam-liquid concentrate into the storage tank. The connection should be provided with strainers of 6 mm mesh, and should have check valves or be so constructed that foam will not be lost from the tank when connection or disconnection is made.

7. The tank should be adequately vented to permit rapid and complete filling without the build-up of excessive pressure and to permit emptying the tank at the maximum design flow rate without danger of collapse. The vent outlets should be directed to the ground to prevent spillage of foam-liquid concentrate on vehicle components.

### 9.19.3.3 Foam-liquid concentrate piping

1. The foam-liquid concentrate piping should be of material resistant to corrosion. Care should be taken that the combinations of dissimilar metals that produce galvanic corrosion are not selected or that such dissimilar metals are electrically insulated. Where plastic piping is used, it should be fabricated from unplasticized resins unless the stipulated plasticizer has been shown not to adversely affect the performance characteristics of the foam-liquid concentrate. The plastic pipe may be reinforced with glass fibers.

2. The foam–liquid concentrate piping should be adequately sized to permit the maximum required flow rate and should be arranged to prevent water from entering the foam tank.

### 9.19.3.4 Foam/liquid pump

This pump should be of the positive displacement or centrifugal type and work independently of the water booster pump (driven by a PTO or other source). The pump should be able to fulfill the characteristics given in Table 9.3 and be able to inject, foam liquid into the water stream at a pressure of 0.7–4 bar above the maximum water pressure, delivered by either the water booster pump or the firewater mains.

The pump should also be able to transfer foam-liquid from drums or storage tank into the foam/liquid tank of the vehicle and vice versa.

The line-up should be provided with a relief valve (set pressure equal to the design pressure of the pump) allowing full flow discharge into the foam liquid tank without overheating the pump or exceeding the specification of the piping system. The pump should be able to fulfill the characteristics given in Tables 9.3 and 9.4.

**Table 9.3 Foam/Liquid Pump Outlet**

| Purpose | Sources | Discharge | Minimum LPM | Discharge Pressure (bar) |
|---|---|---|---|---|
| Delivery foam liquid | Direct from tank | To other foam storage tank | 400 | 2.5 |
| Delivery foam liquid | Suction from drums (suction height 1.5 m) | To storage tank | 120 | 2 |
| Delivery to foam proportioner | Direct from tank | Foam proportioning system | 100–400 | 2 bar above the expected inlet water pressure (8–12 bar) |

**Table 9.4 Characteristics of Foam/Liquid Pump**

| Application | Source of Water Supply | Water Rate in LPM | | Water Press, Bar | | Foam Agent Rate in LPM | | |
|---|---|---|---|---|---|---|---|---|
| | | Min. | Max. | Suction | Discharge | Min. | Max. | |
| | | | | | | | 1% Setting | 6% Setting |
| Foam solution or water | Direct from hydrants (by-passing) the booster pump | 400 | 7800 | 6–12 | | 4 | 78 | 468 |
| Foam solution or water | From hydrants via the booster pump | 400 | 4500 | 6–12 | 11.16 pump diff. head 5 bar | 4 | 45 | 270 |
| Foam solution or water | Suction from open water via the booster pump | 400 | 2400 | 1.5 m suction height | 10 | 4 | 24 | 144 |
| Foam solution or water | Suction from water tank on vehicle via the booster pump | 400 | 2400 | | 10 | 4 | 24 | 144 |
| Foam concentrate to discharge (16 bar) | — | — | — | | | 4 | – | 500 |

The material of the foam pump housing and rotors should preferably be stainless steel type ANSI 304 or 316, with a type ANSI 316 stainless steel shaft.

The manufacturer should give the direction of rotation of the drive shaft, the type of glands and bearings and the materials used as standard.

Proposed alternative materials should be subject to approval by the client. The manufacturer should provide a copy of the pump curves.

### 9.19.3.5 Foam-water tank accessories

The tank volume should be as large as possible, but should at least contain the volume as specified. In any combination of FLC and water the ratio should be 1:6 with a tolerance of 5%.

The tank and all tank components should be of stainless steel or alternative materials, subject to approval.

The tank should be provided internally with sufficient baffles, but baffle sizes and spacing should allow for cleaning and inspection. It should have an expansion dome with a volume of 3% of the tank volume, the dome should be provided with a manhole of minimum diameter 500 mm fitted with a quick release lock.

The tank should also be provided with 2 pressure/vacuum (p/v) valves of sufficient capacity, and with hand-operated ball valves for tank outlet-and-filling. These valves should be readily accessible. The p/v valves should be installed in the middle of the tank to avoid clogging of the valves as a result of the acceleration and braking of the vehicle. The size should be suitable for a filling rate of 1200 LPM. A pen overflow should be fitted to release under the vehicle in the case of overfilling. The location of the overflow should be such that the overflow liquid will not fall on any part of the chassis.

The water tank filling connections should be equipped with a level indicator visible at rear of vehicle and a low-level audible alarm, which will be activated when a level of 10% FLC is reached. The level indicator type should also provide a good indication with dark brown colored foam compound.

Special attention should be given to the design to prevent damage to the tank during filling and possible surging at the point of overflow at high filling rates. Provision should be made to refill the foam liquid tank by foam liquid transporting truck.

### 9.19.3.6 Foam control system

The foam proportioning system should be designed such that foam agent can be added at each individual discharge connection.

It should be possible to manually set the foam percentage at zero and proportionally between 1 and 6% preferably continuously or in 1% steps.

The in-line proportioners should be calibrated in the actual manifold on the vehicle as follows:

- Calibration at water rates of 200 up to 1200 LPM at each delivery;
- foam setting normally 3–6%;
- required accuracy between 0 and plus 0.3%.

### 9.19.3.7 Line-up and piping design of the water/foam system

The line-up should be in accordance with the relevant flow scheme for the specified vehicle. Drain valves, vent valves, and valved flushing connections should be provided to ensure proper flushing of all components.

The size of the piping should be such that the velocity will not exceed 2.8 m/s in the suction lines and 6 m/s in the discharge and return lines.

All components and the piping should have a maximum working pressure of 16 bar and should be able to withstand a test pressure of 1.5 times the maximum working pressure.

The piping, fittings, and other components of the system should be stainless steel.

### 9.19.3.8 Operating and control panel-water/foam

The main operating and control panel should be mounted at the rear of the vehicle.

The width of the panel should be 600–800 mm approximately and consist of the following sections arranged from top to bottom and incorporate as a minimum the following indicators:

1. The panel should be installed at an angle such that a standing operator can easily read the instruments, at an eye level position between 1500—1800 mm. All illuminated lamps with colored lenses should be clearly visible in full sunshine.
2. All elements should be conveniently grouped and clearly identified.
3. The panel should be constructed from oil-resistant material suitable for outdoor tropical sun-exposed conditions.
4. Indicating lamps; should give signal that brakes blocked, power take off (PTO) 1 and/or PTO 2 are engaged.
5. The electrical instruments including the wiring of sections should be installed in a weatherproof box with marine-type enclosures for indicators and lamp fittings;
The applied wiring terminations should be vibration proof.

## 9.19.4 ADJUSTABLE FOAM/WATER MONITOR

The foam/water monitor should be mounted at the rear of the vehicle or over the cab. It should be able to turn 360 degrees horizontally in both directions and rise vertically from −30 degrees depression up to +80 degrees elevation. The monitor should be provided with adjustable deflectors.

When operated as a water jet, the jet should be able to reach the ground at a distance of not more than 8 m at each side of the vehicle. When operated with low-expansion foam the width of the foam blanket should be 4 m minimum at the close throwing distance.

Position setting should be done by a lever or other acceptable methods, however, locking of the monitor in any desired position should be possible.

The foam/water solution discharge rate should be approximately 2000 LPM at a water pressure of 10 bar but may, unless otherwise specified, be up to 4000 LPM. The minimum throw length with foam should be approximately 50 m, while under these conditions no foam should fall on the ground within 20 m.

The monitor should be operable from a fixed platform with swing down type handrails of height 700–900 mm. The water and foam supply should be manually controlled and be operable near the monitor on the platform, a pressure gage should also be fitted near the monitor. The maximum vehicle height normally is not more than 3500 mm, the type of monitor will be specified if required. Hydraulic operation and oscillating type should be supplied if specified.

The monitor and the bearings should be of aluminum bronze material with a stainless steel or aluminum alloy barrel and deflector.

Required foam expansion ratios should be 8–10%.

## 9.19.5 FOAM SYSTEM-BY-PASSING WATER PUMP

Four to six inlets and four to six outlets valved manifold should be provided on either side of the vehicle inlets with instantaneous male and outlets with instantaneous female couplings. At each outlets foam proportioner with control lever should be fixed. In this system when pressure of water is sufficient, water bypassing the water pump, will be mixed by FLC when the FLC pump is engaged by PTO. The FLC pressure should be 0.7–4 bar above the water pressure. When the water pump is required to boost the pressure, water will pass through the water pump.

## 9.19.6 ROUND THE PUMP FOAM PROPORTIONER

In the areas where generally water pressure is not sufficiently high, the water tank mounted on truck is used for initial fire fighting. Foam proportioning system may be of round-the pump type and foam pump is not necessary. If water hydrant is available water passing through the water tank will be boosted by the pump. Differential pressure between suction and discharge through ejector cause the FLC to flow to the suction and boost the required pressure for making foam. A control lever will control percentage of FLC mixing with water. By this system FLC can be used either from FLC tank or through 3 or 4 cm pick-up tube using FLC containers.

### 9.19.6.1 Valves

All hand-operated valves should be of the ball type, either flanged or with wafer type valve bodies installed between flanges. For sizes up to 65 mm valve bodies and trims should be of stainless steel type ANSI 304. Valves 75 mm and larger should be carbon steel to ASTM A 216 WCC or WCB with maximum carbon content of 0.25% and trim of stainless steel type ANSI 304.

The flanges should have CAF gaskets and stud bolts to ASTM A 193 B7 with hexagonal nuts to ASTM A 194 GP 2H.

Material certificates equivalent to DIN50049 type 3.1 B are required for pressure-containing parts. The foam proportioners and the bypass of the foam control valve should be installed at the rear discharge connections, be easily adjustable and with dial settings clearly visible

## 9.20 EXTINGUISHING DRY POWDER SYSTEMS

The fire-fighting vehicle can be equipped with dry powder units, depending on the type of vehicle and its application. The system should consist of the following:

- dry powder tank with charging system
- nitrogen or dry air cylinders for expellent gas and flushing function
- hose reels with powder hose and trigger nozzle
- control inspection and operating panel
- dry powder monitor.

### 9.20.1 DRY POWDER

Urea-based potassium bicarbonate (Monnex) should normally be used.
   *Notes:*
1. A container will hold 70% by weight of "Monnex" compared to other dry powder.
2. Monnex (purple-K) are trade names of potassium bicarbonate base dry powder.

### 9.20.2 POWDER VESSEL DESIGN

The system including pressure vessels and expellent gas cylinders should be designed and manufactured in accordance with the relevant B.S. No. 5430 Part 3.

A formal approval certificate for the vessels is required signed by a pressure vessel Inspecting Authority.

All inlet and outlet connections should be flanged. Each vessel should have a relief valve of sufficient capacity to ensure that the maximum pressure will not exceed the maximum operating pressure by more than 15%. A manhole or inspection nozzle of 150 mm or larger should be provided on the tank. Lifting lugs should also be fitted.

After full discharge the remaining quantity of powder in the vessel should be less than 7% of the charge.

### 9.20.3 LINE-UP AND PIPING DESIGN OF THE DRY POWDER SYSTEM

The line-up should be in accordance with the relevant flow scheme.

All valves should be of the ball type, suitable for dry powder and manually operated. The pipe system, branches, T-pieces, and bends should be smooth and have minimum resistance to the flow of dry powder. The fluid velocity should not exceed 4 m/s. Leak valves and nonreturn valves should be fitted in the nitrogen/compressed air expellant gas and control gas systems so that all switch functions can be carried out correctly.

Each cylinder should be provided with its own valve and be connected to a high-pressure manifold, a manually operated valve should preferably be fitted in the manifold. The pressurized system should be pressure-tested at 1.5 times the maximum working pressure.

A gage should be installed to indicate the pressure.

### 9.20.4 EXPELLANT GAS CYLINDER

Sufficient dry nitrogen or air should be available to empty each powder tank fully and to flush all piping. The working pressure of the dry powder tank should be at least 14 bar with a maximum of 16 bar. The cylinder contents should have a reserve of 30% in order to deal with possible small leakages during intermittent operation and to carry out control function.

Pressure regulator should be so designed that it will automatically reduce the normal cylinder pressure and hold the expellant gas pressure at the designed operating pressure of dry chemical container. Charging time of the vessel should be less than 15 s.

### 9.20.5 POWDER GUN, POWDER HOSE, AND HOSE REELS

The dry powder hose should have a smooth bore of not less than 25 mm diameter and be 30 m long. The safe working pressure of the hose should be two times the working pressure of the powder tank and its bursting pressure three times. The powder gun should have an output of 1.8 kg/s.

The hose reel should have the least possible resistance so that the hose can be unrolled under pressure and not be jammed, a manual rewind mechanism should be provided. The hose reels should have hose gliders and a brake-blocking device.

The system should be such that each powder tank is provided with one hose reel and a valved manifold connection.

The hand nozzle should be of sea-water-resistant aluminum bronze.

### 9.20.6 CONTROL AND OPERATING PANEL POWDER SYSTEMS

A control/inspection and operating panel should be fitted next to or near each powder unit comprising of: flushing valve for each hose, pressure gages for expellant gas working pressure, and a push button to discharge the tank.

### 9.20.7 **PREMIX FOAM SYSTEM**

General Premix foam system is mixture of 6–10% of preferably film forming fluoro protein (AFFF), aqueous film forming foam (FFFP), or alcohol resistance foam liquid mixed with water and used as quick initial fire fighting means for class B fires. The system is in two forms:

1. Gas expelled. The tank should be of rigid high pressure construction pressurized by air or nitrogen and released into two high pressure hose reels terminated to foam-making branch nozzles.
   The tank capacity in this system should be up to 500 L. This premix system is generally used simultaneously with dry chemical fire extinguishers.
2. Pump expelled. In fire trucks with foam liquid pump having suction inlet from the premix tank (60 mm) and outlet to hose reels; by using the pump, premix liquid will be boosted through two hose reels passing through foam making nozzles. Premix foam can also be pressurized by fire truck water pump if specified. When premix liquid is used, the tank also can be refilled either by FLC or premix. The capacity of the tank depends on the class of the vehicle.

#### *9.20.7.1 Twin agent*

Systems of this type combine the rapid fire extinguishing capabilities of dry chemical powder (as well as their ability to extinguish-three dimensional fires) with the sealing and securing capabilities of foam and are of particular importance for protection of flammable liquid hydrocarbon hazards.

This system may be self-contained and the application of each agent is separately controlled so that the agents may be used individually.

The supplier of dry chemical and foam liquid to be used, in the system should confirm that their products are mutually compatible and satisfactory for this purpose. Limitations imposed on either of the agents alone should also be applied to the combined agent system.

Minimum delivery rates for protection of hazard should be the ratio of dry chemical discharge rate and AFFF discharge rate (kg dry chemical-liter/s AFFF). Foam liquid should be in the range of 0.6:1–5:1 liters per second (LPS).

The equipment mounted on the fire truck should be capable of operation for a period of at least 30 s for each agent. For this system twin hose reels must be fixed between dry powder tanks and premix foam tank.

Premix foam system can also be used and pressurized by foam and water pump. Therefore provision should be made for the foam pump suction inlet valve connected to the premix tank. Premix foam system can also be pressurized by nitrogen gas. In this system pressure regulators should be fitted to release the excess pressure from the tank and the capacity of pressurized tank should not exceed more than 500 L.

---

## 9.21 **COMPARTMENT FOR MISCELLANEOUS EQUIPMENT**

### 9.21.1 **HOSE COMPARTMENT**

Hose compartment should be fabricated from noncorrosion material and should be designed to drain effectively and should be smooth and free from projections. Hose compartment must be provided on right side of the vehicle and should not be more than 160 cm above the ground. No other equipment should be mounted or located in hose compartment where it will obstruct the removal of the hose.

## 9.21.2 MISCELLANEOUS EQUIPMENT

The following fire equipment should be carried in major and general purpose fire trucks as mentioned further. They should not be supplied unless specified by client. Selection should be made from the list to suit the types of the vehicles. It may also be necessary to specify other types of equipment when required:

- sixteen lengths of 70 mm diameter 25 m fire hose each with instantaneous 65 mm couplings
- five lengths of 45 mm diameter 25 m fire hose each with instantaneous 65 mm couplings
- four lengths of appropriate size of suction hose with round thread couplings, one of them with metal strainer
- two pairs of suction spanners for suction hoses
- one portable foam/water monitor light alloy chromium (Cr) plated manually operated, mountable on separate base plate (LPM to be specified).
- one suction collecting head round thread with instantaneous male inlet with spring loaded check valve
- two dividing breeching inlets male, and outlets female couplings
- two Nos. 4 cm pick-up tubes with round thread or Storz couplings for filling of foam liquid to the trucks foam tank from drums
- two lengths of 15.4 m hose with round thread or Storz couplings as specified for pumping foam from FLC dispensing vehicle to the fire truck tank;
- four water fog/jet nozzles brass, chromium (Cr) plated adjustable
- 2–10 kg $CO_2$ fire extinguishers
- 2–12 kg dry powder fire extinguishers
- five fireman axes
- four sets of breathing apparatus air type (back-pack) with 4 spare cylinders (to be carried in crew compartment)
- four light alloy foam branches aluminum alloy, anodized, capable of discharging 400 LPM (water/FLC solution) at 7 bars with deflector
- two as above 200 LPM (water/FLC solution) at 7 bars with deflector
- one as above 800 LPM at 7 bar
- one aluminum (2 sections) ladder 8 m (if required)
- four Nos. safety torches
- one first aid kit
- one portable hailer
- two cutting axes
- two 1½ cm cotton rope 25 m
- ten hose couplings and 5 suction coupling gaskets
- three short handle shovels
- one tool box
- one pair of rubber gloves (anti electric shock)
- one portable gas detector
- one crowbar
- one ejector pump
- one polarized battery recharging receptacle mounted on rear of the truck
- one removable search light tripod
- one fire approach or entry suit.

# 9.22 PROPOSED STANDARD SPECIFICATION OF MAJOR FOAM TENDER AND GENERAL PURPOSE FIRE FIGHTING TRUCKS FOR REFINERIES AND OTHER HIGH RISK AREAS

## 9.22.1 MAJOR FOAM TENDER

### 9.22.1.1 Truck

Chassis, engine, brakes, steering, cabin (driver +2) instruments control electrical system, vehicle drive, superstructure, body, equipment cabin, etc. are specified in relevant standards.

### 9.22.1.2 Fire-fighting system

The following fire-fighting system should be mounted on fire truck:

1. water booster pump 6000 LPM at 7 bar
2. FLC booster pump 400 LPM
3. foam-liquid tank 5000 L
4. two 250 kg dry powder extinguishing with one hose reel
5. foam/water monitor 3000 LPM at 10 bar (water/FLC solution) mounted over drivers cabin
6. foam system by-passing water pump six inlets and six outlets manifold as
7. foam ejector proportioning system
8. foam control
9. line-up and piping design of water/foam
10. operating and control panel water/foam
11. miscellaneous equipment.

## 9.22.2 GENERAL PURPOSE FIRE TRUCK

### 9.22.2.1 Truck

Chassis, engine, brakes, steering, extended cabin, electrical, instruments control system, vehicle drive, superstructure, body, equipment, and etc. are specified in Section I.

### 9.22.2.2 Fire-fighting system

The following fire-fighting system as specified in Section II should be mounted:

1. water booster pump multipressure 3000 LPM at 7 bar
2. FLC booster pump 400 LPM
3. water tank 2500 L
4. hose reels
5. twin foam liquid tank 500 L FLC and 500 L premix[a]
6. foam water tank accessories
7. foam control proportioner
8. operating and control panel-water-foam
9. foam/water monitor 2000 LPM at 10 bar
10. foam system bypassing water pump (four inlet and four outlet manifold)

[a]The type of foam used in premixed section to be of AFFF-FFFP or alcohol resistance type as required. The twin tank can also be used as FLC tanks.

**11.** extinguishing dry powder system 2 × 250 kg one hose reel
**12.** miscellaneous equipment.

## 9.23 SPECIFICATION FOR AUXILIARY FIRE-FIGHTING AND EMERGENCY VEHICLES

Auxiliary fire-fighting vehicles comprises of the following units:

**1.** foam liquid dispensing truck
**2.** dry powder extinguishing truck
**3.** twin agent extinguishing truck
**4.** water tender
**5.** emergency service vehicles
**6.** hydraulic boom.

### 9.23.1 FOAM LIQUID DISPENSING TRUCK

Main items include:

**1.** foam concentrated liquid tank 6000–8000 L
**2.** foam transfer pump
**3.** foam monitor
**4.** control panel
**5.** miscellaneous equipment.

### 9.23.2 FOAM LIQUID TANK

Foam concentrated tank to be of 6000–8000 L capacity and with all its components should be made of stainless steel.

The tank should also be provided with 2 p/v relief valves of sufficient capacity, and with hand-operated ball valves for tank outlet and filling. These valves should be readily accessible. The p/v valves should be installed in the middle of the tank to avoid clogging of the valves as a result of the acceleration and braking of the vehicle. The size should be suitable for a filling rate of 1200 LPM. An open overflow discharge should be fitted to release under the vehicle in the case of overfilling. The location of the overflow discharge should be such that the overflow discharge liquid will not fall on any part of the chassis.

The tank filling connections should be provided with strainers. These connections and the tank drain should be fitted with hose couplings and blank caps attached by a chain.

The tank should be equipped with a level indicator (visible at rear of vehicle) and a low-level audible alarm, which will be activated when a level of 10% foam compound concentrate is reached. The level indicator type should also provide a good indication with dark brown colored foam compound.

The size of the nozzles should be such that the velocity will not exceed 2.8 m/s in the suction nozzles and 6 m/s in the supply and return nozzles.

The suction nozzle in the tank should extend to 30 mm inside the tank to avoid sediment build-up in the suction piping.

However, the foam liquid storage tank should be provided with:

- an inlet with an internal filling tube
- an outlet which prevents sludge from overflowing in the pump
- a sump with bolted cover-plate incorporating the sludge drain
- a gage tube to measure the tank contents
- a breather valve to prevent excessive under-pressure or overpressure.

### 9.23.3 FOAM MONITOR

Vehicle should have a foam monitor positioned over the driver's cab. The total foam solution discharge rate should be of 2000 LPM at 10 bar. Manual foam monitor control should be accessible to both driver and crew member.

The foam monitor should be capable of being elevated at least 45 degree above the horizontal and capable to discharge within 6 m in front of the vehicle. The foam monitor should be also capable of being rotated not less than 90 degrees to either side, total traverse not less than 180 degrees.

The monitor should be of manual or hydraulic type as specified.

### 9.23.4 CONTROL PANEL

The control panel should be mounted at the rear of the vehicle and consist of the following:

1. foam liquid level indicator
2. foam discharge and inlet rate
3. water inlet pressure gage
4. foam transfer pump pressure gage
5. foam proportioning levers
6. engine lube oil temperature
7. battery charging current
8. engine temperature
9. engine throttle control lever.

The panel should be installed at an angle as such that standing operator can easily read instruments at an eye level position.

### 9.23.5 EQUIPMENT TO BE CARRIED

The following ancillary fire equipment should be carried in two suitable compartments:

- six lengths of 65 mm standard fire hose
- two 4 cm pick-up tubes with round thread or Storz connections
- four foam branches with water foam liquid solution discharging rate 400 LPM at 7 bar with deflector made of aluminum alloy anodized
- two fog/spray nozzles brass chrome plated, adjustable
- two 4 cm filling hoses with round thread or Storz couplings, each 15 m
- two 12 kg dry powder fire extinguishers.

## 9.23.6 FOAM PROPORTIONING SYSTEM

Six inlet valve, three on each side and six outlet valves at the rear of the vehicle should be provided, all inlet and outlet valves should be of quick opening ball valve made of gun metal chrome plated.

Adjustable 0–6% foam proportioners to be provided at each outlet and foam monitor.

- Foam liquid pump—As specified in Table 9.2.
- Foam control system
  When pressurized water is supplied from hydrant through the vehicle manifold, foam agent is added from the foam liquid pump to the proportioning system and foam/water solution passes through air/foam making branch pipes or foam monitors.
  The foam pump discharge connections can be used to replenish the FLC tank of other fire trucks or other fixed foam systems.

## 9.23.7 DRY POWDER FIRE EXTINGUISHING TRUCK

Main items include:

- the vehicle
- extinguishing dry powder system
- dry chemical powder
- powder vessels
- line-up and piping
- expellant gas
- powder gun, powder hose, and hose reels
- manually operated powder monitor
- control and operating panel.

Chassis, engine, body, and other related equipment should meet the standard specification as given in Section I and powder vessel, line up and piping, expellant gas cylinders, powder gun, powder hose reels, control and operating panel as given in Section 2 Clause 7.15.

### 9.23.7.1 Extinguishing dry powder system

The fire-fighting vehicle can be equipped with dry powder units, depending on the type of vehicle and its application. The capacity depending on risk factor and class of fire should be up to 3000 kg in three stages. The system should consist of the following:

- dry powder tanks with charging system
- nitrogen cylinders for expellant gas and flushing function
- hose reels with powder hose and trigger nozzle
- dry powder monitor
- dry powder manifold, to be connected to external dry powder extinguishing system.

### 9.23.7.2 Dry powder

Urea-based potassium bicarbonate should normally be used. Other types of dry powder may be considered when specified.

### 9.23.8 MANUALLY ADJUSTABLE POWDER MONITOR

If a powder monitor is required, it should be installed on the vehicle in such a way that the powder stream should hit the ground approximately 8 m from each side or in front of the bumper of the vehicle.

The powder monitor should be manually operated and its capacity should be selected between 20 and 50 kg/s with a throw between 30 and 50 m respectively.

The monitor should be horizontally adjustable over 140 degree on each side from the straight forward position. The vertical elevation depends on its installation on the vehicle, but should be at least 90 degree from the most downward position.

An operating handle should be installed at the monitor to open and close the quick acting main pneumatic valve of the powder tank.

The monitor should be equipped with a reliable locking and braking device and a cover on the barrel attached by chain to prevent water entry.

The monitor should be operable from a fixed platform with swing down type handrails.

### 9.23.9 COMBINATION OF DRY POWDER AND PREMIX FOAM (TWIN AGENT)

#### 9.23.9.1 Main items include
- the vehicle
- extinguishing twin agents

#### 9.23.9.2 The vehicle
The vehicle should be suitable for use in areas as specified.

The driving cabin should be tilted type and contain seats for three persons.

#### 9.23.9.3 Superstructure
The nitrogen/air cylinders compartment and foam premix and powder units should be fastened to the chassis by a method which will prevent harmful influence, ensure higher flexibility of the body work superstructure and help to provide a better road grip for the wheels. The electrical system should be dust proof and waterproof and fit for tropical conditions when this is required. Revolving beacon, siren, and two search lights should be provided over the driver's cabin.

### 9.23.10 FLOW CHART

The following information concerning the fire equipment loaded on trucks A and B should be considered as in Table 9.5.

### 9.23.11 EXTINGUISHING POWDER/FOAM INSTALLATION

The vehicle should be equipped with the following depending on the area of fire risk.

- one to four extinguishing powder unit of either 250 or 500 kg, see under "A" and "B" of the aforementioned flow chart
- one to four premix foam unit 250 or 500 L
- twin high-pressure hose 25 mm diameter × 25 m long wound on a hose reel mounted at the rear of the vehicle
- nitrogen cylinders of sufficient capacity for emptying the dry powder and foam tanks.

**Table 9.5 Fire Equipment Loaded on Trucks**

| Extinguishing Capacity | Truck "A" | Truck "B" |
|---|---|---|
| Dry powder | 250 kg Unit | 500 kg Unit |
| Premix foam | 250 L | 500 L |
| Hose | 1 hose reel with 25 m twinned hose for each unit | 1 hose reel with twinned hose for each unit |
| Discharge rate: (powder) | Approx. 2 kg/s | 2 × 3.5 kg/s |
| (Premix) | 200 LPM | 2 × 200 LPM |
| Discharge range: (Powder) (Premix) | | |
| | 10 m | 10 m |
| | 20 m | 20 m |
| Expellant gas system (Powder) | $N_2$ cylinders 20 dm at 150 bar | |

The hose reel should have the least possible resistance and should be such that the hose can be unrolled under pressure and not jammed. It should be possible to operate any length of unrolled hose. The twin hose should have combined trigger nozzles with capacities for discharging the powder and foam.

## 9.24 WATER TENDER

Main items include:

- the vehicle
- water tank
- water pump
- hose reels
- miscellaneous fire-fighting equipment.

### 9.24.1 THE VEHICLE

The chassis, body, engine, vehicle drive, steering, brakes, and electrical system.

The cabin should provide seating for driver plus one crew member. Search lights, siren, and flashing beacon should be provided and the cabin to be of tilted type.

### 9.24.2 WATER TANK

The tank should have the capacity of 6000–8000 L and should be made of galvanized steel and internally protected against corrosion. The tank should be provided with internal traverse baffles and dome having manhole opening with quick release lock. The tank should be provided with:

- an inlet with an internal tube
- an outlet to the pump suction
- sump with bolted cover and drain pipe

- a breather valve to prevent excessive pressure
- a gage tube to measure the water content.

### 9.24.3 WATER PUMP

The booster pump should be of centrifugal multipressure type capable of pumping 2000 LPM at 7 bar and 1000 LPM at 15 bar.

The pump to be driven by engine P.T.O. and should be mounted behind the driver's cab provided with control panel at right side.

One suction connection and two outlet instantaneous valve connections should be provided.

Two high-pressure hose reels also to be fixed, one on each side at the rear. Two separate 6.5 cm filling valved connections should also be provided, one on each side.

### 9.24.4 COMPARTMENT

- Compartment for 5 lengths of 70 mm and 5 lengths of 45 mm hoses with 65 mm standard couplings should be provided on either side.
- Four suction hoses of 3 m length 10 cm dia, two on each side should also be housed, one of the suction hoses with metal strainer.
- Compartment for the equipment should also be provided.
- Five No 20 L foam compound containers preferably (film forming) and two 200 LPM portable foam branches and 2 cm pick up tube connections round thread.
- Two fog/spray control nozzles.
- Two sets of breathing apparatus in sustained position in drivers cab with two spare cylinders.
- Two shovels, two picks, two axes, and a tool kit.
- Two sets of firemen's protective clothing.
- Two 12 kg dry powder fire extinguishers.

## 9.25 EMERGENCY SERVICE VEHICLES

### 9.25.1 EMERGENCY COMBINED RESCUE VEHICLE

Main items include:

- lighting set with reeled cable
- hose reels
- fire-fighting equipment
- cutting saw and tools
- resuscitation and breathing apparatus
- miscellaneous rescue equipment.

#### 9.25.1.1 The vehicle

The vehicle should be of light type, four-wheel drive and with suspension that can be used in rough roads and on soft soil at temperature of $-25°C–60°C$, with the following details:

- the cabin should provide seating for driver plus two, with air conditioning and heating system
- exhaust system should be of spark arrestor type

- hydraulic steering
- the engine can be petrol driven with 150 L fuel tank and 100 L spare tank the fuel system should meet the specification of Clause 6.2.4.3
- towing connection behind the chassis and winch provided in front of the chassis with 30 m pulling wire (2 tons)
- manual transmission
- electrical system 12 V
- warning siren of 95 dB and revolving beacon
- provision to be made for mounting radio operated from the cab
- spot lights (150 mm) on both left and right side over the cab and hand adjustable type for beam selection
- emergency rescue equipment.

### 9.25.1.2 Lighting set

Lighting set if specified should be provided behind the cab, driven by engine PTO with 10–20 kW output at about 25/50 V. One searchlight of 30 cm diameter with suitable halogen bulb should also be provided.

Two sockets for temporary emergency lighting up to 3 kW, one cutting saw with 50 m reel cable if specified.

Cable of 100 m with halogen bulbs should be provided for emergency lighting.

### 9.25.1.3 Rescue equipment

Rescue equipment consisting of cutting, pulling, and other devices as specified by client should be maintained.

### 9.25.1.4 Miscellaneous fire and safety equipment

The following miscellaneous fire and emergency equipment should be carried in the vehicle.

- two dry chemical fire extinguishers of 12 kg
- two sets of breathing apparatus back-pack with two spare cylinders
- one set of oxygen resuscitators with two spare cylinders
- one explosimeter or gas alarm
- one first aid kit
- one 8 m extension aluminum ladder mounted over the vehicle if specified
- one cutting axe
- two shovels, short handles
- two fire man axes
- A tool kit
- two safety torches
- three road stops (reflecting orange/red)
- three reflecting flags
- two automatic flushing beacons (portable)
- loudspeaker.

### 9.25.1.5 Main items and list of equipment

This vehicle is used to carry extra equipment, which might be required for serious emergencies and in addition can being used in rescue operation. Depending on the nature of risk the following list of equipment is to be carried. Suitable compartments should be made available for such equipment and be so arranged, not to obstruct their removal. The engine, chassis, cabin, and superstructure should meet the specifications given in section one.

- fifty lengths of fire hoses 25 m with standard couplings
- ten lengths of fire hoses of 45 mm diameter × 25 m with standard couplings
- six sets of breathing apparatus back-pack type
- twelve Nos. of spare cylinders for above
- two sets of resuscitators with spare cylinders
- six shovels and 6 picks
- two cotton rope 100 m 1½ cm diameter
- sets of gaskets
- two water spray-jet branch pipe 600 LPM at 7 bar
- two portable foam/water monitors
- two collecting and two dividing breechings
- four air foam-making branch pipe 800 LPM
- a complete large size tool box
- ten red flags and road stoppers
- six firemen protective clothing
- six pairs rubber boots
- two gas alarms (transportable detector)
- five sets red beacon
- one first aid box
- two fire approach or entry suits
- two pairs of insulated rubber gloves
- two portable lighting sets (if required)
- two portable lightweight-pumping sets 300–500 LPM (if required) complete with suction hoses.

Special rescue equipment, which should be kept in separate compartment

1. oxy/butane or acetylene cutting tools kit.
2. winch to be provided in front bumper with 50 m pulling wire of 3–4 tons.
3. rescue equipment as specified.

## 9.25.2 FIRE AND EMERGENCY LIGHT VEHICLE

### 9.25.2.1 The vehicle
The vehicle should be of four-wheel drive and have suspension, power, and torque that can be used in rough roads and soft soil in temperature and altitude as specified and with the following details.

- the cabin should provide seating for driver plus one, with air conditioning and heating system
- the engine should be petrol driven well known with 150 L fuel tank
- towing connection mounted behind chassis for towing a trailer of 1½ ton
- manual transmission
- electrical system 2 × 12 V batteries with high efficiency alternator
- warning siren (95 dB) and one revolving beacon
- two spot lights with adjustable beam selection
- radio mounted on dashboard.

### 9.25.2.2 Fire-fighting system

The following first aid fire-fighting system should be mounted on the vehicle:

- fire pump 150–400 LPM at 7 bar
- water (premix) tank 400 L
- one 35 m hose reel terminated to foam-water fog nozzle.

### 9.25.2.3 Miscellaneous fire, safety, and emergency equipment

The following equipment should be carried:

| | |
|---|---|
| 45 mm × 25m standard fire hose | 4 |
| 12 kg dry powder fire extinguisher | 2 |
| Breathing apparatus | 1 |
| Oxygen resuscitator | 1 |
| Suction hose | 4 |
| Tool kit | 1 |
| Foam liquid 20 L container | 2 |
| Cutting axe | 1 |
| Crow bar | 1 |
| Road stop reflector | 1 |
| Reflective flag | 2 |
| Shovel | 1 |
| Rope 1½ cm | 20 m |
| Portable gas detector | 1 |
| Rubber boot | 1 pair |

## 9.25.3 FOAM/WATER HYDRAULIC BOOM

### 9.25.3.1 Chassis

The hydraulic boom should be mounted on an auxiliary chassis at the front or rear of the vehicle chassis and consist of an arm(s), a turntable, and a remote-operated foam water monitor mounted on the upper arm.

### 9.25.3.2 Arms

- The arms should be of profiled tubular steel in order to give the construction a minimum of mass with a maximum of strength and rigidity.
- All rotation points should, when necessary, be provided with grease-lubricated bronze bushes, the grease nipples being easily accessible.
- The lower arm should be mounted on the turntable.
- The movements of the boom should be controlled by hydraulic cylinders.

### 9.25.3.3 Turntable

The turntable should be of a robust steel plate construction. It should run on ball bearings to ensure smooth and continuous rotation of the whole structure. The turntable should operate electrically/hydraulically, but manual operation should also be possible. All electrical equipment installed on the turntable should be suitable and certified intrinsically safe.

### 9.25.3.4 Auxiliary chassis

The boom and turntable should be mounted on an auxiliary steel chassis. The auxiliary chassis should be firmly fixed to the vehicle chassis.

To ensure stability of the boom, the auxiliary chassis should be fitted with four hydraulic jacks, which will relieve the vehicle chassis sufficiently when the boom is in use.

These jacks should consist of telescopic supports and have double-acting hydraulic cylinders for operation of the jacks.

The jacks should extend in pairs, within 20 s by means of manually operated hydraulic valves. These valves should be mounted in such a location that extension of the jacks should be clearly visible at all times.

The pressure pipes of the jacks should incorporate overload releases, which are set so that it is impossible to lift the vehicle chassis off its springs.

The jack cylinders should be provided with nonreturn valves, which will hydraulically lock the jacks when the boom is in use to ensure stability at all times.

### 9.25.3.5 Hydraulic and electrical system

The hydraulic pressure should be obtained through a hydraulic pump drive by a separate air-cooled diesel engine. The diesel engine should be equipped with an electric starting device, an exhaust fitted with a spark arrester of approved type and automatic overspeed protection, the fuel supply should be obtained from the vehicle's fuel tank.

The hydraulic of tank should be large enough to ensure that the oil temperature does not exceed the recommended manufacturer's temperature. During normal operation of the monitor, the hydraulic oil tank should be fitted with a temperature gage. The telescopic movement of booms and turntable should be controlled by means of hydraulic/electric valves installed at the turntable. These valves should regulate smooth speed of movement from zero to maximum. The required electrical energy to actuate the magnetic valves in the valve blocks should be obtained from the vehicle battery.

The hydraulically operated upper arm should be provided with an electronic control device, which allows simultaneous movement of all functions at the same time. The speed of the required movements should be obtained by actuating operating levers. The operating and control panel should be mounted on a dustproof junction box at the left-hand side of the vehicle and should be combined with the water/foam-operating panel. This panel should incorporate the following elements.

### 9.25.3.6 Operating and control panel of the boom

- main operating handles (speed control of all boom movements)
- levers for boom and turntable movements
- levers for jacking movements
- switch for the searchlight on the monitor.

### 9.25.3.7 Operating and control panel of the hydraulic pump unit

- pressure gage, oil system
- start button
- stop button
- control lamp oil pressure engine
- control lamp generator engine
- temperature gage engine
- hand throttle control.

### 9.25.3.8 Safety devices

- To ensure safe operation of the complete hydraulic circuit, a safety valve should be fitted in the pressure piping near to the pump.
- The hydraulic cylinders for the movements of the booms should be provided with "fracture" valves mounted on the cylinders.
- These valves should shut off the oil discharge from the hydraulic cylinders in the event of a pipe failure.
- Driving off with the jacks extended should not be possible. A red light on the dashboard should warn the driver when the jacks are extended. A green light should indicate when the jacks are fully retracted.
- Retracting the jacks should only be possible, when the boom is in its transport position.
- The hydraulic system should be designed so that violent or sudden operation of the levers will be balanced automatically and not affect the smooth movement of the boom.
- The boom should have electric switches or hydraulic valves, which should automatically block movement when the boom has reached its extreme position.
- The hydraulic system should incorporate a hand-operated pump, which will be able to bring the boom back to its transport position in the case of engine failure.
- The electric circuits should each be safeguarded by individual fuses.
- An emergency stop button should be fitted on the operating panel.
- The boom movements should be controlled by operation levers, which should also act as "dead man's switches."
- Warning lights should be mounted on top of the rear jacks.

### 9.25.3.9 Performance of boom and turntable

| | |
|---|---|
| Max. reachable height of monitor | At least 14 m |
| Time to reach max. height | Max. 40 s |
| Movement lower boom | 70 degree approximately |
| Movement upper boom | 90 degree approximately |
| Rotation | 180 degree on each side of the vehicle |

### 9.25.3.10 Water/foam monitor

The water/foam monitor should be mounted at the end of the upper boom. The water or foam solution supply to the monitor should be fed via stainless steel piping and rubber hoses and/or stainless steel rotation joints.

The monitor feed line should be connected to the water/foam system of the vehicle. A hand-operated ball valve should be situated in the monitor feed line.

It should be possible to feed the monitor directly from the hydrant by use of three extra inlets, complete with 6.30 cm ball valves, couplings and blind caps.

The water/foam monitor should be suitable for a straight stream of foam or water, oscillating type monitor up to 2000 LPM solution may be specified.

# 9.26 BRIEF DESCRIPTION AND LIST OF PROPOSED TYPES OF FIRE-FIGHTING TRUCKS

The following fire truck specifications can also be used in high-risk areas such as oil refineries, but the suitability generally depends on availability of fixed fire-fighting system provided. However careful study should be made for selection of the trucks.

Detail specifications for chassis, engine, cabins, electrical system drive, superstructure, brake, cabin, and booster pumps, arc as given in Section 1. And foam-water-dry chemical extinguishers, premix and twin agents in Section 2.

## 9.26.1 TYPES OF FIRE-FIGHTING TRUCKS

---

Truck No. 1: foam-water system + dry powder extinguisher

Booster pump 4000–6000 LPM at 7 bar

Foam liquid tank 1000 L

Fixed monitor 2000 LPM at 10 bar

Dry powder fire extinguisher $2 \times 500$ kg

---

Truck No. 2: foam-water-dry chemical system

Booster pump multi pressure 3000 LPM at 7 bar;

Foam liquid tank 5000 L;

Fixed monitor 2000 LPM at 10 bar;

Dry chemical fire extinguisher twin tank 500 kg each

---

Truck No. 3: crew compartment-foam-water-twin agent systems

Booster pump multi pressure 3000 LPM at 7 bar;

Foam liquid tank 3000 L;

Fixed monitor 2000 LPM at 10 bar;

Dry powder 250 kg with hose reel gas expelled

Premix 250 L with hose reel

---

| Truck No. 4: crew compartment-foam-water-dry powder | |
|---|---|
| Booster pump multi pressure 3000 LPM at | 7 bar |
| Foam liquid tank | 600 L |
| Water tank | 3600 L |
| Dry powder | 500 kg |
| Hose reel (water) | 2 Nos. |
| Hose reel (powder) | 1 No. |

| Truck No. 5: crew compartment-foam-water-premix systems | |
|---|---|
| Booster pump multi pressure 3000 LPM at | 7 bar |
| Foam liquid tank | 400 L |
| Water tank | 2500 L |
| Premix[a] | 500 L |
| Hose reel (water) | 2 Nos. |

[a]*Foam liquid for premix to be of AFFF-FFFP or alcohol resistance type as specified.*

| Truck No. 6: crew compartment foam-water + twin agent system | | |
|---|---|---|
| Booster pump multi pressure | 3000 LPM at 7 bar | |
| Foam liquid tank | 1000 L | |
| Water tank | 3000 L | |
| Twin agent | Water/premix 500 L | |
| | Dry powder 500 kg | Hose reel applicator gas expelled |
| Water hose reel | 2 | |

| Truck No. 7: foam-water system |
|---|
| Booster pump 4000 6000 LPM at 7 bar |
| Foam liquid tank 4000 L |
| Fixed monitors 2000 LPM at10 bar hydraulic or manual |

| Truck No. 8: dry chemical system 4000 kg |
|---|
| 1000 kg dry powder tanks 4 Nos. (in four stages) |
| Dry powder monitor |
| Two hose reels |

| Truck No. 9: dry powder premix twin agents | |
|---|---|
| Dry powder 2000 kg | In 4 stages |
| Premix (AFFF) 2000 L | In 4 stages |
| Twin agent monitor | 1 |
| Hose reel (dry powder) | 1 |
| Hose reel (premix) | 1 |

Truck No. 10: foam dispensing water/foam system

    Foam liquid tank 8000 L

    Foam pump 400 LPM (as specified in Section 1 Clause 7.6.4)

    Foam monitor manual (fixed) 2000 LPM at 10 bar

| Truck No. 11: dry powder 3000 kg in 4 stages | |
|---|---|
| Dry powder monitor | 1 |
| Dry powder hose reel | 2 |

| Truck No. 12: dry powder 3000 kg in 3 stages | |
|---|---|
| Dry powder monitor | 1 |
| Hose reel | 2 |

Truck No. 13: water, foam or premixed system

    Water pump 3000 LPM at 7 bar

    Water or premix tank 2000 L

    Foam tank 2000 L

    Hose reels 2

Truck No. 14: water-foam-twin agent system

    Water pump 3000 LPM at 7 bar

    Foam liquid tank 2000 L

    Twin agent 2 × 250 kg dry powder and 2 × 250 L premix

## 9.27 FIRE FIGHTING PORTABLE, TRAILER, SKID, AND FIXED MOUNTED PUMPS

This standard should apply to portable light weight and trailer skid mounted and fixed pumping units used for fire suppression and other emergency activities. Light weight, trailer, and skid mounted can be carried or towed to the site where they are to be used. This standard establishes requirements for operational characteristics, reliability and service performance of all different types of pumping units, so that a high degree of usefulness to the fire and emergency services can be achieved.

Before purchasing, the client should evaluate the needs and uses of these specifications to assure that the units to be purchased will be equipped in the best way to meet this standard.

### 9.27.1 CLASSIFICATION OF PUMPING UNITS

#### 9.27.1.1 Portable light-weight pumping units

1. *Small volume relatively high pressure.* This pumping unit should be capable of pumping 80 LPM at 13 bar through hose reel of 25 mm and of 4 cm suction inlet.
2. *Medium volume-medium pressure.* This pumping unit should be capable of discharging 250 LPM at 6 bar and 500 LPM at 4 bar of 4 cm discharge outlet while taking suction through 6 cm suction inlet.
3. *Large volume-relatively low pressure.* This pumping unit should be capable of supplying 500 LPM at 4 bar and 1100 LPM at 1½ bar through 6.50 cm discharge outlet while taking suction in 7.60 cm suction inlet.

#### 9.27.1.2 Performance

Portable pumping units should be capable of delivering and maintaining capacity and pressure as specified in this standard by pumping water at any altitude specified above sea level from 3 m lift through 6 cm suction hose with strainers.

The weight of a complete pumping unit including carrying handles and all other components, excluding fuel oil and accessories should not exceed 80 kg for large volume and 70 kg for medium and small.

When starter motor, generator/alternator, and battery are furnished, the weight of such equipment should be added to the weight as specified in the above paragraph.

Centrifugal type pumps with self priming should be used. The body of the pump should be of bronze and stainless steel shafts with reliable seal of carbon or similar should be used. The engine should be of four cycles, easy to start and air cooled and the stop switch should be of nonlocking type. It may be desirable to have the pumping unit mounted on one or, two wheeled dolly so that it can be transported over a considerable distance by one person.

### 9.27.2 PUMP, PUMP CONNECTIONS, AND FITTINGS

Pump should be of the centrifugal type

#### 9.27.2.1 Exception

Pump may be of a type other than centrifugal when specified by the purchaser under special provisions.

Discharge connections of all portable pumping units should be equipped with instantaneous fire hose couplings and suction connections Storz or threaded as specified.

Caps with chains or cables and suitable gaskets should be provided for each suction and discharge connection.

An adapter should be provided that will permit attachment of 3.8 cm hose couplings when the discharge connection is 2.5 cm in size.

An adapter should be provided that will permit attachment of (6.3 cm) hose couplings when the discharge connection is 3.8 cm in size.

Suitable means should be provided for completely draining the pump and its attachment in cold weather.

Pump body should be capable of withstanding a hydrostatic pressure 7 bar (689.5 kPa) above the rated operating pressure.

The pump casing should be capable of being easily disassembled for inspection and replacement of parts.

Pump impeller should be constructed of cast iron, bronze, stainless steel, or copper–nickel alloy.

Pump shaft should be constructed of stainless steel or copper-nickel alloy.

### 9.27.2.2 Priming device

A priming device should be provided. The device should be capable of priming the pump at 3.05 m lift through 6.1 m of suction hose and through the suction inlet within 30 s.

The priming device should be capable of making a vacuum of 43.18 cm of mercury at altitude up to 610 m above sea level.

### 9.27.2.3 Engine

The engine should be air cooled.

If the engine is subject to pump thrust, roller thrust bearing should be provided.

Engine electrical components and ignition exposed to the weather should be protected from water.

A rope, crank, or other manual starter should be provided.

### 9.27.2.4 Exception

An electric starter, starter/generator starter/alternator should be furnished in addition to manual starter, when specified.

A manually adjustable automatic speed control should be furnished that will automatically adjust engine throttle as necessary to maintain the engine speed, and should limit the engine speed load, to the engine manufacturer's recommended minimum speed. The speed controller should hold its position when the engine is operating unattended.

A suitable muffler should be furnished.

A nonlocking stop switch, to stop the engine should be furnished in a readily accessible location.

### 9.27.2.5 Fuel tank

One or more fuel tanks should be furnished.

The fuel tank(s) should be of sufficient size to meet operation of the pumping unit at rated capacity and pressure for at least 2 h without refilling.

### 9.27.2.6 Trailer pumping unit

The pumping unit should be capable of delivering and maintaining capacity and pressure as specified below:

1. Pumping capacity 2000 LPM minimum pressure 7 bar.
2. Pumping capacity 3000 LPM minimum pressure 7 bar.
3. Pumping capacity 4000 LPM minimum pressure 7 bar.

The operating pumping capacity should be ¾ of the above at 10 bar and 2/3 at 15 bar respectively

Unless otherwise specified gasoline driven or diesel engine should be specified in the purchasing order, but diesel engine as prime mover for 4–6 cylinders is preferred.

Centrifugal pumps with automatic priming suitable for tropical sandy weather and salty/dirty water should be considered.

Bronze pump body and impeller, and high-grade stainless steel or copper alloy shaft with carbon seal should be provided.

### 9.27.2.7 Suction and discharge manifold

Suction for type (a) to be of 10 cm diameter with round thread or Storz couplings whichever specified. There should be two delivery connections with instantaneous 6.5 cm female coupling with quick opening valve.

For type (b) and (c), suction inlet should be of 12.5 cm, round thread or Storz couplings whichever specified and there should be four delivery connections with instantaneous female valve of quick opening type.

Suction inlets should have removable or accessible strainer provided inside each external inlet.

All suction inlets should be provided with suitable closures, inlet having male threads should be equipped with caps.

All 6.5 cm outlets should be equipped with valves, which can be opened and closed smoothly and readily at any rated pressure. The flow regulating element of each valve should not change its position under any condition of operation involving discharge pressures to 15 bar, the means to prevent a change in position should be incorporated bar in the operating mechanism and may be manually controlled or automatic. Each discharge valve should be equipped with a drain or bleed-off valve with a minimum 2 cm pipe thread connection for draining or bleeding off pressure from a hose connected to the valve.

### 9.27.2.8 Engine controls

A hand throttle, controlling the fuel supply to the engine and of a type that will hold its set position, should be so located that it can be manipulated from the operator's position with all gages in full view.

When a supplementary heat exchange cooling system is provided, proper valving should be so installed as to permit use of water from the discharge side of the fire pump for the cooling of coolant circulating through the engine cooling system without intermixing.

### 9.27.2.9 Gages and instruments

A pump suction gage should be provided on the left hand side of the gage panel. It should be not less than 10 cm in diameter, and it should read from 75 cm Hg vacuum and from 0 to 20 bar but not more than 40 bar pressure.

A pump discharge pressure gage should be provided located to the right of the suction gage as specified. It should be not less than 10 cm in diameter of a type not subject to damage by vacuum, and should read from zero to not less than 20 bar but not more than 40 bar pressure.

All gages should have 10 mm pipe thread connections and should be mounted so that they are readily visible at the pump operator's position, and so that they are not subject to excessive vibration. They should be suitably enclosed or otherwise protected.

### 9.27.2.10 Priming system

Priming system should be capable of developing vacuum of 70 cm Hg at sea level. An engine pressure and temperature gage should be provided at the pump operating position.

### 9.27.2.11 Starting

Battery starting and alternator charger should be provided. A suitable battery charger should also be provided to keep the engine battery ready for starting.

### 9.27.2.12 Frame

Trailer pump frame should be of stainless steel with four adjustable handles and independent suspension complete with jocky wheel and twin rear mounted probe stand and standard 32.5 cm wheels with mud guards should be supplied.

One detachable flood light with $2 \times 100$ W bulbs and a tripod should be provided.

Sufficient stowage capacity for $4 \times 2$ m length suction hoses together with two delivery hoses, nozzles, spanners and suction strainer etc. should also be provided.

### 9.27.2.13 Instrument

Pump-operating panel should be fitted on the pump at operator's position with the following instruments.

- pump compound suction pressure gage
- pump delivery pressure gage
- engine water temperature gage
- engine oil pressure gage
- battery condition meter
- hours run meter
- revolution per minute (rpm).

### 9.27.2.14 Fuel tank

Fuel tank should be of stainless steel demountable type with capacity for 2 h full load fuel consumption.

## 9.27.3 FIXED PUMPS

Where firewater main is provided with hydrant for fire-fighting operations, fire pumps of sufficient capacity and pressure should be installed to pressurize the water main. The capacity of the pumps depends on the area and risk of major fire. However, 6000 LPM at 10 bar for each pump is the minimum requirement.

Refineries or a larger plant with different multi units are usually subdivided into smaller sections, and each section has a water ring mains with block valves. These valves can be opened to serve water to unit(s) involved in fire.

At least two identical pumps taking suction from open water or storage should be installed; in each zone one electric motor-driven and one diesel engine-driven, the latter serving as a spare.

The power of the drives, both the electric motor and the spare unit should be rated such that it is possible to start these units against an open discharge system which may be pressurized to 3 bar.

The electric motor should be provided with an automatic starting device, which will act immediately after putting the fire alarm system into operation or set water pressure.

The spare unit should be provided with automatic starting facilities, which will act immediately if the electric motor of the original pump is out of order or volume and pressure of water are not sufficient when electric pump is in operation during fire fighting.

Manual starting and stopping of each unit should be possible from a control centre or from the fire station if the latter is permanently manned; it should always be possible at the pump site. Manual starting should be possible without the fire alarm coming into operation.

For a diesel engine, the following additional requirements should be adhered to:

- The capacity of the fuel tank should be such that the engine can operate at full power for at least 24 h. Heat tracing provision should be considered for fuel tank.
- This tank should be installed at such a level that the bottom is at least 0.2 m above the suction valve of the diesel injection pump. The tank should be provided with a level gage and facilities for refilling direct from drums.
- No clutch should be installed between diesel engine and pump.

### 9.27.3.1 Pump connections and facilities

At the suction side of the pumps common strainer facilities should be provided which should be easy to clean.

The discharge line of each pump should be fitted with a check valve, a test valve, a pressure gage, and a block valve with locking device. Each pump should be connected separately to the ring main line.

The test valve should be used for pump testing and be so sized that it will allow a minimum flow of 10% of the maximum pump capacity.

In cases where the pumps are located at a considerable distance from the water distribution system, for example, at a jetty approach, consideration should be given to installing one discharge line only and making provisions for installing, if necessary, a new line in the future or replacing the discharge spool piece after installing a 90 degree elbow.

### 9.27.3.2 Pressure regulation

In order to maintain a system pressure of 10 bar at the most remote location under full flow conditions, the actual discharge pressure of the fire-fighting pumps should normally be well above this figure. If fire-fighting water is required close to the pump area, the high discharge pressure in this area may cause an unsafe situation for personnel handling the hoses or may overstress the fire hoses proper. To keep this pressure within acceptable limits a pressure indicator controller should be installed at the common discharge of the pumps in order to enable adjustment to the pressure required. Local control as well as central panel control is required.

In order to keep the system full of water and permanently under pressure when not in operation two alternatives are possible:

1. A permanent connection to the plant cooling water system should be considered.
2. An electric motor-driven jocky pump. A permanent pressure of approximately 3 bar should be maintained by this pump, which should have a capacity of about 150–180 LPM.

In both cases the pressurizing facilities should be protected against the discharge pressure of the fire-fighting pumps by means of a check valve.

## 9.27.4 **SKID MOUNTED PUMPS**

For oil well fire-fighting operations the most reliable equipment are needed. One of the prime needs to combat the fire is availability of adequate spray water fog to be used for cooling the equipment left at close proximity of radiant heat.

In most cases the source of water is far away from the oil well location and therefore, pipe lines should be laid down from a high pressure water pumping station to the near side of the fire where a large (20000 m$^3$) water pond is erected for water storage. Close to the water storage, a firewater pumping station should be set to feed the firewater mains laid at strategic points. The fire pump requirements are:

### 9.27.4.1 *Pumping units*
Two diesel operated pumps, each with minimum capacity of 12000 LPM at 10 bar is required with the following specifications.

The pumps to be of centrifugal double stage with the most reliable preferably built of cast steel or bronze impeller and stainless steel high tensile shaft with the capacity of 12000 LPM at 10 bar and 30 cm diameter suction and 20–25 cm discharge delivery.

The suction should be of steel pipe terminated to a nonreturn foot valve with strainer.

The engine should be of super charger diesel engine, electric battery starter of 24 V or air pressure start with alternator with sufficient output for charging batteries and minimum of 500 W lighting including flood lights.

The following instrument should be fixed on operating panel of each pumping unit:

1. pressure gage
2. engine oil pressure
3. engine oil temperature
4. engine cooling system temperature
5. start and stopping switches
6. engine tachometer
7. engine speed lever with overspeed protection.

### 9.27.4.2 *Fuel and water storage*
Two overhead tanks in separate locations should be fixed as:

1. Fuel tank with sufficient capacity for 24 h continuous operation of two pumping units.
2. Water storage of 2000 L for pump priming system.
3. A flood light of 24 V-60 W Halogen should be provided over each pumping unit.
4. A walky talky radio should be available to communicate with operational site.

## 9.28 **MATERIAL PROCUREMENT STANDARD**

The purchase of new fire trucks involves a major investment and should be treated as such, thus a large measure of uniformity throughout the procurement should be achieved with due consideration to its economic. The appropriate officials should consult transport authorities and experienced engineers to make thorough study of the needs before a purchase order is processed.

This specification is designed to render access for procurement of various types of fire trucks and pumps as referred in Sections 1–5.

The tests of equipment are important features, which are required to assure that the materials will meet the specified standard. An appropriate list of equipment is prepared and included in the standard specifications, which specifies the relevant fire equipment requirements for each type of fire trucks. Such portable equipment should not be included in a purchasing order except for fire-fighting trucks purchases for new refineries or new fire stations in high-risk areas.

## 9.28.1 CHECKLIST FOR SPECIFICATION PURPOSES

The checklist should be used by the purchaser to ensure that a complete specification of the type of vehicle required will be given to the manufacturer.

- Type and Number of the Vehicle—Country and area of destination.
- Climatic Conditions

| | |
|---|---|
| Minimum local temperature | °C |
| Maximum local temperature | °C |
| Dust area | Yes/no |
| Humidity and corrosive nature | |
| Altitude from sea level. | |
| Tropical subtropical | |

- Chassis
  - make and type of towing hook
  - type of socket for trailer brake
  - type and voltage of connection for lighting of trailer
  - spare wheel and tools: yes/no
  - maximum permissible weights: total weight-axle load.
- Engine
  - gasoline b diesel b
  - on-off buzzer when vehicle is reversing yes/no
  - PTO operating from cabin/at the operating panel manually b hydraulic b
  - electric heating element in the cooling system: yes/no.
- Drivers Cabin
  - tilted b yes/no
  - additional seats
  - air conditioning: yes/no
  - max. head room of vehicle: m.

### 9.28.1.1 Superstructure
- cabinets closed by: rollers/doors
- water hose reels on both sides: yes/no—front/rear.

### 9.28.1.2 Electrical systems
- battery charger: yes/no
- electric supply: 220–240 V, 50 Hz
- color of revolving beacons: red/blue/yellow
- type of cable connection for mobile phone at rear
- tires description.

### 9.28.1.3 Extended cabin
- crew compartment: yes/no
- number of seats in the crew compartment
- make and type of air breathing apparatus
- location and number of suspension.

### 9.28.1.4 Line-up of water/foam
- suction hose couplings: Storz/others
- size of suction couplings: 65/80/125/150 mm.

### 9.28.1.5 Foam/water monitor
- manual-hydraulic
- mounted at rear b over the drivers cab b or other
- max. allowable vehicle passage height: m
- water capacity at 10 bar, LPM
- discharge and throw trajectories to comply with standard.

### 9.28.1.6 Extinguishing powder installation
- powder monitor: yes/no
- capacity of powder monitor : kg/s
- powder delivery manifold: yes/no
- number of delivery connections
- hose couplings on powder manifold: Storz/others
- size of hose couplings
- pressurizing of powder tanks: annually/electrically/pneumatically
- capacity of powder pistol: kg/s
- user's language on operating panel: English/others
- type of powder: monnex/(purple.k) or others.

### 9.28.1.7 Operating panels
- language for identification instruction: English/others

### 9.28.1.8 Painting and coating
- sign writing details to be supplied

### 9.28.1.9 Additional equipment
- requirements should be selected from the list or otherwise specified

### 9.28.1.10 Initial fills of chemicals

- make and type FLC
- quantity/capacity: L
- make and type dry powder
- quantity dry powder: kg.

### 9.28.1.11 Operating manual

- language for instruction books: English/others

### 9.28.1.12 Requisition for trailer

- type of trailer
- voltage of lightings
- height of tow bar from ground level: m
- make and type of tow eye
- make and type of socket for trailer brake
- make and type for lighting socket.

### 9.28.1.13 Performance testing

- road test: 300 km /1500 km
- rough track test : yes/no
- tilt test: yes/no
- other test requirements.

### 9.28.1.14 Quotation requirements

The supplier should include the following information with the quotation:

- technical specification
- vehicle lay-out and arrangement drawing, including a top view
- water foam and dry powder flow schemes
- list of all makes and types of equipment which is purchased from other manufacturers. For each item the manufacturer's documentation on the purchasing specification should be included
- copies of the certified performance curves of the booster pump
- copies of the performance curves of the foam pump
- certified performance data for the monitors and hand nozzles
- test certificate of the dry powder hose
- list of all proposed deviations from the specification and, where applicable, supported with reasons for the deviations
- detailed loading calculations for front and rear axle
- proposed performance testing on the basis of specifications
- list of the proposed color coding for electric wiring
- spare wheel location, if applicable
- list of recommended spare parts for two years' operation
- advice on chassis requirements
- program of in-house quality control during assembly/ construction.

### 9.28.1.15 Miscellaneous

Inspection should include but not be limited to the following:

- welding requirements, in accordance with Standard Specification
- material certificates, DIN50049, etc.
- pump and monitor castings
- all material to be in accordance with the specifications and approved design drawings
- pipe schedule and flange rating
- pressure testing of equipment
- relief valve settings, to be as specified.

Pressure vessels constructed to a design code will be accepted on the evidence of certificates signed by the approved pressure vessel inspecting authority.

### 9.28.1.16 Surveillance during assembly of the vehicle

Surveillance should include but not be limited to the following:

- the manufacturer should check and accept the chassis in accordance with the purchasing specification. Any deviations should be reported to the client's inspector within 6 days after arrival of the chassis, in such cases work should not be allowed to commence without the agreement of the inspector
- Review
- the manufacture's in-house quality control program
- dimensional check
- pump alignment
- piping arrangement/hook-up/couplings
- coating and painting application
- lighting
- electrical installation and cabling
- marking, identification and nameplates
- weatherproofing
- drainage, overflow
- completeness of the vehicle and systems
- additional equipment
- spare parts and special tools
- checking documents and manuals
- visual inspection
- ergonomics and accessibility.

### 9.28.1.17 Performance testing

The following performance test and checks should be carried out.

- Road test—A road test with the fully loaded vehicle over a distance of 300 km on an average type of road. A representative of the chassis supplier should attend this test
  - A brake test in accordance with company's requirements
  - A 2-h rough track test

- A road test of 1500 km followed by a full service-adjustment by the chassis supplier
- When specified in the purchase order

After the road test a second hydrostatic test should be carried out followed by the equipment performance tests as follows.
- Pump balance

During the shop test of pumps with antifriction bearings, operating at rated speed or at any other speed within the specified operating range, the maximum allowable unfiltered root mean square vibration velocity measured on the bearing bracket in any plane with an instrument in accordance with ISO 2954-1975 (E) should not exceed the following value:

| Flow range in % of flow at Best Efficiency Point (BEP) | rms vibration velocity in mm/s |
|---|---|
| 25–49 | 4.5 |
| 50–110 | 3 |

- hydrostatic test of the total system
- calibration of each proportioner
- flexibility test

With a block of 200 mm under one front wheel and the opposite rear wheel, there should be no movement of the cab on the chassis, the lockers should function without restriction and the complete pumping system should be fully operational without additional vibration. The clearance height in the wheel guards during the torsion test above should be at least 50 mm unless otherwise stated by the chassis supplier.

- tilt test only when specified in the purchase order
- measure the wheel loading and deviation from horizontal when fully loaded including the weight of a full crew
- pump performance test
- at 1.5 m suction lift
- at 3.0 m suction lift or at 6.0 m suction height from hydrants with 6–8 bar inlet pressure via the booster pump
- full load pump test for 1 hour uninterrupted
- foam proportioning test
- monitor movement
- water/foam monitor: capacity and throw
- priming time: of the dry powder system
- dry powder gun: capacity and throw
- dry powder monitor: capacity and throw
- dry powder charging time
- quality of produced foam
- hose reels (including rewind mechanism and overrun brakes)

- fog guns
- any other test that may be specified in the requisition.

Certain tests can possibly be waived when equipment has already a "Type approval."

### 9.28.1.18 Documentation

The following certification and documents should be prepared and dispatched before acceptance for shipment:

- water/foam and dry powder flow schemes
- list of all makes and types of equipment which is purchased from other manufacturers. For each item the manufacturer's documentation or the purchasing specification should be included
- copies of the certified performance curves of the booster pump
- copies of the performance curves of the foam pump
- certified performance data for the monitors and hand nozzles
- test certificate of the dry powder hose
- list of all proposed deviations from the specification and, where applicable supported with reasons for the deviations
- detailed loading calculations for front and rear axle
- proposed performance testing on the basis of specifications
- list of the proposed color coding for electric wiring
- spare wheel location, if applicable
- list of recommended spare parts for 2 years operation
- advice on chassis requirements
- program of in-house quality control during assembly/construction.

### 9.28.1.19 Operational instruction and maintenance manuals

Manufacturer should supply five copies of instruction and maintenance manual including trouble shooting instructions with each fire truck and pumping units together with recommended spare part list for 2 years' operation.

### 9.28.1.20 Acceptance tests

Acceptance tests on behalf of purchaser should be as prescribed in purchasing order conducted within 10 days before delivery of each unit in the presence of such person or persons as the purchaser designates in the requirement for delivery. The tests requirement should be conducted in accordance with Chapter 11 1901 N.F.P.A (automotive fire apparatus)

Any inspection and testing in no way relieve the manufacturer of any responsibility for any fire-fighting truck or fire pumps meeting all requirements of this specification and applicable codes.

### 9.28.1.21 Guarantees

Manufacturer should guarantee by letter of acceptance the satisfactory performance of the fire truck and pumping units in accordance with this specification. The manufacturer should also guarantee to replace without charge any or all parts defective due to faulty material, design or poor workmanship for the period of 18 months after shipment.

### 9.28.1.22 Shipping

Adequate shipping support should be provided in order to prevent damages during transit. Provision should be taken to protect the truck and equipment from possible marine exposure.

## 9.29 FIREWATER DISTRIBUTION AND STORAGE FACILITIES

This chapter specifies minimum requirements for water supply for fire-fighting purposes. It is important that all authorities concerned should work together to provide and maintain these minimum water supplies and discussions with municipality fire stations would include not only the water available from the hydrants but also help to assure the continuous and adequate flow of water for fire fighting.

The following items are also included in standards:

- basic for a fire-fighting water system
- firewater pumping facilities
- water tanks for fire protection
- fire hose reel (water) for fixed installation
- water spray fixed system.

Water is the most commonly used agent for controlling and fighting a fire, by cooling adjacent equipment and for controlling and/or extinguishing the fire either by itself or combined as a foam. It can also provide protection for firefighters and other personnel in the event of fire. Water should therefore be readily available at all the appropriate locations, at the proper pressure and in the required quantity.

Firewater should not be used for any other purpose.

Unless otherwise specified or agreed, the company requirements, which are given for major installations such as refineries, petrochemical works, crude oil production areas where large facilities are provided, and for major storage areas should be applied.

In determining the quantity of firewater, that is, "required firewater rate," protection of the following areas should also be considered:

- general process
- storage (low pressure), including pump stations, manifolds and in line blenders, etc.
- pressure storage (LPG, etc.)
- refrigerated storage (LNG etc.)
- jetties
- loading
- buildings
- warehouses.

Basically, the requirements consist of an independent fire grid main or ring main fed by permanently installed fire pumps taking suction from a suitable large capacity source of water such as storage tank, cooling tower basin, river, sea, etc. The actual source will depend on local conditions and is to be agreed with the company.

The water will be used for direct application to fires and for the cooling of equipment. It will also be used for the production of foam.

### 9.29.1 **PUBLIC WATER SYSTEMS**

One or more connections from a reliable public water system of proper pressure and adequate capacity furnishe a satisfactory supply. A high static water pressure should not, however, be the criterion by which the efficiency of the supply is determined.

If this cannot be done, the post indicator valves should be placed where they will be readily accessible in case of fire and not liable to injury. Where post indicator valves cannot readily be used, as in a city block, underground valves should conform to these provisions and their locations and direction of turning to open should be clearly marked.

Adequacy of water supply should be determined by flow tests or other reliable means. Where flow tests are made, the flow in LPM together with the static and residual pressures should be indicated on the plan.

Public mains should be of ample size, in no case smaller than 15 cm (6 in.).

No pressure-regulating valve should be used in water supply except by special permission of the authority concerned. Where meters are used they should be of an approved type.

Where connections are made from public waterworks systems, it may be necessary to guard against possible contamination of the public supply. The requirements of the public health authority should be determined and followed.

Connections larger than 50.8 mm to public water systems should be controlled by post indicator valves of a standard type and located not less than 12.2 m from the buildings and units protected.

## 9.30 **BASES FOR A FIRE-FIGHTING WATER SYSTEM**

A ring main system should be laid around processing areas or parts thereof, utility areas, loading and filling facilities, tank farms and buildings whilst one single line should be provided for jetties and a fire-fighting training ground, complete with block valves and hydrants.

The water supply should be obtained from at least two centrifugal pumps of which one is electric motor-driven and one driven by a fully independent power source, for example, a diesel engine, the latter serving as a spare pump.

The water quantities required are based on the following considerations:

There will be only one major fire at a time. As a recommendation in processing units the minimum water quantity is 200 dm$^3$/s or air foam making and exposure protection. It is assumed that approximately 30% of this quantity is blown away and evaporates; the balance of this quantity, which is 140 dm$^3$/s per processing unit, should be drained via a drainage system.[b]

For storage areas the quantity needed for making air foam for extinguishing the largest cone roof tank on fire and for exposure protection of adjacent tanks.

For pressure storage areas the quantity needed for exposure protection of spheres by means of sprinklers.

---

[b] The quantity of firewater required for a particular installation should be assessed in relation to fire incidents which could occur on that particular site, taking into account the fire hazard, the size, duties, and location of towers, vessels, etc. The firewater quantity for installations having a high potential fire hazard should normally be not less than 820 m$^3$/h and no greater than 1360 m$^3$/h.

For jetties the quantity needed for fighting fires on jetty decks and ship manifolds with air foam as well as for exposure protection in these areas.

The policy for a single major fire, or more, to occur simultaneously should be decided upon by the authorities concerned.

*Note:* The above specification is based on one major fire only.

For new installations the quantities required items mentioned above should be compared, and the largest figure should be adhered to for the design of the fire-fighting system.

The system pressure should be such that at the most remote location a pressure of 10 bar can be maintained during a water take-off required at that location.

Fire-fighting water lines should be provided with permanent hydrants.

Hydrants with four outlets should be located around processing units, loading facilities, storage facilities for flammable liquids, and on jetty heads and berths.

Hydrants with two outlets should be located around other areas, including jetty approaches.

Fire hose reels should be located in each process unit, normally 31–47 m apart at certain strategic points.

The water will be applied by means of hose and branch pipes using jet, spray or fog nozzles, or by fixed or portable monitors preferably with interchangeable nozzles for water or foam jets.

## 9.31  FIREWATER RING MAIN SYSTEM

Firewater ring mains of the required capacity should be laid to surround all processing units, storage facilities for flammable liquids, loading facilities for road vehicles and rail cars, bottle filling plants, warehouses, workshops, utilities, training centers, laboratories, and offices. Normally, these units will also be bounded by service roads. Large areas should be subdivided into smaller sections, each enclosed by fire water mains equipped with hydrants and block valves.

A single firewater pipeline is only acceptable for a fire-fighting training ground. Firewater to jetties should be supplied by a single pipeline provided that it is interconnected with a separate pipeline for water spray systems. The firewater pipelines from the fire pumps to the jetty should be provided with isolating valves, for closing in the event of serious damage to the jetty. These valves should close without causing high surge pressures.

The firewater mains should be provided with full bore valved flushing connections so that all sections and dead ends can be properly flushed out. The flushing connections should be sized for a fluid velocity in the relevant piping of not less than 80% of the velocity under normal design conditions but for not less than 2 m/s.

Firewater mains should normally be laid underground in order to provide a safe and secure system, and which will give in addition, protection against freezing for areas where the ambient temperature can drop below 0°C. When in exceptional circumstances, firewater mains are installed above ground they should be laid alongside roads and not in pipe tracks where they could be at risk from spill fires.

The basic requirements consist of an independent fire grid main or ring main fed by permanently installed fire pumps. The size of ring main and fire pumps should be such as to provide a quantity of water sufficient for the largest single risk identified within the overall installation.

Suction will be from a suitable large capacity source of water such as storage tank, cooling tower basin, river, sea, etc. The actual source will depend on local conditions and is to be investigated. Pump suction lines should be positioned in a safe and protected location and incorporate permanent, but easily cleanable strainers or screening equipment for the protection of fire pumps.

Advantage should be taken where available in obtaining additional emergency water supplies through a mutual aid scheme or by re-cycling but mandatory national or local authority requirements may modify these to a considerable extent.

## 9.32 FIREWATER RING MAIN/NETWORK DESIGN

The firewater mains network pipe sizes should be calculated and based on design rates at a pressure of 10 bar gage at the take-off points of each appropriate section, and a check calculation should be made to prove that pressure drop is acceptable with a blocked section of piping in the network. The maximum allowable flow/velocity in the system should be 3.5 m/s.

Firewater rates should, however, be realistic quantities since they determine the size of firewater pumps, the firewater ring main system and the drainage systems which have to cope with the discharged firewater. If the drainage system is too small or becomes blocked, major hazards such as burning hydrocarbons floating in flooded areas may occur to escalate the fire. Facilities for cleaning should therefore be provided. For large areas such as pump floors, and in pipe tracks, fire stops should be provided to minimize the spillage area. It is assumed that 30% of firewater evaporates or is blown away while extinguishing a fire. This figure should be taken into account for the design of drainage systems.

Under nonfire conditions, the system should be kept full of water and at a pressure of 2–3 bar gage by means of a jockey pump, by a connection to the cooling water supply system, or by static head from a water storage tank. If a jockey pump is used, it should be "spared" and both pumps should have a capacity of 15 $m^3$/h to compensate for leakages.

The firewater ring main systems should be equipped with hydrants. A typical arrangement of a firewater distribution system is shown in Fig. 9.8.

A single water line connected to the ring main system should run along the jetty approach to the jetty deck. This line should be fitted with a block valve located at a distance of about 50 m from the jetty deck.

For small chemical plants, depots, and minor production and treatment areas, etc. for which precise commensurate with the size of risk involved, requirements should be as specified or agreed with the authorities.

## 9.33 FIREWATER PUMPING FACILITIES

Firewater should be provided by at least two identical pumps, each pump should be able to supply the maximum required capacity for a fire water ring main system. Fire- water pumps should be of the submerged vertical type when taking suction from open water, and of the horizontal type when suction is taken from a storage tank.

The firewater pumps should be installed in a location which is considered to be safe from the effects of fire and clouds of combustible vapor, and from collision damage by vehicles and shipping. They should for example, be at least 100 m away from jetty loading points and from moored tankers

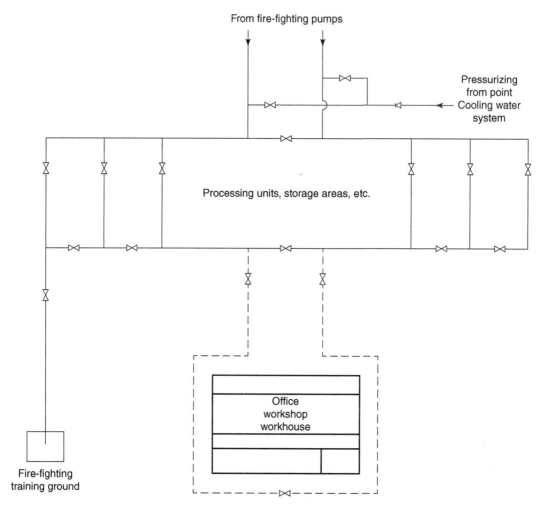

**FIGURE 9.8 Typical Sketch of Fire-Fighting Water Distribution System**

or barges handling liquid hydrocarbons. They should be accessible to facilitate maintenance, and be provided with hoisting facilities.

The main firewater pump should be driven by an electric motor and the second pump, of 100% stand-by capacity, by some other power source, preferably a diesel engine. Alternatively, three pumps, each capable of supplying 60% of the required capacity may be installed, with one pump driven by an electric motor and the other two by diesel engines.

Refinery with over 100 000 barrels a day capacity should have two electric and two diesel pumps.

When the required pump capacity should exceed 1000 m³, two or more smaller pumps should be installed, together with an adequate number of spare pumps. The power of the drives, for both main and standby units should be so rated, that it will be possible to start the pumps against an open discharge with pressure in the firewater ring main system under nonfire conditions, normally at 2–3 bar gage unless otherwise agreed by the relevant authorities. The main firewater pump should be provided with automatic starting facilities, which will function immediately the fire alarm system becomes operational due to one of the following actions:

1. when a fire call point is operated
2. when an automatic fire-detection system is operated
3. when the pressure in the firewater ring main system drops below the minimum required static pressure which is normally 2–3 bar (ga)
4. The stand-by firewater pumps should be provided with automatic starting facilities which will function If the main firewater pumps do not start, or having started, fail to build up the required pressure in the firewater ring main system within 20 s.

Manual starting of each pump unit (without the fire alarms coming into operation) should be possible at the pump, from the control center and, when necessary, from the gate house. Manual stopping of each pump unit should only be possible at the pump.

*Fuel Tank Capacity.* Fuel supply tank(s) should have capacity at least equal to 1 gal per horsepower (5.07 L/kW), plus 5% volume for expansion and 5% volume for sump. Larger capacity tanks may be required and should be determined by prevailing conditions, such as refill cycle and fuel heating due to recirculation, and be subject to special conditions in each case. The fuel supply tank and fuel should be reserved exclusively for the fire pump diesel engine.

The pumps should have stable characteristic curves exhibiting a decrease in head with increasing capacity from zero flow to maximum flow; a relatively flat curve is preferred with a shut off pressure not exceeding the design pressure by more than 15%.

The total water supply within the refinery should be capable of supplying the maximum flow for a period of not less than four to 6 h, consistent with projected fire scenario needs. Where the water system is supplied from a tank or reservoirs, the quantity of water required for fire protection. However, where the tank or reservoir is automatically filled by a line from a reliable, separate supply, such as from a public water system or wells, the total quantity in storage may be reduced by the incoming fill rate.

## 9.34 PLANS

A layout plan should be prepared and approved in every case where new private fire service main is contemplated.

The plan should be drawn to scale and should include all essential details such as:

- Size and location of all water supplies.
- Size and location of all piping, indicating, where possible, the class and type and depth of existing pipe, the class and type of new pipe to be installed and the depth to which it is to be buried.
- Size, type, and location of valves indicate if located in pit or if operation is by post indication key wrench through a curb box. Indicate the size, type, and location of meters, regulators, and check valves.
- Size and location of hydrants, showing size and number of outlets and if outlets are to be equipped with independent gate valves. Indicate if hose boxes and equipment are to be provided and by whom.

- Sprinkler and standpipe risers and monitor nozzles hose reels to be supplied by the system.
- Location of fire department connections, if part of private fire service main system, including detail of connections.
- Location, numbers and size of fire-water pumps installed.

## 9.35 WATER TANKS FOR FIRE PROTECTION

Standards covers elevated tanks on towers or building structures; grade or below grade water storage tanks and pressure tanks.

The size and elevation of the tank should be determined by conditions at each individual property after due consideration of all factors involved. Where tanks are to supply sprinklers.

Whenever possible, standard sizes of tanks and heights of towers should be used as given in standards.

The capacity of the tank is the number of cubic meters available above the outlet opening. The net capacity between the outlet opening of the discharge pipe and the inlet of the overflow should be at least equal to the rated capacity. For gravity tanks with large plate risers, the net capacity should be the number of cubic meters between the inlet of the overflow and the designated low-water level line. For suction tanks, the net capacity should be the number of cubic meters between the inlet of the overflow and the level of the vortex plate.

The standard sizes of steel tanks are: 18.93, 37.85, 56.78, 75.70, 94.63, 113.55, 151.40, 189.25, 227.10, 283.88, 378.50, 567.75, 757.00, 1135.50 and 1892.50 $m^3$ net capacity. Tanks of other sizes may be built (according to NFPA-20).

The capacity of pressure tanks should be as approved by the authority concerned.

The standard sizes of wooden tanks are 18.93, 37.85, 56.78, 75.70, 94.63, 113.55, 151.40, 189.25, 227.10, 283.88 and 378.50 $m^3$ net capacity. Tanks of other sizes may be built.

The standard capacities of COATED-fabric tanks are in increments of 378.5–3785 $m^3$ (According to NFPA 22). Fig. 9.9 shows a typical fire-water storage tank.

**FIGURE 9.9 Shows a Typical Firewater Storage Tank**

### 9.35.1 **EXCEPTION**

The location chosen should be such that the tank and structure will not be subject to fire exposure from adjacent units. If lack of yard room makes this impracticable, the exposed steel work should be suitably fireproofed or protected by open sprinklers.

Fireproofing where necessary should include steel work within 6.1 m of combustible buildings, windows, doors, and flammable liquid and gas from which fire might issue.

1.  When steel or iron is used for supports inside the building near combustible construction or occupancy, it should be fireproofed inside the building, 1.8 m above combustible roof coverings and within 6.1 m of windows and doors from which fire might issue. Steel beams or braces joining two building columns which support a tank structure should also be suitably fireproofed when near combustible construction or occupancy. Interior timber should not be used to support or brace tank structures.
2.  Fireproofing, where required, should have a fire resistance rating of not less than 2 h.
3.  Foundations or footings should furnish adequate support and anchorage for the tower.
4.  If the tank or supporting trestle is to be placed on a building, the building should be designed and built to carry the maximum loads.

Fire water taken from open water is preferred, but if water of acceptable quality for fire-fighting in the required quantity, cannot be supplied from open water, or if it is not economically justified because of distance to install firewater pumps at an open source, water storage facilities should be provided.

Storage facilities may consist of an open tank of steel or concrete or a basin of sufficient capacity. The tank or basin should have two compartments to facilitate maintenance, each containing 60% of the total required capacity and there should be adequate replenishment facilities. A single compartment of 100% capacity is acceptable providing that an alternative source of water, for example, from temporary storage will be available during maintenance periods. The replenishment rate should normally not be less than 60% of the total required firewater pumping capacity.

If a 100% replenishment rate is available, the stored firewater capacity may be reduced if agreed by the N.I.O.C. Authorities that may be considered for replenishment are plant cooling water, open water or below-ground water, provided that it is available at an acceptable distance and in sufficient quantity for a minimum of 6 h uninterrupted fire-fighting at the maximum required rate.

## 9.36 **STEEL GRAVITY AND SUCTION TANKS**

### 9.36.1 **PRESSURE TANKS**

Pressure tanks may be used for limited private fire protection services.

Pressure tanks should not be used for any other purpose unless approved by the authority concerned.

### 9.36.2 **AIR PRESSURE AND WATER LEVEL**

Unless otherwise approved by the relevant authority, the tank should be kept two-thirds full of water, and an air pressure of at least 5.2 bars by the gage should be maintained. As the last of the water leaves the pressure tank, the residual pressure shown on the gage should not be less than zero, and should be sufficient to give not less than 1.0 bars pressure at the highest sprinkler under the main roof of the building

Other pressures and water levels may be required for hydraulically designed systems.

### 9.36.3 LOCATION

Pressure tanks should be located above the top level of sprinklers.

### 9.36.4 HOUSING

Where subject to freezing, the tank should be located in a substantial noncombustible housing. The tank room should be large enough to provide free access to all connections, fittings, and manhole, with at least 457 mm around the rest of the tank. The distance between the floor and any part of the tank should be at least 0.91 m.

The floor of the tank room should be watertight and arranged to drain outside of the enclosure. The tank room should be adequately heated to maintain a minimum temperature of 4.4°C and should be equipped with ample lighting facilities.

### 9.36.5 BURIED TANKS

Where lack of space or other conditions require it, pressure tanks may be buried if the following requirements are satisfied.

For protection against freezing the tank should be below the frost line.

The end of the tank and at least 457 mm of its shell should project into the building basement or a pit in the ground, with protection against freezing. There should be adequate space for inspection and maintenance and use of the tank manhole for interior inspection.

The exterior surface of the tank should be fully coated as follows for protection against corrosion conditions indicated by a soil analysis:

1. An approved cathodic system of corrosion protection should be provided.
2. At least 305 mm of sand should be backfilled around the tank.

The tank should be above the maximum ground water level so that buoyancy of the tank where empty will not force it upward. An alternative would be to provide a concrete base and anchor the tank to it.

The tank should be designed with strength to resist the pressure of earth against it.

A manhole should be located preferably on the vertical center line of the tank end to clear the knuckle while remaining as close as possible to it.

### 9.36.6 TANK'S MATERIAL

Types of materials should be limited to steel, wood, concrete, and coated fabric.

The elevated wood and steel tanks should be supported on steel or reinforced concrete towers.

## 9.37 FIRE HOSE REELS (WATER) FOR FIXED INSTALLATIONS
### 9.37.1 ROTATION

The hose reel should rotate around a spindle so that the hose can be withdrawn freely.

## 9.37.2 **REEL**

The drum or hose support of the first coil of hose should be not less than 150 mm diameter. The fitting by which the hose is attached to the reel should be arranged in such a way that the hose is not restricted or flattened by additional layers of hose being placed upon it.

## 9.37.3 **MANUAL INLET VALVE**

The inlet valve of a manual reel should be a screw down aboveground stop valve or a gate valve.

*Note:* To facilitate ease of installation and maintenance a union should be fitted between the valve and the reel.

The valve should be closed by turning the handle in a clockwise direction. The direction of opening should be permanently marked on the handle, preferably by an embossed arrow and the word OPEN.

## 9.37.4 **REEL SIZE**

Reels should be of sufficient size to carry the length of hose fitted, excluding the nozzle, within the space defined by the end plates. The length of hose fitted should be not more than 45 m for 19 mm internal diameter hose or 35 m for 25 mm internal diameter hose.

## 9.37.5 **RANGE AND WATER FLOW RATE**

The water flow rate should be not less than 24 LPM and the range of the jet should be not less than 6 m. The output of the nozzle, whether plain jet or jet/spray, should comply with the above flow rate except that the range of the spray should be less than 6 m (According to NFPA-22).

## 9.37.6 **LIMITATION OF HOSE IN CERTAIN CIRCUMSTANCES**

Although standards permit up to 45 m of hose on hose reels, frequently there are circumstances in which there is a likelihood of the hose having to be handled by persons of only moderate physical strength.

In such cases, and also when the likely routes for the hose are tortuous, the length and size of hose on the reel should be limited, and the siting and provision of reels should be reviewed with these limitations in mind.

## 9.37.7 **PROVISION**

One hose reel should be provided to cover every 800 m$^2$ of floor space or part thereof.

## 9.37.8 **SITING**

Hose reels should be sited in prominent and accessible positions at each floor level adjacent to exits in such a way that the nozzle of the hose can be taken into every area and within 6 m of each part of an area, having regard to any obstruction.

Where heavy furniture or equipment is introduced into an area, the hose and nozzle should be capable additionally of directing a jet into the back of any recess formed.

In exceptional circumstances consideration should also be needed as to the desirability of siting hose reels in such a way that if a fire prevents access to one hose reel site, the fire can be attacked from another hose reel in the vicinity.

## 9.37.9 COORDINATING SPACES FOR HOSE REELS

- The spaces required for most types of hose reels and their location in relation to floor or ground level for "horizontal" hose reels are not given as these are considered to be special installations.
- The figures indicate the range of acceptable choices from the point of view of dimensional coordination. First preferences are indicated by a thick blob and second preferences are indicated by a smaller blob.

*Notes:*
1. The basic space accommodates:
   **a.** the reel and valve
   **b.** the hanging loop of hose
   **c.** the guide or necessary space for proper withdrawal of the hose
   **d.** the component case (if any)
2. The space sizes have been based on the normal arrangement where the water supply is fed upward. Downward or sideways feeds should be treated as special installations.

### 9.37.9.1 Minimum requirement of water supply for hose reels

As a minimum, the water supply to hose reels should be such that when the two topmost reels in a building or unit are in use simultaneously, each will provide a jet of approximately 6 m in length and will deliver not less than 0.4 LPS (24 LPM). For example, when a length of 30 m of hose reel tubing is in use with a 6.5 mm nozzle, a minimum running pressure of 1.5 bar (1 bar = $10^5$ N/m$^2$ = 100 kPa) will be required at the entry to each reel and similarly for a 4.5 mm nozzle where a minimum running pressure of 4 bar will be required.

## 9.38 WATER SPRAY FIXED SYSTEMS FOR FIRE PROTECTION

Water spray is applicable for protection of specific hazards and equipment, and may be installed independently of or supplementary to other forms of fire protection systems or equipment.

## 9.38.1 HAZARDS

Water spray protection is acceptable for the protection of hazards involving:

- gaseous, liquid flammable and toxic materials
- electrical hazards such as transformers, oil switches, motors, cable trays, and cable runs
- ordinary combustibles such as paper, wood, and textiles
- certain hazardous solids.

### 9.38.2  USES

In general, water spray may be used effectively for any one or a combination of the following purposes:

1. extinguishment of fire
2. control of burning
3. exposure protection
4. prevention of fire.

### 9.38.3  LIMITATIONS

There are limitations to the use of water spray, which should be recognized. Such limitations involve the nature of the equipment to be protected, the physical and chemical properties of the materials involved and the environment of the hazard.

Other standards also consider limitations to the application of water (slopover, frothing, electrical clearances, etc.)

### 9.38.4  ALARMS

- The location, purpose, and type of system should determine the alarm service to be provided.
- An alarm, actuated independently of water flow, to indicate operation of the detection system should be provided on each automatically controlled system.
- Electrical fittings and devices designed for use in hazardous locations should be used where required by the standard.

### 9.38.5  FLUSHING CONNECTIONS

A suitable flushing connection should be incorporated in the design of the system to facilitate routine flushing as required.

### 9.38.6  WATER SUPPLIES

It is of vital importance that water supplies be selected which provide water as free as possible from foreign materials.

### 9.38.7  VOLUME AND PRESSURE

The water supply flow rate and pressure should be capable of maintaining water discharge at the design rate and duration for all systems designed to operated simultaneously.

For water supply distribution systems, an allowance for the flow rate of hose streams or other fire protection water requirements should be made in determining the maximum demand.

Sectional control shut-off valves should be located with particular care so that they will be accessible during an emergency case.

When only a limited water source is available, sufficient water for a second operation should be provided so that the protection can be reestablished without waiting for the supply to be replenished.

## 9.38.8 SOURCES

The water supply for water spray systems should be from reliable fire protection water supplies, such as:

- connections to waterworks systems
- gravity tanks (in special cases pressure tanks)
- fire pumps with adequate water supply.

## 9.38.9 FIRE DEPARTMENT CONNECTION

One or more fire department connections should be provided in all cases where water supply is marginal and/or where auxiliary or primary water supplies may be augmented by the response of suitable pumper apparatus responding to the emergency. Fire department connections are valuable only when fire department pumping capacities equal maximum demand flow rate.

Careful consideration should be given to such factors as the purpose of the system, reliability, and capacity and pressure of the water system. The possibility of serious exposure fires and similar local conditions should be considered. A pipeline strainer in the fire department connection should be provided if indicated by 13.8.5. Where a fire department connection is required, suitable suction provisions for the responding pumper apparatus should be provided.

## 9.38.10 WORKMANSHIP

Water spray system design, layout, and installation should be entrusted to none but fully experienced and responsible parties. Water spray system installation is a specialized field of sprinkler system installation, which is a trade in itself.

Before a water spray system is installed or existing equipment remodeled, complete working plans, specifications, and hydraulic calculations should be prepared and made available to interested parties.

## 9.39 DENSITY AND APPLICATION
### 9.39.1 EXTINGUISHMENT

- Extinguishment of fires by water spray may be accomplished by surface cooling, by smothering from steam produced, by emulsification, by dilution, or by various combinations thereof. Systems should be designed so that, within a reasonable period of time, extinguishment should be accomplished and all surfaces should be cooled sufficiently to prevent "flashback" occurring after the system is shut off.
- The design density for extinguishment should be based upon test data or knowledge concerning conditions similar to those that will apply in the actual installation.

A general range of water spray application rates that will apply to most ordinary combustible solids or flammable liquids is from 6.1 LPM/m$^2$ to 20.4 LPM/m$^2$ of protected surface.

*Note:*
There is some data available on water application rates needed for extinguishment of certain combustibles or flammables; however, much additional test work is needed before minimum rates can be established.

Each of the following methods or a combination of them should be considered when designing a water spray system for extinguishment purposes:

1. surface cooling
2. smothering by steam produced
3. emulsification
4. dilution
5. other factors.

## 9.39.2 CABLE TRAYS AND CABLE RUNS

When insulated wire and cable or nonmetallic tubing is to be protected by an automatic water spray (open nozzle) system maintained for extinguishment of fire which originates within the cable or tube (ie, the insulation or tubing is subject to ignition and propagation of fire), the system should be hydraulically designed to impinge water directly on each tray or group of cables or tubes at the rate of 6.1 LPM/m$^2$ on the horizontal or vertical plane containing the cable or tubing tray or run.

## 9.39.3 EXCEPTION

Other water spray densities and methods of application should be used if verified by tests and acceptable to the authority.

Automatic detection devices should be sufficiently sensitive to rapidly detect smoldering or slow-to-develop flames. When it is contemplated that spills of flammable liquids or molten materials will expose cables, nonmetallic tubing, and tray supports, design of protection systems should be in accordance with that recommended for exposure protection.

When electrical cables or tubing in open trays or runs are to be protected by water spray from fire or spill exposure, a basic rate of 12.2 LPM/m$^2$ of projected horizontal or vertical plane area containing the cables or tubes should be provided.

Water spray nozzles should be arranged to supply water at this rate over and under or to the front and rear of cable or tubing runs and to the racks and supports.

Where flame shields equivalent to 1.6 mm thick steel plate are mounted below cable or tubing runs, the water density requirements may be reduced to 6.1 LPM/m$^2$ over the upper surface of the cable or rack. The steel plate or equivalent flame shield should be wide enough to extend at least 152 mm beyond the side rails of the tray or rack in order to deflect flames or heat emanating from spills below cable or conduit runs.

Where other water spray nozzles are arranged to extinguish, control or cool exposing liquid surfaces, the water spray density may be reduced to 6.1 LPM/m$^2$ over the upper surface, front or back of the cable, or tubing tray or run.

Fixed water spray systems designed for protecting cable or tubing and their supports from heat of exposure from flammable or molten liquid spills should be automatically actuated.

## 9.39.4 CONTROL OF BURNING

A system for the control of burning should function at full effectiveness until there has been time for the flammable materials to be consumed, for steps to be taken to shut off the flow of leaking material, for the assembly of repair forces, etc. System operation for hours may be required.

Nozzles should be installed to impinge on the areas of the source of fire, and where spills may travel or accumulate. The water application rate on the probable surface of the spill should be at the rate of not less than 20.4 LPM/m$^2$.

Pumps or other devices which handle flammable liquids or gases should have the shafts, packing glands, connections, and other critical parts enveloped in directed water spray at a density of not less than 20.4 LPM/m$^2$ of projected surface area.

### 9.39.5 EXPOSURE PROTECTIONS

1. The system should be able to function effectively for the duration of the exposure fire, which is estimated from a knowledge of the nature and quantities of the combustibles and the probable effect of fire-fighting equipment and materials. System operation for hours should be required.
2. Automatic water spray systems for exposure protection should be designed to operate before the formation of carbon deposits on the surfaces to be protected and before the possible failure of any containers of flammable liquids or gases because of the temperature rise. The system and water supplies should, therefore, be designed to discharge effective water spray from all nozzles within 30 s following operation of the detection system.
3. The densities specified for exposure protection contemplate minimal wastage of 2.0 LPM/m$^2$.

### 9.39.6 SPRAY NOZZLES

Care should be taken in the application of nozzle types. Distance of "throw" or location of nozzle from surface should be limited by the nozzle's discharge characteristics.

Care should also be taken in the selection of nozzles to obtain waterways, which are not easily obstructed by debris, sediment, sand, etc. in the water.

### 9.39.7 SELECTION

The selection of the type and size of spray nozzles should be made with proper consideration given to such factors as physical character of the hazard involved, draft or wind conditions, material likely to be burning, and the general purpose of the system.

### 9.39.8 POSITION

Spray nozzles may be placed in any position necessary to obtain proper coverage of the protected area. Positioning of nozzles with respect to surfaces to be protected, or to fires to be controlled or extinguished, should be guided by the particular nozzle design and the character of water spray produced.

The effect of wind and fire draft on very small drop sizes or on larger drop sizes with little initial nozzle velocity should be considered, since these factors will limit the distance between nozzle and surface, and will limit the effectiveness of exposure protection, fire control or extinguishment. Care should be taken in positioning nozzles that water spray does not miss the targeted surface and reduce the efficiency or calculated discharge rate LPM/m$^2$. Care should also be exercised in placement of spray nozzles protecting pipelines handling flammable liquids under pressure, where such protection is intended to extinguish or control fires resulting from leaks or ruptures.

### 9.39.9 **STRAINERS**

Main pipeline strainers should be provided for all systems utilizing nozzles with waterways less than 9.5 mm and for any system where the water is likely to contain obstructive material.

Mainline pipeline strainers should be installed so as to be accessible for flushing or cleaning during the emergency.

Individual strainers should be provided at each nozzle where water passageways are smaller than 3.2 mm.

Care should be taken in the selection of strainers, particularly where nozzle waterways are less than 6.5 mm in dimension. Consideration should be given to size of screen perforation, to volume available for accumulation without excessive friction loss and the facility for inspection and cleaning.

### 9.39.10 **VESSELS**

These rules for exposure protection contemplate emergency relieving capacity for vessels, based upon a maximum allowable heat input of 18,930 $W/m^2$ of exposed surface area. The density should be increased to limit the heat absorption to a safe level in the event required emergency relieving capacity is not provided.

Water should be applied to vertical or inclined vessel surfaces at a net rate of not less than 9.2 LPM/$m^2$ of exposed uninsulated surface.

Where run-down is contemplated for vertical or inclined surfaces the vertical distance between nozzles should not exceed 3.7 m.

The horizontal extremities of spray patterns should at least meet.

Spherical or horizontal cylindrical surfaces below the vessel equator cannot be considered wettable from rundown.

Where projections (manhole flanges, pipe flanges, support brackets, etc.) will obstruct water spray coverage, including run-down or slippage on vertical surfaces, additional nozzles should be installed around the projections to maintain the wetting pattern which otherwise would be seriously interrupted.

Bottom and top surfaces of vertical vessels should be completely covered by directed water spray at an average rate of not less than 9.2 LPM/$m^2$ of exposed uninsulated surface. Consideration should be given to slippage but on the bottom surfaces the horizontal extremities of spray patterns should at least meet.

Special attention should be given to distribution of water spray around relief valves and around supply piping and valve connection projections.

Uninsulated vessel skirts should have water spray applied on one exposed (uninsulated) side, either inside or outside, at a net rate of not less than 9.2 LPM/$m^2$.

### 9.39.11 **STRUCTURES AND MISCELLANEOUS EQUIPMENT**

Horizontal, stressed (primary) structural steel members should be protected by nozzles spaced not greater than 3 m on centers (preferably on alternate sides) and of such size and arrangement as to discharge not less than 4.1 LPM/$m^2$ over the wetted area (Fig. 9.10).

The wetted surface of structural member-beam or column is defined as one side of the web and the inside surface of one side of the flanges as shown above.

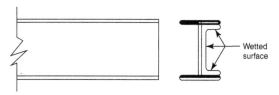

**FIGURE 9.10 The Wetted Surface of Structural Member-Beam or Column**

Vertical structural steel members should be protected by nozzles spaced not greater than (3 m) on centers (preferably on alternate sides) and of such size and arrangement as to discharge not less than $9.2$ LPM/m$^2$ over the wetted area (Fig. 9.10).

Metal pipe, tubing, and conduit runs should be protected by water spray directed toward the horizontal plane surface projected by the bottom of the pipes or tubes.

Nozzles should be selected to provide essentially total impingement on the entire horizontal surface area within which pipes or tubes are or could be located.

For single-level pipe racks, water spray nozzles should discharge onto the underside of the pipe at a plan view density of $9.2$ LPM/m$^2$.

For two-level pipe racks, water spray nozzles should discharge onto the underside of the lower level at a plan view density of $8.2$ LPM/m$^2$ and additional spray nozzles should discharge onto the underside of the upper level at a plan view density of $6.1$ LPM/m$^2$.

For three-, four-, and five-level pipe racks, water spray nozzles should discharge onto the underside of the lowest level at a plan view density of $8.2$ LPM/m$^2$ and additional spray nozzles should discharge onto the underside of alternate levels at a plan view density of $6.1$ LPM/m$^2$. Water spray should be applied to the underside of the top level even if immediately above a protected level.

For pipe racks of six or more levels, water spray nozzles should discharge onto the underside of the lowest level at a plan view density of $8.2$ LPM/m$^2$ and additional spray nozzles should discharge onto the underside of alternate levels at a plan view density of $4.1$ LPM/m$^2$. Water spray should be applied to the underside of the top level even if immediately above a protected level.

Water spray nozzles are to be selected and located such that extremities of water spray patterns should at least meet and the discharge should essentially be confined to the plan area of the pipe rack.

Spacing between nozzles should not exceed 3 m and nozzles should be no more than 0.8 m below the bottom of the pipe level being protected.

Consideration should be given to obstruction to the spray patterns presented by pipe supporting steel.

Where such interferences exist, nozzles should be spaced within the bays.

*Exceptions:*

1. Water spray protection with the same density as specified previously may be applied to the top of pipes on racks where water spray piping cannot be installed below the rack due to possibility of physical damage or space is inadequate for proper installation.
2. Vertically stacked piping may be protected by water spray directed at one side of the piping at a density of $6.1$ LPM/m$^2$. Table 9.6 provides more details of information.

**Table 9.6  Protection of Metal Pipe, Tubing, and Conduit**

| Number of Rack Levels | Plan View Density at Lowest Level | | Plan View Density at Upper Level(s)[a] | | Levels Requiring Nozzles |
|---|---|---|---|---|---|
| | gpm/ft.$^2$ | (LPM)/m$^2$ | gpm/ft.$^2$ | (LPM)/m$^2$ | |
| 1 | 0.25 | 9.2 | N/A | N/A | All |
| 2 | 0.20 | 8.2 | 0.15 | 6.1 | All |
| 3, 4, or 5 | 0.20 | 8.2 | 0.15 | 6.1 | Alternate |
| 6 or more | 0.20 | 8.2 | 0.10 | 4.1 | Alternate |

$^a$*The table values contemplate exposure from a spill fire.*

### 9.39.11.1 Transformers

Transformer protection should contemplate essentially complete impingement on all exterior surfaces, except underneath surfaces, which in lieu thereof may be protected by horizontal projection. The water should be applied at a rate not less than 9.2 LPM/m$^2$ of projected area of rectangular prism envelope for the transformer and its appurtenances and not less than 6.1 LPM/m$^2$ on the expected nonabsorbing ground surface area of exposure. Additional application is needed for special configurations, conservator tanks, pumps, etc. Spaces greater than 305 mm in width between radiators, etc. should be individually protected.

Water spray piping should not be carried across the top of the transformer tank, unless impingement cannot be accomplished with any other configuration and provided the required distance from the live electrical components is maintained.

In order to prevent damage to energized bushings or lightning arrestors, water spray should not envelop this equipment by direct impingement, unless so authorized by the manufacturer or manufacturer's literature.

### 9.39.11.2 Belt conveyers

#### 9.39.11.2.1 Drive unit

Water spray system should be installed to protect the drive rolls, the take-up rolls, the power units and the hydraulic-oil unit. The rate of water application should be 9.2 LPM/m$^2$ of roll and belt.

Nozzles should be located to direct water spray onto the surfaces to extinguish fire in hydraulic oil, belt, or contents on the belt. Water spray impingement on structural elements should be such as to provide protection against radiant heat or impinging flame.

#### 9.39.11.2.2 Conveyer belt

A water spray system should be installed to automatically wet the top belt, its contents, and the bottom return belt. Discharge patterns of water spray nozzles should envelop, at a rate of (9.2 LPM/m$^2$) of top and bottom belt area, the structural parts, and the idler-rolls supporting the belt. Water spray system protection should be extended onto transfer belts, transfer equipment and transfer buildings beyond each transfer point.

Or, systems for the protection of adjacent belts or equipment should be interlocked in such a manner that the feeding belt water spray system will automatically actuate the water spray system protecting the first segment of the downstream equipment.

Special consideration should be given to the interior protection of the building, gallery, or tunnel housing the belt conveyer equipment.

Also, the exterior structural supports for galleries should be protected from exposure such as fires in flammables located adjacent to the galleries.

The effectiveness of belt conveyer protection is dependent upon rapid detection and appropriate interlocks between the detection system and the machinery.

## 9.40 FIRE AND EXPLOSION PREVENTION

The system should be able to function effectively for a sufficient time to dissolve, dilute, disperse, or cool flammable or hazardous materials. The possible duration of release of the materials should be considered in the selection of duration times.

The rate of application should be based upon experience with the product or upon test.

### 9.40.1 SIZE OF SYSTEM

Separate fire areas should be protected by separate systems. Single systems should be kept as small as practicable, giving consideration to the water supplies and other factors affecting reliability of the protection. The hydraulically designed discharge rate for a single system or multiple systems designed to operate simultaneously should not exceed the available water supply.

Separation of fire areas should be by space, fire barriers, diking, special drainage, or by a combination of these. In the separation of fire areas consideration should be given to the possible flow of burning liquids before or during operation of the water spray systems.

### 9.40.2 AREA DRAINAGE

Adequate provisions should be made to promptly and effectively dispose of all liquids from the fire area during operation of all systems in the fire area. Such provisions should be adequate for:

1. water discharged from fixed fire protection systems at maximum flow conditions
2. water likely to be discharged by hose streams
3. surface water
4. cooling water normally discharged to the system.

There are four methods of disposal or containment:

1. grading
2. diking
3. trenching
4. underground or enclosed drains.

The method used should be determined by:

1. the extent of the hazard
2. the clear space available
3. the protection required.

Where the hazard is low, the clear space is adequate, and the degree of protection required is not great, grading is acceptable. Where these conditions are not present, consideration should be given to dikes, trenching, underground, or enclosed drains.

## 9.41 VALVES

### 9.41.1 SHUT-OFF VALVES

Each system should be provided with a shut-off valve so located as to be readily accessible during a fire in the area the system protects or adjacent areas, or, for systems installed for fire prevention, during the existence of the contingency for which the system is installed.

Valves controlling water spray systems, except underground gate valves with roadway boxes, should be supervised open by one of the following methods:

- central station, proprietary, or remote station alarm service
- local alarm service which will cause the sounding of an audible signal at a constantly attended point
- locking valves open
- sealing of valves and approved weekly recorded inspection when valves are located within fenced enclosures under the control of the owner.

### 9.41.2 AUTOMATICALLY CONTROLLED VALVES

Automatically controlled valves should be as close to the hazard as accessibility during the emergency will permit, so that a minimum of piping is required between the automatic valve and the spray nozzles.

Remote manual tripping devices, where required, should be conspicuously located where readily accessible during the emergency and adequately identified as to the system controlled.

### 9.41.3 HYDRAULIC CALCULATION

Table 9.7 shows details of process unit firewater requirements.

| Table 9.7 Process Unit Firewater Requirements | | |
|---|---|---|
| | **Minimum Firewater Rate** | |
| **Type of Process Unit** | **m³/h** | **US gpm** |
| Atmospheric distillation, vacuum, or combination units with up to 15,900 m³/d (100,000 bbl/d) throughput treating plants; asphalt stills; others | 598 | 2500 |
| Atmospheric distillation, vacuum, or combination units with 15,900 m³/d (100,000 bbl/d) or higher throughput; catalytic cracking units | 900 | 4000 |
| Light-end units containing volatile oils and hydrogen, such as reformers, catalytic desulfurizers, and alkylation units | 900 | 4000 |
| Lube oil units and blending facilities | 454 | 2000 |

## 9.42 FIXED ROOF TANKS CONTAINING HIGH-FLASH LIQUIDS

When the stored product has a closed-cup flash point of 65°C (150°F) or higher, a fixed roof tank can be considered relatively safe. Then water for foam extinguishment is not required, provided the following conditions are met:

- If the product is heated, there must be no possibility of the storage temperature exceeding either the flash point or 93°C (200°F).
- There must be no possibility of hot oil streams entering the tank at temperatures above 93°C (200°F) or their flash point.
- Cutter stock having a flash point below the storage temperature must never be pumped into the tank for blending purposes.
- Sufficient firewater should be available to cool exposed adjacent tankage in the event of ignition. Then the tank should be pumped out or allowed to burn out.
- The product should not be crude oil with boil over characteristics. If the product were crude, the fire would have to be extinguished before the heat wave reached water at the tank bottom.
- Storage temperatures between 93°C (200°F) and 121°C (250°F) should be avoided, as water lenses or water at the tank bottom may reach boiling temperature at any time, resulting in a serious frothover.

If the product is heated above 121°C (250°F) foam extinguishment cannot be accomplished and slopover will occur if foam is applied.

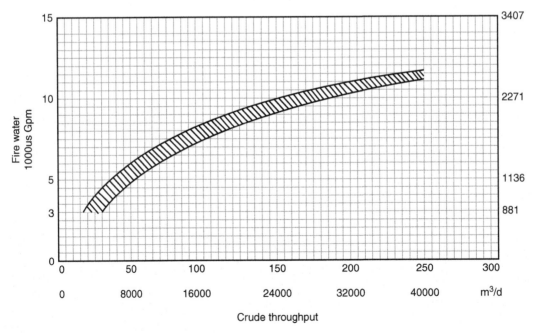

**FIGURE 9.11 Firewater Versus Crude Throughput**

## 9.42.1 FLOATING ROOF TANKS

Floating roof tanks are considered virtually ignition proof, except for rim fires. Thus, there should be sufficient firewater to cool the shell and extinguish a rim fire.

### 9.42.1.1 Pressure storage

The water requirement for cooling pressure storage spheres or drums may exceed the maximum cone roof tank fire-water requirement when spheres are of large diameter, or when a number of spheres or drums are closely spaced. However, when adjacent spheres are not over 15 m (50 ft.) in diameter and are at least 30 m (100 ft.) apart, shell-to-shell, cooling of these spheres may be disregarded.

### 9.42.1.2 Low-pressure refrigerated storage

Water for monitor cooling streams should be available to cool the tank shell if it is exposed to fire. Water must not be applied directly to refrigerated LP-gas or cryogenic flammable liquid spills or spill fires, since much more rapid vapor evolution or increased fire intensity will result.

### 9.42.1.3 Cooling water for exposed tankage

During a tank or sphere fire, cooling streams may be needed for adjacent tankage. However, this does not apply to process units where flammable liquids are not contained in sufficient volume to generate enough heat to require cooling adjacent tankage. Allow at least two 57 m³/h [250 US gallons per minute (gpm)] cooling streams, or a total of 114 m³/h (500 US gpm), for each adjacent, unshielded fixed or floating roof tank within the following limits:

Within 15 m (50 ft.) of a burning tank or sphere of any size, regardless of wind direction.

Within one tank diameter and within a quadrant that will require the maximum amount of cooling water for tanks that fall within the quadrant.

Within 45 m (150 ft.) of a sphere and within the most congested quadrant.

### 9.42.1.4 FireWater capacity versus crude throughput

Fig. 9.11 shows firewater capacity, in cubic meters per hour (thousands of US gpm), plotted against crude throughput, in cubic meters per day (thousands of barrels per calendar day). The curve represents an average of data collected from plants all over the world. This curve can be used as a guide to the required quantity when calculating refinery firewater capacity by the method prescribed in this section.

Firewater should be obtained from an unlimited source, such as a natural body of water. When this is not possible, the supply should always be available in a storage tank or reservoir.

# SOLIDS HANDLING SYSTEMS AND DRYERS

The handling of solids plays an important role in the vast number of industries, particularly in chemicals and petrochemical process industries. Bulk materials handling comprises a wide range of techniques incorporating a number of handling modes. Attempts are made here to highlight some of the important modes specifically utilized in OGP process plants by providing a basis on which the basic conveying requirements, and parameters, are given for proper selection and the design of the systems.

"Process design of offsite facilities for OGP processes," is broad and contain various subjects of paramount importance. This chapter intends to cover essential process requirements, and governing the selection of a proper handling system for bulk materials, with specific concerns to capacity requirements, material characteristics, process requirements, and flow properties of solids. Classification codes summarizing the behavior of bulk solids, process design considerations from the views of operating conditions and other process design information and criteria to the extent specified herein are covered.

## 10.1 MOST SALIENT FEATURES

The selection of a specific handling system for bulk materials requires full knowledge of the physical and chemical properties of the materials to be handled. Material characteristics and the prevailing plant conditions, play an important role in determining the flow behavior of the product, thereby influencing the type of equipment selected. The most salient features for selection of a suitable handling system are:

1. *Capacity*. Capacity requirement is a prime factor in conveyor selection. Belt conveyors, which can be manufactured in relatively large sizes, to operate at high speeds, deliver large tonnage economically. On the other hand, screw conveyors become extremely cumbersome as they get larger and cannot be operated at high speeds without creating serious abrasion roblems.
2. *Distance*. Length of travel is definitely limited for certain type of conveyers. With high-tensile strength belting, the length limit on belt conveyors can be a matter of miles. Air conveyors are limited to 300 m, vibrating conveyors to 100 m. In general as the length of travel increases, the alternatives become fewer.
3. *Lift*. Lifting can usually be handled most economically by vertical or inclined bucket elevators, but when lift and horizontal travel are combined other conveyors should be considered. Conveyers that combine several directions of travel in a single unit are generally more expensive, but since they require a single drive, this feature often compensates for the added base cost.
4. *Materials*. Material characteristics, chemical, and physical characteristics should be considered especially flow ability, abrasiveness, friability, and lump size. Effects of chemicals, moisture, and

oxidation effects from exposure to atmosphere can be harmful to the material being conveyed, or to the conveyor's material. Certain types of conveyors lend themselves to such special requirements better than the others.

5. *Processing*. Processing requirements can be met by some conveyors with little or no change in design. For example, screw conveyors are available for a wide variety of processing operations such as mixing, dewatering, heating, and cooling. A continuous flow conveyor may provide a desired cooling for solids simply because it puts the conveyed material into direct contact with heat conducting metals.

6. *Flow properties of solids*. The flow characteristics of bulk solid materials depend on their physical and chemical properties. The main characterizing factors to be considered in selection of solids conveying systems are:

   a. *Product grouping*. The flow ability of the product yields two main categories of bulk solids and they are specified below:

   – *Group I*. This group includes free-flowing materials, that is, non-cohesive products; those that do not undergo any plastic deformation when subjected to high pressures. When the load is removed, the particles return to their original condition in terms of both shape and flow characteristics.

   – *Group II*. All products that undergo plastic deformation when subjected to external pressure, that is, cohesive products for which the degree of deformation is strongly influenced by both temperature and moisture. When the load is removed, the particles do not regain their original shape, thereby yielding poor flow condition.

## 10.1.1 FLUIDIZATION CHARACTERISTICS

The ability of the material to fluidize and whether the product has an affinity to trap air or gas is of major importance when designing and selecting handling and storage systems for solids.

## 10.1.2 FLOW FUNCTION

Flow functions for both short and long residence time in silos and storage bins shall be taken into account in design and selection. Additional complications occur when the product is stored at elevated temperatures or when humidity could influence the moisture content of the product.

## 10.1.3 IMPORTANT FLOW FEATURES

For the successful operation of any materials' handling system the flow of solids from bins and silos must be controlled. The following important features should be considered in selection and design of such storage facilities.

## 10.1.4 FACTORS INFLUENCING FLOW

Three essential factors must be considered when designing a storage hopper or bin:

1. *Geometric form of the hopper*. The elements which must be considered include:
   a. cone angle θ (theta)

**b.** size of outlet
**c.** shape (circular or rectangular)
**d.** hopper construction material.

2. *Product characteristics*
   **a.** particle size and shape
   **b.** particle size distribution
   **c.** particle density and bulk density
   **d.** cohesiveness of the product
   **e.** fluidizability
   **f.** floodability
   **g.** deaeration characteristics.

3. *Additional factors*
   **a.** influence of humidity
   **b.** temperature of product and process
   **c.** storage time
   **d.** ambient conditions.

## 10.2 CONVEYING OF BULK SOLIDS

### 10.2.1 CONVEYOR SELECTION

Main guidelines and prime factors of this specification must be considered in the course of evaluating and selection. However, it is advisable to check with the manufacturer to be sure that the application is proper.

Conveyor selection must be based on the characteristics of a material as conveyed. For instance if packing or aerating can occur in the conveyor, the machine's performance will not meet expectations if calculations are based on an average mass per cubic meter. Storage conditions, variations in ambient temperature and humidity, and storage methods may all affect conveying characteristics. So, such factors should also be carefully considered before making a final conveyor selection.

### 10.2.2 MECHANICAL CONVEYORS

Mechanical conveying techniques are the most widely used form of materials handling in chemical and petrochemical industries. Mechanical conveyors have distinct advantages in terms of the ability to effect accurate control in the monitoring of material from one process to another.

Under the standard specification, some basic features of the various types of mechanical conveyors as well as the safety and environmental considerations are covered.

### 10.2.3 SELECTION OF MECHANICAL CONVEYORS

#### 10.2.3.1 Belt conveyors

Belt conveyors are the most widely used and versatile mode of mechanical conveying systems employed to transport materials horizontally or on an inclined either up or down. Fig. 10.1, represents a typical belt-conveyor arrangement, with the following main components of the system:

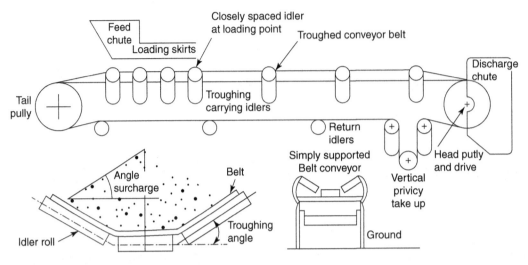

**FIGURE 10.1 Typical Belt-Conveyor Arrangement**

1. the belt, which forms the moving and supporting surface on which the conveyed material rides.
2. the idler, which form the supports for the carrying and return strands of the belt.
3. the pulleys, which support and move the belt and control its tension.
4. the drive, which imparts power to one or more pulleys to move the belt and its load.
5. the structure, which supports and maintains the alignment of the idlers and pulleys, and supports the driving machineries.

### 10.2.3.2 Requirements
Unless otherwise specified, all conveyers, drives, supports, and electricals including the control panel and other materials necessary to complete the conveying system shall be furnished by the vendor.

The vendor shall furnish all appliances, special tools, and accessories that are necessary or incidental to the proper installation and safe operation of the equipment, even though these items may not be included in the drawings, specifications, or data sheets.

A summary of utility requirements including electric power, plant and instrument air, cooling water, steam, etc. shall be submitted with the vendor's quotation.

The vendor shall submit recommendations for the following three categories of spare parts:

1. Erection and pre-commissioning.
2. Commissioning and initial operation.
3. 2 years of normal operation.

### 10.2.3.3 Design
The conveying system shall be designed for 24 h continuous operation at the rated output in the specified environment.

General arrangement drawings of conveying system showing location of gravity take-up, bents, support locations, etc. shall be prepared by the vendor.

The vendor shall be responsible for guaranteeing the performance of the belt conveyor system. The data sheets and drawings shall indicate minimum requirements, but these shall in no way relieve the vendor from his responsibility for providing a system capable of meeting the required performance.

Construction materials shall be selected and specified in accordance with service condition and handling material specification.

Belt conveyors shall be CEMA Grade 2, designed for the material conveying temperature.

The following operating conditions shall be considered in the design of the belt conveyor :

1. Service condition is the first step which should be considered. The method by which the conveyor will be fed, the point where loading takes place, and where the material will be discharged.
2. Surroundings which involve such conditions as high temperature or corrosive atmosphere can affect the belt, machinery, and structure.
3. Continuous service may require extremely high quality components and even specially designed equipment for servicing while the belt is in operation.
4. Belt width and operating speed are functions of the bulk density and lump size of the material, using the narrowest possible belt for a given lump size and operating it at maximum speed can result in lowest cost.
5. For detailed design requirement of belt conveyors, reference is given to ISO 5048, 1989, "Continuous Mechanical Handling Equipment Belt Conveyors."

The following criteria shall be considered in the design of the belt conveyor:

1. belt conveying capacity, belt incline, and belt loading points shall clearly be determined in design data sheets.
2. design and engineering of belt conveyors must be directed toward keeping the belt in operating condition.
3. the belt transfer points should be reduced to a minimum to cut degradation, dust, and cost.
4. all belt lines should be elevated to a specified level above ground to ease inspection, maintenance, and cleanup.
5. clearances above roadways and rail lines should permit the passage of cranes and other mobile equipment, as well as fire control vehicles. Minimum vertical clearance for passage above/below other facilities shall be as per Table 10.1.

**Table 10.1 Minimum Vertical Clearance**

| Location | Minimum clearance (m) |
|---|---|
| 1. Above major roads open to unrestricted traffic (such as periphery of process unit area limits) | 6.0 |
| 2. Within process unit areas: above internal roadways provided for access of maintenance and firefighting equipment | 3.5 |
| 3. Above walkways and elevated platforms | 2.0 |
| 4. Under any low level piping in paved or unpaved areas | 0.6 |

**6.** care should be taken to control dust emissions. The amount of dust released depends upon the physical characteristics of the bulk material and the manner in which the material is handled. An enclosure should be placed around the transfer to control the dust emission.

**7.** in addition to requirements, the following design information shall be submitted for the company's review:

    **a.** dimensioned outline drawings showing equipment physical arrangement and elevation

    **b.** completed specification sheets giving manufacture, size, type, or model of specific equipment to be furnished

    **c.** materials of construction

    **d.** calculations and data necessary to support and interpret the calculations

    **e.** detailed drawing and data showing:

       – device for cleaning the conveying surface of the belt

       – length of each type and size of belt to be furnished for field splices

       – type and location of idlers

       – seal and lubrication sign of idlers.

For general definitions and nomenclatures of the terms used see ISO 2148, 1974, "Continuous Handling Equipment Nomenclatures."

### 10.2.3.4 Screw conveyors

The screw conveyor is one of oldest and most versatile conveyor types, it consists of a helicoid or sectional flight rotating in a stationary trough. Power to convey is transmitted through the pipe or shaft and is limited by the allowable size of this member.

The allowable loading and screw speed are limited by characteristics of the material. Light, free-flowing, non-abrasive materials fill the trough deeply, permitting a higher rotating speed than with heavier and more abrasive materials. The manufacturer's recommendations should be considered in determining the allowable loading.

In addition to their conveying ability, screw conveyors can be adapted to a wide variety of processing operations. Almost any degree of mixing can be achieved with a screw conveyor.

There is a diversified number of uses for screw conveyors, such as controlled heating or cooling, mixing, and blending. This part only applies to a screw conveyor used in horizontal or inclined position (up to approximately 20°) for a regular, controlled, and continual supply of the bulk materials.

Required power is made up of three components:

Ph: power necessary for the progress of the material
Pn: drive power of the screw conveyor at no load
Psr: power due to inclination.

A variety of screw conveyors can be used for handling of solid materials, depending on the characteristics of the material to be conveyed. Typical screw conveyors with relevant nomenclature are given in ISO 2148, 1974.

### 10.2.3.5 Field of application

The design calculations and requirements shall only be used for a screw conveyor, used in a horizontal or inclined position.

*Note*: Excluded from this section are the special screws for following special fields of application:

1. extracting screws
2. calibrating screws
3. mixing screws
4. moistening screws
5. inclined screws (above 20°)
6. vertical screws.
    a. The nominal capacity of a screw conveyor IV shall be obtained from the equation given under British Standard BS 4409, Part, 3 1982 and ISO 7119, 1981.
    b. In selection of the trough-filling coefficient the conditions specified in BS 4409, Part 3, 1982 or ISO 7119, 1981 shall be considered.
    c. The screw needs to have a smooth flight surface in order to reduce friction and power consumption.
    d. The peripheral speed of the screw should not be too high so as to prevent the material being thrown upwards which will spoil its transport. The peripheral speed should be chosen as a function of the screw diameter D, the physical properties of the material and the filling coefficient.

### 10.2.3.6 Drive power of loaded screw
The total drive power of the loaded screw in kilowatts is given by the formula in BS 4409, Part 3.

### 10.2.3.7 Drive power due to progress of material $P_H$
The power necessary for the progress of the material $P_H$ should be obtained from the formulas given in BS 4409, Part 3.

### 10.2.3.8 Drive power of conveyor at no load $P_N$
The drive power $P_N$ is very low compared to the power required for the progress of the material. This value is proportional to the diameter $D$ and length of the screw $L$, but In practice it is given in kilowatts and should be obtained from the formula given under BS 4409.

### 10.2.3.9 Power due to inclination $P_{st}$
The power $Pst$ in kilowatts, will be the product of the capacity $W$ by the height $H$ and by the acceleration due to gravity $g$. This value should be obtained from the formulas given under Clause 8.3 of British Standard BS 4409, Part 3.

### 10.2.3.10 Total power necessary for the shaft of the screw conveyor
The total power necessary is the sum of the various powers and should be obtained from the formula given under British Standard BS 4409, Part 3.

### 10.2.3.11 Design requirements
Screw conveyor enclosure shall be ANSI/CEMA, No. 350 Class III E for indoor or sheltered service, and Class IV E for outdoor applications.

Enclosures shall be designed to allow access over the entire length of the conveyor.

Screw conveyors shall be sized based on a trough loading of no greater than 30% of trough volume with the specified lowest material bulk density $\rho$ (rho) exhibited during transport by the equipment.

Design information to be submitted for review by the company shall include but not be limited to the following:

1. dimensioned outline drawings showing equipment physical arrangement and elevations
2. completed specification sheets giving manufacturer, size, type of model of specific equipment
3. material of construction
4. calculations and data necessary to support and interpret the calculations.

## 10.2.4 CONTINUOUS HANDLING EQUIPMENT OR PNEUMATIC CONVEYORS

Pneumatic conveyors are commonly used to transport dry granular or powdered materials, both vertically and horizontally, to remote plant areas that would be hard to reach economically with mechanical conveyors.

Pneumatic systems can be completely enclosed to prevent product contamination, material loss, and dust emission. Furthermore, some materials are better protected from degradation when they are conveyed using an inert gas or dried air. Conveying rates are comparable to those for most mechanical conveyors.

### 10.2.4.1 Advantages and disadvantages of pneumatic conveyors
Pneumatic conveying offers the following advantages:

1. *Flexibility.* The system can follow any route; a change in flow direction is achieved by simple addition of a pipe bend; it is relatively easy to extend pipeline length.
2. *Pollution-free transport.* Because solids are contained in a closed conduit, toxic materials can be transported safely and the product is also protected from environmental contamination.
3. *Reduction in manpower and minimal maintenance.* Pneumatic conveying systems have virtually no moving parts, require minimal maintenance, and have relatively modest manpower requirements.
4. *Control.* With the latest feeding techniques, pneumatic conveying systems can be easily integrated into modern chemical plants employing sophisticated process control.
5. *Range of products transported.* Pneumatic conveying techniques have been successfully applied to a wide range of products. Modern handling techniques have facilitated the transportation of flakes, powders, and granules.

The pneumatic conveyors may also offer certain disadvantages as:

1. *Higher power consumption.* When compared with other mechanical handling systems.
2. *Limited distances.* Pneumatic handling techniques are generally limited to the indoor transportation of materials, although there are systems operating which move products over a maximum distance of 4000 m.
3. *Limited throughput.* Pneumatic conveying techniques are best suited to handling products at rates of hundreds, rather than thousands of tons per hour.

### 10.2.4.2 Types of pneumatic conveying systems
Pneumatic conveyors are classified according to following basic types:

1. Pressure system
   a. In pressure systems (Fig. 10.2a) material is dropped into the air stream (at above atmospheric pressure) by a rotary air-lock feeder. The velocity of the stream maintains the bulk material in

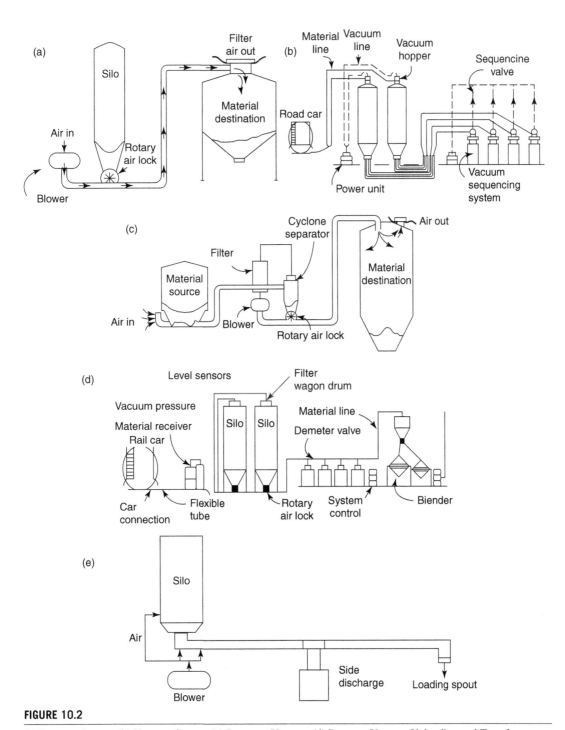

**FIGURE 10.2**

(a) Pressure System (b) Vacuum System (c) Pressure–Vacuum (d) Pressure Vacuum Unloading and Transfer
(e) Fluidizing System

suspension until it reaches the receiving vessel, where it is separated from the air by means of cyclone or air filter.

   **b.** Pressure systems are used for free flowing materials of almost any particle size, up to 6.35 mm. These systems are favored when one source must supply several receivers. Conveying air is usually supplied by positive displacement blowers.

**2.** Vacuum systems

   **a.** Vacuum systems (Fig. 10.2b) are characterized by material moving in an air stream of pressure less than ambient. The advantages of this type are that, all the pumping energy is used to move the product, and that material can be sucked into the conveyor line without the need of a conveying feeder.

   **b.** Vacuum systems are typically used when flows do not exceed 7000 kg/h. The equivalent conveyor length is less than 305 m. They are widely used for finely divided materials.

   **c.** Vacuum systems are designed in particular for flows under 7.6 kg/min. used to transfer materials short distances from storage bins or bulk containers to process units. These type of conveyors are widely used in plastics and other processing operations where the variety of conditions requires flexibility in choosing pickup devices, power sources, and receivers.

**3.** Pressure–vacuum systems

   **a.** Combined pressure–vacuum systems (Fig. 10.2c) combine the best of both the pressure and vacuum methods. A vacuum is used to induce material into the conveyor and move it a short distance to a separator. Air passes through a filter and into the suction side of a positive displacement blower.

   **b.** The most typical application is the combined bulk vehicle unloading and transferring to product storage (Fig. 10.2d).

**4.** Fluidizing systems

   **a.** Fluidizing systems generally convey pre fluidized, finely divided non-free flowing materials over short distances, such as from storage bins to the entrance of a main conveying system.

   **b.** Fluidizing is accomplished by means of a chamber in which air is passed through a porous membrane that forms the bottom of the conveyor (Fig. 10.2).

   **c.** Prefluidizing has the advantage of reducing the volume of conveying air needed, consequently less power is required.

   **d.** The most common application of this type of conveyor is the well-known railroad air slide covered hopper car.

## 10.2.5 **EFFECTS OF MATERIAL PROPERTIES**

The properties of a material decide whether or not it can be successfully conveyed. Material properties are used in determining the type of system required and the design details of auxiliary equipment.

   Material properties should be properly analyzed and sorted according to the system capacity, power demand, and other effects to be considered. A summary of how material properties affect the design of a pneumatic conveyor is shown in Table 10.2.

## 10.2.6 **DESIGN CONSIDERATIONS**

The energy needed to operate the system shall be assessed:

**Table 10.2  Effects of Material Properties on Selection of Pneumatic Equipment**

| Property | Used to Determine |
|---|---|
| Relative density (specific gravity) | Power and air requirements |
| Settled specific mass | Power and air requirements |
| Aerated specific mass | Bin and hopper volumes; feeder capacities |
| Sieve analysis-complete spectrum down to +325 mesh and average particle size in micrometers (microns) through 325 mesh | Dust collection requirements; feeder clearances; type of seals at points of air pressure differential; type of bearing protection required, power requirements |
| Abrasiveness relative to a commonly known abrasive material | Type of conveying system materials of construction; type of feeder; type of seals at points of air pressure differential; type of bearing protection required; power requirements |
| Moisture content and rate of deliquescence | Type of conveying system; dust collection requirements; storage turnover requirements; air drying and bin venting requirements, type of feeder; necessity for flow inducers in bins and hoppers |
| pH or corrosiveness relative to some well-known acid | Materials of construction; air drying and venting requirements |
| Tackiness relative to some well-known sticky material | Type of conveying system; air drying and venting requirements; type of dust collection system |
| Aeration and deaeration characteristics | Type of conveying system; type of flow inducers in bins and hoppers; type of level indicator, design of chutes from bins and hoppers; requirements for deareation |
| Angle of repose | Design of bins and hoppers; type of flow inducers in bins and hoppers |
| Toxicity | Type of dust collection; type of flow inducers in bins and hoppers |
| Temperature limitations | Requirements for cooling conveying air; insulation requirements for bins, dust collectors and ducts |
| Crystal structure or form of particles | Type of conveying system; type of feeder; form of piping system |
| Absorption of odors | Location and type of conveying air filters |

1. to ensure that the most efficient system is chosen
2. to make an account of the utility requirements that must be furnished, that is, electric power and compressed air.

The actual brake horsepower (in kilowatts) necessary to develop the required air horsepower (in kilowatts) depends on the volumetric and mechanical efficiencies of the air mover. In estimating power requirements, the following efficiencies may be used:

1. 76% for reciprocating compressors
2. 67% for sliding-vane compressors
3. 65% for positive displacement blowers with lobed impellers.

**4.** 70% for positive displacement blowers with meshed screws.
**5.** 64% for fans with radial blades.

In any air mover used for pneumatic conveying, the volume of air entering the pipeline must equal the mass rate of flow required to maintain the material to air ratio of the conveying system under standard conditions.

All of the calculation procedures and specific design equations used by the vendor for determining:

**1.** volume of air entering the air mover
**2.** power consumption
**3.** pressure drop.

Should be provided for the company's review and all design basic data and information should be included in data sheets relevant to specific equipment and conveying lines. Table 10.3 presents air velocities for some conveying system.

Since pneumatic conveyors and their components are subject to continual improvements by a fast-changing supplier industry, the vendor manufacturers should be invited to submit alternative designs.

## 10.3 HANDLING BULK MATERIALS IN PACKAGES AND CONTAINERS

Main factors that should be considered during the selection process are:

Product properties with respect to its handling process to (CEMA) codes.

If a product is defined to be poisonous, flammable, or oxidizing, special consideration must be given to potential hazards. The container must then meet the regulations for transportation as a Dangerous Article, set forth by internationally accepted bodies and organizations such as the International Air Transportation Association (IATA) and Civil Aeronautics Board (CAB), and such others.

**Table 10.3 Air Velocities Needed to Convey Solids and Various Bulk Densities**

| Bulk Density (kg/m³) | Air Velocity (m/min) | Bulk Density (kg/m³) | Air Velocity (m/min.) |
|---|---|---|---|
| 160 | 884 | 1120 | 2347 |
| 240 | 1094 | 1200 | 2438 |
| 320 | 1256 | 1280 | 2515 |
| 400 | 1402 | 1360 | 2591 |
| 480 | 1539 | 1440 | 2652 |
| 640 | 1780 | 1600 | 2804 |
| 720 | 1882 | 1680 | 2880 |
| 800 | 1981 | 1760 | 2957 |
| 880 | 2072 | 1840 | 3118 |
| 960 | 2179 | 1920 | 3200 |
| 1040 | 2270 | | |

Transport and storage of a highly degradable pelletized product in bags may present the risk of breakage, especially if the bags are shipped across the country or must frequently be rehandled. So, a particular attempt should be made for container selection on the basis of straightforward analysis of practical limitations on container applicability.

Materials that pack under pressure should not be packaged in containers that are stored in such manner as to foster consolidation.

### 10.3.1 SELECTION OF CONTAINER

The cost of the product and the cost of the container must be factored into economic evaluations.

*Note*: As a general guidance, products with low prices are seldom found in the more expensive containers. Such containers are generally associated with a comparatively expensive product, or with special conditions that justify their higher costs.

### 10.3.2 PAPER BAG

A paper bag is the most versatile and economical package available for storing and handling a wide variety of powdered, granular, prilled, or lump bulk products and has the following advantages:

1  It can be manufactured in a variety of sizes and strengths.
2  Full bags can be dense-packed and stacked for efficient use of storage space; empty bags require very little space.
3  Outer surface bags can be specially treated to resist abrasion, scuffing, and insect infestation.
4  Special barrier sheets can be built into the bag to prevent passage of moisture and odors, and to resist penetration by grease and oil.

The following basic factors should be considered when sizing the paper bags:

1. what mass of bags is most convenient for the customers
2. the product bulk density before and after bagging
3. the bag size, which fits the available pallets
4. the kind of filling equipment
5. pallet size which best fits the truck, rail units, or other transport means
6. the height of stacked pallets with concern on the safety measures
7. the best bag-loading pattern.

### 10.3.3 FIBER DRUMS

Fiber drums are rigid containers, which are used widely for handling and storing bulk-solid products with the following foremost advantages and disadvantages:

1. light mass
2. economy
3. readiness for filling with no setup or assembly required
4. they can be stacked several tiers high
5. one disadvantage of fiber drums is that empty units require a great deal of storage area.

In fabrication of fiber drums by the manufacturers, they must follow united freight classification regulations or such similar regulations which are acceptable to the company and subject to company's approval.

Fiber drums should be ordered according to their inside diameter, wall thickness, and overall outside height. Specifications should define wall construction, types of ends, and any special barrier treatment.

### 10.3.4 STEEL DRUMS

Steel drums are used for certain dry products requiring strength, water tightness, weatherability, and general ruggedness. In this regard, specification given by Department of Transportation (DOT) is recommended..

Steel drums must bear a code indicating the metal gage, volume capacity, maker's name, date of manufacturer, and so forth.

## 10.4 STORAGE OF SOLIDS IN BULK

### 10.4.1 OUTDOOR BULK STORAGE (PILE)

A hydrophilic-products processing plant needs a large storage area to keep its products dry during the low demand season. Long experience achieved in many plants proved that hydrophilic material, such as ammonium nitrate, can be safely stored outside in bulk. This approach may be applied to similar bulk materials.

The outdoor bulk storage bins are considered as easy and economical to construct and simple in operation.

### 10.4.2 DESIGNING AND OUTDOOR STORAGE BIN

In design phase many factors have to be considered such as: storage capacity, rate of stockpiling, operation, the materials' physical and chemical properties, shape of the stockpile, area required, available capital investment, operating cost, product protection, etc.

## 10.5 SOLID–LIQUID SEPARATORS

This section covers minimum requirements for the process design (including criteria for type selection) of solid–liquid separators used in the production of the oil and/or gas, refineries and other gas processing and petrochemical plants.

Typical sizing calculation together with introduction for proper selection is also given for guidance.

In this section, process aspects of three types of most frequently used solid–liquid separators are discussed more or less in detail. These three types are:

1. filters
2. centrifuges
3. hydrocyclones.

## 10.5.1 **SOLID–LIQUID SEPARATOR TYPES**

Solid–liquid separator types often used in OGP processes are discussed in this part:

- filters
- centrifuges
- hydrocyclones
- gravity settlers.

### 10.5.1.1 *Separation principles*

Solid–liquid separation processes are generally based on either one or a combination of "gravity settling," "filtration" and "centrifugation," principles.

The principles of these kinds of mechanical separation techniques are briefly described in the following clauses. Note that as a general rule, mechanical separation occurs only when the phases are immiscible and/or have different densities.

### 10.5.1.2 *Mechanical separation by gravity*

Solid particles will settle out of a liquid phase if the gravitational force acting on the droplet or particle is greater than the drag force of the fluid flowing around the particle (sedimentation). The same phenomenon happens for a liquid droplet in a gas phase and immiscible sphere of a liquid immersed in another liquid.

Rising of a light bubble of liquid or gas in a liquid phase also follows the same rules, that is, results from the action of gravitational force (floatation).

Stokes' law applies to the free settling of solid particles in the liquid phase.

### 10.5.1.3 *Mechanical separation by momentum*

Fluid phases with different densities will have different momentum. If a two-phase stream changes direction sharply, greater momentum will not allow the particles of the heavier phase to turn as rapidly as the lighter fluid, so separation occurs. Momentum is usually employed for bulk separation of the two phases in a stream. Separation by centrifugal action is the most frequently technique used in this field.

### 10.5.1.4 *Mechanical separation by filtration*

Filtration is the separation of a fluid-solid or liquid gas mixture involving passage of most of the fluid through a porous barrier, which retains most of the solid particulates or liquid contained in the mixture.

Filtration processes can be divided into three broad categories, cake filtration, depth filtration, and surface filtration.

### 10.5.1.5 *Patterns of filtration process*

Regarding the flow characteristic of filtration, this process can be carried out in the three following forms:

1. *Constant-pressure filtration.* The actuating mechanism is compressed gas maintained at constant pressure.
2. *Constant-rate filtration.* Positive displacement pumps of various types are employed.
3. *Variable-pressure, variable-rate filtration.* The use of a centrifugal pump results in this pattern.

## 10.5.2 LIQUID FILTERS

Filtration is the separation of particles of solids from fluids (liquid or gas) or liquid from liquid gas mixture by use of a porous medium. This section deals only with separation of solids from liquid, that is, "liquid filtration."

### 10.5.2.1 Mechanisms of filtration

Three main mechanisms of filtration are cake filtration, depth filtration, and surface filtration. In cake filtration, solids form a filter cake on the surface of the filter medium. In depth filtration, solids are trapped within the medium using either cartridges or granular media such as sand or anthracite coal. Surface filtration, also called surface straining, works largely by direct interception. Particles larger than the pore size of the medium are stopped at the upstream surface of the filter.

### 10.5.2.2 Types of liquid filters

Considering the flow characteristics, three types of filtration processes exist, constant pressure, constant rate, and variable pressure–variable rate. Regarding the manner of operation, filtration may be continuous or batch.

Filter presses and vacuum drum filters are well known examples for batch and continuous filters respectively.

Most commonly used types of liquid filters may be named as follows:

1. strainers
2. screens
3. cartridge filters
4. candle filters
5. sintered filters
6. precoat filters
7. filter presses
8. rotary drum filters
9. rotary disk filters
10. belt filters
11. leaf filters
12. tipping pan filters.

### 10.5.2.3 Filter media

There are many different types of filter media available and all have an important role in filtration.

The range includes: paper, natural and synthetic fibers, powders, felt, plastic sheet and film, ceramic, carbon, cotton yarn, cloth, woven wire, woven fabric, organic and inorganic membranes, perforated metal, sintered metals, and many other materials. These may be generally divided into four groups, general media, membrane type media, woven wire, and expanded sheet media.

### 10.5.2.4 General types of media

The main advantage of paper is its ability to remove finer particles and its main disadvantage is its limited mechanical strength. Examples of paper are: filter sheets, natural fabrics, synthetic fabrics either monofilament or multifilament felts, needle felts, bounded media, wool resin electrostatic media, mineral wools, diatomaceous earth, perlite, silica hydrogels, glass fiber, charcoal cloth, carbon fiber, anthracite and ceramic media are some types of filter media in this group. Applications of filter cloths including some advantages and disadvantages of this type of media are shown in Table 10.4.

### 10.5.2.5 Membrane filters

Particles with diameters from smaller than 0.001 μm up to 1 μm can be filtered by microfiltration, ultrafiltration, reverse osmosis, dialysis, electrodialysis processes using porous, microporous, and non-porous membranes.

Membranes may be made from polymers, ceramic, and metals.

**Table 10.4  Applications of Filter Cloths**

| Material | Suitable for | Maximum Service Temp. (°C) | Principal Advantage(s) | Principal Disadvantage(s) |
|---|---|---|---|---|
| Cotton | Aqueous solutions, oils, fats, waxes cold acids and volatile organic acids | 90 | Inexpensive | Subject to attack by mildew and fungi |
| Jute wool | Aqueous solutions<br>Aqueous solutions and dilute acids | 85<br>80 | Easy to seal joints in filter presses<br>High strength or flexibility | High shrinkage, subject to moth attack in store. Absorbs water; not suitable for alkalis |
| Nylon | Acids, petrochemicals, organic solvents, alkaline suspensions | 150 | Easy cake discharge. Long life | Not suitable for alkalis |
| Polyester (terylene) | Acids, common organic solvents, oxidizing agents. | 100 | Good strength and flexibility | May become brittle |
| PVC | Acids and alkalis | Up to 90 | Initial shrinkage | Heat resistance poor |
| PTFE | Virtually all chemicals | 200 | Extreme chemical resistance | High cost |
| Polyethylene | Acids and alkalis | 70 | Excellent cake discharge | Soften at moderate temperatures |
| Polypropylene | Acids, alkalis, solvents (except aromaties and chlorinated hydrocarbons) | 130 | Easy cake discharge | |
| Dynes | Acids, alkalis, solvents, petrochemicals | 110 | Low-moisture absorption | |
| Orlon | Acids (including chromic acid), petrochemicals | Over 150 | Suitable for a wide range of chemical solutions, hot or cold (except alkalis) | |
| Vinyon | Acids, alkalis, solvents, petroleum products | 110 | | Lacks fatigue strength for flexing |
| Glass fiber | Concentrated hot acids, chemical solutions. | 250 | | Abrasive resistance poor |

### 10.5.2.6 Woven wire

Woven wire cloth is widely used for filtration and is available in an extremely wide range of materials and mesh sizes.

It can be woven from virtually any metal ductile enough to be drawn into wire form, preferred materials being phosphor bronze, stainless steel of the nickel/chrome type-AISI 304, 316, 316L, and monel.

Woven wire cloth is described nominally by a mesh number and wire size, that is, N mesh M mm (or swg). Mesh numbers may range from 2 (two wires per 25.4 mm or 1 in.) up to 400. Fine mesh with more than 100 wires per lineal 25.4 mm (in.) is called gauze. Woven wires may also be described by aperture opening, for example:

- coarse-aperture opening 1–12 mm;
- medium-aperture opening 0.18–0.95 mm (180–950 μm);
- fine-aperture opening 0.020–0.160 mm (20–160 μm).

### 10.5.2.7 Expanded sheet and nonwoven metal mesh

Perforated metal sheets, drilled plates, milled plates and expanded metal mesh are examples of this type of filter media.

Most of the strainers, air and gas filters, etc. are usually made using the type of filters media. Predictable and consistent performance is the main characteristic of it, which results from the controllability of the size of screen opening by the manufacturer.

### 10.5.2.8 Filter rating

Filters are rated on their ability to remove particles of a specific size from a fluid, but the problem is that a variety of very different methods are applied to specifying performances in this way. Quantitative figures are only valid for specific operating or test conditions.

### 10.5.2.9 Absolute rating

The absolute rating, or cut-off point of a filter refers to the diameter of the largest particle, normally expressed in micrometers (μm), which will pass through the filter. It therefore represents the pore opening size of the filter medium. Filter media with an exact and consistent pore size or opening thus, theoretically at least, should have an exact absolute rating.

Certain types of filter media, such as papers, felts, and cloths, have a variable pore size and thus no absolute rating at all. The effective cut-off is largely determined by the random arrangement involved and the depth of the filter. Performance may then be described in terms of nominal cut-off or nominal rating.

### 10.5.2.10 Nominal rating

A nominal filter rating is an arbitrary value determined by the filter manufacturer and expressed in terms of percentage retention by mass of a specified contaminant (usually glass beads) of given size. It also represents a nominal efficiency figure, or more correctly, a degree of filtration.

### 10.5.2.11 Mean filter rating

A mean filter rating is a measurement of the mean pore size of a filter element. It establishes the particle size above which the filter starts to be effective.

### 10.5.2.12 Beta (ß) ratio

The beta ratio is a rating system introduced with the object of giving both filter manufacturer and user an accurate and representative comparison amongst filter media. It is determined by a multi-pass test, which establishes the ratio of the number of upstream particles larger than a specific size to the number of downstream particles larger than a specified size, that is,

$$\beta_x = \frac{N_u}{N_d}$$

where, $\beta_x$ is beta rating (or beta ratio) for contaminants larger than $x$ µm $N_u$ is number of particles larger than $x$ µm per unit of volume upstream $N_d$ is number of particles larger than the $x$ µm per unit of volume downstream.

### 10.5.2.13 Filter efficiency for a given particle size

Efficiency for a given particle size ($E_x$) can be derived directly from the ratio by the following equation:

$$E_x = \frac{\beta_{x-1}}{x} \times 100$$

where, $E_x$ is filter efficiency for particles with $x$ micrometer diameter size $\beta_x$ (beta) is the rating or B ratio of filter, (dimensionless); see Table 10.5 $x$ is the particle size (m).

*Example*: If a filter has a $\beta_5$ rating of 100, this would mean that the filter is capable of removing 99% of all particles of greater size than 5 µm.

**Table 10.5  Filter Rating**

| $\beta$ Value at $x$ mm $\beta_\lambda$ | Cumulative Efficiency µ% Particles $x$ µm | Stabilized Downstream Count $x$ µm Where Filter is Challenged Upstream With 1,000,000 Particles $x$ µm |
|---|---|---|
| 1.0 | 0 | 1,000,000 |
| 1.5 | 33 | 670,000 |
| 2.0 | 50 | 500,000 |
| 10 | 90 | 100,000 |
| 20 | 95 | 50,000 |
| 50 | 98.0 | 20,000 |
| 75 | 98.7 | 13,000 |
| 100 | 99.0 | 10,000 |
| 200 | 99.5 | 5000 |
| 1000 | 99.90 | 1000 |
| 10,000 | 99.99 | 100 |

### 10.5.2.14 Filter efficiency (separation efficiency)

As noted previously the nominal rating is expressed in terms of an efficiency figure. Efficiency is usually expressed as a percentage and can also be derived directly from the beta ratio as this is consistent with the basic definition of filter efficiency which is:

Number of emergent particles = 1 − number of incident particles × 100 (%)

### 10.5.2.15 Filter permeability

Permeability is the reciprocal expression of the resistance to flow offered by a filter. It is normally expressed in terms of a permeability coefficient, but in practice, permeability of a filter is usually expressed by curves showing pressure drop against flow rate.

## 10.5.3 FILTER SELECTION

### 10.5.3.1 Factors to be considered in filter selection

Three major factors, which should be considered in filter selection are performance, capital, and operating costs and availability. Performance and some other important factors are discussed in the following sections.

### 10.5.3.2 Performance

Filter performance may be determined by the "cut-off" achieved by the filter and/or other methods. The most meaningful figure now widely adopted is the "beta rating" associated with a particle size and efficiency figure.

### 10.5.3.3 Filter size

The size of filter needs to be selected with regards to the acceptable pressure drop and time required between cleaning or element replacement. This is closely bound up with the type of element and the medium employed.

Where space is at a premium, the overall physical size can also be a significant factor.

### 10.5.3.4 Surface versus depth filters

Surface type filters generally have relatively low permeability. To achieve a reasonably low-pressure drop through the filter, the element area must be increased so that the velocity of flow through the element is kept low.

### 10.5.3.5 Compatibility

Other essential requirements from the filter element are complete compatibility with the fluid and system. Compatibility with the fluid itself means freedom from degradation or chemical attack or a chemically compatible element. At the same time, however, 'mechanical compatibility' is also necessary to ensure that the element is strong enough for the duty involved and also free from migration.

### 10.5.3.6 Contamination levels

Contamination level may also affect the type of filter chosen for a particular duty, thus an oil bath filter for example may be preferred to a dry element type in a particularly dust laden atmosphere (eg, internal combustion engines operating under desert conditions) due to its large dust holding capacity.

### 10.5.3.7 Prefiltering

Particularly where fine filtering is required, the advisability or even necessity of prefiltering should be considered. In fact, with any type of filter which shows virtually 100% efficiency at a particle size substantially lower than the filtering range required, prefiltering is well worth considering as an economic measure to reduce the dirt load reaching the filter depending on the level of contamination involved.

## 10.5.4 LIQUID FILTER SELECTION GUIDE

### 10.5.4.1 Selection of media

Basic types of fluid filters are summarized in Tables 10.6 while Table 10.7 presents a basic selection guide.

It must be emphasized, however, that such a representation can only be taken as a general guide.

Particular applications tend to favor a specific type of filter and element or range of elements. Furthermore, filtering requirements may vary considerably. Thus, instead of being a contaminant, the residue collected by the filter may be the valuable part, which needs to be removed easily (necessitating the use of a type of filter which builds up a cake). Equally, where the residue collected is contamination, ease of cleaning or replacement of filter elements may be a necessary feature for the filter design.

**Table 10.6  Basic Types of Fluid Filters**

| Type | Media | Remarks |
|------|-------|---------|
| Surface | 1. Resin-impregnated paper (usually pleated). | Capable of fine (nominal) filtering |
| | 2. Fine-wove fabric cloth (pleated or 'star' form) | Low permeability |
| | 3. Membranes | Low resistance than paper |
| | 4. Wire mesh and perforated metal | Ultra-fine filtering |
| | | Coarse filtering and straining |
| Depth | 1. Random fibrous materials | Low resistance and high dirt capacity |
| | 2. Felts | Porosity can be controlled/graduated by manufacture Provide both surface and depth filtering. Low resistance |
| | 3. Sintered elements | Sintered metals mainly, but ceramics for high temperature filters |
| Edge | 1. Stacked discs. | Paper media are capable of extremely fine filtering |
| | 2. Helical wound ribbon | Metallic media have high strength and rigidity |
| Precoat | Diatomaceous earth, pearlite powered volcanic rock, etc. | Form filter beds deposited on flexible semi-flexible or rigid elements |
| | | Particularly suitable for liquid clarification |
| Adsorbent | 1. Activated clays | Effective for removal of some dissolved contaminants in water, oils, etc. Also used as precoat or filter bed material |
| | 2. Activated charcoal | Particularly used as drinking water filters |

**Table 10.7 General Selection Guide for Fluid Filters**

| Element | Sub micrometer (under 1) | Ultra-Fine (1–2.5) | Very fine (2.5–5) | Fine (5–10) | Fine/Medium (10–20) | Medium (20–40) | Coarse (10–20) |
|---|---|---|---|---|---|---|---|
| Perforated metal | | | | | | | X |
| Wire mesh | | | | | | | X |
| Wire gauze | | | | | | X–X | X |
| Pleated paper | | | | | X–X | X–X | |
| Pleated fabric | | | | | | X–X | |
| Wire wound | | | | | X | X–X | |
| Wire cloth | | | | X | X–X | X–X | X |
| Sintered wire cloth | | | | X | X–X | | |
| Felt | | | | | | X–X | |
| Metallic felt | X | X–X | X | X–X | X–X | | |
| Edge type, paper | | | X–X | X–X | | X | X |
| Edge type, ribbon element | | | | | X | X–X | X |
| Edge type, nylon | X | X–X | | X–X | X–X | X–X | X |
| Microglass | | | X–X | Limited application for liquids | | | |
| Mineral wool | X | X–X | | | | | |
| Ceramic | | | X–X | X–X | | X–X | X |
| Filter cloths | X | X–X | | X | X–X | | |
| Membrane | | X | X–X | | | X–X | |
| Sintered metal | | | X–X | X–X | X–X | | |
| Sintered PTFE | | | | X–X | X–X | X–X | X |
| Sintered polythene | | | | | | | X |

As a rough or primary selection procedure, the following steps may be followed:

1. Find the particle size range, either from design data or use Table 10.5.
2. Find suitable filter media using Tables 10.4 and 10.6.
3. Considering other process factors, find the proper filtration type and filter medium from Tables 10.6 and 10.8.

### 10.5.4.2 Selection of filter type

When the filter medium type is fixed, filter type selection should be performed based upon the process requirements like the allowable pressure drop, physical size, cleaning period, cleaning method, the

**Table 10.8  General Guide to Contaminant Sizes**

| Contaminant | Particle Size (mm) | | | | | |
|---|---|---|---|---|---|---|
| | Under 0.01 | 0.01–0.1 | 0.1–1 | 1–10 | 10–100 | 100–1000 |
| Hemoglobin | X | | | | | |
| Viruses | X | X | | | | |
| Bacteria | | | X–X | | | |
| Yeasts and fungi | | | X–X | X | | |
| Pollen | | | | X | | |
| Plant spores | | | | X–X | X | |
| Inside dust | X | X–X | X–X | X–X | X | |
| Atmospheric dust | | | | X–X | X–X | |
| Industrial dusts | | | | X | X–X | X–X |
| Continuously suspended dusts | X | X–X | X–X | | | |
| Oil mist | | X | X–X | X–X | | |
| Tobacco smoke | | X–X | X–X | X | | |
| Industrial gases | | | | X–X | | |
| Aerosols | X | X–X | X–X | X | | |
| Powdered insecticides | | | | X–X | | |
| Permanent atmospheric pollution | X | X–X | X–X | | | |
| Temporary atmospheric pollution | | | | X–X | X–X | X–X |
| Contaminants harmful to machines | | | | X | X–X | X–X |
| Machine protection normal | | | | X | X–X | X–X |
| Machine protection maximum | | | | X–X | X–X | X–X |
| Silt control | | | | X | | |
| Partial silt control | | | | | X | |
| Chip control | | | | | X | |
| Air filtration, primary | | | | | X–X | X–X |
| Air filtration, secondary | | | | X–X | | |
| Air filtration, ultra-fine | | X | X–X | | | |
| Staining particle range | | | X–X | X | | |

value of the residue and the actions which should be taken on it, etc. Then there are the factors which dictate whether a continuous or a batch filter should be chosen.

Other important factors to be taken into consideration are cost and maintainability of the filter.

# 10.6 CENTRIFUGES

Centrifugal separation is a mechanical means of separating the components of a mixture by accelerating the material in a centrifugal field.

Commercial centrifuges can be divided into two broad types, sedimentation centrifuges and centrifugal filters.

## 10.6.1 SEDIMENTATION CENTRIFUGES

Sedimentation centrifuges remove or concentrate particles of solids in a liquid by causing the particles to migrate through the fluid radially toward or away from the axis of rotation, depending on the density difference between particles and liquid.

In commercial centrifuges the liquid-phase discharge is usually continuous.

## 10.6.2 SEDIMENTATION BY CENTRIFUGAL FORCE

A solid particle settling through a liquid in a centrifugal-force field is subjected to a constantly increasing force as it travels away from the axis of rotation. It therefore never reaches a true "terminal" velocity.

However, at any given radial distance $r$ the settling velocity of a sufficiently small particle is very nearly given by the Stokes' Law relation.

### 10.6.2.1 Centrifugal filters

The centrifugal filter supports the particulate solids phase on a porous septum, usually circular in cross section, through which the liquid phase is free to pass under the action of the centrifugal force. The density of the solid phase is important only for calculation of the mass loading in the available volume of the basket. A more important parameter is the permeability of the filter cake under the applied centrifugal force.

Centrifugal filtration is often applied to batch production on fine, slow draining solids, but it is better suited to handle medium to coarse particles that require fair to good washing and a low residual liquid content.

## 10.6.3 SELECTION OF CENTRIFUGES

Table 10.9 indicates the particle size range to which the centrifuge types are generally applicable.

Table 10.10 summarizes the several types of commercial centrifuges, their manner of liquid and solid discharge, their unloading speed, and their relative maximum (pumping) capacity. When either the liquid or the solid discharge is not continuous, the operation is said to be cyclic.

Cyclic or batch centrifuges are often used in continuous processes by providing appropriate upstream and downstream surge capacity.

Table 10.9  Classification of Centrifuges by Size of Dispersed Particles

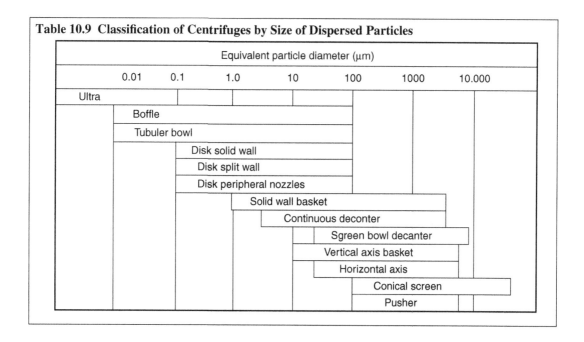

*Note*: Unless operating data on similar material are available from other sources, continuous centrifuges should be selected and sized only after tests on a centrifuge of identical configuration.

### 10.6.3.1 Power requirement

Typical energy demand values for sedimentation centrifuges handling dilute slurries, in joules per liter of feed, are, for tubular and disk, 950–9500, for nozzle-discharge disk, 1900–11,500, and for helical-conveyer decanters, 2800–14,300. Nozzle-discharge centrifuges typically consume 54,000–1,44,000 kJ/1000 kg of solids discharged through the nozzles.

Typical values for centrifugal filters handling moderately concentrated feeds, in kilojoules per ton (1000 kg) of dry solids, are, for automatic batch (constant speed), 10,800–36,000, for automatic batch (variable speed), 18,000–90,000, for "pusher" centrifuges, 7,200–27,000, and for vibrating and oscillating conical-screen machines, 1,080–36,000.

### 10.6.3.2 Required data for selection

For preliminary screening as to the suitability of the application for centrifugal separation and tentative selection of suitable centrifuge types, the following information is needed:

1. nature of the liquid phase(s)
   a. temperature
   b. viscosity at operating temperature
   c. density at operating temperature
   d. vapor pressure at operating temperature
   e. corrosive characteristics

**Table 10.10  Typical Specifications for Metal (Membrane) Filters**

| Max. Pore Size of Filters (Diameter in mm) | Open Pore Area per 1 dm² (Approx. Values) | Models Available | |
| | | Discs (Max. Diameter in mm) | Squares (Max. Edge Length in mm) |
| --- | --- | --- | --- |
| 2.0 | 4.6 mm² | 140 | 150 |
| 3.0 | 10.5 mm² | 140 | 150 |
| 5.0 | 29.0 mm² | 140 | 300 |
| 8.0 | 74.0 mm² | 140 | 300 |
| 10.0 | 1.2 cm² | 140 | 300 |
| 15.0 | 3.3 cm² | 140 | 500 |
| 20.0 | 5.8 cm² | 140 | 500 |
| 25.0 | 9.0 cm² | 140 | 500 |
| 30.0 | 13.0 cm² | 140 | 500 |
| 40.0 | 23.6 cm² | 140 | 500 |
| 50.0 | 39.0 cm² | 140 | 500 |
| 70.0 | 32.0 cm² | 140 | 500 |
| 80.0 | 38.0 cm² | 140 | 500 |
| 90.0 | 48.0 cm² | 140 | 500 |
| 100.0 | 44.0 cm² | 140 | 500 |

    **f.** fumes are noxious, toxic, inflammable, or none of these
    **g.** contact with air is not important, is undesirable, or must be avoided.
**2.** nature of the solids phase
    **a.** particle size and distribution
    **b.** particles are amorphous, flocculant, soft, friable, crystalline, or abrasive
    **c.** particle size degradation is unimportant, undesirable, or highly critical
    **d.** concentration of solids in feed
    **e.** density of solids particles
    **f.** retained mother liquid content, _____% is tolerable, _____% is desired
    **g.** rinsing to further reduce soluble mother liquor impurities is unnecessary or required.
**3.** quantity of material to be handled per batch or per unit time.

## 10.7  HYDROCYCLONES

Hydrocyclones are used for solid–liquid separations; as well as for solid classification, and liquid–liquid separation. It is a centrifugal device with a stationary wall, the centrifugal force being generated by the liquid motion.

### 10.7.1 VARIABLES AFFECTING HYDROCYCLONE PERFORMANCE

#### *10.7.1.1 Effects of process variables*

The ability of this gravity-force machine to effect an adequate solids/liquid separation is governed by Stokes' Law. Specifically, the ease of separation is directly proportional to the suspended particle diameter, squared times the relative density (specific gravity), differential between the solid and the liquid phases, and inversely proportional to the viscosity of the continuous liquid phase.

#### *10.7.1.2 Effect of mechanical design characteristics*

The ability of a hydrocyclone to meet required solids/liquid separation needs is governed by the design variables of the equipment itself. These variables include cone diameter, overall body length, as well as the dimensions of the feed, apex, and vortex openings. Regardless of the "Stokes" data available and the equipment design formulas that may exist, the suitability of a hydrocyclone to a given process must depend upon existing information known from past experience or upon results developed by laboratory or field testing.

## 10.8 FILTER MEDIA SPECIFICATIONS

Table 10.11 shows typical filter media specifications.

Tables 10.12–10.16 and figures provide more information about various aspects of the solid handling process

### 10.8.1 PRESSURE DROP CHARACTERISTICS OF STRAINERS

Please check Figs. 10.3 and 10.4 for estimation and characteristics of pressure drop.

**Table 10.11 Principal Weaves for Wire Cloths**

| Name | Characteristics | Absolute rating range $\mu m$ | Remarks |
|---|---|---|---|
| Square (plain or twilled) | Largest open area and lowest flow resistance | 20–300+ | Most common type of weave. |
| | Aperture size is the same in both directions | | Made in all grades from coarse to fine. |
| Plain Dutch single weave | Good contaminant retention properties with low flow resistance | 20–100 | Openings are triangular. |
| Reverse plain Dutch weave | Very strong with good contaminant retention | 15–115 | |
| Twilled Dutch double weave | Regular and consistent aperture size | 6–100 | Used for fine and ultra-fine filtering. |

**Table 10.12  A Typical List of Wire Cloth Specifications**

| Aperture (μm) | Wire (mm) | Diameter (mm) | Open Area % | Mech Wire | Diameter (in.) |
|---|---|---|---|---|---|
| 25 | 0.025 | 0.025 | 25 | 500 | 0.0010 |
| 28 | 0.028 | 0.025 | 28 | 480 | 0.0010 |
| 32 | 0.032 | 0.028 | 28 | 425 | 0.0011 |
| 38 | 0.038 | 0.025 | 36 | 460 | 0.0010 |
| 40 | 0.04 | 0.032 | 31 | 425 | 0.0012 |
| 42 | 0.042 | 0.036 | 29 | 400 | 0.0014 |
| 45 | 0.045 | 0.036 | 31 | 350 | 0.0014 |
| 50 | 0.05 | 0.036 | 34 | 325 | 0.0014 |
| 56 | 0.056 | 0.040 | 34 | 115 | 0.0016 |
| 63 | 0.063 | 0.040 | 47 | 300 | 0.0016 |
| 75 | 0.075 | 0.036 | 46 | 270 | 0.0014 |
| 75 | 0.075 | 0.053 | 36 | 250 | 0.0021 |
| 80 | 0.08 | 0.050 | 38 | 230 | 0.0020 |
| 85 | 0.085 | 0.040 | 46 | 200 | 0.0016 |
| 90 | 0.09 | 0.050 | 41 | 200 | 0.0020 |
| 95 | 0.095 | 0.045 | 46 | 200 | 0.0018 |
| 100 | 0.1 | 0.063 | 38 | 180 | 0.0025 |
| 106 | 0.106 | 0.05 | 46 | 180 | 0.0020 |
| 112 | 0.112 | 0.08 | 34 | 150 | 0.0032 |
| 125 | 0.125 | 0.09 | 34 | 165 | 0.0035 |
| 140 | 0.14 | 0.112 | 31 | 130 | 0.0045 |
| 150 | 0.15 | 0.10 | 36 | 120 | 0.0040 |
| 160 | 0.16 | 0.10 | 38 | 100 | 0.0040 |
| 180 | 0.18 | 0.14 | 32 | 100 | 0.0055 |
| 200 | 0.2 | 0.125 | 38 | 100 | 0.0050 |
| 200 | 0.2 | 0.14 | 35 | 80 | 0.0055 |
| 224 | 0.224 | 0.16 | 34 | 80 | 0.0065 |
| 250 | 0.25 | 0.16 | 37 | 75 | 0.0065 |
| 280 | 0.25 | 0.22 | 31 | 65 | 0.009 |
| 315 | 0.315 | 0.20 | 37 | 60 | 0.008 |
| 400 | 0.4 | 0.22 | 42 | 50 | 0.009 |
| 400 | 0.4 | 0.25 | 38 | 50 | 0.010 |
| 435 | 0.425 | 0.38 | 36 | 40 | 0.011 |

| Table 10.12 A Typical List of Wire Cloth Specifications* (*cont.*) | | | | | |
|---|---|---|---|---|---|
| Aperture (μm) | Wire (mm) | Diameter (mm) | Open Area % | Mech Wire | Diameter (in.) |
| 500 | 0.5 | 0.20 | 51 | 40 | 0.008 |
| 500 | 0.5 | 0.25 | 44 | 36 | 0.010 |
| 500 | 0.5 | 0.32 | 37 | 36 | 0.012 |
| 560 | 0.56 | 0.28 | 44 | 33 | 0.011 |
| 560 | 0.56 | 0.36 | 37 | 30 | 0.014 |
| 630 | 0.63 | 0.25 | 51 | 30 | 0.010 |
| 630 | 0.63 | 0.28 | 48 | 28 | 0.011 |
| 630 | 0.63 | 0.40 | 37 | 30 | 0.016 |
| 710 | 0.71 | 0.32 | 48 | 28 | 0.012 |
| 710 | 0.71 | 0.45 | 37 | 25 | 0.018 |
| 800 | 0.8 | 0.32 | 51 | 25 | 0.012 |
| 800 | 0.8 | 0.5 | 38 | 22 | 0.020 |
| — | 1 | 0.36 | 54 | 22 | 0.0014 |
| — | 1 | 0.61 | 38 | 20 | 0.0025 |
| — | 1.25 | 0.4 | 57 | 18 | 0.0016 |
| — | 1.6 | 0.5 | 58 | 16 | 0.0020 |
| — | 2 | 0.56 | 61 | 16 | 0.0022 |
| — | 2.5 | 0.71 | 61 | 12 | 0.0028 |
| — | 3.15 | 0.8 | 64 | 10 | 0.0022 |
| — | 4 | 1.0 | 64 | 8 | 0.04 |
| — | 5 | 1.25 | 64 | 6 | 0.05 |
| — | 6.3 | 1.25 | 70 | 5 | 0.05 |
| — | 7.1 | 1.4 | 70 | 4 | 0.055 |
| — | 8 | 2 | 64 | 3 | 0.08 |
| — | 10 | 2.5 | 64 | 3 | 0.10 |
| — | 12.5 | 2.8 | 67 | 2 | 0.11 |
| — | 16 | 1.2 | 69 | — | 0.12 |

## 10.8.2 SPECIFICATIONS OF CENTRIFUGES AND CAPACITY CHECKING FOR HYDROCYCLONES

## 10.8.3 CAPACITY CHECKING FOR SIZED HYDROCYCLONES

If the pressure differential for a hydrocyclone separation unit is fixed by the process conditions and a properly sized device has been selected, the capacity of the unit can be determined as shown in Fig. 10.5 When a hydrocyclone does not have adequate capacity over the pressure range indicated to handle a given problem, multiple hydrocyclones are manifolded in parallel.

**Table 10.13 Perforated Mesh Sieves**

| Nominal Width of Aperture (Side of Square) | | Plate Thickness BG | Aperture Tolerances | | | |
|---|---|---|---|---|---|---|
| | | | Average | | Maximum | |
| mm | in | BG | % | Units | % | Units |
| 101.60 | 4 | 10 | 0.20 | 80 | 0.50 | 200 |
| 88.90 | 3½ | 10 | 0.20 | 70 | 0.49 | 170 |
| 76.20 | 3 | 12 | 0.20 | 60 | 0.50 | 150 |
| 69.85 | 2¾ | 12 | 0.20 | 55 | 0.51 | 140 |
| 63.50 | 2½ | 14 | 0.20 | 50 | 0.52 | 130 |
| 57.15 | 2¼ | 14 | 0.20 | 45 | 0.53 | 120 |
| 50.80 | 2 | 16 | 0.20 | 40 | 0.50 | 100 |
| 47.63 | 17/8 | 16 | 0.21 | 40 | 0.53 | 100 |
| 44.45 | 1¾ | 16 | 0.20 | 35 | 0.51 | 90 |
| 41.28 | 15/8 | 16 | 0.21 | 35 | 0.55 | 90 |
| 38.10 | 1½ | 16 | 0.20 | 30 | 0.53 | 80 |
| 34.93 | 13/8 | 16 | 0.22 | 30 | 0.58 | 80 |
| 31.75 | 1¼ | 16 | 0.24 | 30 | 0.56 | 70 |
| 28.58 | 11/8 | 16 | 0.26 | 30 | 0.62 | 70 |
| 25.40 | 1 | 16 | 0.25 | 25 | 0.60 | 60 |
| 22.23 | 7/8 | 16 | 0.23 | 20 | 0.69 | 60 |
| 19.05 | ¾ | 16 | 0.27 | 20 | 0.80 | 60 |
| 15.88 | 5/8 | 16 | 0.32 | 20 | 0.80 | 50 |
| 12.70 | ½ | 16 | 0.40 | 20 | 1.00 | 50 |
| 9.53 | 3/8 | 18 | 0.53 | 20 | 1.06 | 40 |
| 7.94 | 5/16 | 18 | 0.58 | 18 | 1.16 | 36 |
| 6.35 | ¼ | 18 | 0.60 | 15 | 1.20 | 30 |
| 4.76 | 3/16 | 20 | 0.64 | 12 | 1.33 | 25 |

## 10.9 DRYERS

Drying is an important unit operation concept in which water and other volatile liquids can be separated from solids and semisolid materials and from gases and liquids. Drying is most commonly used in OGP process plants for removal of water or solvents from solids by thermal means, dehydration of gases by condensation, adsorption or absorption, and drying of liquids by fractional distillation, or adsorption of fluids.

**Table 10.14  Perforated Metal Data**

| Size of Hole | | |
|---|---|---|
| **mm** | **in.** | **½ Open Area** |
| Round hole | | |
| 0.38 | 0.015 | 10 |
| 0.55 | 0.0215 | 20 |
| 0.70 | 0.0275 | 30 |
| 0.80 | 0.0315 | 32 |
| 1.09 | 0.043 | 25 |
| 1.40 | 0.049 | 25 |
| 1.50 | 0.055 | 32 |
| 1.5 | 0.059 | 37 |
| 1.64 | 0.065 | 36 |
| 1.75 | 0.069 | 19 |
| 2.16 | 0.085 | 33 |
| 2.45 | 0.097 | 36 |
| 2.85 | 0.112 | 50 |
| Square hole (parallel) | | |
| 1.50 | 0.059 | 44 |
| 3.17 | 0.125 | 44 |
| 6.00 | 0.236 | 54 |
| 6.35 | 0.256 | 44 |
| 7.00 | 0.273 | 41 |
| 9.52 | 0.375 | 44 |
| 11.00 | 0.437 | 49 |
| 12.70 | 0.500 | 44 |
| 19.05 | 0.750 | 56 |
| 25.40 | 1.00 | 44 |
| Square hole (alternate) | | |
| 1.75 | 0.069 | 32 |
| 3.17 | 0.125 | 32 |
| 4.75 | 0.187 | 44 |
| 6.75 | 0.250 | 44 |
| 7.93 | 0.312 | 64 |

(*Continued*)

**Table 10.14  Perforated Metal Data (*cont.*)**

| Size of Hole | | ½ Open Area |
|---|---|---|
| **mm** | **in.** | |
| 9.53 | 0.375 | 56 |
| 11.10 | 0.437 | 60 |
| 12.70 | 0.500 | 53 |
| 19.05 | 0.750 | 56 |
| 25.40 | 1.0 | 57 |
| Diamond squares | | |
| 4.75 | 0.178 | 36 |
| 9.52 | 0.375 | 49 |
| 12.70 | 0.500 | 48 |
| 15.87 | 0.625 | 42 |
| 19.05 | 0.750 | 44 |
| 25.40 | 1.0 | 43 |
| Round end slots | | |
| 10.00 × 0.50 | 0.394 × 0.019 | 13 |
| 10.00 × 1.00 | 0.394 × 0.039 | 23 |
| 10.00 × 1.50 | 0.394 × 0.059 | 32 |
| 20.00 × 2.00 | 0.787 × 0.059 | 34 |
| 10.00 × 2.00 | 0.394 × 0.079 | 30 |
| 30.00 × 2.00 | 0.787 × 0.079 | 30 |
| 13.00 × 2.50 | 0.518 × 0.098 | 28 |
| 20.00 × 2.50 | 0.787 × 0.098 | 31 |
| 12.00 × 3.00 | 0.427 × 0.118 | 38 |
| 20.00 × 3.00 | 0.787 × 0.118 | 47 |
| 25.00 × 3.50 | 0.984 × 0.117 | 38 |
| Square and slots (parallel) | | |
| 10.00 × 0.40 | 0.394 × 0.016 | 14 |
| 10.00 × 0.56 | 0.394 × 0.022 | 19 |
| 10.00 × 0.76 | 0.394 × 0.03 | 25 |
| 20.00 × 1.10 | 0.812 × 0.043 | 33 |
| 20.32 × 1.44 | 0.800 × 0.057 | 29 |
| 19.05 × 1.59 | 0.730 × 0.0625 | 27 |
| 13.00 × 3.50 | 0.511 × 0.089 | 37 |

**Table 10.14  Perforated Metal Data (*cont.*)**

| Size of Hole | | ½ Open Area |
|---|---|---|
| **mm** | **in.** | |
| 20.00 × 3.35 | 0.787 × 0.128 | 41 |
| 19.84 × 1.96 | 0.781 × 0.150 | 41 |
| 19.05 × 4.75 | 0.730 × 0.187 | 45 |
| 15.87 × 6.35 | 0.625 × 0.250 | 47 |
| 20.00 × 8.00 | 0.787 × 0.314 | 49 |
| Diagnrial slots | | |
| 12.29 × 0.50 | 0.484 × 0.020 | 14 |
| 12.29 × 0.62 | 0.484 × 0.024 | 19 |
| 11.91 × 0.73 | 0.469 × 0.029 | 12 |
| 11.91 × 1.07 | 0.469 × 0.042 | 25 |
| 20.62 × 1.09 | 0.812 × 0.043 | 27 |
| 9.90 × 2.38 | 0.390 × 0.093 | 27 |
| 11.91 × 3.17 | 0.469 × 0.125 | 37 |
| 12.70 × 1.96 | 0.500 × 0.156 | 36 |
| 12.70 × 1.04 | 0.500 × 0.041 | 28 |
| 20.00 × 2.00 | 0.787 × 0.078 | 29 |
| 11.50 × 1.50 | 0.454 × 0.059 | 24 |
| 19.05 × 3.17 | 0.750 × 0.059 | 40 |
| Triangular holes | | |
| 3.17 | 0.125 | 26 |
| 5.00 | 0.197 | 15 |
| 6.50 | 0.256 26 | |
| 9.52 × 11.11 | 0.375 × 0.437 | 16 |
| Oral holes | | |
| 7.00 × 3.00 | 0.276 × 0.118 | 32 |
| 9.00 × 4.25 | 0.354 × 0.167 | 38 |
| 9.00 × 5.00 | 0.354 × 0.197 | 45 |
| 14.00 × 6.00 | 0.551 × 0.236 | 46 |
| 13.50 × 7.00 | 0.531 × 0.276 | 45 |

**Table 10.15 Specifications and Performance Characteristics of Typical Sedimentation Centrifuges**

| Type | Bowl Diameter (mm) | Speed (r/min) | Maximum Centrifugal | Liquid (m3/h) | Solid (kg/h) | Typical Motor Size (Kw) |
|---|---|---|---|---|---|---|
| Tubular | 45 | 50.000 | 62.400 | 0.021–0.056 | | |
| | 105 | 15.000 | 13.200 | 0.023–2.3 | | 1.49 |
| | 25 | 15.000 | 15.900 | 0.045–4.5 | | 2.24 |
| Disk | 178 | 12.000 | 14.300 | 0.023–2.3 | | 0.246 |
| | 330 | 7.500 | 10.400 | 1.14–11.4 | | 4.74 |
| | 610 | 4.000 | 5.500 | 4.5–45 | | 5.6 |
| Nozzle discharge | 254 | 10.000 | 14.200 | 4.3–9 | 91–910 | 14.9 |
| | 406 | 6.250 | 8.900 | 5.68–34 | 360–3600 | 29.8 |
| | 685 | 4.200 | 6.750 | 9–90 | 910–10.000 | 93.2 |
| | 762 | 3.300 | 4.600 | 9–90 | 910–10.000 | 93.2 |
| Helical conveyer | 152 | 8.000 | 5.500 | To 4.5 | 27–227 | 3.73 |
| | 356 | 4.000 | 3.380 | To 17 | 453–1360 | 14.9 |
| | 457 | 3.500 | 3.130 | To 11.4 | 453–1360 | 11.2 |
| | 635 | 3.000 | 3.190 | To 50.8 | 2270–11000 | 112 |
| | 813 | 1.800 | 1.470 | To 56.8 | 2720–9100 | 44.7 |
| | 1016 | 1.600 | 1.450 | To 58 | 9100–163000 | 74.6 |
| | 1372 | 1.000 | 770 | To 170 | 18150–54400 | 112 |
| Knife discharge | 508 | 1.800 | 920 | | 0.028 | 14.9 |
| | 915 | 1.200 | 740 | | 0.115 | 22.4 |
| | 1727 | 900 | 780 | | 0.574 | 29.8 |

In drying, material is transferred from one phase to another, which is complicated by the need to transfer heat and mass simultaneously, but in the opposite direction. Heat is transferred first, usually in different external heat-transfer mode such as: convection, conduction, radiation, dielectric heating etc. Then mass transfer occurs, involving the removal of surface moisture and movement of internal moisture to the surface. Many dryers employ more than one of these modes. Nevertheless, most industrial dryers are characterized by one that predominates: heat transfer mechanism.

Industrial dryers may be classified according to the physical characteristics of the material being dried, the method of transferring the thermal energy to wet product, the source of the thermal energy, the method of physical removal of the solvent vapor, and the method of dispersion (in the case of wet solids) in the drying operation.

As a consequence of dryer specialization, the selection of the type of dryer appropriate to the specific product to be dried becomes a critical step in the specification and design of the processing plant.

Table 10.16 Centrifugal Filters Classified by Flow Pattern

| Flow pattern | Fixed-bed type force** | Centrifugal (m³) | Basket capacity* (under lip ring) |
|---|---|---|---|
| Liquid: continuous (interrupted for discharge of solids) Solids: batch | Vertical axis Manual unload Container unload Knife unload Horizontal axis Knife unload | 1200 550 1800 1000 | 0.453 0.566 0.453 0.566 |
| Flow pattern | Moving-bed type | Centrifugal force** | Solids capacity (kg/h)*** |
| Liquid: continuous Solids: continuous | Conical scree Wide angle Differential scroll Axial vibration Torsional vibration Oscillating | 2400 1800 600 600 600 | 68,000 136,000 |
| | Cylindrical scree Differential Reciprocating | 600 600 | 36,000 27,000 |

*Reduce by 1/3 for volume of processed solids ready to be discharged.
**Nominal maximum centrifugal force ($w^2 r/g$) developed, usually less in larger sizes,
***Nominal maximum capacity of largest sizes, subject to reduction as necessary to meet required performance on a given application.

The choice of the wrong type of dryer can lead to inefficient operation, reduced product quality, and loss of profit.

This section is intended to cover minimum requirements for the process design of dryers used in oil, gas, and petrochemical process plants.

Although, as a common practice, dryers are seldom designed by the users, but are brought from companies that specialize in design and fabrication of drying equipment, the scope covered herein is for the purpose of establishing and defining general principles of the drying concept and mechanism, dryer classification and selection and to provide an accumulation of design information and criteria required for proper selection, design, and operation of solid, liquid, and gaseous drying equipment (dryers).

## 10.9.1 WET SOLID DRYERS

In the drying process the goal of many operations is not only to separate a volatile liquid, but also to produce a dry solid of specific size, shape, porosity, texture, color, or flavor. So, a good understanding of liquid and vapor mass transfer mechanism prior to design work is strongly recommended.

In the drying of wet solids, the following main factors, which essentially are used in the process design calculation of dryers should be defined in accordance with mass and heat transfer principles, process conditions and drying behavior:

1. drying characteristics
2. constant-rate period

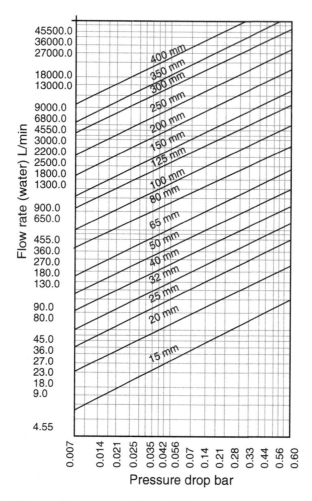

**FIGURE 10.3 Estimation of Pressure Loss in Typical Y-Type Strainers**

Note: This chart is based on water of relative density (specific gravity) 1.0 and viscosity 2–3 cSt. Screens are clean and are 40 × 40 woven wire mesh.

3. falling-rate period
4. moisture content
5. diffusion concept.

## 10.9.2 DRYING CHARACTERISTICS

The drying characteristics of wet solids is best described by plotting the average moisture content of material against elapsed time measured from the beginning of the drying process. Fig. 10.6 represents a typical drying-time curve. The experimental estimation of this curve must be made before one can begin the design calculations. The influence of the internal and external variables of drying on the drying-time curve should be determined in order that an optimal design can be developed.

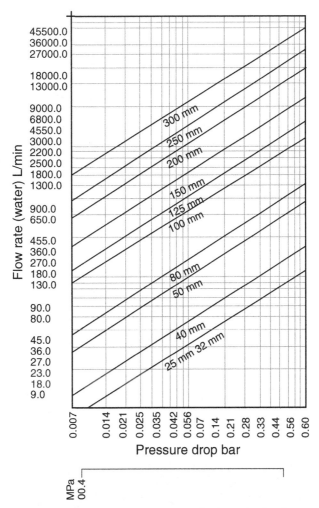

**FIGURE 10.4 Typical Pressure Drop Characteristics for Basket With 3 mm Perforations**

The drying-rate curve (Fig. 10.7) is derived from the drying-time curve by plotting slopes of the latter curve against the corresponding moisture content. The distinctive shape of this plot, shown in Fig. 10.2, illustrates the constant-rate period, terminating at the critical moisture content, followed by the falling-rate period. The variables that influence the constant-rate period are the so-called external factors consisting of gas mass velocity, thermodynamic state of the gas, transport properties of the gas, and the state of aggregation of the solid phase changes in gas temperature, humidity, and flow rate will have a pro-found effect on the drying rate during this period. The controlling factors in the falling-rate period are the transport properties of the solids and the primary design variable is temperature.

The characteristic drying behavior in these two periods is markedly different and must be considered in the design. In the context of economics, it shall be costlier to remove water in the falling-rate period than it is to remove it in the constant-rate period, accordingly, it is recommended that the length of the constant-rate period is extended with respect to falling rate as much practicable.

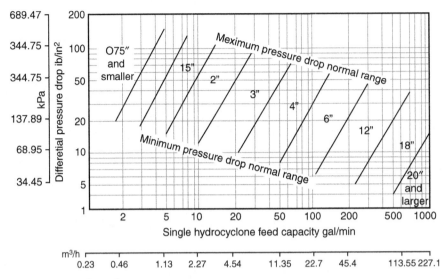

**FIGURE 10.5 Hydrocyclone Capacity**

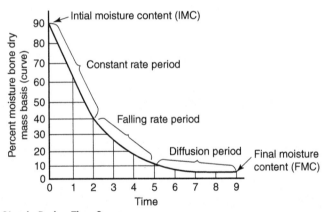

**FIGURE 10.6 Typical Classic Drying-Time Curve**

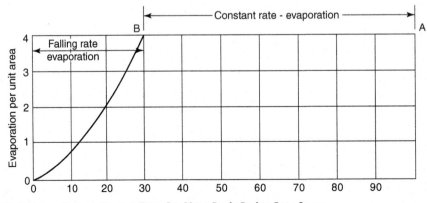

**FIGURE 10.7 Moisture Content, Percent Bone Dry Mass Basis Drying-Rate Curve**

### 10.9.2.1 Constant-rate period

In Fig. 10.2, the horizontal segment AB, which pertains to the first major drying period, is called the constant-rate period. During this period, the solid is so wet that a continuous film of water exists over the entire drying surface, and this water acts as if the solid were not there. If the solid is nonporous, the water removed in this period is mainly superficial water on the solid's surface.

The evaporation from a porous material is subject to the same mechanism as that from a wet-bulb thermometer.

### 10.9.2.2 Critical moisture content

The critical moisture content is the average material moisture content at which the drying rate begins to decline. A prototype drying test should be conducted to determine the critical moisture content. In Fig. 10.2, point B represents the constant-rate termination and marks the instant when the liquid water on the surface is insufficient to maintain a continuous film covering the entire drying area. The critical point (B) occurs when the superficial moisture has evaporated. In porous solids point B of Fig. 10.2 is reached when the rate of evaporation becomes the same as obtained by the wet-bulb evaporative process.

## 10.9.3 DETERMINING OF DRYING TIME

The three following methods are generally used in order of preference for determining of dryingtime:

1. Conduct tests in a laboratory dryer simulating conditions in the commercial machine, or obtain performance data directly from the commercial machine.
2. If the specific material is not available, obtain drying data on similar material by either of the above methods. This is subject to the investigator's experience and judgment.
3. Estimate drying time from theoretical Equation or any such appropriate theoretical formula.

When designing commercial equipment, tests are to be conducted in a laboratory dryer that simulates commercial operating conditions. Sample materials used in the laboratory tests should be identical to the material found in the commercial operation. The result from several tested samples should be compared for consistency. Otherwise, the test results may not reflect the drying characteristics of the commercial material accurately.

When laboratory testing is impractical, commercial drying can be based on the equipment manufacturer's experience as an important source of data.

Since estimating drying time from theoretical equations are only approximate values, care should be taken in using this method.

When selecting a commercial dryer, the estimated drying time determines what size machine is needed for a given capacity. If the drying time has been derived from a laboratory test, the following should be considered:

- In a laboratory dryer, considerable drying may be the result of radiation and heat conduction. In a commercial dryer, these factors are usually negligible.
- In a commercial dryer, humidity conditions may be higher than in a laboratory dryer. In drying operations with controlled humidity, this factor can be eliminated by duplicating the commercial humidity condition in the laboratory dryer.
- Operating conditions are not as uniform in a commercial dryer as in a laboratory dryer.

- Because of the small sample used the test material may not be representative of the commercial material. Thus, the designer must use experience and judgment to correct the test drying time to suit commercial conditions.

### 10.9.4 PSYCHOMETRY

Before drying can begin, a wet material must be heated to such a temperature that the vapor pressure of the liquid content exceeds the partial pressure of the corresponding vapor in the surrounding atmosphere. The effect of atmospheric vapor content of a dryer on the drying rate and material temperature is conveniently studied by construction of a psychometric chart (Fig. 10.8).

---

## 10.10 CLASSIFICATION OF INDUSTRIAL DRYING

Industrial dryers may be classified according to the following categories:

1. *Method of operation.* This category refers to the nature of the production schedule. For large-scale production the appropriate dryer is of the continuous type with continuous flow of the material into and out of the dryer. Conversely, for small production requirements, batch-type operation is generally desired.
2. *Physical properties of material.* The physical state of the feed is probably the most important factor in the selection of the dryer type. The wet feed may vary from a liquid solution, a slurry, a

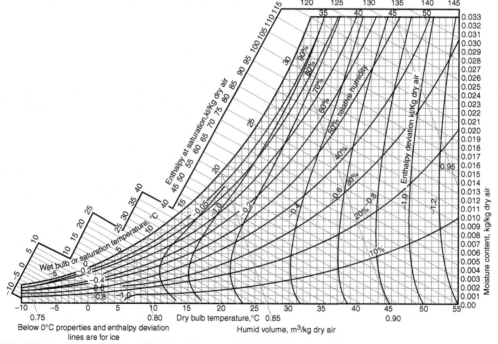

**FIGURE 10.8** Psychometric Chart [Air-Water Vapor at 101.325 kPa (=1 atm.)]

paste, or filter cake to free-flowing powders, granulations, and fibrous and non-fibrous solids. The design of the dryer is greatly influenced by the properties of the feed; thus dryers handling similar feeds have many design characteristics in common.

*Note*: A comprehensive classification of commercial dryers based on properties of materials handled, is given in Perry's *Chemical Engineering Handbook*.

**3.** *Conveyance*. In many cases, the physical state of the feed dictates the method of conveyance of the material through the dryer; however, when the feed is capable of being preformed, the handling characteristics of the feed may be modified so that the method of conveyance can be selected with greater flexibility. Generally, the mode of conveyance correlates with the physical properties of the feed.

**4.** *Method of energy supply*. Where the energy is supplied to the material by convective heat transfer from a hot gas flowing past the material, the dryer is classified as a convection type. Conduction-type dryers are those in which the heat is transferred to the material by the direct contact of the latter with a hot metal surface.

**5.** *Cost*. Cost effect of dryer selection influences the classification of industrial drying. When capacity is large enough, continuous dryers are less expensive than batch units. Those operating at atmospheric pressure cost about 1/3 as much as those at vacuum. Once through air dryers are one-half as expensive as reciprocable gas equipment. Dielectric and freeze dryers are the most expensive and are justifiable only for sensitive and specialty products. In large scale drying, rotary, fluidized bed, and pneumatic conveying dryers cost about the same.

**6.** *Special process features*. Special characteristics of the drying material together with particular features of the product are carefully considered in classifying dryer and selection of dryer type. Hazardous, heat sensitive, quality sensitive products, and cost effects can clearly dictate process consideration in classifications.

## 10.11 SELECTION OF DRYER

The choice of the best type of dryer to use for a particular application is generally dictated by the following factors:

**1.** the nature of the product, both physical and chemical
**2.** the value of the product
**3.** the scale of production
**4.** the available heating media
**5.** the product quality consideration
**6.** space requirements
**7.** the nature of the vapor (toxicity, flammability)
**8.** the nature of the solid (flammability, dust explosion hazard, toxicity).

For application of factors, in selection of process, a systematic procedure involving the following steps is recommended:

**1.** *Formulating of drying case as completely as possible*. In this step, the specific requirements and variables should be identified; thus, the important information derived can be summarized as:
   **a.** the product and its purity;

**b.** initial and final moisture content;

**c.** range of variation of initial and final moisture content;

**d.** production rate and basis.

2. *Collecting all available data related to the case.* In this step, the previous experience related to drying of particular product of interest or of a similar material should be investigated.

3. *Physical and chemical properties to be established.* The physical and chemical properties of feed and product including physical state of feed (filter cake, granulations, crystals, extrusions, briquettes, slurry, paste, powder, etc.) including size, shape, and flow characteristics; chemical state of the feed (pH, water of crystallization, chemical structure, degree of toxicity of vapor or solid, corrosive properties, inflammability of vapor or solid, explosive limits of vapor); and physical properties of dry product (dusting characteristics, friability, flow characteristics, and bulk density). Finally, available drying data in the form of prior laboratory results, pilot-plant performance data, or full-scale plant data on the drying of similar materials should be obtained.

4. Defining of critical factors, constraints and limitations associated with particular product and with available resources:

   **a.** Any particular hazards related to the handling of the product (wet or dry) should be specifically and quantitatively identified.

   **b.** Any characteristics of the product that present potential problems should be recognized.

   **c.** Degree of uniformity of drying will work as an important consideration in the selection process.

5. *Making a preliminary identification of the appropriate drying systems.* In this step, an identification of several dryer types that would appear to be appropriate should be made. This can be accomplished by simply comparing the properties and critical factors identified in Steps 3 and 4 with the characteristic features of the industrial dryers classified previously.

6. *Selection of optimal drying system and determining its cost effectiveness.* This step is followed on the basis of forging, and the optimal dryer type is identified and the appropriate design calculations or experimental programs can be conducted. Thus, the ultimate choice is usually that which is dictated by minimum total cost. However, it should be noted that a detailed economic analysis might lead to a selection based on maximum profit rather than minimum cost.

When selecting a dryer, there are several questions that need to be answered for all types of dryers. Rotary dryers will be used to illustrate problems because they dry more material than any other dryer. A few of the problems are as follows:

What type of dryers can handle the feed? If the feed is liquid, dryers such as spray, drum, or one of the many special dryers that can be adapted to liquids may be used.

If the feed is quite sticky, it may be necessary to recycle much of the product in order to use a certain type of dryer. The best solution to the feed problem is to try the material in a pilot unit. The pilot unit for a spray dryer needs to be near the size of the production unit as scale-up is quite difficult in this case.

Is the dryer reliable? Is the dryer likely to cause shut downs of the plant, and what performance history does this unit have in other installations? How long is the average life of this type of dryer?

How energy efficient is this type of dryer? For example, a steam tube dryer may have an efficiency of 85% while a plain tube type of rotary dryer may have an efficiency of only 50%.

However, production of the steam entails additional costs so the plain tube may be more efficient in overall production.

The higher the temperature of inlet gas stream, the higher the efficiency of the dryer in general. A fluid-bed dryer has a high back-mix of gas so it is possible to use a fairly high entering gas temperature.

Any dryer can use recycled stack gas to lower the inlet gas temperature and thus obtain a high efficiency for dryer. However, if there is any organic material in the stack gas, it may be cracked to form a very fine carbonaceous particulate, which is almost impossible to remove from the stack. Recycle also increases the dew point of the incoming gas, which lowers the drying potential of the dryer. This lowering of the potential is quite important when drying heat-sensitive material. What type of fuel can be used for heating? Direct heating is usually the most efficient unit, and natural gas and LPG are the best fuels. However, both gases are getting more expensive and in many cases will not be available. The next best fuel is light fuel oil, which can be burned readily with a "clean" stack.

This material is expensive, and in some cases may be in short supply. The third best fuel is heavy fuel oil, which is usually available, but this oil requires special burners and may not give a sufficiently clean stack. Coal is dusty and hard to handle.

The stack gas usually is too contaminated for use in most installations.

- Does the dryer have a dust problem? Steam tube units use very low air flow and have minor dust problems, while a plain tube uses high air rates and may have serious dust problems. In some cases the stack dust removal devices may cost more than the dryer.
- How heat sensitive is the material to be dried? Most materials have a maximum temperature that can be used without the product deteriorating. This temperature is a function of the time of exposure as the thermal deterioration usually is a rate phenomenon. Wet material can stand much higher temperatures in the gas due to the evaporation cooling.
- As an example: A rotary dryer working with alfalfa can use 760°C entering gas in a cocurrent unit. A countercurrent unit at this temperature would burn the alfalfa. As the temperature of the entering gas determines the efficiency of the dryer, concurrent dryers, on the average, are more efficient than countercurrent dryers.
- What quality of product will be obtained from the dryer? Freeze drying usually will give an excellent product, but the cost is prohibitive in most cases. A dryer needs to balance quality against cost of production of a satisfactory product.
- What space limitations are placed on the installation? There are certain height limitations in some buildings, and floor space may be limited or costly.
- Maintenance costs are often a major consideration. If moving parts either wear out or break down due to material "balling-up" or sticking, the plant may be shut down for repair, and repairs cost money. If this is a problem, a record should be kept of the performance of the unit. It may be possible to get this information from a plant, which is using this particular unit on a similar product.
- What is the labor cost? A tray dryer has high labor costs, but it is the best dryer in many cases where only small amounts of material need to be handled.
- Is a pilot unit available which can be used to get data to design the needed production facility? Nearly all new products need pilot plant data for a satisfactory design of a dryer.

- In the case of spray drying an industrial size unit needs to be used. Drum and rotary units and most other dryers can be scaled-up with sufficient success from laboratory-sized units.
- What is the capital investment for the dryer and all the accessories?
- What is the power requirement for the dryer? A deep fluid-bed dryer needs hot gas at a higher pressure than most other dryers: 0.47 m³/s of gas requires approximately 0.75 kW per 102 mm of water pressure.
- What quantity of product is desired? For larger production a spray or rotary dryer should be considered. Rotary and spray dryers handle most large production demands, but in small production plants other dryers are often more economical.
- Can the dryer perform over a wide range of production rates and still give a satisfactory product in an efficient manner?
- Is a sanitary dryer needed? A sanitary dryer is one that has no grooves or corners that can trap product, and hence can be easily cleaned. If no corrosion can be allowed, most of the units should be made of stainless steel.
- Once the above points have been examined, it is possible to select a few types of dryers that appear to be the best for the particular operation. Sufficient information and data should be obtained on these dryers to determine the size needed. Firm quotations should be obtained from the manufacturers. The most economical dryer now can be selected on the basis of quality of product and capital and operating costs.

## 10.12 POLYMER DRYERS

1. Polymer dryers may be classified and selected according to the mode of heat transfer, that is, direct-heat and indirect heat dryers. Dryers combining both heat-transfer modes are often used for polymer drying.
2. Radiant-heat dryers are not commonly used because most polymers are heat sensitive to some degree and material temperature is difficult to control under radiant sources.
3. Within broad ranges, polymer dryers may be classified on the basis of material residence time as:
   a. *Short resident time*. Spray dryers, pneumatic conveyors, drum dryers, and thin-film belt dryers, when the material residence time is less than 1 min.
   b. *Medium residence time*. Continuous-fluid-bed dryers, vibrating fluid-bed dryers, steam-tube dryers, and direct-heat rotary dryers; when the residence time is up to 1 h.
4. *Long residence time*. Batch fluid-bed dryers, batch or continuous-tray dryers, rotating-shelf dryers, hopper dryers, vacuum rotary, and rotating dryers; when the residence times vary from one to several hours.
   a. Short residence-time dryers are usually employed only for solutions and fine particle slurries during constant rate drying. The longer residence-time dryers are used for materials containing bound moisture and for operations involving capillary or diffusional drying. Solids flow control is difficult in continuous fluidized-bed and rotary dryers.
   b. A classification of polymer dryers according to adiabatic and non-adiabatic processes is a general guideline for selecting a specific kind of equipment for particular product. However, a general classification for the purpose of choosing the correct dryer for a specific process is not suggested. Classifications are useful for review to ensure that all feasible alternatives are considered early in the selection process.

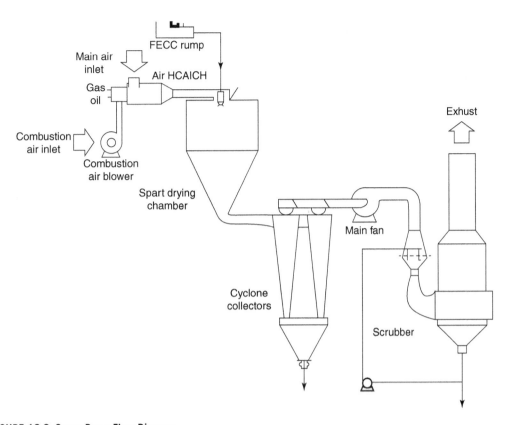

**FIGURE 10.9 Spray Dryer Flow Diagram**

    **c.** The specific operating characteristics of various dryers used for some important polymer drying and polymer grade by a competent vendor is given herein below for further useful review and consideration in the selection process.

**1.** Poly vinyl chloride (PVC)

    **a.** Emulsion-grade PVC is dried in spray dryers (Fig. 10.9). A spray dryer which is a direct-heat adiabatic dryer, is the first choice for this polymer grade. Centrifugal disk spray machines are usually chosen, because they are scalable to higher capacities and do not require high-pressure pumps. Cocurrent flow of spray gas and product permits a high inlet gas temperature.

    **b.** Suspension-grade PVC can be centrifuged to a dry-basis moisture content of 25–35%. The cocurrent rotary dryer is still the most commonly chosen option and is installed in the manner depicted in Fig. 10.10. Dry product leaving the system carries less than 0.2% of moisture. A controlling system installed to measure the temperature loss to indicate the dryer is approaching overload. Cocurrent gas–solid flow is employed in such a way that the gas of the highest temperature contacts the wettest polymer, and overheating of dry product is avoided.

    **c.** An alternative to the cocurrent rotary dryer is the two-stage arrangement of a pneumatic conveying dryer followed by a fluid-bed dryer shown in Fig. 10.11. This setup is tailored to accommodate the two drying periods, or phases, which are characteristic of several

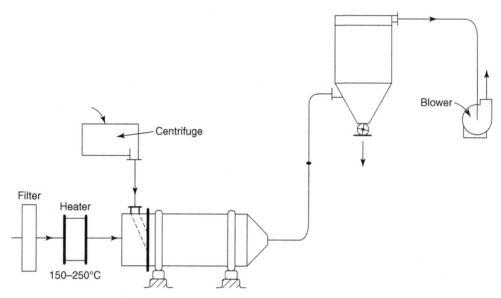

**FIGURE 10.10 Rotary Drying of Suspension-Grade (Poly Vinyl Chloride)**

commodity polymers. A representative drying profile is shown in Fig. 10.12. During only a few seconds residence time, a properly sized pneumatic conveying dryer easily removes the surface moisture. A fluid-bed dryer with a residence time of about 30 min completes the drying process at a relatively low temperature during falling-rate drying of capillary moisture. Benefits include the reduced likelihood of adhesion of wet particles in the conveyor and longer residence time in the fluid bed, which allows a lower drying temperature, uniform product quality, and easy scale-up.

    **d.** A third suspension-grade PVC drying arrangement employs a single fluid-bed, which combines direct with indirect-heat transfer by use of internal, indirect-heat, plate coil heating surface (Fig. 10.13). This method minimizes dust recovery and gas-handling costs by reducing gas consumption to only that needed for fluidization and vapor removal, whereas most of the energy needed for evaporation is transferred indirectly from the heating surface. Total energy required is about 45% of that used by the cocurrent rotary dryer and 55% of that needed by the pneumatic conveyor-fluid bed combination. Residence time and plug-flow in the indirect heat fluid bed are controlled by arranging the plate coils to form internal baffles and plug-flow channels.

**2.** Polyproplene and high-density polyethylene (HDPE)

    **a.** These polymers may be wet with water or an organic solvent. They are dried after centrifugation, and product temperature must not exceed 100–110°C, therefore, liquid vapor pressure has an overriding influence on dryer selection.

    **b.** A direct-heat rotary dryer was used earlier, but now has proved to be a poor choice for organic solvent service. Large, expensive gas-tight rotary seals are needed between each end of the rotating dryer cylinder and its stationary end enclosures. Continuous maintenance is needed to ensure precise sealing.

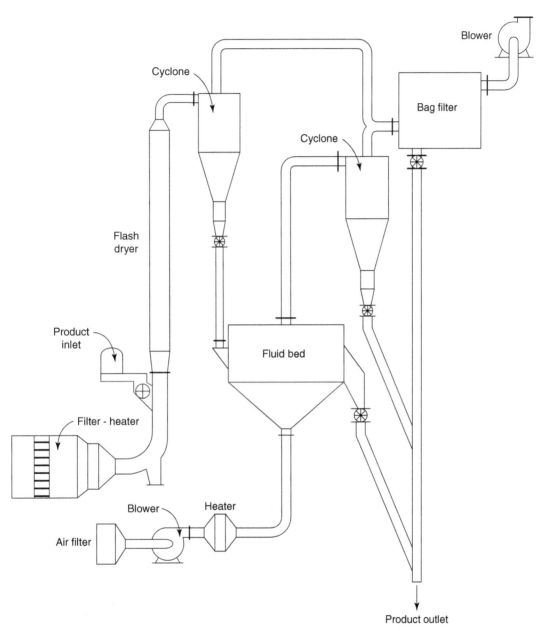

**FIGURE 10.11 Two-Stage Drying System (Pneumatic Conveying-Fluid-Bed Dryer) for Suspension-Grade (Poly Vinyl Chloride)**

   **c.** Two-stage paddle agitator type dryers (Fig. 10.14), are the preferred alternative. These paddle dryers are preferable to the rotary dryer because their cylinders are stationary. Shaft seals are very small compared to rotary cylinder seal. The first paddle dryer removes all surface liquid under constant drying conditions. This stage is characterized by intense agitation, deagglomeration,

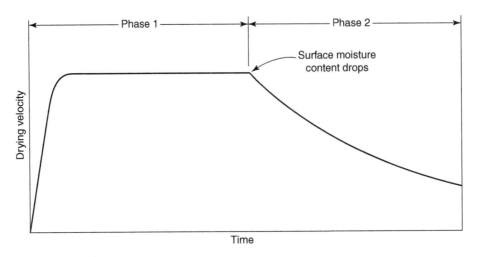

**FIGURE 10.12  Drying Requirements of Polymers**

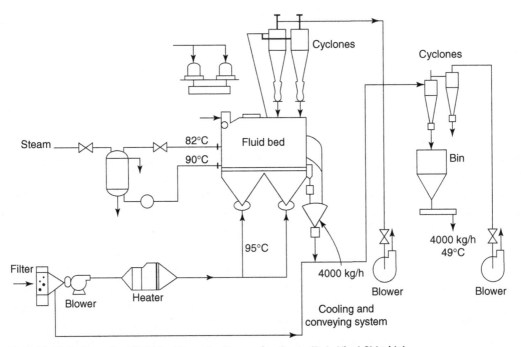

**FIGURE 10.13  Indirect-Heat Fluid-Bed Dryer for Suspension-Grade (Poly Vinyl Chloride)**

rapid heat transfer, and short residence time. The second stage is designed for the removal of bound liquid and combines moderate agitation with a long residence time and a small temperature differential. Each drying stage includes an independent gas recycle and solvent-recovery system.

    **d.** A combination of pneumatic conveying-fluid bed dryers (Fig. 10.15) incorporating closed-circuit inert gas recirculation is also employed. In these types, again constant-rate drying is

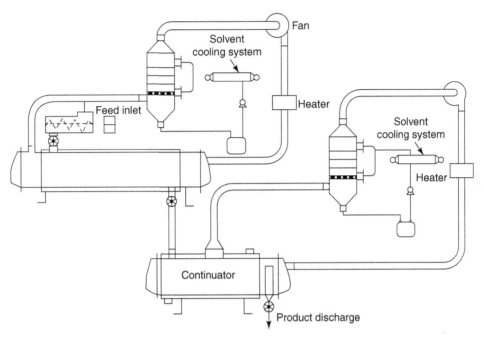

**FIGURE 10.14  Two-Stage Drying of High Density Polyethylene and Polypropylene**

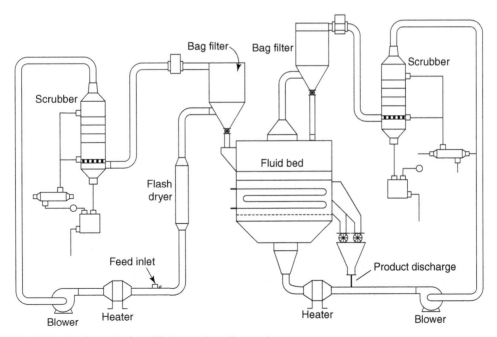

**FIGURE 10.15  Two-Stage Drying of Polypropylene Homopolymer**

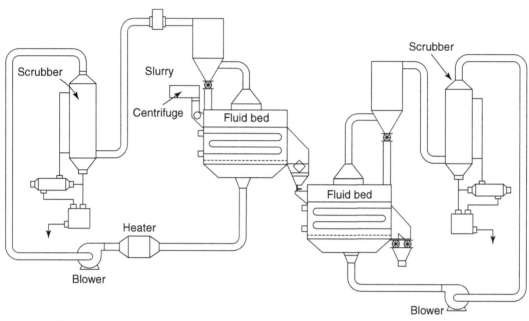

**FIGURE 10.16 Two-Stage Fluid-Bed Dryer for Polypropylene Homopolymer**

separated from falling-rate drying which allows the use of higher gas temperature and solvent partial pressure in the first stage.
  **e.** Multistage fluid-bed dryers have been used successfully for Polypropylene (PP) and High Density Polyethylene (HDPE) (Fig. 10.16). As with the aforementioned paddle-dryer system, energy efficiency will be improved by the use of indirect heat plate coils in the fluid beds, especially in the first stage.
  **f.** Efficiency may also be improved by installing three or more stage drying systems. Fluid beds are vulnerable in situations where feed properties cannot be controlled specifically for dryer performance. Fluid beds are susceptible to defluidization if feed is sticky or cohesive such as olypropylene copolymers.
**3.** Acrylonitrile-butadine-styrene (ABS) polymers
  **a.** The drying characteristics of ABS Polymers vary with changes in composition. The usual requirement is to dry a centrifuge cake from 50% moisture to less than 1.0%. A product temperature of 100°C is about the maximum permissible. The pneumatic conveyor yields good thermal efficiency and is suitable for fine particles. The rotary dryer has a longer residence time and is suitable for these particles.
  **b.** Using a two-stage dryer with an arrangement similar to that shown in Fig. 10.17, with or without closed circuit gas recycle is a third choice which is free from those disadvantages employed for pneumatic conveyor and rotary types. In this type, each stage is designed for intense mechanical agitation, and particle lumps and agglomerates formed in the centrifuge are broken apart as drying proceeds. A product moisture content as low as 0.3% can be obtained in this manner.

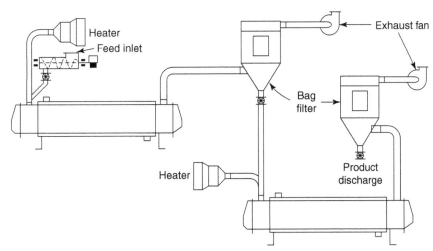

**FIGURE 10.17  Two-Stage Paddle-Agiator Dryer for Acrylonitrile-Butadiene-Styrene Polymer (ABS)**

**c.** A fourth alternative is the two-stage, pneumatic conveying fluid-bed dryer. Closed circuit inert-gas systems are installed on most new ABS polymer dryers to minimize polymer oxidation and the escape of styrene monomer, and increase the thermal efficiency of the dryer. A closed circuit, inert-gas indirect-heat disk dryer for ABS is illustrated in Fig. 10.18.

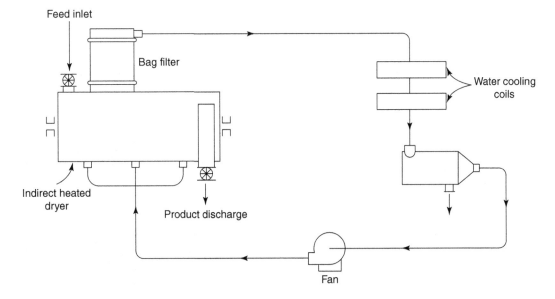

**FIGURE 10.18  Indirect-Heat Disk Dryer for Abs Polymer**

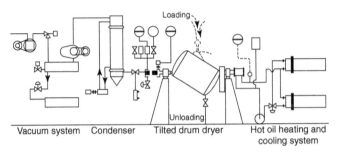

**FIGURE 10.19 Flow Sheet of a Drying Plant for Nylon and Polyester Chips with Heating and Cooling System Dust Collector and Vacuum Unit**

4. Drying of hydroscopic polymers
   a. Nylon and polyester are prominent examples of hydroscopic polymers drying polyester pellets before solid-stage polymerization is carried out in what is generally called a pellet-dryer. Both nylon and polyester absorb moisture from the atmosphere during handling and storage. Presence of moisture will cause discoloration and viscosity degradation, in a melting, extrusion, molding, and spinning process.
   b. Nylon may absorb 0.5–1.0% moisture and should be dried to less than 0.2% before melting. Because nylon is susceptible to oxidation and discoloration at elevated temperatures, most nylon pellet dryers are provided with closed-circuit inert-gas circulation. When dried with dehumidified air, the temperature should never exceed 80°C.
   c. Polyester absorbs up to 0.5% moisture and must be dried to 0.005% to avoid viscosity loss during the melting process. Polyester does not degrade in air and does not polymerize below 180°C, it may be safely dried in dehumidified air.
   d. For small productions, batch drum-type and double-cone rotary vacuum dryers are employed (Fig. 10.19). Internal pressure is 0.1–1.0 kPa (0.75–7.5 mmHg) when drying nylon, and less than 0.1 kPa (0.75 mmHg) for polyester. Jacket temperature is maintained with steam or hot oil at the desired final polymer temperature. Batch drying time for nylon and polyester is 8–24 h, depending on the batch and dryer sizes. In larger pellet dryers, the drying rate is limited by heat transfer.
   e. Dryer heating surface to working volume ratios is low and varies inversely with nominal shell diameters. Installation of internal, heated tubes, or plate coils in larger dryers alleviates the deficiency of heat transfer, but not sufficiently as it is the limiting feature of most rotating vacuum dryers.
   f. Continuous drying is the preferred method to avoid atmospheric exposure. Nylon and polyester are dried in fluid-beds, mechanically agitated hoppers, or simple moving-bed hoppers, where circulating dehumidified and heated air or inert gas through the bed heats the polymer and removes the moisture.
   g. A moving-bed, hopper-dryer arrangement for polyester pellets is typically illustrated in Fig. 10.20.
   h. When polyester is dried in a rotating vacuum dryer a separate crystallization step is usually not necessary because the heating rate is so low that crystallization takes place gradually over a period of several hours. In the hopper, temperature is controlled at 150–180°C by dehumidified air or inert gas with a dew point below −40°C.

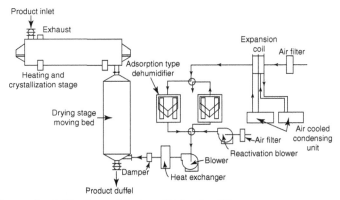

**FIGURE 10.20 Continuous Crystallization and Drying of Polyester Chips**

## 10.13 **COMPRESSED AIR DRYER**

Compressed air may be dried by:

**1.** absorption
**2.** adsorption
**3.** compression
**4.** cooling
**5.** combination of compression and cooling.

*Note*: Mechanical drying methods and combined compression and cooling are used in large-scale operations. They are generally more expensive than those employing desiccants and are used when compression of the gas is a necessary step in the operation or when its cooling is required.

### 10.13.1 **RATING PARAMETERS AND REFERENCE CONDITIONS**

Reference standard conditions and rating parameters are both necessary in defining the performance of an air dryer and in comparing one make up dryer with another.

The reference conditions in Table 10.17 and performance rating parameters in Table 10.18, are to ISO 7183, and shall form invariable and variable parts of this statement respectively.

### 10.13.2 **SPECIFICATION**

Important specification data together with relevant explanatory notes, essentially required in the period of design, enquiry, and purchase and also for the use, when specifying and inspecting of compressed air dryers are tabulated in Table 10.19. For detailed specification and testing procedure see ISO 7183.

In addition to the reference conditions and the performance rating parameters, some other important performance data which should be concluded in process design of compressed air dryers and is required for performance comparisons of the vendors'/manufacturers' proposals is tabulated in Table 10.20.

**Table 10.17 Reference Conditions**

| Quantity | Unit | Value | | Tolerance |
|---|---|---|---|---|
| | | **Option A** | **Option B** | |
| Inlet temperature | °C | 35 | 38 | ±1 |
| Inlet pressure | bar | 7 | 7 | ±7% |
| Inlet pressure dew point | °C | 35 | 38 | ±2 |
| Cooling air inlet temperature | °C | 25 | 38 | ±3 |
| Cooling water inlet temperature | °C | 25 | 30 | ±3 |
| Ambient air temperature | °C | 25 | 38 | ±3 |

*Note: The choice between A and B will be influenced by the intended geographical location of the equipment.*

**Table 10.18 Performance Rating Parameters**

| Quantity | Unit | Value |
|---|---|---|
| Outlet pressure dew point | °C | As specified |
| Outlet air flow | L/s or m³/s | As specified |
| Pressure drop across dryer | bar | As specified |
| Frequency of electrical power supply | Hz | As specified |

## 10.14 ADSORPTION DRYERS

The majority of industrial gases and liquids require some level of water removal between initial processing and final intended use. Unit operations and processes typically employed in drying industrial fluids include the following:

- distillation (including azeotropic and extractive distillation)
- mechanical separation
- adsorption (including liquid desiccants as dehydration media)
- adsorption (including solid desiccant materials).

Drying with adsorbent discloses a number of the advantages on comparison with fractional distillation, wet scrubbing, or other processes, which necessitates its paramount importance use in OGP process plants.

These advantages include:

1. lower capital and operating costs
2. high reliability because adsorption performance is relatively unaffected by changes in flow rate or composition
3. eliminates problems caused by azeotrope formation

**Table 10.19  Specification**

| Item | Description | Symbol | Unit | Remarks | Explanatory notes |
|---|---|---|---|---|---|
| 1 | Compressor type | — | — | | State the type of compressor(s) (eg, displacement or turbo compressor), the type of lubrication (nonlubricated, minimum lubrication or oil flooded) and the type of coolant (air, water, oil). See ISO 5388 |
| 2 | Mode of operation of compressor plant | — | — | Continuous/intermittent | Details should be given of the operating intervals (on periods) and the position of the compressed air dryer in the compressed air pipework system |
| 3 | Volume of air receiver | $V$ | L, m$^3$ | | State the volume of the air receiver |
| 4 | Air volume flow rate related to the intake conditions in compliance with 4.10.1 | $q_{v1}$ | L/s or m$^3$/s | | The maximum compressed air volume flow accepted by the dryer under the reference conditions including air required for regeneration, pressurizing or cooling purposes |
| 5 | Effective (gage) pressure of the compressed air | $p_1$ | bar | | The inlet air pressure shall be stated |
| 6 | Temperature of compressed air | $t_1$ | °C | | The temperature of compressed air at the inlet of the dryer will affect its performance and shall be stated. |
| 7 | Pressure dew point of compressed air | tpd1 | °C | | If the dryer is installed immediately following the compressor aftercooler, the compressed air may be assumed to be saturated. However, the humidity of the air should be measured if the dryer is installed downstream of the air receiver or in the pipework remote from the aftercoolers |
| 8 | Pressure drop across dryer | $p$ | bar | | — |
| 9 | Oil presence in compressed air | — | g/m$^3$ | | The supplier should state the type and amount of compressor lubricant that can be expected at the dryer inlet. |
| 10 | Aggressive components in air | — | — | | Any pollution of incursive (aggressive) contaminants should be stated |
| 11 | Coolant | — | — | Water/air | |
| 12 | Coolant temperature | $tc_1$ | °C | | The coolant temperature shall be measured |

*(Continued)*

**Table 10.19 Specification (*cont.*)**

| Item | Description | Symbol | Unit | Remarks | Explanatory notes |
|------|-------------|--------|------|---------|-------------------|
| 12.1 | Coolant quality | — | — | | Any aggressive component in the coolant should be stated |
| 12.2 | Coolant pressure | — | bar | | |
| 13 | Position of air dryer | — | — | Before/after air receiver | When designing and specifying the air dryer the position of the air receiver is important and shall be stated |
| 14 | Dryer location | — | — | Indoors/out-doors | It is necessary to state the location of the dryer (eg, indoors, outdoors, hazardous area) |
| 15 | Ambient conditions (maximum and minimum) | — | — | | Any special ambient conditions shall be stated in the enquiry |
| 16 | Power available | — | — | | To include supply voltage, frequency and number of phases |

**Table 10.20 Data for Performance Comparisons**

| Description | Symbol | Unit | Explanatory notes |
|-------------|--------|------|-------------------|
| Types of compressed air dryer | — | — | Specific details with regard to operation and design/type of the compressed air dryer should be given as well as a specification of the equipment included in the delivery |
| Mode of operation of compressed air dryer | — | — | Details should be provided of the mode of operation of the compressed air dryer, for example, continuous operation, on/off operation (for refrigeration dryers) alternating operation (in the case of adsorption dryers) as well as automatic, semi automatic or manual |
| Cycle time | — | s | — |
| Air volume flow rate related to the intake condition | $q_{v2}$ | L/s or m³/s | The volume of air delivered by the dryer under the reference conditions, that is, after maximum bleed air, pressurizing air and cooling air flows have been deducted |
| Mass flow of compressed air (if required) | $q_{m2}$ | kg/s | If required, the manufacturer of the dryer should calculate in the mass of flow from the volume flow and state the value to the tender |
| Temperature of dried compressed air | $t_2$ | °C | The temperature shall be measured |

**Table 10.20 Data for Performance Comparisons (*cont.*)**

| Description | Symbol | Unit | Explanatory notes |
|---|---|---|---|
| Pressure drop across dryer | $p$ | bar | If the dryer is delivered with integral filters, they shall be included in the pressure drop |
| Highest pressure dew point under operating condition | $t_{pd}$ | °C | The maximum pressure dew point shall be stated for operating conditions |
| Nominal pressure dew point as requested by purchaser | $t_{pd}$ | °C | — |
| Coolant flow | $q_{v\,c2}$ | L/s | — |
| Energy requirements: | | | — |
| Electric power at dryer terminals including all components (this includes cooling air fans), max. and average | p | kW | — |
| Bleed air; dump losses, etc., max. and average | $q_{v\,loss}$ | L/s | — |
| Steam consumption | — | L/s(or kg/h) | — |
| Steam condition | | | — |
| Pressure | — | bar | — |
| Temperature | — | °C | — |
| Water (for cooling according to coolant temperature which is used at any heat exchanger of dryer) | $q_v$ | L/s | Pressure, quality inlet temperature and temperature should be stated |
| Noise level of air dryer | — | dB | — |

*Note: For source of Specification Data reference is made to ISO 7183.*

4. low maintenance because corrosion is not a problem
5. simple process control and response, resulting in easy startup, shutdown, and a virtually unlimited turndown ratio
6. handling and disposal problems associated with corrosive liquid chemicals are not a factor with inert solid desiccants
7. fully automatic, unattended operation possible
8. very low dew point attainable.

## 10.14.1 SOLID DESICCANT

### 10.14.1.1 Characteristics

Adsorbents used for removing water from a fluid stream are known as "solid desiccant." The characteristics of solid desiccants vary significantly depending on their physical and chemical properties. Many known solids have some ability to adsorb, but relatively few are commercially important. Some of the qualities that make a solid adsorbent commercially important are:

1. available in large quantity
2. high capacity for the gases and liquids to be adsorbed
3. high selectivity
4. ability to reduce the materials to be adsorbed to a low concentration
5. ability to be regenerated and used again
6. physical strength in the designed service
7. chemical inertness.

## 10.14.2 CRITERIA FOR SOLID DESICCANT SELECTION

In order to make the proper selection of solid desiccant, the following criteria should be considered.

### 10.14.2.1 Cycled capacity

The equilibrium loading is also known as the equilibrium capacity. This capacity gradually decreases during repeated adsorption regeneration cycles, essentially because of desiccant fouling and degradation. Consideration must be given to a desiccant's capacity over a long period of use rather than its capacity when freshly manufactured.

### 10.14.2.2 Ability to reach the required outlet moisture specification
### 10.14.2.3 Susceptibility to deactivation in specific service

Ability to exclude certain side reactions as well as to maintain chemical inertness in the stream being dried is important (eg, certain types of desiccant materials perform better than others in olefinic or acidic service).

### 10.14.2.4 Cost

The initial cost of the desiccant, the operating cost, the recharge cost as related to change-out frequency, and the initial capital equipment cost should be evaluated in desiccant selection.

### 10.14.2.5 Pressure drop

Pressure drop is a function of desiccant particle size and type (eg, beads or pellets), and is important on both adsorption and regeneration legs of the cycle.

### 10.14.2.6 Regeneration capability

The quantity and quality of regeneration gas available, as well as the temperature available to remove the moisture from the "loaded" desiccant.

### 10.14.2.7 Service

The availability and capability of a desiccant supplier to provide needed service is very important in view of the complex processing that is often required.

*Note*: The order given herein above, does not necessarily dictate the relative priority that mainly depends on the user's particular circumstance.

## 10.14.3 DESIGN BASIS

Design and optimization of the adsorption process is a complex task; the vendor's/manufacturers advice shall save much time and effort. However, in order to design an optimum adsorption system, the design engineer must have an accurate design basis data, information and the variations and upsets, which may occur, in the processing stream. This type information shall also be required by adsorbent manufacturers in order to provide recommendations on specific applications. As a minimum, the following information should be available:

1. *Type of fluid*. Physical state (gas or liquid composition) and water level.
2. *Operating conditions*. Flow rate, temperature, and pressure.
3. *Outlet water specification*
4. *Preferred adsorption cycle time*. This time should be integrated into the operation and be consistent with the needs of the system. Switching vessels every 24 h or with change in operator shifts every 8 h is a fairly common way to designate this time.
5. *Regeneration*. The available fluid, its composition, quantity available, pressure, and maximum temperature available for regeneration, as well as contaminant levels (especially water concentrations), must be known.
6. *Existing equipment*. In certain circumstances it is necessary or desirable to replace one type of adsorbent with another as processing conditions change. In most cases the same equipment can be used, but careful consideration must be given to interior vessel volume, vessel configuration, and number, and adsorption and regeneration system flow.

### 10.14.4 STANDARD CONFIGURATION OF ADSORBER

Vertical cylindrical vessels filled with adsorbent are the simplest fixed-bed adsorption system. Cylindrical adsorption vessels are usually arranged in two-bed or three-bed systems. Also, multiples of these basic systems, containing three, four, five, or more units, are not uncommon. As mentioned previously, one bed in the dual-bed system is adsorbing or drying. While the other is desorbing or regenerating (Fig. 10.21), in a three-bed system, one of the following three basic piping configurations is employed (Fig. 10.22a–c):

1. *Two beds on parallel adsorption, one bed regenerating*. This system is usually employed where a minimum pressure drop is required on adsorption, or where the use of small-diameter, multibed systems reduces vessel costs. In this arrangement, more efficient adsorption is obtained because flow is slit in half and, therefore, mass transfer zone size per vessel is reduced.
2. *Two beds on series adsorption, one bed regenerating*. This system is usually employed when mass transfer zones are long. Each bed "moves" sequentially from:
   a. trim, or downstream, adsorption to
   b. lead, or upstream, adsorption and then to
   c. regeneration.

A bed spends 1/3 of its cycle time in each of the three positions. The trim bed is long enough to contain a mass transfer zone, and it guards against water breakthrough into sensitive downstream equipment. In the lead position, nearly all of the adsorbent becomes loaded to equilibrium capacity.

3. *One bed on adsorption, two beds regenerating in series*. This system is usually employed where there is little regeneration gas available. Each bed "moves" from:
   a. adsorption to
   b. heating and then to
   c. cooling.

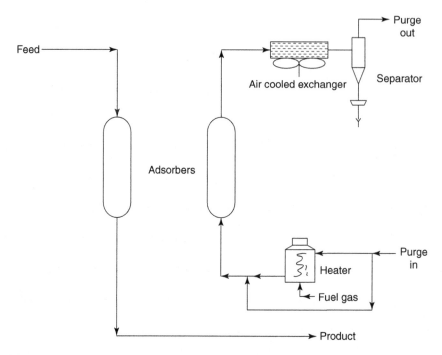

**FIGURE 10.21 Dual Bed System**

Again, a bed spends 1/3 of its cycle time in each of the three positions. In one arrangement, clean purge gas flows first to the bed to be cooled, next to a heater, then to the bed to be heated and desorbed, and finally to discharge. Many bed combinations are possible with the optimum arrangement being dictated by the basic processing constraints and economics. Three-bed systems offer many benefits to meet unique processing needs. However, they require more valves and more complicated piping than the dual-bed system. In some situations a one-bed system may be the only vessel required. This is usually the case in intermittent or batch-type operation where adsorption drying is not required on a continuous basis (Fig. 10.23).

## 10.14.5 DESIGN CRITERIA AND CALCULATIONS

### 10.14.5.1 Flow velocity

Flow velocity, pressure drop, and adsorber bed diameter are all related. When any one of these parameters is fixed along with cycle time, the other two are also fixed. A limitation on pressure drop is usually the key parameter, and is generally the basis for fixing the other two. However, typical superficial linear velocities through beds of adsorbent are in the order of 10–20 m/min, for gases and 0.3–0.6 m/min, for liquids.

### 10.14.5.2 Bed diameter

Vessel costs tend to increase dramatically with diameter. This becomes more significant as the operating pressure (and consequently, wall thickness) goes up.

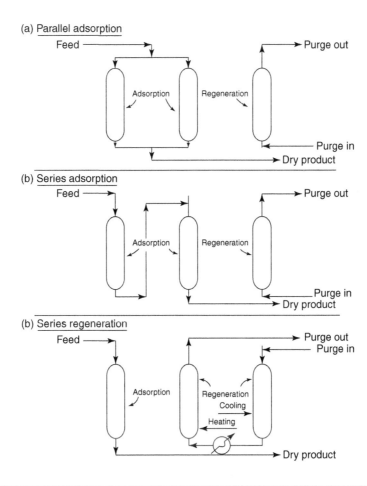

**FIGURE 10.22**

(a–c) Multiple Bed Systems

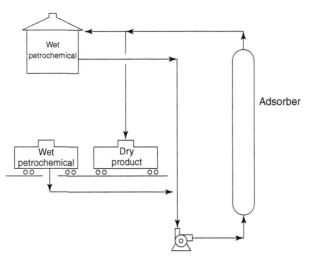

**FIGURE 10.23 Single Bed System**

The minimum diameter for an adsorber bed is set by pressure drop limitations. A pressure drop analysis is required for each of the steps in the adsorption cycle, including the pressurizing and depressurizing steps.

### 10.14.5.3 Pressure drop

It will be necessary during various stages of dryer evaluation to determine the fixed-bed pressure drop in order to check fluidization limits, pressure drop variation with changes in fluid flow rate, utilization of existing equipment, etc.

The pressure drop through packed adsorbent beds may be determined by using the modifier, Ergun correlation, which has proved to be very reliable.

The Ergun equation for the calculation of pressure drop in adsorbent beds is in good agreement with numerous pressure drop measurements on commercial adsorption units for both gas-phase and liquid-phase operation.

The following form of this equation is suitable for calculating pressure drop through adsorbent beds:

$$\frac{P}{L} = \frac{f_T \cdot C_T \cdot G}{\rho \cdot D_P}$$

where, $C_T$ is pressure drop coefficient, in $(m \cdot h^2/m^2)$; $D_P*$ is effective particle diameter, in (m); $f_T$ is friction factor; G is superficial mass velocity, in $(kg/h.m^2)$; L is distance from bed entrance, in (m); P is pressure drop, in $(kg/m^2)$; $\rho$ (rho) is fluid density, in $(kg/m^3)$; P/L is pressure drop per unit length of bed, in $[(kg/m^2)/m]$.

*Note*: The friction factor, $f_T$, is determined from Fig. 10.19 which has $f_T$ plotted as a function of modified Reynold's number:

Modified $Re = D_P \cdot G/\mu$

Where, $\mu$ (mu) is fluid viscosity, in (kg/m h).

Notes:

1. The pressure drop coefficient, $C_T$ is determined from Fig. 10.24
2. $C_T$ is plotted as a function of external void fraction, $\varepsilon$ (epsilon)
3. The suggested values for $\varepsilon$ and $D_P$ for various sizes of adsorbents are:

| Sizes of Adsorbents | External Void Fraction, $\varepsilon$ | Effective Particle Diameter, $D_p$ (mm) |
|---|---|---|
| 0.32 mm pellets | 0.37 | 3.72 |
| 0.16 mm pellets | 0.37 | 1.86 |
| 14 × 30 mesh granules | 0.37 | 1.00 |

### 10.14.5.4 Adsorption equipment

General guidelines and design criteria for auxiliary equipment of adsorption system such as blowers, heaters, heat exchangers, pumps, compressors, piping, valving, and insulation are given in relevant referenced standards and other standards which should be considered in process design of adsorption system. The adsorber vessel design, however, requires some attention to detail to achieve optimum desiccant performance.

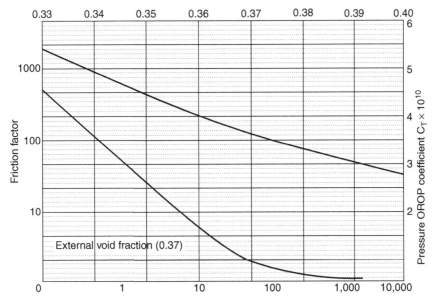

**FIGURE 10.24** Modified Reynolds number $D_p \cdot G/\mu$ Friction Factor, $f_t$ And Pressure Drop Coefficient, $C_t$ for Modified Ergun Equation

### 10.14.5.5 Adsorber vessel design

Fig. 10.25 details one of the many acceptable adsorption vessel designs including its, bed support system, nozzles, baffles, bed support media etc. all require special design review and consideration as:

1. *Bed support system.* The support system should be designed to hold the mass of the desiccant material, forces exerted by process pressure drop, and a substantial safety factor. A tight seal of specified mesh screen should be provided against the vessel walls. The I beams fastened to the supports and the vessel wall, should be free to move slightly during process heating and cooling of the system.
2. *Nozzles.* The inlet and outlet nozzles should be placed on the axis of the vertical vessel, to obtain proper flow distribution. The guidelines for the distance between the nozzle and the bed are:
   a. outlet nozzle, two pipe diameters;
   b. inlet nozzle, 5–6 pipe diameters;
   c. the ratio of the vessel diameter to the pipe diameter also has an effect. The larger this ratio is, the greater the distance should be from the bed to the nozzle.
3. *Baffles.* For proper flow, distribution baffles should be installed in the inlet and outlet nozzles. Preferred baffle type and design shall be based on the vendor/manufacturer's experience. The goals, however, of any baffling should include:
   a. ensuring low pressure drops past the baffle;
   b. preventing direct impingement on the desiccant bed;
   c. breaking up the flow into several directions not merely redirecting the entire flow to another direction.
4. Bed support media
   a. A hard, mechanically strong, inert, high-density, inexpensive bed support that can take thermal cycling is desirable above and below the desiccant bed. The material on top acts as

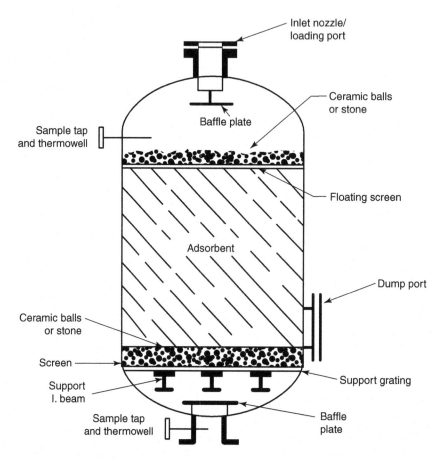

**FIGURE 10.25 Adsorber Vessel Design**

a guard layer, flow distribution media for the gas, and prevents desiccant particle movement caused by possible eddy currents from uneven flow distribution. The material is a relatively large size to minimize pressure drop and its own movement. A depth of 100–150 mm of 25–40 mm size material is typically required for the top support layer. A floating screen between the support media and the desiccant bed can be used to prevent migration into the desiccant bed.

**b.** Support media is placed at the bottom of the bed in many systems to a depth of about 80 mm. Usually a 6–10 mm size material is necessary to prevent desiccant particles from slipping between the large support media. This material provides some additional baffling and is less likely to fall through small open spaces in the mechanical bed support.

**5.** *Thermal wells.* Should be placed for process requirements of temperature measure including inlet and outlet flows, hot gas into and out of the vessel and temperature near the wall of the vessel. In addition to the pressure taps, sample taps should be provided when occasionally measuring the pressure drop across the vessel is required.

### 10.14.5.6 Equipment vendor/adsorbent manufacturers consultation

It is recommended that prior to package-equipment selection and design, the equipment vendor and the adsorbent manufacturer should be consulted by the contractor/licensor, since their technical staff can provide considerable experience and input into the final process and mechanical design and equipment selection. The adsorbent manufacturer, who typically works with the equipment vendor in setting final specifications, can assist in integration of the desiccant unit into the process scheme for optimum efficient performance.

### 10.14.5.7 Specification form for a dryer

Below are some specifications for a typical dryer.

| 1. Operation | Mode | Batch/continuous |
|---|---|---|
| | Operating cycle | ___ h |
| 2. Feed | a. Material to be dried | ___ |
| | b. Feed rate | ___ kg/h |
| | c. Nature of feed | Solution/slurry/sludge/granular/fibrous/sheet/bulky |
| | d. Physical properties of solids: | |
| | – Initial moisture content | ___ kg/kg |
| | – Hygroscopic-moisture content | ___ kg/kg |
| | – Heat capacity | ___ kJ/kg°C |
| | – Bulk density, wet | ___ kg/m$^3$ |
| | – Particle size | ___ mm |
| | e. Moisture to be removed: | |
| | – Chemical composition | ___ |
| | – Boiling point at 1 bar | ___ °C |
| | – Heat of vaporization | ___ MJ/kg |
| | – Heat capacity | ___ kJ/kg°C |
| | f. Feed material is | Scaling/corrosive/toxic/abrasive/explosive |
| | g. Source of feed | ___ |
| 3. Product | a. Final moisture content | ___ kg/kg |
| | b. Equilibrium-moisture content at 60% r.h | ___ kg/kg |
| | c. Bulky density | ___ kg/m$^3$ |
| | d. Physical characteristics granular/flaky/ fibrous/powdery/sheet/bulky | |
| 4. Design restraints | a. Maximum temperature | |
| | – When wet | ___ °C |

| | | |
|---|---|---|
| | – When dry | ___ °C |
| | b. Manner of degradation | ___ |
| | c. Material-handling problems | |
| | – When wet | ___ |
| | – When dry | ___ |
| | d. Will flue-gases contaminate product? | ___ |
| | e. Space limitations | ___ |
| 5. Utilities | a. Steam available at | ___ bar pressure ($10^6$ N/m$^2$) |
| | – Maximum quantity | ___ kg/h |
| | – Costing | ___ S/kg |
| | b. Other fuel | ___ |
| | – At | ___ kg/h |
| | – With heating value | ___ MJ/kg |
| | – Costing | ___ S/kg |
| | c. Electric power | ___ V |
| | – Frequency | ___ hz |
| | – Phases | ___ |
| | – Costing | ___ S/kWh |
| 6. Present method of drying | | ___ |
| 7. Rate-of-drying data under constant external conditions: | | ___ |
| Or data from existing plant | | ___ |
| 8. Recommended materials of construction | Parts in contact with wet material | ___ |
| | Parts in contact with vapors | ___ |

# LOADING AND UNLOADING FACILITIES

The loading and unloading facilities in the oil and gas processing industries vary with the size and complexity of the plant and the location and requirements of the consumers. Because of seasonal and other variations and product distribution, loading facilities shall be quite flexible and its capacity may far exceed normal plant production.

This chapter covers minimum requirements for process design and engineering of loading and unloading facilities for road tankers in the oil and gas processing industries.

It should be noted that the scope of this chapter is liquid applications and road tankers. Furthermore, in this chapter the unloading part is limited to probable discharges of the products remaining in the tankers that arrive for loading.

## 11.1 TRUCK LOADING AND UNLOADING

### 11.1.1 LOADING

This section is limited to provision of, process design of new facilities for loading of bulk road vehicles at normal installations for different products. For this reason, the designs shown include features which will not be necessary in all situations; and when new facilities are planned it is recommended that the simplest facilities that will efficiently perform the filling operation should be constructed. These requirements can also be used for the modernization and/or extension of existing loading facilities for road tankers.

Specifying the yearly average loading capacity, the size of tanker and loading assembly may be fixed and pump capacity will be calculated.

It should be noted that in case there is freedom in tanker size and/or loading assembly then economical evaluation shall be considered for such selections.

#### 11.1.1.1 Loading facilities in the context of the overall distribution system

The importance of bulk vehicle loading facilities as part of the total distribution complex must be fully realized when plans are made for the construction of new facilities, or the modernization and extension of existing arrangements. It is therefore necessary to examine the operation of the distribution system in order to optimize both its efficiency and the size of the loading facilities. The latter are an integral part of the distribution system and should not be studied in isolation; changes in the system and/or operating procedures can have a considerable effect upon vehicle loading requirements. In this context the objective must be to optimize the number of loading bays, and product loading spouts per bay, in relation to the overall distribution system, capital investment, and operating expenditure.

First, the cost of one's own and the contractor's vehicles should be assessed for the time spent (vehicle standing charges) while:

- queuing for a loading bay
- waiting for a loading arm while in the bay
- being loaded in the bay

Second, for existing installations the traffic flow must be studied to establish the present arrival patterns of vehicles at the loading facilities and hence the peak loading periods. The types of delivery such as urban, country, and over long distances, will influence arrival patterns.

Application of simple methods planning techniques to these operations will show whether efficiency can be improved by changes in:

- working hours
- shift patterns
- staggered starting times
- night loading
- dispatching and delivery systems

The objective being to improve utilization of existing facilities and of the existing road transport fleet.

For new installations the above information may not be available. In such cases an operational system must be established in which the various factors mentioned are carefully considered in relation to practice in the local industry, and in consultation with the designers.

### 11.1.1.2 Environmental conservation

It is the policy of OGP industries to conduct their activities in such a way that proper regard is paid to the conservation of the environment. This not only means compliance with the requirements of the relevant legislation, but also constructive measures for the protection of the environment, particularly in respect of avoidance/containment of spillages.

### 11.1.1.3 Vapor recovery system

The recovery of product vapors such as gasoline is of interest for economic, safety and environmental reasons. In most locations where bulk lorries are loaded, the total gasoline vapor emissions have not been considered a significant factor affecting the quality of the local environment. Nevertheless, at the design stage, system should be reviewed to see if it becomes necessary to install a vapor collection system return line for poisonous, hazardous, and high vapor pressure products [RVP > 0.34 bar (abs)].

However, it is essential to minimize the generation, and hence the emission of vapors during loading by eliminating the free fall of volatile products and reducing jetting and splashing.

In areas where action has been required by national authorities to minimize vapor emissions at loading facilities, bulk vehicles may have to be filled with a closed vapor system; this entails the following modifications to loading arrangements:

#### 11.1.1.3.1 Top Loading

As the majority of loading facilities in service are top loading, the best solution would be to replace (or modify) the existing loading arms so that when volatile products are loaded, the manhole is sealed and vapors are diverted into a vapor return system. The latter may be either integral with the loading arm or a vapor manifold on the vehicle connected to all the tank compartments which would be similar to the system described in Section 11.1.1.3.2.

### 11.1.1.3.2 Bottom Loading
Bulk vehicles equipped for bottom loading require a pipe connection from the vapor emission vent of each compartment into a vapor recovery manifold, which should terminate in a position which is easily accessible from ground level for use at both the loading bay or retail outlets as required. The coupling connections for liquid and vapor must be different types.

### *11.1.1.4 Reduction of vapor emissions*
Apart from installing a full vapor recovery system, considerable reduction in vapor emissions can be achieved by avoiding free fall and splashing of volatile products in top and bottom filling operations, as follows.

### 11.1.1.4.1 Top Filling
The loading arms should be designed to reach the end compartments of a vehicle tank in such a manner that the down pipe can penetrate vertically to the bottom of the compartment.

### 11.1.1.4.2 Bottom Filling
It may be necessary to fit deflectors in the vehicle tank at the point of entry of the product into the compartment.

Such measures have the following advantages:

1. minimizing the hazard of static electricity
2. minimizing the amount of vapor formation
3. reducing product losses
4. reducing the fire risk: the concentration of vapor emanating from the compartments will be dissipated faster to below the explosive limit

### *11.1.1.5 Spillage control*
The main items to be considered at the loading facilities are the provision of:

- emergency shut-off valve to prevent or reduce spillage due to overfilling, hose failure, etc.
- emergency push-button switch to stop the pumps, activate an alarm, and close all flow control and block valves on the island
- adequate drainage and interception arrangements

### *11.1.1.6 Health and safety*
Loading facilities are labor intensive (because of numbers of driving personnel) and vulnerable because of emission of vapors. It is the most likely source of accidents in a depot and hence particular attention needs to be paid to working conditions.

### *11.1.1.7 Static electricity*
To minimize the hazard of static electricity it is essential first, to ensure that the vehicle tank and loading equipment are at the same potential. This should be arranged by providing a bonding interlock system connecting the vehicle tanks to the loading rack and product flow-control valves.

Second, maximum safe flow rates in the loading system should be considered.

## 11.1.2 **LOADING SYSTEMS**

Ideally, the loading system should be able to fill all compartments of the vehicle without needing to move the vehicle. The spacing between loading systems at the loading island should allow the loading arms or hoses to be operated independently, without interference between each other, or meter heads, and with minimum obstruction of access for the operator.

### 11.1.2.1 *Choice of loading system-top or bottom*

The first criteria for selection of loading system is the volatility characteristics of the product. If RVP (Reid Vapor Pressure) of the product at 38°C is higher than 0.55 bar (abs) in summer or 0.83 bar (abs) in winter then bottom loading shall be used.

The second aspect is the requirements to restrict emissions from a specific product which dictates to use bottom loading.

Besides the above-mentioned limitations, the relative merits of top and bottom loading system are summarized in Table 11.1.

## 11.1.3 **CONTROL SYSTEM**

### 11.1.3.1 *Control of product flow*

11.1.3.1.1 Filling by Volume

Measurement of product volume governs the amount of product filled into each compartment and this is normally arranged by flow through a positive displacement meter. Slowing down and stopping the flow is usually controlled by a preset quantity control device which represents the first line of control. In the event of any emergency, for example, malfunction of the mechanism, or incorrect setting of the preset, etc. the possibility of a spillover occurs, and a second line of control is necessary. Methods of achieving this are as follows:

*11.1.3.1.1.1* Top Filling. The fitting of a "deadman control" in the form of a "hold-open" valve also enables the operator—when filling through an open manhole—to watch the level of the product and to stop the flow immediately in any emergency. The valve-operating lever (or control rod) must be located so that the filler can see the product in the compartments at high level, while avoiding the vapor plume emitted from the manhole. However, the temptation to tie the hold-open valves in the open position, has resulted in spillovers.

This factor, together with the necessity for operators to stand on vehicles while fillings, has led to the increasing use of liquid-level control equipment as a positive secondary means of stopping product flow in an emergency.

Where two or more compartments are required to be filled at the same time, liquid level control equipment is strongly recommended as a secondary means of stopping the flow of product.

*11.1.3.1.1.2* Bottom Filling. With all loading operations at ground level, and vehicle manhole covers remaining closed, the use of an overfill protection system based upon liquid-level detection equipment becomes essential.

The liquid-level control equipment should be linked into an interlock system which covers bonding of the vehicle, and access to the products by means of controls on the loading arms. This enhances safety and provides the basis for an automatic control system.

**Table 11.1  The Relative Merits of Top and Bottom Loading**

| Safety Features | Bottom Loading | Top Loading |
|---|---|---|
| Worksite | Ground level | On platform. Can be made safe by provision of guard rails and access ramps to vehicles, but at extra cost. |
| Vapor emissions (no vapor recovery) | Closed manhole covers gives rise to small pressure build-up to operate the vents resulting in marginally less vapor emission. | Open manhole covers therefore slightly greater vapor emission. |
| Control of product flow assuming meter preset does not work | Reliance on overspill protection equipment. | Positive visual control by loader assuming "hold-open" valve is correctly used. Two-arm loading requires overspill protection when the conditions are the same as for bottom loading. |
| Product handling equipment | Arms, and particularly hoses filled with product are heavier to handle. Generally, hose diameters should be limited to DN80 (3 in.). | Care is needed to ensure that the down-pipe of loading arms, is correctly positioned in each compartment. DN100 and DN150 (2 and 6 in.) diameter counter-balanced arms are easily handled. |
| Electrostatic precautions | Flow rates restricted to 75% of that for equivalent top loading system. | |
| Environmental conservation | | |
| Vapor recovery (loading bay) | Vehicles must be fitted with a vapor recovery manifold connecting each compartment; of sufficient capacity to cope with simultaneous loading of 2, 3, or 4 compartments. | Each product loading arm must be fitted with a vapor sealing head so that vapors are diverted into a vapor recovery system; either (a) on loading arm or (b) manifold provided for gasoline deliveries to retail outlets. Care must be taken to position collar seal in fill opening. Liquid level sensing equipment must be fitted on loading arms or in each vehicle tank compartment. |
| Vapor recovery (service stations) | | |
| Performance | | |
| Preparation for loading (normal) | Vehicles already equipped with vapor return manifold for use when loading. | Vehicles must be fitted with vapor return manifold. |
| Preparation for loading (vapor return) | Removal of caps and connecting couplings is contained within small operating envelope. | Greater area of operation because of positioning of manhole covers. |
| Loading arrangement | Additional coupling connection to vapor manifold. (No significant difference between systems.) | Care must be taken to position arm/vapor head in fill opening. (No significant difference between systems.) |
| Product flow rates | Simultaneous loading of two or more compartments more easily arranged. | |

*(Continued)*

**Table 11.1  The Relative Merits of Top and Bottom Loading (*cont.*)**

| Safety Features | Bottom Loading | Top Loading |
|---|---|---|
| Costs | | |
| Capital costs | 25% slower per compartment than equivalent top handling system because of electrostatic hazard in certain filling operations. | Additional structure and safety equipment for working platform. |
| | 1. Approximately 17% more loading space is required than that of an equivalent top-loading gantry. Additional cost for greater roof area. | |
| | 2. (a) All vehicle compartments must be fitted with loading dry-break couplings. | |
| | (b) To minimize over-filling risk, vehicles must be fitted with liquid level sensing equipment. | |
| | (c) Deflectors must be fitted to foot valves to minimize jetting and turbulence. | |
| | (d) Additional product handling equipment on islands. Depending upon by group's requirements, this may be about 30–50% more. | |
| Maintenance costs | The additional equipment above will require to be maintained/replaced. Out-of-service time of vehicles for maintenance may be increased. | Maintenance of working platform and safety features. |
| Constraints | | |
| Vehicle accommodation | Can more easily accept range of vehicle capacities and heights (present and future). | Less flexible than bottom loading arrangement. |
| Compatibility with competitors and contractors vehicles | All vehicles likely to use loading bays must be fitted with suitable equipment. Industry agreement to adopt similar practices should be encouraged. | More flexible. |
| Compartment outlets full or empty | Possible need to persuade authorities to change law to permit outlet pipes filled with product, otherwise drainage must be arranged with consequent measurement and operational problems. | No problem. |
| Sophistication | Less flexible operation. Increased maintenance. Need for greater control of maintenance. | More flexible operation. |

### 11.1.3.1.2 Filling by Mass

Where the weighbridge is positioned at the loading bay, the filling can be controlled by a preset mechanism operating in two stages before cutting off at the total loaded mass. Only one compartment can be loaded at a time with this method. The requirement of secondary protection against overfilling is met:

- For bottom loading; as for Section 11.1.3.1.1.2.
- For top loading: use of a "hold-open" type valve on loading arm with operator standing on gantry platform (NOT VEHICLE) in a position to observe compartment being loaded. For single (or large compartments) it may be desirable to fit liquid-level control equipment if the driver/loader has other things to do on the loading platform.

### *11.1.3.2 Automation*

An interlock system whereby product will not flow unless and until:

- the vehicle is properly earthed or bonded
- the loading arm is in its correct position

Measurement of product flow into vehicle compartments should be through a positive displacement meter. This enables systems to be developed which capture the data for the product and quantity loaded into a specific vehicle which is required to identify itself before product will flow.

### *11.1.3.3 Provision for automation*

The basic equipment which must be available on the loading island comprises:

- an earth interlock system
- a positive displacement meter with preset unit and/or two-stage product flow-control valve, at each product loading point
- a meter pulse unit transmitting per unit volume
- means for taking temperature into account, for example:
  - temperature compensating meters
  - thermometer pocket in product lines for measuring temperature by resistance thermometers or temperature recorders

Cables for transmission of data on product and flow quantities must be run in separate wiring conduits and not in the same conduit carrying power, lighting and control valve cables.

## 11.1.4 PROCESS DESIGN PARAMETERS

The individual factors that contribute to the total cost of loading vehicles are:

- the cost of the loading facilities (capital charges for bays, structures, pumps, lines, meters, weighbridges, etc.)
- the cost of vehicle time while occupying the loading bay and while queuing for a loading bay, or waiting for a loading arm while in the bay (vehicle standing charges)
- vehicle capacities and dimensions
- shift patterns, including staggered starts and night loading. In this context the method of operation can be single or double shift patterns, or 24-h service, or a combination of these

Having established the likely future pattern of vehicle arrivals during peak hours, the extent and cost of alternative loading methods and loading rates can be determined and costed, and the economic balance obtained between the cost of vehicle queuing delays and the cost of providing extra loading facilities which will reduce or eliminate them.

### 11.1.4.1 Peak demand

Any loading facility should be designed to meet the forecast loading demand during peak periods. To calculate the facilities required, it is necessary to determine the quantity to be loaded in the peak hour for each product, at the same time establishing the quantities required for each multiproduct vehicle loading combination; and to forecast the future peak demands on which the size (number of loading bays) will be based.

After establishing the total number of loading bays, the effect of major sensitivities should be studied, in particular the reduction of loading bays by one (or more) on the waiting time for all vehicles, and vice versa, in order to ensure that an economic optimum for the whole system is chosen.

### 11.1.4.2 Product flow rates

Flow rates are generally restricted by safety precautions (ie, prevention of excessive static electricity generation), also the economic size of pumps, pipework, and measuring equipment.

As regards safety precautions, concerning static electricity on flow rates, the rate of flow should not normally exceed the figures as given in Table 11.2.

### 11.1.4.3 Simultaneous loading using two or more arms/hoses

Considerable benefit can be achieved by loading a vehicle using two or more loading arms or hoses simultaneously. The additional cost of meters or loading arms, etc. is usually well compensated by the savings from reduced vehicle time in the bay, and in a reduction in the number of loading bays required.

In the case of top loading, the simultaneous use of two or more arms will result in the need for additional equipment to prevent overfilling which may not be necessary for single arm operation. The cost and other consequences arising from such equipment must be taken into account in the economic comparison.

### 11.1.4.4 Calculation of number of spouts and pumping capacity

The determination of the optimum number of spouts for loading facilities is important because it directly affects capital costs of the facilities on the one hand, and operating costs of vehicle fleet on the other hand.

| Table 11.2 Flow Rate Limitation for Static Electricity | | | |
|---|---|---|---|
| **Product** | **Maximum Loading Rate (m³/h)** | | |
| | **DN80 (3 in.)** | **DN100 (4 in.)** | **DN150 (6 in.)** |
| Top loading | 108 | 144 | 216 |
| Bottom loading | 78 | 105 | |

Loading rates and the number of spouts required for each product varies with:

1. truck size
2. number of loading hours per day
3. number of loading days per week
4. time required for positioning, look-up, and depositioning of truck and
5. size of loading assemblies

### 11.1.4.5 Heating for loading arms

When heated pipelines are used, the pipework up to and including the final valve on the loading arm should be heated. Since heating is often required only in cold weather or during start up, it is economical to consider using thermostatically controlled flame/explosion-proof electric heating.

### 11.1.5 EQUIPMENT

Typical equipment required for a truck loading operation is shown in Fig. 11.1.

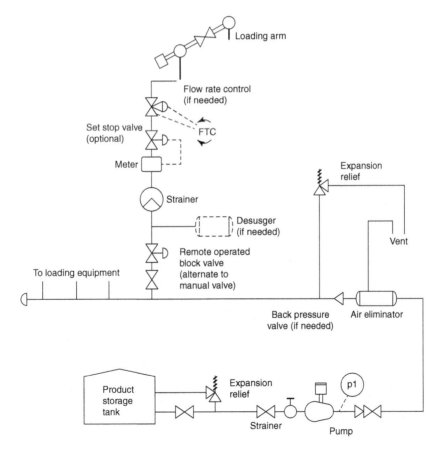

**FIGURE 11.1  Typical Schematic Diagram of Usual Equipment Needed for Tank-Truck Loading**

Note: The symbols shown in the figure have only illustration value.

Air eliminators are used to disengage air and other vapors which would affect the accuracy of metering. Disengaging of vapor is done at about 2 bar (ga) and if there is not at least this amount of static head difference between the air eliminator and the loading spout discharge, a back pressure valve must be provided. This may be a swing-type check valve.

Desurgers are installed in some installations to decrease hydraulic shock resulting from quick shut-off. Strainers are provided to keep dirt and other foreign particles out of the meters, which are normally of the positive-displacement recording type. Set stop valves are used to stop product flow automatically at a predetermined quantity set on the set-stop counter of the meter. These valves can be used in remote-controlled systems and can also serve as a remotely operated block valve to prevent unauthorized withdrawal of product.

Rate-of-flow controllers are self-contained flow-indicating control valves used to prevent over-speeding and wear of meters. The flow indicator is usually a pitot venturi, and a straight meter run of at least six pipe diameters is recommended when the controller is downstream of a strainer, globe valve, or short-radius elbow. The loading arm is a mass or spring balanced assembly of pipe and swing joints which will reach various points on trucks of a range of heights. A controlled closing loading valve is included in the assembly. This decreases the flow rate rapidly to a small percentage of capacity, after which shut-off is slow to prevent shock.

### 11.1.5.1 Pumps

Pumps and loading devices shall be sized to provide rates of flow appropriate to the capacity of the facility. Extreme care shall be taken to ensure that the rates of flow are such that the operator can follow the course of loading and unloading at all times and have adequate time to shut down the facility before the tank or tanks are emptied completely or before they are filled beyond their maximum filling height.

Transfer systems shall be designed such that dangerous surge pressures cannot be generated when the flow in either direction is stopped.

Provision may be made for forced or natural circulation of cold liquid through the loading facility when it is not in service to minimize relief problems and thermal recycling.

The pumps should have flat head capacity characteristics to provide a reasonably constant discharge pressure under varying delivery and discharge conditions. Usual pump differentials are 2.5–3 bars without major changes in static head.

### 11.1.5.2 Flow indicators

Sight flow indicators are not required for large installations, but they may be desirable in some instances in which small quantities of liquid transfer are involved. However, care must be taken to ensure that such equipment is adequately designed for the pressure to which it may be subjected. Either the flapper type or rotor type is satisfactory. The flapper type must be properly installed with respect to direction of flow because it also serves as a check valve.

### 11.1.5.3 Pressure gages

Pressure gages shall be located in a sufficient number of places in the liquid and vapor lines to allow the operator to have a constant check on operating pressure, differentials, and so forth to ensure safe operation.

### 11.1.5.4 Emergency shut-off valves

Emergency shut-off valves shall incorporate all of the following means of closing:

1. Automatic shut-off through thermal (fire) actuation. When fusible elements are used they shall have a melting point not exceeding 120°C.
2. Manual shut-off from a remote location.
3. Manual shut-off at the installed location.

Installation practices for emergency shut-off valves shall include the following considerations:

1. Emergency shut-off valves shall be installed in the transfer line where hose or swivel piping is connected to the fixed piping of the system. Where the flow is only in one direction, a back-flow check valve may be used in place of an emergency shut-off valve if it is installed in the fixed piping downstream of the hose or swivel piping.
2. Emergency shut-off valves shall be installed so that the temperature sensitive element in the thermally actuated shut-off system is not more than 1.5 meters in an unobstructed direct line from the nearest end of the hose or swivel-type piping connected to the line in which the valve is installed.
3. The emergency shut-off valves or back-flow check valves shall be installed in the plant piping so that any break resulting from a pull will occur on the hose or swivel piping side of the connection while retaining intact the valves and piping on the plant side of the connection. This may be accomplished by the use of concrete bulkheads or equivalent anchorage or by the use of a weakness or shear fitting.

### 11.1.5.5 Metering equipment used in loading and unloading

When liquid meters are used in determining the volume of liquid being transferred from one container to another, or to or from a pipeline, such and accessory equipment shall be installed in accordance with the procedures stipulated by the API "Manual of Petroleum Measurement Standards" and Recommended Practice 550.

### 11.1.5.6 Hoses and arms

Hoses and arms for transfer shall be suitable for the temperature and pressure conditions encountered. Hoses shall be provided for the service and shall be designed for a bursting pressure of not less than five times the working pressure. The hose working pressure shall be considered as the greater of the maximum pump discharge pressure or the relief valve setting.

Provisions shall be made for adequately supporting the loading hose and arm. When determining counter masses, ice formation on uninsulated hoses or arms shall be considered.

Flexible pipe connections shall be capable of withstanding a test pressure of one and one-half times the design pressure for that part of the system.

## 11.2 TRUCK UNLOADING

Many points which have been referred to under the subject of loading, are applied here as well. Furthermore, in the following section specific reference is made to discharging of some quantities of products possibly remaining in the tankers before loading again.

## 11.2.1 DISCHARGING UNLOADED PRODUCTS

Vehicles may sometimes return for loading with a quantity of product remaining on board. Attempts should be made to minimize this, and if it occurs, to check the quantity and grade and then to "load on to." Where this cannot be done, the product must be offloaded, and tanks, pipelines, and pumps provided as required. Offloading facilities should be located at a separate bay to avoid congestion at the loading bays; a typical arrangement is illustrated in Fig. B.3.

Such facilities can also be used if vehicle flushing and draining is required, or for special grade changing procedures.

Since quantities to be offloaded should be small, offloading rates do not need to be as fast as loading rates, and rates of about 50–80 m³/h are usual. With suitable manifolding, the pump used occasionally to pump out product from the offloading tanks may be used to speed offloading from the vehicle; otherwise gravity discharge into an underground tank is acceptable.

### 11.2.1.1 Typical loading systems flow schemes
Figs. 11.2 and 11.3 show the typical loading systems (Figs. 11.4–11.8).

### 11.2.1.2 Hose specifications
#### 11.2.1.2.1 Hoses for Road and Rail Tankers for Petroleum Products
This part specifies requirements for rubber and plastic hoses and assemblies for carrying gasoline, kerosene, fuel and lubricating oils, including aviation fuels with an aromatic hydrocarbon content of not more than 50% at temperature up to 80°C. All types are suitable for use with a vacuum not exceeding 0.5 bar.

#### 11.2.1.2.2 Types and Classes
***11.2.1.2.2.1*** Types.  Hoses are designated as follows.

| | |
|---|---|
| Type A | Rough bore externally armored hose principally for gravity discharge with a maximum working pressure of 3 bar. |
| Type AX | Rough bore composite hose principally for gravity discharge with a maximum working pressure of 3 bar. |
| Type B | Rough bore externally armored hose with a maximum working pressure of 7 bar. |
| Type BX | Rough bore composite hose with a maximum working pressure of 7 bar. |
| Type C | Smooth bore hose with smooth or corrugated exterior principally for gravity discharge with a maximum working pressure of 3 bar. |
| Type D | Smooth bore hose with smooth or corrugated exterior with a maximum working pressure of 7 bar. |
| Type E | Smooth bore reeling hose with a maximum working pressure of 7 bar. |
| Type F | Smooth bore reeling hose of controlled dilation for metered delivery with a maximum working pressure of 7 bar. |

***11.2.1.2.2.2*** Classes.  Types A, AX, B, and BX are divided into the following two classes:

Class 1. For aviation and other uses.
Class 2. For nonaviation use.

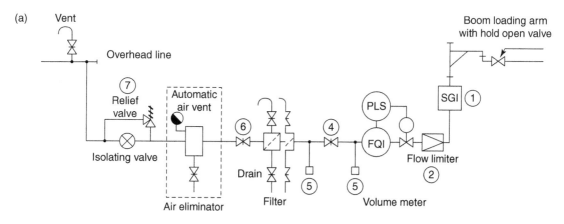

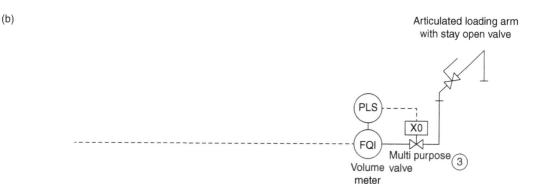

**FIGURE 11.2  Typical Loading Systems**

(a) Loading of bulk road vehicles by meter-mechanically controlled. (b) Loading of bulk road vehicles by meter-automatically controlled.

Notes:

1. Sight glass should be incorporated only where required by local regulations.
2. Flow limiter protects the meter if one pump is used for more than 1 m.
3. Multipurpose solenoid-operated flow-control valve:
   a. protects the meter if one pump is used for more than 1 m
   b. operated by meter quantity preset control
   c. shuts off product flow if actuated by overfill prevention system
   d. interlocked with bonding of vehicle to loading equipment
4. Gate valve, block, and bleed.
5. Meter test proving point, self-sealing coupling.
6. Valve assists filter draining/cleaning.
7. Relief valve relieves thermal expansion pressure.
8. For black oils:
   a. Air-eliminating equipment is not required.
   b. Loading line equipment including the loading arm may have to be heated.
   c. Boom type arms with hydraulically operated valves may be required.

(a)

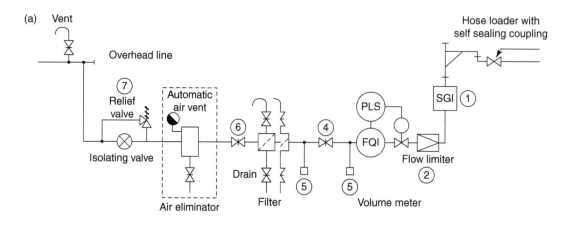

(b)

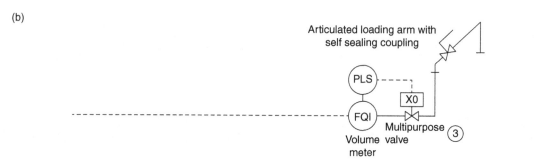

**FIGURE 11.3 Typical Loading Systems**

(a) Loading of bulk road vehicles by meter-mechanically controlled. (b) Loading of bulk road vehicles by meter-automatically controlled.

Notes:

1. Sight glass should be incorporated only where required by local regulations.
2. Flow limiter protects the meter if one pump is used for more than 1 m.
3. Multipurpose solenoid-operated flow-control valve:
   a. protects the meter if one pump is used for more than 1 m
   b. operated by meter quantity preset control
   c. shuts off product flow if actuated by overfill prevention system
   d. interlocked with bonding of vehicle to loading equipment
4. Gate valve, block, and bleed.
5. Meter test proving point, self-sealing coupling.
6. Valve assists filter draining/cleaning.
7. Relief valve relieves thermal expansion pressure.
8. For black oils:
   a. Air-eliminating equipment is not required.
   b. Loading line equipment including the loading arm may have to be heated.
   c. Boom type arms with hydraulically operated valves may be required.

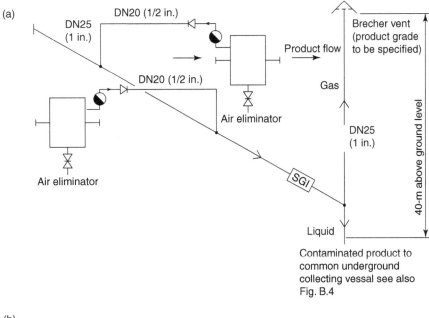

(a)

DN20 (1/2 in.)

DN25
(1 in.)

DN20 (1/2 in.)

Product flow

Brecher vent
(product grade
to be specified)

Gas

Air eliminator

DN25
(1 in.)

SGI

40-m above ground level

Air eliminator

Liquid

Contaminated product to
common underground
collecting vessal see also
Fig. B.4

(b)

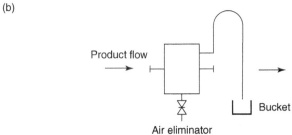

Product flow

Air eliminator

Bucket

**FIGURE 11.4  Typical Pipeline Collection Systems for Air Eliminators**

(a) Scheme for open outlets and (b) scheme for open outlets.

Notes:

1. Instead of a central collecting tank small individual tanks may be installed on each gantry stand.
2. Draining on each filter meters and vehicle tanks also flushings from vehicle tanks should be collected in a small tank trolley for eventual disposal by appropriate down grading.
3. Where common collecting lines are used these should be segregated to high or low flash point products etc

(a)

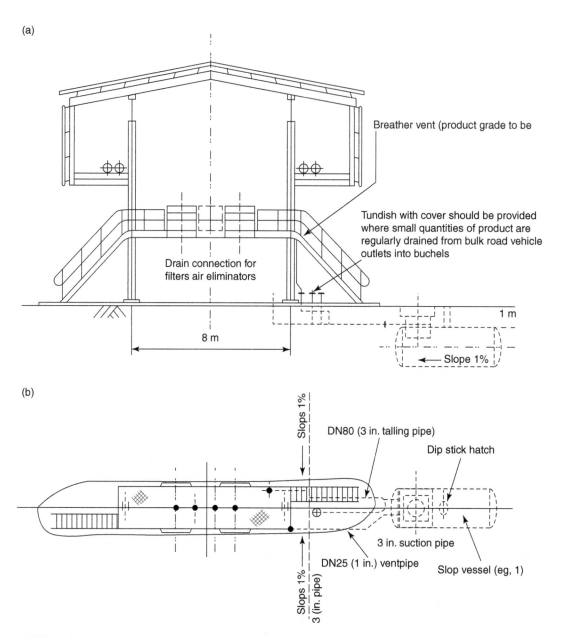

Breather vent (product grade to be

Tundish with cover should be provided where small quantities of product are regularly drained from bulk road vehicle outlets into buchels

Drain connection for filters air eliminators

1 m

8 m

Slope 1%

(b)

Slops 1%

DN80 (3 in. talling pipe)

Dip stick hatch

3 in. suction pipe

DN25 (1 in.) ventpipe

Slop vessel (eg, 1)

Slops 1%
3 (in. pipe)

**FIGURE 11.5  Typical Underground Vessel for Slops, With Closed Drainage Flow Scheme**

(a) Box from gantry and (b) underground vessel for slops.

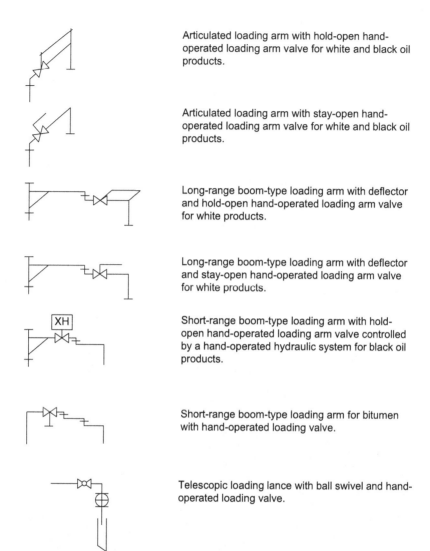

Articulated loading arm with hold-open hand-operated loading arm valve for white and black oil products.

Articulated loading arm with stay-open hand-operated loading arm valve for white and black oil products.

Long-range boom-type loading arm with deflector and hold-open hand-operated loading arm valve for white products.

Long-range boom-type loading arm with deflector and stay-open hand-operated loading arm valve for white products.

Short-range boom-type loading arm with hold-open hand-operated loading arm valve controlled by a hand-operated hydraulic system for black oil products.

Short-range boom-type loading arm for bitumen with hand-operated loading valve.

Telescopic loading lance with ball swivel and hand-operated loading valve.

**FIGURE 11.6 Symbols for Bulk Road Vehicle Loading Arms (Top Loading)**

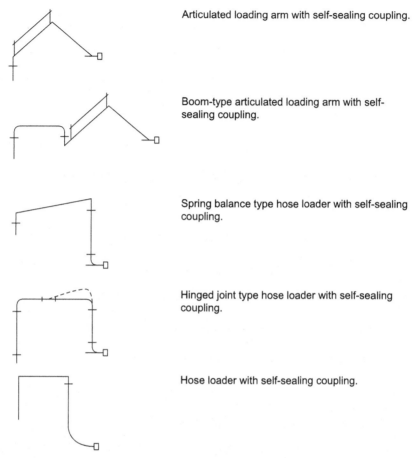

Articulated loading arm with self-sealing coupling.

Boom-type articulated loading arm with self-sealing coupling.

Spring balance type hose loader with self-sealing coupling.

Hinged joint type hose loader with self-sealing coupling.

Hose loader with self-sealing coupling.

**FIGURE 11.7 Symbols for Bulk Road Vehicle Loading Arms and Hoses (Bottom Loading)**

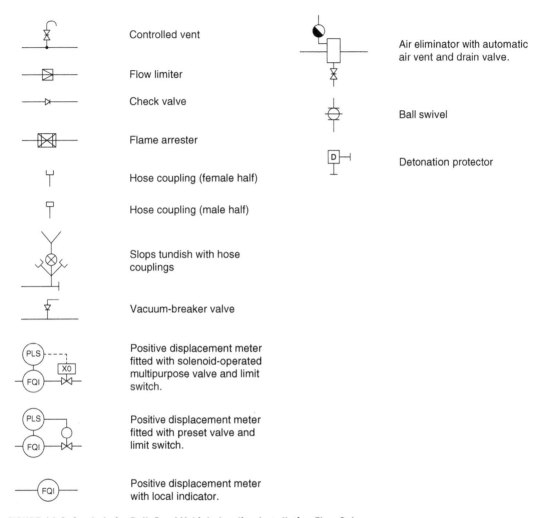

| | |
|---|---|
| | Controlled vent |
| | Flow limiter |
| | Check valve |
| | Flame arrester |
| | Hose coupling (female half) |
| | Hose coupling (male half) |
| | Slops tundish with hose couplings |
| | Vacuum-breaker valve |
| | Positive displacement meter fitted with solenoid-operated multipurpose valve and limit switch. |
| | Positive displacement meter fitted with preset valve and limit switch. |
| | Positive displacement meter with local indicator. |

| | |
|---|---|
| | Air eliminator with automatic air vent and drain valve. |
| | Ball swivel |
| | Detonation protector |

**FIGURE 11.8 Symbols for Bulk Road Vehicle Loading Installation Flow Schemes**

**Table 11.3  Nominal Bores and Tolerances**

| Types A, AX, B, BX, C, and D (mm) | Types E and F (mm) | Permissible deviations (mm) |
| --- | --- | --- |
| — | 25 | ±1.25 |
| 32 | 32 | ±1.25 |
| 38 | 38 | ±1.5 |
| 51 | 51 | ±1.5 |
| 63 | — | ±1.5 |
| 76 | — | ±2.0 |
| 102 | — | ±2.0 |

**Table 11.4  Pressure Ratings**

| Pressure | Types A, AX, and C (bar) | Types B, BX, D, E, and F (bar) |
| --- | --- | --- |
| Maximum | | |
| Working | 3 | 7 |
| Proof | 4.5 | 10.5 |
| Minimum burst | 12 | 28 |

## 11.2.1.2.3 Dimensions and Tolerances

***11.2.1.2.3.1*** Bore.  The bore of the hose shall comply with the nominal dimensions and tolerances given in Table 11.3 when measured in accordance with BS 5173: Section 101.1.

### 11.2.1.3 Pressure requirements

The maximum working pressure, proof pressure and minimum burst pressure of hoses shall be as given in Table 11.4.

### 11.2.1.4 Performance requirements

Besides pressure requirements, the hose shall have resistance to vacuum of up to 0.5 bar. It shall also have sufficient resistance to materials to be handled.

Note: For further details see BS 3492 (Figs. 11.9–11.11).

Typical truck dimensions

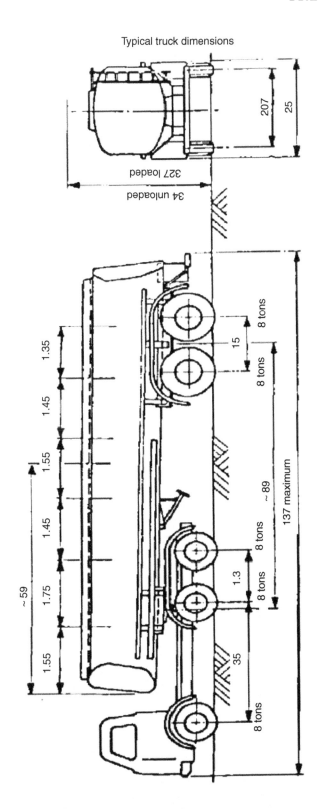

Dimension in meters

Total weight:    30 tons
Pay load:       26 tons
Compartments: 6, 5000 L
Turning radius:  0.5 m
(outer)

**FIGURE 11.9 Typical 38 Ton Bulk Road Vehicle**

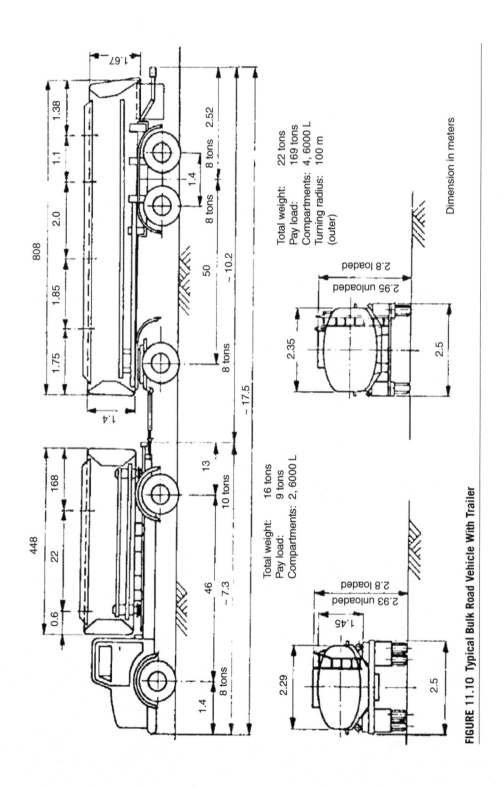

**FIGURE 11.10 Typical Bulk Road Vehicle With Trailer**

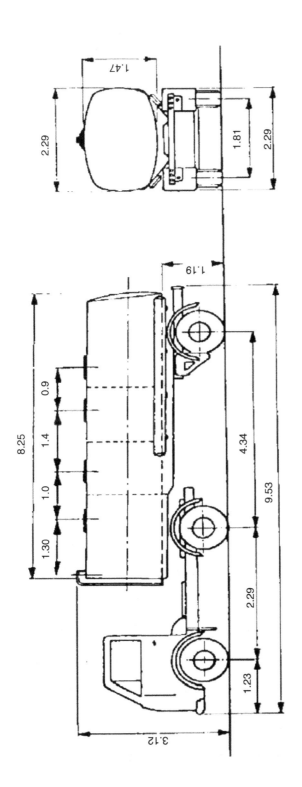

Total weight:      18.5 tons
Pay load:          16.500 tons
Compartments:   2, 5500 L
                 2, 27,500 L
Turning radius:    8 m

**FIGURE 11.11  Typical Bulk Road Vehicle—16.5 m³ Capacity**

# ADDITIONAL LIST OF READING ON RETROFITTING

BSI (British Standards Institution)

BS 5351                    "Specification for Steel Ball Valves for the Petroleum, Petrochemical and Allied Industries," 1986

BS 6843                    "Classification of Petroleum Fuels," 1988

API (American Petroleum Institute)

API-RP-550                 "Manual on Installation of Refinery Instruments and Control Systems," 4th Ed., Jul. 1985

API Standard 616           "Gas Turbines for the Petroleum, Chemical and Gas Industry Services," Ed.1998

API (American Petroleum Institute)

API 672                    "Packaged, Internally Geared Centrifugal Air Compressors for General Refinery Services," 2nd Ed., Apr. 1988.

API 680                    "Packaged Reciprocating Plant and Instrument Air Compressors for General Refinery Services," 1st Ed., Oct. 1987.

API (American Petroleum Institute)

API Std. 650               "Welded Steel Tanks for Oil Storage", 10th Ed., Nov. 1998

NACE (National Association of Corrosion Engineers)

TPC Publication 1, 1984    "Cooling Water Treatment Manual", 3rd Ed.,

ANSI (American National Standard Institute)

ANSI B 16.5                "Pipe Flanges and Flanged Fittings, NPS 1/2 through NPS 24," Ed. 2003

ANSI (American National Standards Institute)

B16.5                      "Steel Pipe Flanges and Flanged Fittings"

B31.3                      "Petroleum Refinery Piping"

ISA (Instrument Society of America)

ISA-S-7.3                  "Quality Standard for Instrument Air"

ISA-RP-7.7                 "Recommended Practice for Producing Quality Instrument Air"

ASME (American Society of Mechanical Engineers)

Section VIII, Div.1 "Boiler and Pressure Vessel Code"

BSI (British Standards Institution)

| | |
|---|---|
| BS1515-Part I | "Fusion Welded Pressure Vessels" |
| BS1655 | "Flanged Automatic Control Valves for the Petroleum Industry" |

ASME (American Society of Mechanical Engineers)

| | |
|---|---|
| PTC-4.1 | "Heat Loss Abbreviated Method" |
| ASME | "Boiler and Pressure Vessel Code, Section I" |

ASTM (American Society for Testing and Materials)

| | |
|---|---|
| | "Special Technical Publication No. 148" |
| ASTM D-1066 | "Standard Practice for Sampling Steam" |
| ASTM D-1125-50T | "Standard Test Method for Electrical Conductivity and Resistivity of Water" |

NAFM (National Association of Fan Manufacturers)

"Standard Test Code"

ASME/ANSI (American Society of Mechanical Engineers)

| | |
|---|---|
| B16.5 | "Steel Pipe Flanges and Flanged Fittings" |
| B31.3 | "Process Piping" |
| B31.1 | "Power Piping" |
| Sec. 1 | "Rules for Construction of Power Boilers" |

BSI (British Standard Institution)

| | |
|---|---|
| BS 799 | "Oil Burning Equipment," Parts 3 and 4 |
| BS 1560 | "Steel Pipe Flanges and Flanged Fittings for the Petroleum Industry" |
| BS 2790 | "Design and Manufacture of Shell Boilers of Welded Construction" |
| BS 3059 | "Specification for Steel Boiler and Superheater Tubes" |

ISO (International Organization for Standardization)

| | |
|---|---|
| 831 | "Rules for Construction of Stationary Boilers" |

BSI (British Standards Institution)

| | |
|---|---|
| BS 4485 | "British Standard Specification for Water Cooling Towers" |
| Part 1: 1969 | "Glossary of Terms" |
| Part 2: 1988 | "Methods for Performance Testing" |

Part 3: 1988                        "Thermal Design Principles"

Part 4: 1975                        "Structural Design of Cooling Towers"

CTI (Cooling Tower Institute, USA)

CTI Bulletin                        "Nomenclature for Industrial Water Cooling Tower"

NCL-109                             "Acceptance Test Code"

ASME (American Society of Mechanical Engineers)

   "ASME Test Code"

API (American Petroleum Institute)

API Standard 610                    "Centrifugal Pumps for Petroleum, Petrochemical and Natural Gas Industries"

API Standard 618                    "Reciprocating Compressors for Petroleum, Chemical and Gas Industry Services"

API Standard 619                    "Positive Displacement Compressors for Petroleum, Chemical and Gas Industry Services"

API Standard 660                    "Shell-and-Tube Heat Exchangers for General Refinery Services"

API Standard 661                    "Air-Cooled Heat Exchangers for General Refinery Services"

API Specification 11 P (Spec. 11 P)

             "Specification for Packaged Reciprocation Compressors for Oil and Gas Production Services"

ASME (American Society for Mechanical Engineers)

ASME Standard, No. 120              Prepared by Pannel IV of Joint ASME-ASTM-NEMA Committee on "Gas Turbine Lubrication System," Clause 6.0, "Coolers", Louisville, Ky, May 1957

ASME Code, Section VIII, Division 1 Part 2, Appendix A

TEMA (Tubular Exchanger Manufacturers Association, Inc)

TEMA Class R Heat Exchangers

ASME (The American Society of Mechanical Engineers)

   "ASME Code," Section VIII, Division 1

API (American Petroleum Institute)

API RP-550, Section 8.5,           "Winterizing", 4th Ed., Oct. 1980

ANSI (American National Standards Institute) / ASME (American Society of Mechanical Engineers)

PTC 39.1                            "Performance Test Codes for Condensate RemovalDevices for Steam Systems"

ANSI (American National Standards Institute) / FCI (Fluid Controls Institute)

69-1                                "Pressure Rating Standards for Steam Traps"

85-1                                "Standards for Production and Performance Tests for Steam Traps"

ABMA (American Boiler Manufactures' Association)

ASME (American Society of Mechanical Engineers)

AWWA (American Water Works Association, Inc)

|  |  |
|---|---|
|  | "Water Treatment Plant Design", 1971 |
| Manual M21 | "Ground Water", Latest Edition |

API (American Petroleum Institute)

|  |  |
|---|---|
|  | "API Glossary of Terms Used in Petroleum Refining", 2nd Ed., 1962 |

GPSA (Gas Processors Suppliers Association)

|  |  |
|---|---|
|  | "Engineering Data Book", Section 18, 1987 |

USPHS (US Public Health Service)

|  |  |
|---|---|
|  | "Drinking Water Standards", No. 956, 1962 |

BSI (British Standards Institution)

| | |
|---|---|
| BS 65: 1991 | "Specification for Vitrified Clay Pipes, Fittings and Ducts, Also Flexib Mechanical Joints for Use Solely with Surface Water Pipes and Fittings" |
| BS 143 &1256: 2006 | "Specification for Malleable Cast Iron and Cast Copper Alloy Threaded Pipe Fittings" |
| BS 416-1: 1990 | "Discharge and Ventilating Pipes and Fittings, Sand-Cast or Spun in Cast Iron Part 1 Specification for Spigot and Socket Systems" |
| BS 437: 2008 | "Specification for Cast Iron Drain Pipes, Fittings and their Joints for Socketed and Socketless Systems" |
| BS 460: 2002 + A2: 2007 | "Cast Iron Rainwater Goods Specifications" |
| BS EN 512: 1995 + A: 2001 | "Fiber-Cement Products- Pressure Pipes and Joints" |
| BS EN 545: 2010 | "Ductile Iron Pipes, Fittings, Accessories and their Joints for Water Pipelines- Requirements and Test Methods" |
| BS EN 588-1: 1997 | "Fibre-Cement Pipes for Sewers and Drains – Part 1: Pipes, Joints and Fittings for Gravity Systems" |
| BS EN 777.1: 2011 | "Specification for Masonry Units Part 1: Clay Masonry Units" |
| BS EN 10224: 2002 | "Non-Alloy Steel Tubes and Fittings for the Conveyance of Water an Aqueous Liquids- Including Water for Human Consumption Technical Delivery Conditions" |
| BS EN 10346: 2009 | "Continuously Hot – Dip Coated Steel Flat Products-Technical Delivery Conditions" |
| BS 1169: 1953 | "Rubber Sealing Rings for Domestic Preserving Jars for Fruit and Vegetables" |
| BS EN 13101: 2002 | "Steps for Under Ground man Entry Chambers" |
| BS EN ISO 1461: 2009 | "Hot Dip Galvanized Coatings on Fabricated Iron and Steel Articles-Specifications and Test Methods (ISO1461:2009)" |

| BS 3506: 1969 | "Specifications for Unplasticized PVC Pipe for Industrial Uses" |
| BS 4211: 2005 + A1: 2008 | "Specification for Permanently Fixed Ladders" |
| BS 4622: 1970 | "Specifications for Gray Iron Pipes and Fittings" |
| BS 4660: 2000 | "Thermoplastics Ancillary Fittings of Nominal Sizes 110 and 160 for below Ground Gravity Drainage and Sewerage" |
| BS 4942: 1981 | "Short Link Chain for Lifting Purposes" |
| BS EN 1401-1: 2009 | "Plastic Piping Systems for Non-Pressure Underground Drainage and Sewerage Unplasticized Poly (Vinyl Chloride) (PVC-U) Part 1: Specification for Pipes, Fittings and the System" |
| BS EN 1917: 2002 | "Concrete Manholes and Inspection Chambers, Unreinforced, Steel Fibre and Reinforced" |
| BS EN 877: 1999 + A1: 2006 | "Cast Iron Pipes and Fittings, their Joints and Accessories for the Evacuation of Water form Buildings Requirements, Test Methods and Quality Assurance" |

DIN (Deutsches Institut Für Normung e.V.)

| DIN 1230: 1986 | "Vitrified Clayware for Sewers" |
| DIN 4032: 1981 | "Concrete Pipes and Fittings" |
| DIN 19534: Pt. 1: 1979 | "Pipes and Fittings of Unplasticized Rigid PVC with Plug Socket for Sewerage Pipes and Lines" |

ISO (International Organization for Standardization)

| ISO 49: 1997 | "Malleable Cast Iron Fittings Threaded to ISO 7-1" |
| ISO 4435: 1991 | "Unplasticized Poly (Vinyl Chloride) (PVC-U) Pipes and Fittings for Buried Drainage and Sewerage Systems Specifications" |
| ISO 6708: 1995 | "Pipework Components Definition and Selection of DN (Nominal Size)" |
| ISO 7186: 2011 | "Ductile Iron Products for Sewage Applications" |
| ISO 8283-1: 1991 | "Plastics Pipes and Fittings Dimensions of Sockets and Spigots for Discharge Systems Inside Buildings" |
| | Part 1: Unplasticized Polycvinyle Chloride (PVC-U) and Chlorinated Polycviny Chlorid (PVC-C) |

NFC (National Fire Codes) NFPA

| Section 20 | "Centrifugal Fire Pumps" |

NFPA (National Fire Protection Association)

| N.F.P.A. 1901 Chapter 11 | "Atootive Fire Apparatus" |

BSI (British Standard Institution)

| Supplement to BS 5000 | "Pressure Vessels" |
| BS-336-6391 | "Fire Hose Couplings and Ancillary Equipment" |
| BS-5430 Part 3 | |

ANSI (American National Standards Institute)

    ANSI-B 16.5                     "Pipe Flanges and Fittings"

ISO (International Organization for Standardization)

    ISO-2954                       "Mechanical Vibration of Rotating Reciprocating and Machinary Equipment"

DIN (Deutsches Institute Fur Normung e.V.)

    DIN-50049                     "Documents for Material Tests"

    DIN-14690                     "Fire Fighting Equipment"

    DIN-49462/49463              "Splash Proof and Water Tight Multiple Socket Outlets"

    DIN-45639                     "Inside Noise of Motor Vehicle"

IEC (International Electrotechnical Commission)

    IEC-529 (IP-Code)              "Degree of Protection Provided by Enclosures"

ASME (American Society of Mechanical Engineers)

    Section II                        "Material Specification Part A – Ferrous Materials"

BSI (British Standards Institution)

    BS 4409, Part 1, 1991         "Screw Conveyors, Specification for Fixed Trough Type"

    BS 4409, Part 3, 1982         "Screw Conveyors, Method for Calculating Driver Power"

ISO (International Organization for Standardization)

    ISO 7119                       "Continuous Mechanical Handling Equipment for Loose Bulk Materials Screw Conveyors," 1st Ed., 1981

    ISO 5048                       "Continuous Mechanical Handling Equipment Belt Conveyors with Carrying Idlers-Calculation of Operating Power and Tensile forces," 2nd Ed., 1989

    ISO 2148                       "Continuous Handling Equipment Nomenclatures," 1st Ed., 1974

ANSI/CEMA (American National Standard Institute/Conveyor Equipment Manufacturers Association)

    350 Class III E                 "Screw Conveyors," 1st Ed., 1971

    350 Class IV E                 "Package Handling Conveyors," 1st Ed., 1970

ISO (International Standard Organization)

    7183, 1986                     "Compressed Air Dryers-Specifications and Testing," Section 4, 1st Ed., 15 Mar. 1986

    5388, 1981                     "Standard Air Compressors, Safety Rules and Code of Practice," 1st Ed., 1981

API (American Petroleum Institute)

| | |
|---|---|
| Bul. 1003 | "Precautions Against Electrostatic Ignition During Loading of Tank Motor Vehicles" Oct., 1975 |
| RP 550, 1986 | "Manual on Installation of Refinery Instruments and Control Systems" |
| API Std. | "Manual of Petroleum Measurement Standards," Chapter 6.2, 1st Ed., Oct. |
| | "Loading Rack and Tank Truck Metering Systems" 1983, Reaffirmed Mar. 1990 |

BSI (British Standards Institution)

| | |
|---|---|
| BS SP 3492 | "British Standard Specification for Road and Rail Tanker Hoses and Hose Assemblies for Petroleum Products, Including Aviation Fuels" 1987 |
| BS 5173 | "Method of Test for Rubber and Plastic Hoses and Assemblies" Sec. 101.1, 1985 |

# Glossary of Terms

**Adiabatic drying** The drying process described by a path of content adiabatic cooling temperature on the psychometric chart.

**Adsorbate** The molecules that condense on the adsorbent surface, for example, water in the case of drying.

**Adsorbate loading** The concentration of adsorbate on adsorbent, usually expressed as kilogram adsorbate per kilogram adsorbent.

**Adsorbent** A solid material which demonstrates adsorption characteristics.

**Adsorption** The phenomenon whereby molecules in the fluid phase spontaneously concentrate on a solid surface without undergoing any chemical change.

**Adsorption selectivity** The preference of a particular adsorbent material for one adsorbate over another on the basis of certain characteristics of the adsorbate such as polarity or molecular mass.

**After cooler** After-cooler is a species of surface condenser in which compressed air/gas is cooled after compression.

**After cooling** It involves cooling of air in a heat exchanger following the completion of compression to (1) reduce the temperature and (2) liquefy condensate vapors.

**Air flow** It is total quantity of air, including associated water vapor flowing through the tower.

**Air heater or air preheater** A heat-transfer apparatus through which combustion air is passed and heated by a medium of higher temperature, such as the products of combustion, steam, or other fluid.

**Ambient temperature** It is the temperature level of atmosphere in the environment of the equipment installation.

**Ambient wet bulb temperature** Ambient wet-bulb temperature is wet-bulb temperature of air-measured windward of the tower and free from the influence of the tower.

**Approach** Approach is the difference between recoiled water temperature and nominal inlet air wet-bulb temperature.

**Aqueous film-forming foam** Also known as AFFF, it is a mixture of fluorocarbon and hydrocarbon surfactants.

**Atomizer** A device used to reduce a fluid to a fine spray. Atomization means are normally either steam, air or mechanical.

**Back pressure** The pressure on the outlet or downstream side of a flowing system. In an engine, the pressure which acts adversely against the piston, causing loss of power.

**Basin kerb** It is the top level of the retaining wall of the cold water basin; usually the datum point from which tower elevation points are measured.

**Bed expansion** The effect produced during backwashing; the resin particles become separated and rise in the column. The expansion of the bed, due to the increase in the space between resin particles, may be controlled by regulating the backwash flow.

**Blowdown** The continuous or intermittent removal of some of the water in the boiler or cooling water system to reduce the concentration of dissolved and/or suspended solids.

**Brake kilowatt** The actual power input at the crankshaft of the compressor drive.

**Bunker "C" fuel oil** A heavy residual fuel oil used by ships, industry, and for large-scale heating installations. In industry, it is often referred to as Grade No. Fuel Oil.

**Burner** A device for the introduction of fuel and air into a boiler at the desired velocities, turbulence, and concentration to establish and maintain proper ignition and combustion. Burners are classified by the types of fuel fired such as, oil, gas, or a combination of gas and oil. A secondary consideration in classifying burners is the means by which combustion air is mixed with the fuel.

**Burner management system** The portion of a boiler control system which is associated with the supply of fuel to the burners. This includes the complete fuel train, safety shut-off valves, fuel pressure and temperature limits, burner starting and sequencing logic, and annunciation of trouble signals.

**Capillary flow**   The flow of liquid through the interstices and over the surfaces of a solid caused by liquid–solid molecular attraction.

**Carryover**   The moisture and entrained solids forming the film of steam bubbles, a result of foaming in a boiler. Carryover is caused by a faulty boiler water condition.

**Cell**   Cell is the smallest subdivision of a cooling tower bounded by exterior walls and partition walls which can function as an independent unit in regards to air and water flow.

**Cell height**   The distance from the basin kerb to the top of fan deck, but not including fan stack.

**Cell length**   The dimension parallel to the longitudinal axis and the plane where louvers are usually placed.

**Cell width**   The dimension perpendicular to the tower longitudinal axis and usually at right angles to the louver area.

**Centrifugal pumps**   A pump in which the pressure is developed principally by the action of centrifugal force.

**Chlorine requirement**   The amount of chlorine, expressed in milligrams per kilograms, required to achieve the objectives of chlorination under specified conditions.

**Chlorine residual**   The amount of available chlorine present in water at any specified period, subsequent to the addition of chlorine.

**Circulating water flow**   Circulating water flow is the quantity of hot water flowing into the tower.

**Cold-water basin (basin pond)**   A device underlying the tower to receive the cold water from the tower, and direct its flow to the suction line or sump.

**Column anchor**   Column anchor is a device for attaching the tower structure to the foundation; it does not include the foundation bolt.

**Combustion**   Combustion is the rapid oxidation of fuel accompanied by production of heat.

**Combustion control system**   The portion of a boiler control system associated with the control and maintenance of air/fuel mixtures throughout the operating range of the burner and during changes in firing rate.

**Company/employer/owner**   Refers to one of the related affiliated companies of the petroleum industries of Iran such as National Iranian Oil Company (NIOC), National Iranian Gas Company (NIGC), National Petrochemical Company (NPC), etc. as parts of the Ministry of Petroleum.

**Concentration**   The increase of impurities in the cooling water due to the evaporative process.

**Concentration ratio**   The ratio of the impurities in the circulating water and the impurities in the make-up water.

**Condensation**   The constituent of air or gas when liquefied due to a certain reduction in the coolant temperature against the air/gas inlet temperature.

**Constant-rate period**   The drying period during which the rate of liquid removal per unit of drying surface is constant.

**Contractor**   Refers to the persons, firm, or company whose tender has been accepted by the employer and includes the contractor's personnel representative, successors, and permitted assigns.

**Controlled-volume pump**   A reciprocating pump in which precise volume control is provided by varying the effective stroke length. Such pumps are also known as proportioning, chemical injection, or metering pumps. (1) In a packed-plunger pump, the process fluid is in direct contact with the plunger. (2) In a diaphragm pump, the process fluid is isolated from the plunger by means of a hydraulically actuated flat or shaped diaphragm.

**Cooling range (range)**   The difference between the hot water temperature and the recooled water temperature.

**Cooling system**   A self-contained, closed cooling water system, capable of taking the heat transmitted to the heating surface, to the extent specified by the manufacturer.

**Critical diameter**   The diameter of particles larger than which will be eliminated in a sedimentation centrifuge.

**Critical moisture content**   The moisture content of the material at the end of the constant-rate period. The critical moisture content is not a unique property of the material, but is influenced by its physical shape as well as the conditions of the drying process.

**Cycle time**   The amount of time allocated for one bed in an adsorption system to complete adsorption to a predetermined outlet specification level and to be reactivated.

**Damper**   A device for introducing a variable resistance for regulating the volumetric flow of gas or air.

**Desiccant**  An adsorbent that shows primary selectivity for the removal of water. All adsorbents are not necessarily desiccants.

**Desiccant fouling**  Material adsorbed from the carrier stream may not be desorbed satisfactorily on regeneration. Some reaction may also occur on the adsorbent leading to products that are not desorbed. These reaction products may inhibit efficient adsorption and obstruct or "foul" the capacity of the active surface.

**Design basis**  A good design basis requires a sound knowledge of the stream to be processed, the desired outlet specification, and the way system will be operated. The design conditions on which an adsorption system is based are not necessarily the actual operating conditions, nor the least or most stringent operating conditions.

**Dew point**  Temperature, referred to as a specific pressure (degrees Celsius), at which the water vapor begins to condensate.

**Dew point (air)**  The temperature at which condensate will begin to form if the air is cooled at constant pressure. At this point, the relative humidity is a percentage.

**Discharge stack**  The part of the shell or casing of a forced draught tower, through which the outlet air is finally discharged. (See "fan stack" for induced draught towers and "shell" for natural draught towers.)

**Distribution basin**  The elevated basin used to distribute hot water over the tower packing.

**Distribution header**  The pipe or flume delivering water from the inlet connection to lateral headers, troughs, flumes, or distribution basins.

**Distribution system**  Those parts of a tower beginning with the inlet connection which distribute the hot circulating water within the tower to the point where it contacts the air.

**Down spout**  A short vertical pipe or nozzle used in an open distribution system to discharge water from a flume or lateral on to a splash plate.

**Downcomer**  A tube or pipe in a boiler or waterwall circulating system through which fluid flows downward between headers.

**Draft**  The negative pressure (vacuum) of the flue gas measured at any point in the boiler, expressed in millimeters of water column (kilopascals).

**Drift eliminator**  A system of baffles located in the tower designed to reduce the quantity of entrained water in the outlet air.

**Drift loss**  The water lost from the tower as liquid droplets entrained in the outlet air.

**Drivers**  A piece for imparting motion to another piece either directly or indirectly.

**Dry gas**  A gas which does not contain fractions that may easily condense under normal atmospheric conditions.

**Dry powder**  Fire extinguishing agent in a fine form primarily of sodium bicarbonate or urea-based potassium bicarbonate (monnex or purple K) with added material to produce water repellency and free-flowing characteristics.

**Dry-bulb thermometer**  An ordinary thermometer, especially one of the two similar thermometers of a psychometer whose bulb is unmoistened.

**Drying rate**  The amount of water (kg) removed per square meter of drying area per hour. Or the volume flow rate of condensed gas at Standard Reference Atmosphere Condition of an absolute pressure of. kPa (bar) and a temperature of degree Celsius.

**Economizer**  A series of tubes located in the path of the flue gases. Feedwater is pumped through these tubes on its way to the boiler in order to absorb waste heat from the flue gas.

**Effective volume**  The volume within which space the circulating water is in intimate contact with the air flowing through the tower.

**Efficiency, fuel**  Fuel efficiency refers to the heat absorbed divided by the net heat of combustion of the fuel as heat input, expressed as a percentage.

**Efficiency, thermal**  Thermal efficiency refers to the total heat absorbed divided by total heat input, expressed as a percentage.

**Energy conservation**  It is saving in power consumption, as roughly estimated, each degree Celsius decrease in gas temperature between the stages shall result in 1% in power consumption.

**Equilibrium loading** The loading of an adsorbate on the given adsorbent, usually expressed in kilogram of adsorbate per hundred kilogram of adsorbent when equilibrium is achieved at a given pressure, temperature, and concentration of the adsorbate.

**Equilibrium moisture content** The amount of moisture, in the solid that is in thermodynamic equilibrium with its vapor in the gas phase, for given temperature and humidity conditions. The material cannot be dried below its corresponding equilibrium moisture content.

**Excess air** Excess air is the amount of air above the stoichiometric requirement for complete combustion expressed as a percentage.

**FLC** Foam–liquid concentrate.

**Falling-rate period** The part of drying time where the drying rate varies in time.

**Fan** It is a rotary machine which propels air continuously. This is used for moving air in a mechanical draught tower and is usually of the axial-flow propeller type. The fan may be of induced draught or forced draught application.

**Fan casing** Those stationary parts of the fan which guide air to and from the impeller. In the case of an induced draught fan, the casing may form the whole or part of the fan stack.

**Fan deck** The surface enclosing the top of an induced draught tower, exclusive of any distribution system which may also form a part of the enclosure.

**Fan drive assembly** The components for providing power to the fan, normally comprising driver, drive shaft, and transmission unit, and primary supporting members.

**Fan duty (static)** The inlet volume dealt with by a fan at a stated fan static pressure.

**Fan duty (total)** The inlet volume dealt with by a fan at a stated fan total pressure.

**Fan power** The power input to the fan assembly, excluding power losses in the driver.

**Fan stack** A cylindrical or modified cylindrical structure enclosing the fan in induced draught towers.

**Fan static pressure** The difference between the fan total pressure and the fan velocity pressure.

**Fan total pressure** The algebraic difference between the mean total pressure at the fan outlet and the mean total pressure at the fan inlet.

**Fan velocity pressure** The velocity pressure corresponding to the average velocity at the fan outlet, on the basis of the total outlet area without any deductions for motors, fairings, or other bodies.

**Fan-stack height** The distance from the top of the fan deck to the top of the fan stack.

**Filling installations** Facilities for truck loading from entering time up to leaving.

**Film packing** An arrangement of surfaces over which the water flows in a continuous film throughout the depth of the packing.

**Filter** A piece of unit operation equipment by which filtration is performed.

**Filter medium** The "filter medium" or "septum" is the barrier that lets the liquid pass while retaining most of the solids; it may be a screen, cloth, paper, or bed of solids.

**Filtrate** The liquid that passes through the filter medium is called the filtrate.

**Fire-fighting crew** Professional fire fighters.

**Fire-fighting truck** Applies to fire-fighting vehicle, fire engine, and automotive fire apparatus.

**Fluoroprotein foam compound** Conventional protein foam modified by the addition of fluorocarbon surfactants.

**Free moisture content** The liquid content that is removable at a given temperature and humidity. Free moisture may include both bound and unbound moisture, and is equal to the total average moisture content minus the equilibrium moisture content for the prevailing conditions of drying.

**Fuel gas** Any gas used for heating.

**Gantry** A framework on a loading island, under or besides which one or two loading bays with some articulated loading arms/hoses are arranged.

**Heat load** The rate of heat removal from the circulating water within the tower.

**Heating surface** The surface which transmits heat directly from the heating medium to the cooling medium.

**Heating value of a fuel** The caloric, thermal, or heating value of a fuel is the total amount of heat generated by the complete combustion of a unit quantity of fuel, expressed as kilojoules per kilogram for liquid fuels and millijoule per newton meter for gas fuels.

**High-risk areas** Includes refineries, petrochemical plants, production facilities gas plants, and other related installation where provision of manned or retained fire stations is approved by IPI.

**Hot-water temperature** The temperature of circulating water entering the distribution system.

**Ignitor** A term used in industry to denote the device that provides the proven ignition energy required immediately to light the pilot flame.

**Inhibitor** A substance, the presence of which in small amounts in a petroleum product prevents or retards undesirable chemical changes from taking place in the product, or in the condition of the equipment in which the product is used. In general, the essential function of inhibitors is to prevent or retard oxidation or corrosion.

**Inlet air** The air flowing into the tower; it may be a mixture of ambient air and outlet air.

**Inlet air-wet bulb temperature** It is the average wet-bulb temperature of the inlet air; including any recirculation effect. This is an essential concept for purposes of design, but is difficult to measure.

**Inlet pressure** The lowest air pressure in the inlet piping to the compressor and may be expressed as either gage pressure or absolute pressure.

**Inlet temperature** The temperature of liquid coolant entering the heating surface at a specified point in the inlet piping.

**Inlet temperature** The air temperature at the inlet flange of the compressor.

**Intercooler** A species of surface condenser placed between the two consecutive cylinders of a multistage compressor so that the heat of compression generated in the first stage cylinder may be removed (in part or whole) from the compressed air/gas, as it passes through the next stage cylinder's inter-cooler.

**Intercooling** The cooling of air between stages of compression: (1) to reduce the temperature; (2) to reduce volume to be compressed in the succeeding stage; (3) to liquefy condensable vapors; and (4) to save energy.

**Liquefied petroleum gas (LPG)** Light hydrocarbon material, gaseous at atmospheric temperature and pressure, held in the liquid state by pressure to facilitate storage, transport, and handling. Commercial liquefied petroleum gas consists of propane, butane, or a mixture thereof.

**Liquid coolant system** The coolant system by which the heating surfaces are cooled by liquid.

**Loading arm/hose** A piping or hose arrangement for filling in a truck.

**Loading bay** An inlet for trucks to stay under product loading.

**Loading facilities** Facilities consist of pumping and filling installations.

**Loading island** A raised area over which loading arms/hoses and related facilities are installed.

**Louvres** Louvres is members installed in a tower wall, to provide openings through which air enters the tower; usually installed at an angle to the direction of airflow to the tower.

**Low heating value (LHV)** The high heating value minus the latent heat of vaporization of the water formed by burning the hydrogen in the fuel.

**Make-up** It is water added to the circulating water system to replace water loss from the system by evaporation, drift, purge, and leakage.

**Manufactured gas** All gases made artificially or as by-products, as distinguished from natural gas; applied particularly to a utility sendout.

**Maximum allowable working pressure (MAWP)** The maximum continuous pressure for which the manufacturer has designed the equipment (or any part to which the term is referred) when handling the specified fluid at the specified temperature.

**Maximum suction pressure** The highest allowable suction pressure to which the pump is subjected during operation.

**Mazut** A Russian name for distillation residues used largely as fuel oil; also spelled "masut" or "mazout."

**Mesh** The "mesh count" (usually called "mesh"), is effectively the number of openings of a woven wire filter per millimeter, measured linearly from the center of one wire to another mm from it, that is, mesh = $l/(w + d)$.

**Moisture content**  The ratio of water and water vapor by mass to the total volume (gram per cubic meter).

**Motor-rated power**  It is nameplate power rating of the motor driving the fan.

**Mud or lower drum**  A drum or header-tube pressure chamber located at the lower extremity of a water tube boiler convection bank which is normally provided with a blowoff valve for periodically blowing off sediment collecting in the bottom of the drum.

**Multistage reciprocating compressor**  The compressor in which the compression, when a perfect gas or air is entropically compressed, the gas inlet temperature as well as the amount of work spent is the same at each stage.

**Multiple feed**  The combination of two or more pumping elements with a common driver.

**Natural gas**  Naturally occurring mixtures of hydrocarbon gases and vapors, the more important of which are methane, ethane, propane, butane, pentane, and hexane.

**Net positive suction head required (NPSHR)**  It is the NPSH in meters, determined by the vendor testing, usually by water. NPSHR is measured at the suction flange and corrected to the datum elevation. NPSHR is the minimum NPSH at rated capacity required to prevent a head drop of more than 1% due to cavitation within the pump.

**Nominal inlet air-wet bulb temperature**  The arithmetical average of the measurements taken within 1 m of the air inlets and between 1 and 5 m above the basin kerb elevation on both sides of the cooling tower.

**Nominal tower dimensions**  The dimensions used to indicate the effective size of cells, or cooling tower. In the horizontal plane, they refer to the approximate width and length of packed areas, and in the vertical plane to the height above the basin kerb level.

**Normal cubic meters per hour (Nm/h)**  Refers to capacity at normal conditions (kPa and °C) and relative humidity expressed as a percentage.

**Normal operating point**  The point at which usual operation is expected and optimum efficiency is desired. This point is usually the point at which the vendor certifies that performance is within the tolerances stated in this standard.

**Open area**  It is defined as a percentage of the whole area of a woven wire filter, is shown by (Fo).

**Outlet air**  The mixture of air and its associated water vapor leaving the tower. (See airflow.)

**Outlet air wet-bulb temperature**  The average wet-bulb temperature of the air discharged from the tower.

**Outlet temperature**  The temperature of liquid coolant discharged from the heating surface at specified point in the outlet piping.

**Overflow**  The stream being discharged out of the top of a hydrocyclone, through a protruding pipe, is called "overflow." This stream consists of bulk of feed liquid together with the very fine solids.

**PTO**  Power take off.

**Packing (filling)**  The material placed within the tower to increase heat and mass transfer between the circulating water and the air flowing through the tower.

**Partial pressure**  Absolute pressure exerted by any component in a mixture (millibar).

**Plenum**  The enclosed space between the eliminator and the fan stack in induced draught towers, or the enclosed space between the fan and the packing in forced draught towers.

**Pressure part**  A component that contains pressurized water or steam or a mixture of the two.

**Purge (blow down)**  It is the water discharged from the system to control concentration of salts or other impurities in the circulating water.

**Purging**  The displacement of one material with another in process equipment; frequently, displacement of hydrocarbon vapor with steam or inert gas.

**Rated capacity**  Rated process capacity specified by the company to meet process conditions with no negative tolerance (NNT) permitted. *Note*: The acceptable standard for reciprocating compressor industry, tolerance of ±percent is applicable to capacity. Because of this tolerance on capacity, the manufacturer will increase the required capacity by 1% prior to sizing the compressor.

**Rated discharge temperature**  The highest predicted (not theoretical adiabatic) operating temperature resulting from the rated service conditions.

**Rated speed in rotations (revolutions) per minute**  The highest speed required to meet any of the specified operating conditions.

**Recirculation (recycle)**  Recirculation is that portion of the outlet air which re-enters the tower.

**Recirculation rate**  The flow of cooling water being pumped through the entire plant cooling loop.

**Recooled water temperature**  The average temperature of the circulating water entering the basin.

**Refinery gas**  Any form or mixture of gas gathered in the refinery from the various units.

**Regenerant**  The solution used to restore the activity of an ion exchanger. Acids are employed to restore a cation exchanger to its hydrogen form; brine solutions may be used to convert the cation exchanger to the sodium form. The anion exchanger may be regenerated by treatment with an alkaline solution.

**Relative humidity**  The ratio of actual pressure of existing water vapor to maximum possible pressure of water vapor in the atmosphere at the same temperature, expressed as a percentage.

**Relative humidity (relative vapor pressure)**  Ratio of the partial pressure of water vapor (millibar) to its saturation pressure (millibar) at the same temperature.

**Residual fuel oil**  Topped crude oil or viscous residuum in refinery operations.

**Rinse**  The operation which follows regeneration; a flushing out of excess regenerant solution.

**Saturation pressure**  Total pressure at which moist air at a certain temperature can coexist in equilibrium with a plane surface of pure condensed phase (water or ice) at the same temperature (millibar).

**Seal chamber pressure**  The highest pressure expected at the seals during any specified operating condition and during start up and shut down. In determining this pressure, consideration should be given to the maximum suction pressure, the flushing pressure and the effect of internal clearance changes.

**Shell**  Shell is that part of a natural draught tower which induces airflow.

**Shop-assembled boilers**  Water tube boilers, wholly or partly assembled in the manufacturer's workshop, requiring no further fabrication work on the pressure parts and shipped as one unit. It should be noted that such boilers are sometimes referred to by suppliers as packaged boilers.

**Slurry**  A free-flowing mixture of solids and liquids.

**Splash packing**  An arrangement of horizontal laths or splash bars which promotes droplet formation in water falling through the packing.

**Splash plate**  It is used in an open distribution system to receive water from a downspout and to spread water over the wetted area of the tower.

**Spout**  An outlet for loading through an arm or a hose, identical with "loading point."

**Spray nozzle**  It is used in a pressure distribution system to break up the flow of the circulating water into droplets, and effects uniform spreading of the water over the wetted area of the tower.

**Standard air**  Air at degree Celcius dry bulb temperature, percent relative humidity, and kilopascal, which has a density of kilograms per cubic meter.

**Standard air**  Dry air having density of kilograms per liter, at degree Celsius and atmosphere (mmHg).

**Standard cubic meter per hour (Sm/h)**  Refers to the flow rate at any location corrected to a pressure of kilopascals and a temperature of degree Celsius with a compressibility factor and in a dry condition.

**Sump (basin sump or pond sump)**  It is a lowered portion of the cold-water basin floor for draining down purposes.

**Synthetic resin**  Amorphous, organic, semisolid, or solid material derived from certain petroleum oils among other sources; approximating natural resin in many qualities and used for similar purposes.

**Tower pumping head**  Tower pumping head is the head of water required at the inlet to the tower, measured above the basin kerb to deliver the circulating water through the distribution system.

**Underflow**  The stream containing the remaining liquid and the coarser solids, which is discharged through a circular opening at the apex of the core of a hydrocyclone is referred to as "underflow."

**Vapor concentration (absolute humidity)** The ratio of water vapor by mass to the total volume (gram per cubic meter).

**Water loading** It is circulating water flow expressed in quantity per unit of packed plan area of the tower.

**Wet-bulb temperature** The temperature taken on the wet-bulb thermometer, the one whose bulb is kept moist while making determinations of humidity. Because of cooling that results from evaporation, the wet-bulb thermometer registers a lower temperature than the dry-bulb thermometer.

**Working pressure** The maximum continuous pressure for which the manufacturer has designed the equipment (or any part to which the term refers).

# Further Readings

Abdel-Aal, H.K., Aggour, M., Fahim, M.A., 2003. Petroleum and Gas Field Processing. Chemical Industries. Marcel Dekker, New York, USA.

Ahrendts, J., 1977. Die Exergie Chemisch Reaktionsfähiger Systemevol. 43VDI-Verlag, Düsseldorf, Germany.

Alveberg, L.J., Melberg, E.V. (Eds.), 2013. Facts 2013 the Norwegian Petroleum Sector. Ministry of Petroleum and Energy and Norwegian Petroleum Directorate.

Ashour, I., Al-Rawahi, N., Fatemi, A., Vakili-Nezhaad, G., 2011. Applications of Equations of State in the Oil and Gas Industry. pp. 165–178 (Chapter 7).

Aspen Technology, 1999. Aspen Plus and Modelling Petroleum Processes. Aspen Technology, Burlington, USA.

Aspen Technology, 2004. Aspen Hysys 2004.2 e User Guide. Aspen Technology, Cambridge, USA.

Bahadori, A., 2009a. Minimize vaporization and displacement losses from storage containers. Hydrocarbon Process. 88 (6), 83–84.

Bahadori, A., 2009b. Estimating water-adsorption isotherms. Hydrocarbon Process. 88 (1), 55–56.

Bahadori, A., 2010. A method for prediction of scale formation in calcium carbonate aqueous phase for water treatment and distribution systems. Water Quality Res. J. Canada 45 (3), 379–389.

Bahadori, A., 2011a. Rapid estimation of n-alkanes surface tension using a simple predictive tool. SPE Projects Facilities Const. J. 6 (4), 173–178, SPE-140741-PA.

Bahadori, A., 2011b. Prediction of salinity of salty crude oil using arrhenius-type asymptotic exponential function and Vandermonde matrix. SPE Projects Facilities Const. J. 6 (1), 27–32, SPE-132324-PA.

Bahadori, A., 2011c. Simple method for estimation of effectiveness in one tube pass and one shell pass counter-flow heat exchanger. Appl. Energ. 88, 4191–4196.

Bahadori, A., 2011d. Prediction of compressed air transport properties at elevated pressures and high temperatures using simple method. Appl. Energ. 88 (4), 1434–1440.

Bahadori, A., 2011e. Estimation of potential precipitation from an equilibrated calcium carbonate aqueous phase using simple predictive tool. SPE Projects Facilities Const. 6 (4), 158–165, SPE-132403-PA.

Bahadori, A., 2011f. Prediction of saturated air dew point at elevated pressures using simple arrhenius-type function. Chem. Eng. Technol. 34 (2), 257–264.

Bahadori, A., 2011g. Estimation of combustion flue gas acid dew point during heat recovery and efficiency gain. Appl. Thermal Eng. 31, 1457–1462.

Bahadori, A., 2011h. Development of predictive tool for the estimation of true vapor pressure of volatile petroleum products. J. Energ. Sources Part A.

Bahadori, A., 2011i. A simple method for the estimation of performance characteristics of cooling towers. J. Energ. Inst. 84 (2), 88–93.

Bahadori, A., Mokhatab, S., 2009. Simple methodology predicts optimum pressures of multistage separators. Petrol. Sci. Technol. 27 (3), 315–324.

Bahadori, A., Vuthaluru, H.B., 2009a. Simple methodology for sizing of absorbers for TEG gas dehydration systems. Energy 34 (2009), 1910–1916.

Bahadori, A., Vuthaluru, H.B., 2009b. New method accurately predicts carbon dioxide equilibrium adsorption isotherms. Int. J. Greenhouse Gas Control 3 (2009), 768–772.

Bahadori, A., Vuthaluru, H.B., 2009c. Prediction of bulk modulus and volumetric expansion coefficient of water for leak tightness test of pipelines. Int. J. Pressure Vessels Piping 86, 550–554.

Bahadori, A., Vuthaluru, H.B., 2009d. Predicting emissivity of combustion gases. Chem. Eng. Progress 105 (6), 38–41.

Bahadori, A., Vuthaluru, H.B., 2010a. Simple Arrhenius-type function accurately predicts dissolved oxygen saturation concentrations in aquatic systems. Proc. Safety Environ. Protect. 88 (5), 335–340.

Bahadori, A., Vuthaluru, H.B., 2010b. A method for estimation of recoverable heat from blowdown systems during steam generation. Energy 35 (8), 3501.

Bahadori, A., Vuthaluru, H.B., 2010c. Estimation of performance of steam turbines using a simple predictive tool. Appl. Thermal Eng. 30 (2010), 1832–1838.

Bahadori, A., Vuthaluru, H.B., 2010d. Estimation of maximum shell-side vapour velocities through heat exchangers. Chem. Eng. Res. Design 88, 1589–1592.

Bahadori, A., Vuthaluru, H.B., 2010e. Estimation of energy conservation benefits in excess air controlled gas-fired systems. Fuel Process. Technol. 91 (10), 1198–1203.

Bahadori, A., Vuthaluru, H.B., 2010f. Predictive tools for the estimation of downcomer velocity and vapor capacity factor in fractionators. Appl. Energ. 87, 2615–2620.

Bahadori, A., Vuthaluru, H.B., 2010g. Prediction of methanol loss in vapor phase during gas hydrate inhibition using arrhenius-type functions. J. Loss Prevent. Proc. Ind. 23 (3), 379–384.

Bahadori, A., Vuthaluru, H.B., 2010h. Simple equations to correlate theoretical stages and operating reflux in fractionators. Energy 35 (2010), 1439–1446.

Bahadori, A., Vuthaluru, H.B., 2010i. Novel predictive tools for design of radiant and convective sections of direct fired heaters. Appl. Energ. 87 (2010), 2194–2202.

Bahadori, A., Vuthaluru, H.B., 2010. A simple method for the estimation of thermal insulation thickness. Appl. Energ. 87 (2010), 613–619.

Bahadori, A., Vuthaluru, H.B., 2010. Estimation of displacement losses from storage containers using a simple method. J. Loss Prevent. Proc. Ind. 23 (2010), 367–372.

Bahadori, A., Vuthaluru, H.B., 2010. Novel predictive tool for accurate estimation of packed column size. J. Nat. Gas Chem. 19 (2), 146–150.

Bahadori, A., Vuthaluru, H.B., 2010. A simple correlation for estimation of economic thickness of thermal insulation for process piping and equipment. Appl. Thermal Eng. 30, 254–259.

Bahadori, A., Vuthaluru, H.B., 2010. Prediction of silica carry-over and solubility in steam of boilers using simple correlation. Appl. Thermal Eng. 30 (2010), 250–253.

Bahadori, A., Vuthaluru, H.B., 2011. Estimation of saturated air water content at elevated pressures using simple predictive tool. Chem. Eng. Res. Design 89, 179–186.

Bahadori, A., Zeidani, K., 2008. New equations predicting the best performance of electrostatic desalter. Petrol. Sci. Technol. 26 (1), 40–49.

Bahadori, A., Vuthaluru, H.B., Mokhatab, S., 2009. Simple correlation accurately predicts aqueous solubility of light alkanes. J. Energ. Sources Part A 31 (9), 761–766.

Bejan, A., 2006. Advanced Engineering Thermodynamics, 3rd ed. John Wiley & Sons, New York, USA.

Bejan, A., Tsatsaronis, G., Moran, M., 1996. Thermal Design & Optimization. John Wiley & Sons, New York, USA.

Bothamley, M., 2004. Offshore processing options for oil platforms. In: Proceedings of the SPE Annual Technical Conference and Exhibition. Society of Petroleum Engineers, Houston, USA, pp. 1–17 (Paper SPE 90325).

Brodyansky, V.M., Sorin, M.V., Le Goff, P., 1994. The Efficiency of Industrial Processes: Exergy Analysis and Optimization. Elsevier, Amsterdam, The Netherlands.

Danish Energy Agency, 2011. Danmarks olie og gas produktion. Tech. Rep. København, Denmark: Energistyrelsen. Available from: www.ens.dk/DA-DK/UNDERG RUNDOGFORSYNING/OLIE_OG_GAS/RAPOLIEGAS/

de Oliveira, Jr, S., Van Hombeeck, M., 1997. Exergy analysis of petroleum separation processes in offshore platforms. Energ. Convers. Manage. 38 (15–17), 1577–1584.

Dhole, V.R., Linnhoff, B., 1993. Total site targets for fuel co-generation, emissions, and cooling. Comput. Chem. Eng. 17, 101–109.

Dimian, A.C., 2003. Integrated Design and Simulation of Chemical Processes. Elsevier, Amsterdam, The Netherlands.

Elmegaard, B., Houbak, N., 2005. DNA—a general energy system simulation tool. In: Amundsen, J., (Ed.), Proceedings of SIMS 2005—46th Conference on Simulation and Modeling. Tapir Academic Press, Trondheim, Norway, pp. 43–52.

Fazlollahi, S., Maréchal, F., 2013. Multi-objective, multi-period optimization of biomass conversion technologies using evolutionary algorithms and mixed integer linear programming (MILP). Appl. Therm. Eng. 50 (2), 1504–1513.

Gerber, L., Gassner, M., Maréchal, F., 2011. Systematic integration of LCA in process systems design: application to combined fuel and electricity production from lignocellulosic biomass. Comput. Chem. Eng. 35 (7), 1265–1280.

Grassmann, V.P., 1950. Zur allgemeinen definition des wirkungsgrades. Chemie Ingenieur Technik 4 (1), 77–80.

Hammond, G.P., 2007. Industrial energy analysis, thermodynamics and sustainability. Appl. Energ. 84, 675–700.

Hancock, W.P., 1983. Development of a reliable gas injection operation for the North Sea's largest-capacity production platform, Statfjord A. J. Petrol. Technol. 35 (11), 1963–1972.

Ingeniøren/bøger., 2000. Pumpe Ståbi. København. Ingeniøren A/S 2000, Denmark.

Jones, D.S.J.S., Pujadó, P.R. (Eds.), 2006. Handbook of Petroleum Processing. Springer, Dordrecht, The Netherlands.

Kelly, S., Tsatsaronis, G., Morosuk, T., 2009. Advanced exergetic analysis: approaches for splitting the exergy destruction into endogenous and exogenous parts. Energy 34 (3), 384–391.

Klemes, J., Varbanov, P., 2012. Heat integration including heat exchangers, combined heat and power, heat pumps, separation processes and process control. Appl. Therm. Eng. 43, 1–6.

KLM Technology Group Utilities Standards #03-12 Block Aronia, Jalan Sri Perkasa 2, Taman Tampoi Utama, 81200 Johor Bahru, Malaysia.

Kloster, P., 1999. Energy optimization on offshore installations with emphasis on offshore combined cycle plants. In: Proceedings of the Offshore Europe Conference. Society of Petroleum Engineers, Aberdeen, UK, pp. 1–9 (Paper SPE 56964).

Kloster, P., 2000. Reduction of emissions to air through energy optimisation on offshore installations. In: Proceedings of the SPE International Conference on Health, Safety, and the Environment in Oil and Gas Exploration and Production. Society of Petroleum Engineers, Stavanger, Norway, pp. 1–7 (Paper SPE 61651).

Knopf, F., 2011. Modeling Analysis and Optimization of Process and Energy Systems. John Wiley & Sons, Hoboken, USA.

Kotas, T.J., 1980a. Exergy concepts for thermal plant: first of two papers on exergy techniques in thermal plant analysis. Int. J. Heat Fluid Flow 2 (3), 105–114.

Kotas, T.J., 1980b. Exergy criteria of performance for thermal plant: second of two papers on exergy techniques in thermal plant analysis. Int. J. Heat Fluid Flow 2 (4), 147–163.

Kotas, T.J., 1995. The Exergy Method of Thermal Plant Analysis. Krieger Publishing, Malabar, USA.

Lazzaretto, A., Tsatsaronis, G., 2006. SPECO: a systematic and general methodology for calculating efficiencies and costs in thermal systems. Energy 31, 1257–1289.

Li, K.J., 1996. Use of fractionation column in an offshore environment. In: Proceedings of the SPE Annual Technical Conference and Exhibition. Society of Petroleum Engineers, New Orleans, USA, pp. 1–11 (Paper SPE 49121).

Lieblein, S., 1957. Analysis of Experimental Low-Speed Loss and Stall Characteristics of Two-Dimensional Compressor Blade Cascades. NACA, Washington, DC, USA, RM E57A28.

Linnhoff, B., Eastwood, A., 1997. Overall site optimization by pinch technology. Process. Saf. Environ. Prot. Trans. Inst. Chem. Eng. Part B 75 (Suppl.), S138–S144.

Lyons, W.C., Plisga, J.G. (Eds.), 2004. Standard Handbook of Petroleum & Natural Gas Engineering. Gulf Professional Publishing, Burlington, USA.

Manning, F.S., Thompson, R.E., 1991a. Oilfield Processing of Petroleum: Crude Oilvol. 2PennWell Books, Tulsa, USA.

Manning, F.S., Thompson, R.E., 1991b. Oilfield Processing of Petroleum: Natural Gasvol. 1PennWell Books, Tulsa, USA.

Marechal, F., Kalitventzeff, B., 1996. Targeting the minimum cost of energy requirements: a new graphical technique for evaluating the integration of utility systems. Comput. Chem. Eng. 20 (suppl. 1), S225–230.

Maréchal, F., Kalitventzeff, B., 1998. Energy integration of industrial sites: tools, methodology and application. Appl. Therm. Eng. 18 (11), 921–933.

Maréchal, F., Kalitventzeff, B., 1999. Targeting the optimal integration of steam networks: mathematical tools and methodology. Comput. Chem. Eng. 23 (suppl. 1), S133–S136.

Margarone, M., Magi, S., Gorla, G., Biffi, S., Siboni, P., Valenti, G., 2011. Revamping, energy efficiency, and exergy analysis of an existing upstream gas treatment facility. J. Energy Resour. Technol. 133, 012001-1–012001-9.

Mattson, C., Messac, A., 2005. Pareto frontier based concept selection under uncertainty with visualization. OPTE Optim. Eng. 6 (1), 85–115.

Meyer, L., Tsatsaronis, G., Buchgeister, J., Schebek, L., 2009. Exergo environmental analysis for evaluation of the environmental impact of energy conversion systems. Energy 34 (1), 75–89.

Molyneaux, A., 2002. A Practical Evolutionary Method for the Multi-Objective Optimisation of Complex Integrated Energy Systems Including Vehicle Drivetrains. PhD thesis, École Polytechnique Fédérale de Lausanne.

Moran, M.J., 1998. Fundamentals of exergy analysis and exergy-based thermal systems design. In: Bejan, A., Mamut, E. (Eds.), Proceedings of the NATO Advanced Study Institute on Thermodynamic Optimization of Complex Energy Systems. NATO Science Series, vol. 69, Kluwer Academic Publishers, Neptun, Romania, pp. 73–92.

Morris, D.R., Szargut, J., 1986. Standard chemical exergy of some elements and compounds on the planet Earth. Energy 11 (8), 733–755.

Muir, D.E., Saravanamuttoo, H.I.H., Marshall, D.J., 1989. Health monitoring of variable geometry gas turbines for the Canadian Navy. ASME J. Eng. Gas Turbines Power 111 (2), 244–250.

Nemet, A., Klemes, J., Kravanja, Z., 2012. Optimising a plant economic and environmental performance over a full lifetime. Chem. Eng. Trans. 29, 1435–1440.

Nguyen, T.V., Pierobon, L., Elmegaard, B., 2012. Exergy analysis of offshore processes on North Sea oil and gas platforms. In: Proceedings of CPOTE 2012—The 3rd International Conference on Contemporary Problems of Thermal Engineering. Gliwice, Poland; 2012. pp. 1–9 (Paper 45).

Nguyen, T.V., Pierobon, L., Elmegaard, B., Haglind, F., Breuhaus, P., Voldsund, M., 2013. Exergetic assessment of energy systems on North Sea oil and gas platforms. Energy 62, 23–36.

Nguyen, T.V., Jacyno, T., Breuhaus, P., Voldsund, M., Elmegaard, B., 2014. Thermodynamic analysis of an upstream petroleum plant operated on a mature field. Energy 68, 454–469.

Nord, L.O., Bolland, O., 2013a. Steam bottoming cycles offshore – challenges and possibilities. J. Power Technol. 92 (3), 201–207.

Nord, L.O., Bolland, O., 2013b. Design and off-design simulations of combined cycles for offshore oil and gas installations. Appl. Therm. Eng. 54, 85–91.

Norwegian Ministry of Petroleum and Energy, 2012. Facts 2012 – the Norwegian petroleum sector. Tech rep, Norwegian Petroleum Directorate, Oslo, Norway.

Norwegian Petroleum Directorate, 2012. Standards relating to measurement of petroleum for fiscal purposes and for calculation of $CO_2$—tax.

Norwegian Petroleum Directorate, 2013. Petroleum Resources on the Norwegian Continental Shelf.

Oen, A., Vik, E., 2000. Hydrogen sulfide forecasting under PWRI (Produced Water Reinjection). In: Proceedings of the SPE International Conference on Health, Safety, and the Environment in Oil and Gas Exploration and Production. Stavanger, Norway, pp. 1–6 (Paper SPE 61245).

Oljedirektoratet. Faktasider. URL, factpages.npd.no/factpages/Default.aspx? culture¼nb-no; 2013.

Pearson, K., 1895. Note on regression and inheritance in the case of two parents. Proc. Roy. Soc. Lond. 58, 240–242.

Peng, D.Y., Robinson, D.B., 1976. A new two-constant equation of state. Ind. Eng. Chem. Fundam. 15 (1), 59–64.

Peng, D.Y., Robinson, D.B., 1976. A new two-constant equation of state. Ind. Eng. Chem. Fundam. 15 (1), 59–64.

Petrakopoulou, F., Tsatsaronis, G., Morosuk, T., Carassai, A., 2012. Conventional and advanced exergetic analyses applied to a combined cycle power plant. Energy 41 (1), 146–152.

Pierobon, L., Kandepu, R., Haglind, F., 2012. Waste heat recovery for offshore applications. In: Proceedings of the ASME 2012 International Mechanical Engineering Congress and Exposition, ASME. Energy, Parts A and B, vol. 6, pp. 503–512.

Plisga, G.J., Lyons, W.C. (Eds.), 2004. Standard Handbook of Petroleum & Natural Gas Engineering. 2nd ed. Burlington, USA, Gulf Professional Publishing.

Puntervold, T., Austad, T., 2008. Injection of seawater and mixtures with produced water into North Sea chalk formation: impact on wettability, scale formation and rock mechanics caused by fluid-rock interaction. J. Petrol. Sci. Eng. 63 (1e4), 23–33.

Renon, H., Prausnitz, J.M., 1968. Local compositions in thermodynamic excess functions for liquid mixtures. AIChE J. 14 (1), 135–144.

Rivero, R., 2002. Application of the exergy concept in the petroleum refining and petrochemical industry. Energ. Convers. Manag. 43 (9e12), 1199–1220.

Rivero, R., Rendon, C., Monroy, L., 1999. The exergy of crude oil mixtures and petroleum fractions: calculation and application. Int. J. Appl. Thermodyn. 2 (3), 115–123.

Rodriguez, I., Hamouda, A.A., 2010. An approach for characterization and lumping of plus fractions of heavy oil. SPE Reserv. Eval. Eng. 13 (2), 283–295.

Rohde, D., Walnum, H., Andresen, T., Nekså, P., 2013. Heat recovery from export gas compression: analyzing power cycles with detailed heat exchanger models. Appl. Therm. Eng. 60 (1–2), 1–6.

Rosen, M., Dincer, I., 1999. Exergy analysis of waste emissions. Int. J. Energ. Res. 23 (13), 1153–1163.

Saravanamuttoo, H.I.H., Rogers, G.F.C., Cohen, H., Straznicky, P., 2008. Gas Turbine Theory, 6th ed. Pearson Prentice Hall, Upper Saddle River, USA.

Sato, N., 2004. Chemical Energy and Exergy: an Introduction to Chemical Thermodynamics for Engineers, 1st ed. Elsevier, Amsterdam, The Netherlands.

Satter, A., Iqbal, G.M., Buchwalter, J.L., 2008. Practical Enhanced Reservoir Engineering: Assisted with Simulated Software. PennWell Books, Tulsa, USA.

Schwartzentruber, J., Renon, H., 1989. Extension of UNIFAC to high pressures and temperatures by the use of a cubic equation of state. Ind. Eng. Chem. Res. 28 (7), 1049–1055.

Schwartzentruber, J., Renon, H., Watanasiri, S., 1989. Development of a new cubic equation of state for phase equilibrium calculations. Fluid Phase Equilib. 52, 127–134.

Sciubba, E., 2001. Beyond thermoeconomics? The concept of extended exergy accounting and its application to the analysis and design of thermal systems. Exergy Int. J. 1 (2), 68–84.

Sciubba, E., 2003. Extended exergy accounting applied to energy recovery from waste: the concept of total recycling. Energy 28 (13), 1315–1334.

Siemens, 2011. SGT-500 industrial gas turbine. Tech rep, Siemens Industrial Turbomachinery AB, Finnspong, Sweden.

Soave, G., 1972. Equilibrium constants from a modified Redlich-Kwong equation of state. Chem. Eng. Sci. 27 (6), 1197–1203.

Soave, G., 1993. 20 years of Redlich-Kwong equation of state. Fluid Phase Equilib. 82, 345–359.

Song, T.W., Kim, T.S., Kim, J.H., Ro, S.T., 2001. Performance prediction of axial flow compressors using stage characteristics and simultaneous calculation of interstage parameters. In: Proceedings of the Institution of Mechanical Engineers, Institution of Mechanical Engineers, vol. 215 (Part A). London, UK, pp. 89–98.

Spina, P.R., 2002. Gas Turbine performance prediction by using generalized performance curves of compressor and turbine stages. In: Proceedings of the ASME Turbo Expo 2002 Power for Land, Sea, and Air (GT2002), vol. 2. Amsterdam, The Netherlands, pp. 1073–1082.

Standard Norge, 2005. NORSOK Standard and Fiscal Measurement Systems for Hydrocarbon Gas. Tech. Rep. Standards, Norway.

Standard Norge, 2007. NORSOK Standard and Fiscal Measurement Systems for Hydrocarbon Liquid. Tech. Rep. Standards, Norway.

Stodola, A., 1924. Dampf- und Gasturbinen, 6th ed. Springer, Berlin, Germany.

Stortinget, 1990. Lov om avgift på utslipp av $CO_2$ i petroleumsvirksomhet på kontinentalsokkelen Tech. Rep. 72. Finansdepartementet.

Stortinget, 1996. Lov om petroleumsvirksomhet [petroleumsloven]. Tech. Rep. 72. OED (Olje- og energidepartementet).

Sulzer Pumps, 2010. Centrifugal Pump Handbook, 3rd ed. Elsevier, New York, USA.

Svalheim, S.M., 2002. Environmental regulations and measures on the Norwegian continental shelf. In: Proceedings of the SPE International Conference on Health, Safety and Environment in Oil and Gas Exploration and Production. Society of Petroleum Engineers, Kuala Lumpur, Malaysia, pp. 1–10 (Paper SPE 73982).

Svalheim, S.M., King, D.C., 2003. Life of field energy performance. In: Proceedings of the SPE Offshore Europe Conference, July. Society of Petroleum Engineers, Aberdeen, UK, pp. 1–10 (Paper SPE 83993).

Szargut, J., 1989. Chemical exergies of the elements. Appl. Energ. 32 (4), 269–286.

Szargut, J., 1998. Exergy in the thermal systems analysis. In: Bejan, A., Mamut, E. (Eds.), Proceedings of the NATO Advanced Study Institute on Thermodynamic Optimization of Complex Energy Systems. NATO Science Series, vol. 69, Kluwer Academic Publishers, Neptun, Romania, pp. 137–50.

Szargut, J., Morris, D., Steward, F., 1988. Exergy Analysis of Thermal, Chemical, and Metallurgical Processes. Hemisphere, New York, USA.

Taylor, J., 1997. An Introduction to Error Analysis and the Study of Uncertainties in Physical Measurements, 2nd ed. University Science Books, Sausalito, USA.

Templalexis, I., Pilidis, P., Pachidis, V., Kotsiopoulos, P., 2011. Development of a one-dimensional streamline curvature code. ASME J. Turbomach. 133 (1), 011003.

Tock, L., Maréchal, F., 2012. H2 processes with $CO_2$ mitigation: thermo-economic modeling and process integration. Int. J. Hydrogen Energ. 37 (16), 11785–11795.

Traupel, W., 1977. Thermische Turbomaschinen, 3rd ed. Springer, Berlin, Germany.

Tsatsaronis, G., 1993. Thermoeconomic analysis and optimization of energy systems. Prog. Energ. Combust. Sci. 19, 227–257.

Tsatsaronis, G., Park, M.H., 2002. On avoidable and unavoidable exergy destructions and investment costs in thermal systems. Energ. Convers. Manag. 43, 1259–1270.

Tsatsaronis, G., Winhold, M., 1985. Exergoeconomic analysis and evaluation of energy conversion plants—A new general methodology. Energy 10 (1), 69–80.

Tsonopoulos, C., Heidman, J., 1985. From Redlich-Kwong to the present. Fluid Phase Equilib. 24 (1), 1–23.

Vanner, R., 2005. Energy use in offshore oil and gas production: trends and drivers for efficiency from 1975 to 2025. PSI Working Paper, Policy Studies Institute, University of Westminster, London, UK.

Vik, E.A., Dinning, A.J., 2009. Produced Water Re-Injection and the Potential to Become an Improved Oil Recovery Method. Aquatem A/S, Tech. Rep. Oslo, Norway.

Vik, E.A., Bakke, S., Stang, P., Edvardsson, T., Gadeholt, G., 2000. Integrated environmental risk management of future produced water disposal options at Draugen on the Norwegian Continental Shelf. In: Proceedings of the SPE International Conference on Health, Safety and the Environment in Oil and Gas Exploration and Production. Society of Petroleum Engineers, Stavanger, Norway, pp. 1–9 (Paper SPE 61179).

Vik, E.A., Janbu, A., Garshol, F., Henninge, L.B., Engebretsen, S., Oliphant, D., et al., 2007. Nitrate based souring mitigation of produced water, side effects and challenges from the Draugen produced water re-injection pilot.

In: Proceedings of the SPE International Symposium on Oilfield Chemistry. Society of Petroleum Engineers, Houston, USA, pp. 1–11 (Paper SPE 106178).

Voldsund, M., Ertesvåg, I.S., Røsjorde, A., He, W., Kjelstrup, S., 2010. Exergy analysis of the oil and gas separation processes on a North sea oil platform. In: Favrat, D., Maréchal, F. (Eds.), Proceedings of ECOS 2010—The 23rd International Conference on Efficiency, Cost, Optimization, Simulation and Environmental Impact of Energy Systems. Power plants & industrial processes, vol. IV. p. 468.

Voldsund, M., He, W., Røsjorde, A., Ertesvåg, I.S., Kjelstrup, S., 2012. Evaluation of the oil and gas processing at a real production day on a North sea oil platform using exergy analysis. Desideri, U., Manfrida, G., Sciubba, E. (Eds.), Proceedings of ECOS 2012 and the 25th International Conference on Efficiency, Cost, Optimization, Simulation and Environmental Impact of Energy Systems, vol. II, Firenze University Press, Perugia, Italy, pp. 153–166.

Voldsund, M., Ertesvåg, I.S., He, W., Kjelstrup, S., 2013. Exergy analysis of the oil and gas processing a real production day on a North Sea oil platform. Energy 55, 716–727.

Wall, G., 1988. Exergy flows in industrial processes. Energy 13 (2), 197–208.

Walnum, H., Nekså, P., Nord, L., Andresen, T., 2013. Modelling and simulation of CO2 (carbon dioxide) bottoming cycles for offshore oil and gas installations at design and off-design conditions. Energy 59, 513–520.

Whitson, C.H., Brulé, M.R., 2000. Phase Behavior. In: SPE monograph seriesvol. 20 Society of Petroleum Engineers, Richardson, USA.

# Subject Index

## Y

Printed in the United States
By Bookmasters